Dear S+O,

Thanks!

Oliv

Institutions, Discourse and Regional Development

The Scottish Development Agency and the Politics of Regional Policy

P.I.E.-Peter Lang

Bruxelles · Bern · Berlin · Frankfurt am Main · New York · Oxford · Wien

Henrik HALKIER

Institutions, Discourse and Regional Development

The Scottish Development Agency and the Politics of Regional Policy

"Regionalism & Federalism"
No. 8

Presses Interuniversitaires Européennes
Brussels, 2006
1 avenue Maurice, B-1050 Brussels, Belgium
info@peterlang.com; www.peterlang.com

ISSN 1379-4507
ISBN 90-5201-275-X
US ISBN 0-8204-6653-0
D/2006/5678/03
Printed in Germany

Bibliographic information published by "Die Deutsche Bibliothek"
"Die Deutsche Bibliothek" lists this publication in the "Deutsche Nationalbibliografie"; detailed bibliographic data is available in the Internet at <http://dnb.ddb.de>.

CIP available from the British Library, GB
and the Library of Congress, USA.

Contents

PART II
CORPORATE DEVELOPMENTS

PART III
DEVELOPING POLICIES

Preface

Writing a book, you accumulate innumerable debts. This particular book has relied on the kind assistance of a large number of interviewees who recounted their experience of regional policy in Scotland and provided access to unpublished materials – in itself a fascinating journey through oak-panelled rooms, 1970s office buildings and the occasional pub which sets the standards for future research. Extensive periods in Scotland would not have been possible without financial support from Den Obelske Familiefond, the Aarhus School of Business, the Danish Social Science Research Council, and the Faculty of Humanities at Aalborg University – nor, indeed, the flexibility of colleagues with regard to teaching and administration.

In the process of researching and writing, Scotland has become a home away from home: facilities have been provided on numerous occasions at the University of Strathclyde, first by the Department of Economics and then by the European Policies Research Centre, and – equally important – the friendly hospitality of John Scouller, Richard Brooks, Kim Swales, Eric Rahim, Frank Stephens, John Bachtler, Sandra Taylor, and Phil Raines ensured that the prolonged outings also contained seriously unserious elements. In a bizarre twist of fate my obsession with Scottish regional policy lead my beloved Grethe to embark on and complete her PhD in marine biology at the University of Stirling, something which made the inimitable Scot & Olivia an unmissable part of the Scottish experience.

The research process has benefited greatly from the tireless efforts of research librarians in Aarhus, Aalborg, Glasgow, Edinburgh and London, the audiotyping of Ellen Madsen & Lis Rejnert at the Aarhus School of Business, and the research assistance of Fiona McLaren and Colin McCallum at Strathclyde University. The production of the manuscript has been facilitated by the many lucid comments by colleagues in Denmark and Scotland on individual chapters, and so, without implicating them in the final result, thanks are due to Poul Rind Christensen, Bill Stevenson, Morten Pilegaard, John Scouller, Craig Campbell, Neil Hood, Frank Kirwan, Chris Moore, Brian Hogwood, Johan Smed, Kim Swales, Ulf Hedetoft, Bente Halkier, David May, Charlotte Damborg, and Poul Thøis Madsen – and to the proofreaders Iben Kirkegaard, Julie Larsen, Grethe Fallesen, Karin Jørgensen, Helle Weiergang, Karina Andersen, and Marianne Høgsbro.

The final version of the book has benefited from being produced in the congenial and stimulating interdisciplinary environments of the International Studies programme and the *SPIRIT* doctoral school at Aalborg University, and, across the North Sea, at the EPRC in Glasgow. Over the years the Regional Studies Association has provided intellectual inspiration and networking opportunities at events all over Europe which, in turn, has resulted in collaborative research projects, and thanks are due to my collaborators Mike Danson, Charlotte Damborg, Bo Svensson, Anders Östholl, and Anette Therkelsen for discussing issues of concepts and interpretation time and time again.

Writing, it would seem, has a strong social and interactive dimension, but it also requires quiet spaces and long misty days. I would like to thank friends and loved ones who have tolerated absences of body and mind over the years, in Aarhus, in Aalborg and in Scotland.

Aalborg, August 2005
Henrik Halkier

List of Tables

List of Figures

List of Abbreviations

BOP	Business Opportunities Programme
CBI	Confederation of British Industry
CBIS	Confederation of British Industry, Scotland
CoSIRA	Council for Small Industries in Rural Areas
COSLA	Convention of Scottish Local Authorities
DBRW	Development Board for Rural Wales
DCC	Development Consultative Committee
DTI	Department of Trade and Industry
FDI	Foreign direct investment
GDP	Gross domestic product
GEAR	Glasgow Eastern Area Renewal project
HCPD	House of Commons Parliamentary Debates
HIDB	Highlands and Islands Development Board
HOID	Head Office Investment Directorate
IBB	Invest in Britain Bureau
IDC	Industrial Development Certificate
IDS	Industry Department for Scotland
IRC	Industrial Reorganization Corporation
LIS	Locate in Scotland
MBO	Management buyout
NEB	National Enterprise Board
RDA	Regional Development Agency
RDO	Regional Development Organisation
SBD	Small Business Division
SCDI	Scottish Council Development & Industry
SCRI	Scottish Council Research Institute
SDA	Scottish Development Agency
SDAINV	Industrial investment database (cf. Chapter 8)
SDASECT	Sectoral initiatives database (cf. Chapter 9)
SEDCOR	Strathclyde Economic Development Corporation

SEPD	Scottish Economic Planning Department
SICRAS	Small Industries Council for Rural Areas in Scotland
SIEC	Scottish Industrial Estates Corporation
SME	Small and medium-sized enterprise
SNP	Scottish National Party
STUC	Scottish Trades Union Congress
UDC	Urban Development Corporation
WDA	Welsh Development Agency
3i	Investors In Industry

Part I

Setting the Scene

CHAPTER 1

Introducing Aims and Methods

1.1. A Certain Unease

Picture this. In the 1960s and 1970s British regional policy relied on strong and effective policy instruments such as central government regulation and financial incentives that redirected economic activity from the prosperous South to problem regions in the North and West. In Scotland these UK-wide measures were supplemented by central government which in 1975 established a semi-autonomous body, the Scottish Development Agency (SDA), that invested in local firms and used public ownership to stimulate economic growth. After 1979 the Conservative governments of Margaret Thatcher cut back central government regional incentives to a bare minimum, and the SDA was made to rely on 'soft' policy instruments such as advisory services aimed at persuading private firms to become more efficient. A radical change from a 'strong' to a 'weak' form of regional policy in the sense that the pressure on private firms to comply with public priorities were reduced significantly in the 1980s, and consequently the influence of public policies on the spatial distribution of economic activity decreased.

This 'radical change' interpretation of regional policy in Britain has alarmed some observers worried about the increasing impotence of government *vis-à-vis* private firms and the market, while others have applauded the introduction of a less 'interventionist' approach. But underneath the conflicting normative evaluations we find the very same story about the strategies and incentives employed in British regional policy in general and by the SDA in particular. From the 1970s to the 1980s

- policies changed radically,
- policy changes weakened public influence on private firms,
- policy changes had been instigated by central government, and
- policy changes reflected the neo-liberal ideology of the Thatcher government.

The widespread consensus could of course simply reflect that this was 'what really happened', but since the 'radical change' interpretation bears strong resemblance to the views officially endorsed by successive Conservative governments, the 'factual' consensus could perhaps also be a product of the pervasiveness of Thatcherite political discourse. As regional policy is essentially about giving private firms incentives to behave in accordance with the spatial goals of public policy, this form of government intervention sat uncomfortably with the declared intention of new-right Conservatives to 'roll back the state' and 'free the market', and thus the continued existence of regional policy measures in the 1980s needed a plausible explanation, and claiming that the new policies were 'weaker' and working 'with the grain of the market' could have been one way of mediating between ideological discourse and policy practice.

The general unease created by the possible origins of consensus is not dispelled when the 'radical change' interpretation is subjected to closer scrutiny. From a historical perspective, research in other policy areas has demonstrated that despite the radical rhetoric of the Conservative government a high degree of continuity can be found in many areas of public policy in Britain in the 1980s, something which suggests that this could have been the case in regional policy as well. From a regional studies perspective some have argued that the new forms of regional incentives are not only able to influence patterns of economic development but can indeed be more effective than the traditional financial ones, while others have observed that the changes in Britain would seem to fit well into broader European trends and hence questioned whether they are in fact a legacy of Thatcherism or simply reflect a broader paradigmatic shift in regional policy away from central government financial incentives towards 'softer' policy instruments employed by regionally-based development bodies. And from a policy analysis perspective, the 'radical change' interpretation have tended to compare different policies by means of fairly broad dichotomies such as 'liberalism/interventionism' and 'weak/strong' policy instruments, thereby avoiding a more detailed analysis of the strategies and incentives involved in the interaction between public and private actors. Evidence is, in other words, gradually emerging which would seem to question key assumptions in the hitherto prevailing and officially endorsed 'radical change' interpretation of regional policy in Scotland and Britain in the 1970s and 1980s.

The present book is an attempt to overcome the limitations of much of the existing literature on the development of regional policy in Britain in general and the SDA in particular by developing a comprehensive understanding which is both conceptually coherent and sustained by empirical analysis. In order to achieve this ambition, it is paramount that the scope of the study is expanded *significantly. Firstly*, in terms of the historical and geographical context, the examination of Scottish and British matters must be situated in the context of general European trends in regional policy in order to be able to place conceptual and explanatory schemes into a broader perspective. *Secondly*, in order to fully explore the politics of regional policy, the empirical analysis of regional policies must not only be able to illuminate in detail the incentives and strategies employed in order to identify the changing capacity of public actors to influence decision-making in private firms, but must also situate these exchanges in their institutional and discursive context. And *thirdly*, a unitary conceptual framework must be developed, capable of analysing both 'traditional' and 'new' forms of regional policy, and of understanding developments on both sides of the alleged 'great British divide' of 1979. Expanding the scope of the study along these lines clearly makes the road towards uncovering the changing nature of the SDA and British regional policy a longer and more winding one, involving an extensive dialogue with often unconnected literatures in various fields of academic endeavour. But apart from, hopefully, improving the understanding of the specific historical and geographical focus of the present book, this may, in turn, also result in a text that could be of interest outwith the British Isles as a conceptual and historical contribution to the analysis of new paradigms in regional development and public policy.

1.2. Aims and Designs

The purpose of this book is to analyse the development of the SDA in the context of British regional policy from 1975 to 1991 with particular emphasis on clarifying the nature and origins of continuity and change. This sets two central tasks for the empirical enquiry, namely

1. to establish the development of the SDA and its policies in order to determine the balance between continuity and change with regards to objectives, implementation and interaction with private firms
2. to illuminate the relationship between the SDA and its political environment, including its sponsors in the Scottish Office as a territorial department of central government, other actors in the British regional policy network, and the influence of ideological and territorial forms of political discourse.

Although the text can be seen as interdisciplinary because it draws on insights and arguments from a broad range of academic traditions, the overall structure of the enquiry is clearly historical, with importance given to change over time in institutions, discourse, and policies. Moreover, the specific context of this enquiry is the intersection of two broader historical trends which both would seem to involve a reconfiguration of the relationship between public and private actors: the rise of Thatcherism and neo-liberalism in Britain on the one hand, and the emergence of a new bottom-up paradigm in European regional policy which emphasised the role of non-financial policy instruments on the other. In a situation characterised by complexity and flux it is of course particularly important to be able to employ a coherent conceptual framework capable of spanning conceptually across different policy paradigms and accounting for the dynamic relationship between organisations, policies and political discourse. As such a framework was found not to be readily available, theoretical reflection and conceptual clarification will constitute a major preparatory task in its own right, and here the discussion will proceed on the basis of a general acceptance of central tenets of the so-called 'new institutionalism' while the elaboration of a conceptual framework for the analysis of the politics and paradigms of regional development will be inspired mainly by traditions within the study of public policy, political discourse, and networks and inter-organisational relations. We have, in other words, a heavily contextualised case study which necessitates a good deal of 'concept crunching' in dialogue with a range of more general literatures – and hence uncovering the history of the SDA may take on a wider significance.

This research design has, like many of the policy decisions retraced in the following, resulted from an attempt to make the most of unfavourable circumstances on the basis of limited resources while continuing to pursue long-standing goals. The relationship between institutions and economic development, both in general and in non-metropolitan settings in particular, has been a central interest in my research for nearly two decades,[1] but these issues could of course have been pursued in others ways than through a contextualised case study of one particular region – and even when this approach had been decided upon, another region could have been chosen. The two most obvious alternative research designs would have been to undertake a survey of a large number of regions and development bodies, or to carry out a comparative study of a smaller number of regions. However, a survey-type design would have pointed towards a focus on a limited number of predefined issues[2] – with the added risk of reproducing existing simplistic conceptual dichotomies – and hence would have made it difficult to capture key aspects of the policy process and the evolving relationship between institutions, policy instruments and discourses of regional

1 See e.g. Halkier 1986, 1987, 1990b, 1991c.

2 For examples of this, see e.g. Halkier & Danson 1997, Wong 2000.

development. The idea of comparing a small number of regions, internationally or on a purely British basis, would seem to be much more attractive because this would have afforded more opportunities for exploring the political processes in each locality while maintaining the possibility of benefiting from a systematic comparison between regions and/or countries, but still the perceived difficulties of managing four detailed case studies covering the politics of regional policy from design via implementation to economic effects in the absence of a coherent conceptual foundation[3] made this solution less attractive.

In short, the nature of the existing literature on regional policy clearly pointed towards the need for an extensive effort with regard to reconceptualisation, and hence one explorative case-study would seem to be an obvious solution – but why Scotland? On the one hand regional policy in England was essentially limited to the traditional centralised top-down paradigm until the late 1990s and did not entail the same coexistence between different approaches to regional development as Scotland or Wales,[4] and hence an English case study would be much less interesting from a wider European perspective where bottom-up policies had become a prominent feature already in the 1980s.[5] On the other hand a wider and more innovative range of bottom-up regional policies seemed to have developed in Scotland than in Wales,[6] and in order to locate the case study in a British region where different European policy paradigms coexisted, Scotland would certainly seem to be the best choice.

The result is a book that brings an interdisciplinary perspective to bear on the rise and transformation of an organisation which was famed across Europe for its innovative policies, yet surrounded by political controversy in Scotland. By undertaking an in-depth study of development of the SDA in its British and European context, the text would seem to have several layers that could appeal to different audiences:

- Those specifically interested in *Scottish politics* should appreciate an in-depth discussion of an important pre-devolution example of 'preferential regionalism' towards Scotland, as well as a critical account of the subsequent development of an organisation that has often been interpreted in rather simplistic manners as reflecting broarder trends

3 The dual Anglo-German regional comparisons at the heart of Jeffrey Anderson's stimulating book *The Territorial Imperative* (1992) are made possible by the combination of a robust conceptual framework and focus on a particular aspect of the policy process.

4 See e.g. Benneworth 2001, cf. Hudson 1989a.

5 Cf. the discussion in Section 2.1.

6 Compare e.g. Morgan (1994, 1998) on Wales with the work on Scotland by Moore & Booth (1989) and Danson *et al.* (1990b, 1992).

in British politics while the entrepreneurial capacity of the SDA to pursue its aims through innovative measures under different discursive disguises have been given less attention.

- Readers interested in *British politics* would see the book as a contribution to the growing literature on the uneven relationship between political discourse and policy change in the Conservative 1980s, focusing on key dimensions of economic policy and territorial management within the British union state.
- From a *regional studies* perspective systematic book-length attention has seldom been given to neither the study of regional development activities across different European policy paradigms nor, indeed, the politics of regional policy, i.e. why particular strategies have been adopted or rejected and the role in such processes of policy entrepreneurship and discursive strategies by semi-autonomous bodies like the SDA.
- And finally from a broader *theoretical* perspective those grappling with similar conceptual and analytical issues in other areas of public policy may be inspired, positively or negatively, by the attempt, within an overarching institutionalist perspective, to incorporate new developments with regard to inter-organisational relations and policy instruments and to operationalise the analysis of discourse by drawing on contributions to institutional and conceptual history.

Having said that, when reading the text, three delimitations should be borne in mind. In *geographical terms* the focus on the SDA implies that the discussion of Scottish matters takes up far more space than the British context, although developments on both levels are retraced on the basis of analysis of original source material. Throughout the book Scotland is, by the way, referred to as a region within the UK, reflecting the focus of the text on public policy and Scotland's position within the British union state; but that this region at the same time is also considered to be a nation is one of the fascinating complexities that will be discussed in some detail later (Section 4.1) and, indeed, something that will play a role in the subsequent analysis of the origins and development of the SDA. As regards the *temporal limits* of the study, the core of the investigation covers the period in which the SDA existed, i.e. from its setting up in December 1975 to the merger with the Scottish branch of the Training Agency in April 1991 which produced Scottish Enterprise, a new body with an enhanced policy remit and a drastically revamped organisation. The process leading up to the creation of the SDA will of course be analysed in order to establish the political environment into which the Agency was originally inserted, as will the origins of the Scottish Enterprise proposal because it is likely to reflect on the previous Agency experience – but the extensive practical preparations for the organisational merger from the publication of the Scottish Enterprise white paper in December 1988 till the new body eventually started operating will only be covered in so far as they have direct implications for the work of the SDA or give insight into the changing political agenda for regional development in Scotland. And finally in terms of *analytical perspective*, it must be stressed that the focus is very much on the politics of regional policy not just in terms of identifying problems but also in relation to the often fiercely contested choices of strategies, policy instruments, and organisational settings. Changing directions of regional policy, e.g. the current focus on cluster development initiatives or SDA-type regional development agencies, cannot be reduced to technocratic adoption of 'best practice' but must be understood in the context of wider political preferences with regard to e.g. degrees of decentralisation and the relationship between public and private actors. This also implies that while existing evaluations and the perceived success (or otherwise) of particular policies will of course be taken into account because they form part of the political debate, the book does not claim to undertake a systematic evaluation of the economic impact of the SDA and its policy programmes.

1.3. An Overview of the Text

The purpose of the first part of the text is to identify the issues, both historical and conceptual, which the book will be addressing. Following this introduction, Chapter 2 provides an overview of the literature in three central areas, starting out with general trends in the development of regional policy in Europe and Britain, then turning to Thatcherism as political discourse and policy, and finally focusing specifically on interpretations of the history of the SDA. The exposition will in particular highlight two central problems in the existing literature. On the one hand that on a general level the paradigmatic shift from 'top-down' to 'bottom-up' policies has been replicated in the academic literature as two separate research traditions with few lines of communications in between them, and thus the development of a coherent unitary analytical framework has been severely impeded. On the other hand – and perhaps not entirely unrelated to the first point – the 'radical change' interpretation of British, and indeed Scottish, regional policy has ex- or implicitly been accepted in much of the literature despite its rather problematic underlying assumptions and/or shallow analytical foundations.

Chapter 3 begins the work of reconstruction by establishing the methodological and conceptual basis upon which the historical analysis will proceed. As one of the problems in the existing literature on regional policy would seem to be a fairly one-dimensional understanding of the relationship between public and private actors, the text outlines some fundamental institutionalist assumptions about institutions, actors and discourse which are then applied in the ensuing elaboration of a conceptual framework for the study of the development of regional policy. First the notion of regional policy is defined, setting it apart from other forms of public policy with a spatial dimension and arguing that accounting for changing policy paradigms requires a broad definition that is able to include e.g. regionally-based forms of industrial policy. Then elements of a conceptual framework are elaborated on the basis of an institutionalist approach to policy analysis, stressing in particular the importance of the institutional and discursive political environment in which policies are made and the interaction between policy design and implementation with regard to the relationship between public bodies and private firms. Having identified a range of key variables, these are then integrated within a coherent conceptual approach to the analysis of regional policy, spanning across policy paradigms and covering individual organisations, their interaction, and the wider institutional and discursive environment in which they operate. Finally, this approach is used to generate ideal types of the characteristics of paradigms which are relevant for the empirical analysis of the SDA and British regional policy, namely the traditional central-government led 'top-down' approach and the new regionally-based 'bottom-up' approach revolving around semi-autonomous regional development agencies (RDAs).

Chapter 4 sets the scene by providing detailed historical background for the political and economic nature of 'the Scottish problem' that will be explored in the ensuing empirical analysis, a complex web of issues that in a sense was both the *raison d'être* of the SDA and the condition under which it operated. The text starts by introducing the pre-devolution political system of the British union state, including the dual role of the Scottish Office as the arm of central government in Edinburgh and the voice of Scotland in London, and the politics of unionism and nationalism that sets Scottish politics apart from the English two-party system. Then the economic legacies of the industrial past and the problems of structural adjustment, and finally the development of British regional policy from the 1930s till the mid-1970s is surveyed, focusing especially on aims, methods and perceived results of what was predominantly a traditional form of top-down policy.

The analysis of the development of the SDA begins at the corporate level of the organisation. Chapter 5 is a historical enquiry into the political origins of the organisation, discussing the sources of the notion of an RDA for Scotland, the specific circumstances that eventually led to the adoption of a Labour government proposal by the British parliament in 1975, and the possible legacies which the political debate on the proposal could have for the future development of the Agency. Chapter 6 provides an analysis of the development of the resources – authority, information, organisation and finance – of the SDA, identifying general features of the relationships between the Agency and its political sponsors in the Scottish Office, and comparing findings with evidence from other parts of the British regional policy network. Chapter 7 studies the development of the strategic statements for the SDA, both those produced by the Agency itself and those entailed in the regulations of its political sponsors at the Scottish Office, in order to identify the regional problems targeted and the remedial strategies chosen. These objectives and strategies are then set against a reinterpretation of the broader British background which has often been seen as a shift from an interventionism to neo-liberalism, both with regard to economic strategy in general and regional policy in particular, and on the basis of this the various analyses on the corporate level are drawn together in order to produce an assessment of the organisational and strategic aspects of the development of the SDA. This will provide some first indications about the relationship between continuity and change, the respective roles of the arm's-length body and its sponsor department, the degree to which the Agency was in line with developments on the British level, and what areas of SDA activity that were particularly prone to becoming politically controversial.

The link between the official strategies of an organisation and the activities actually performed may at the best of times be a tenuous one, and this is of course no less so for a heterogenous regional development agency operating in a period where the relationship between public and private actors was high on the political agenda, and the empirical analysis therefore proceeds to consider individual policy areas in which the SDA has been engaged, with each of the chapters focusing on a particular form of public intervention that played a prominent role in the policy profile of the Agency. Each of the chapters is structured in three parts, looking first at *policy design* – origins, sponsorship relations, organisational and informational resources, aims and methods – and then considering *implementation* – instruments, modes of operation, impacts on firms and institutions – before finally comparing the Scottish case with parallel initiatives in the wider *British regional policy network*. Taken together the five chapters cover the vast majority of SDA activities, with the length of individual chapters reflecting primarily the political significance of particular policies, and controversy is constantly recurring also in the detailed discussions of concrete activities because also e.g. the way in which particular policy instruments are employed or the symbols involved in organisational reorganisation have been part and parcel of the politics of regional policy in Scotland. The relatively long Chapter 8 examines the development of the most controversial activity of the SDA, namely direct investments in private firms through equity and loan capital, which started out as a 'flagship' activity of the Agency but was officially downgraded by the incoming Conservative government to a limited auxiliary role in order to prevent it from becoming a form of 'nationalisation through the back door'. Chapter 9 focuses on the other major innovative activity of the SDA, the attempt to develop strategies for particular sectors of Scottish industry; an approach to regional development the wisdom of which gradually became more and more disputed. Chapter 10 investigates another controversial high-profile activity, namely the efforts to persuade foreign investors to locate in Scotland. Visibility would always be ensured by the possibility of announcing the sudden creation of a large number of new jobs, and controversy surrounded both the competition between Scotland and other parts of Britain for foreign investment, and the long-term wisdom of the strategy, allegedly making the region over-dependent on externally controlled low-grade assembly jobs. In contrast to this Chapter 11 examines the development of one of the more low-key forms of Agency activity, namely provision of advice to mainly small and medium-sized enterprises (SMEs); a form of public policy ill-suited to provide photo opportunities for Chief Executives and Scottish Office ministers, but nonetheless a persistent feature of the attempt to support the growth of new and small firms. Chapter 12 considers the development of infrastructure and area-based initiatives, activities that accounted for a significant share of total SDA expenditure but still remained relatively low-key, except for the occasional outburst of territorial politics when it came to prioritising between different parts of Scotland. Finally, Chapter 13 concludes the analysis of SDA policies by comparing the characteristics of the changing policy profile with the overall strategies discussed in Chapter 7, and by returning to the original questions concerning the balance between continuity and change, the capacity of the Agency to influence private firms, the relationship between the arm's-length body and its political sponsors in the Scottish Office, and the position of SDA activities *vis-à-vis* comparable activities conducted elsewhere through other actors within the British regional policy network.

Having analysed SDA activities and the interaction of the Agency with its sponsor department, Chapter 14 turns attention to the discursive environment of the organisation, focusing on the political debate about the SDA and its activities in relation to the perceived weaknesses and strengths of the Scottish economy. The text combines the analysis of both party-political exchanges and contributions from key interest organisations within a chronological framework that follows the public debate from the heated disputes of the late 1970s and early 1980s, via the strangely quiet mid-1980s, right through to the debates about the Scottish Enterprise proposal which ultimately led to the Agency being merged into a new RDA with powers unparalleled elsewhere in Britain.

Finally, Chapter 15 contains the overall conclusion with regard to the development of the SDA in the British and European context, addressing the issues raised in Chapter 1 on the basis of the results of the theoretical deliberations and empirical historical enquiries, before suggesting a number of more general points raised by the Scottish case with regard to the conceptualisation of new trends in regional development initiatives in general and the politics of regional policy in particular.

1.4. Sources and Methods

In the following the main sources on which the book will be based will be introduced, providing an overview of the different types of materials around which the argument will be built.

The first parts of the text – presenting the issues, developing the conceptual framework and providing historical background – rely overwhelmingly on a critical examination of existing work, and the depths in which the literature is covered vary according to the role played by different areas of research in the overall structure of the book. With regard to Scottish matters, the intention has been to have access to all relevant texts, something that is not unrealistic given the relatively limited size of the literature. Concerning developments on the British level, the ambition has been slightly lower, namely to cover all major works on regional policy, and significant contributions to the related areas of industrial and urban policy – plus, of course, British postwar history and politics in general and the 1975-91 period in particular. In order to develop a broader historical and conceptual perspective on the development of regional policy, experience from outside the British Isles has been pursued through targeted reading on European cases, especially from Scandinavia and to a lesser extent Germany and France. The selection of general literature pertaining to institutions and public policy reflects a long-standing preoccupation with the relationship between economy, politics and social institutions, an interest which clearly stretches across disciplinary boundaries.

The empirical analysis of the development of the SDA in the context of British regional policy is primarily based on documentary sources, supplemented by a critical reexamination of the results of existing research. It is of course paramount to be able to draw on contemporary sources that were either directly part of the policy process or which can provide detailed reports of developments, but in order to be able to follow the various stages of the policy process in greater detail it has been necessary to complement this material with sources generated as part of the research process, both by creating new accounts of developments through interviews with key actors and through the creation of new documentary sources on the basis on unpublished data. All in all four groups of sources are particularly prominent, namely:

- SDA, government and other official texts produced within the policy network,
- documents from interest organisations and political parties,
- media reports, and
- interviews with persons involved in the making and implementation of regional policy in Scotland.

As regards SDA and government material, a systematic search has been conducted aimed at obtaining a complete coverage of relevant *published* sources, including acts of parliament, government white papers and reports, the proceedings and papers of parliament and its committees, reports of the National Audit Office, and – not least – the extensive annual reports produced by the SDA and the Department of Trade and Industry in London. If the sole object of investigation had been to meticulously reconstruct the negotiations between the Agency and the Scottish Office, it would of course have been essential to have extensive access to internal files of the SDA and the Scottish Office, but given the much broader focus of the present study it was decided only to supplement with *unpublished* material in selected key areas. Accordingly, through the kind cooperation of persons involved with the SDA *written sources* and *unpublished data* have been obtained that relates to two areas of particular interest to the present study. *Firstly*, the development of the Agency's corporate strategies and policy-making process were greatly illuminated by access to a number of internal documents, including the papers of the Development Consultative Committee, relating especially to corporate strategy and policy development. *Secondly*, on the basis of unpublished information a database covering all the SDA's major industrial investments from 1975 onwards has been constructed, making it possible for the first time ever to undertake a detailed analysis of the development of this important and politically contentious activity.

The study of the politics of regional policy has, in addition to the sources mentioned above, also relied on printed and other material from interest organisations and political parties. Again a very extensive coverage has been achieved as access to published material like annual reports and board/committee papers and policy documents was kindly granted by the Scottish Trades Union Congress and the Scottish Council Development & Industry. Moreover, support in locating more or less 'grey' printed material was given by the Glasgow Chamber of Commerce, the Scottish Engineering Employers' Association, the Scottish Conservative Party, the Scottish National Party, and the Scottish Council of the Labour Party.[7] The public debate has been covered by means of newspapers and business magazines: the major indexed Scottish broadsheet, the Glasgow Herald, has been systematically searched from 1970 onwards, together with business magazines like *Business Scotland* and *Scottish Business Insider*.

Last but not least additional material has been generated by conducting an extensive series of interviews with persons associated with the SDA in order to illuminate aspects of the policy-making process, e.g. the perceptions of majors actors of strategies, implementation, and inter-organisational relations, that would otherwise have been more difficult to reach. Altogether 33 semi-structured interviews have been undertaken with chairmen, board members, directors and other senior managers of the Agency plus key actors from its political environment such as government ministers, Scottish Office civil servants, and representatives of interest organisations and political parties. Few of these approaches were unsuccessful, and thus a very considerable part of the decision-makers in Scottish regional policy in the period from 1975 to 1991 were interviewed, generating more than 800 pages of original source material based on taped and audiotyped conversations. All interviews were undertaken in 1990-91, a point in time where it had been decided that the SDA would merge with the Scottish arm of the Training Agency to form Scottish Enterprise, and perhaps the fact that the organisation was soon to become history also increased the enthusiasm of the interviewees for recounting their version of the development of the Agency and brought forward a flood of detailed memories about 'the good/bad old days' where the SDA was undisputedly a central actor in the Scottish regional policy network.

All in all the core of the book is built around documentary sources, some of which have been constructed as part of the research process and hence never exploited before. Quite apart from the possible relevance of the text from a comparative or conceptual perspective, this in itself should ensure that the study should be able to shed new light on the development of the SDA in the context of British regional policy.

This overview of key sources has underlined that analysis of texts will be crucial for the empirical study, simply because they are often the only means through which information about policy processes in the past can be obtained, and thus texts will not only be analysed in relation to other texts in order to corroborate controversial facts,[8] but also in

7 The Scottish branch of the employers organisation CBI, the Confederation of British Industry, regretted that they were not able to be of assistance because relevant files did not exist due to a rather informal pattern of organisation.

8 A traditional, and important, preoccupation within history (see e.g. Dahl 1980 pp. 78ff, Clausen 1963 pp. 84ff).

order to establish their relation to broader discursive patterns.[9] Analysis of texts has traditionally been a method central to the study of history[10] and has, more recently, also become an integrated part of the armoury of qualitative methods in the study of politics and policy,[11] but unlike many writings within other text-based disciplines this book will follow the conventional approach in research traditions whose interests lie beyond the text itself in attempting to limit the use of micro-analytical terminology in order to avoid further burdening of a text that is already long and steeped in unavoidable jargon.

---

9 Ifversen 2000b, cf. the discussion of discourse in Chapter 3.

10 See e.g. Bloch 1954, Clausen 1963, Dahl 1980, Carr 1987, Jenkins 1991, Ifversen 2000a, 2000b.

11 E.g. Elklit 1987, Humlum 1990, Andersen 1999, Jørgensen & Phillips 1999.

CHAPTER 2

Reviewing Literatures

For a study of the development of the SDA in the context of British regional policy, the existing literature represents both a rich source of inspiration and an analytical challenge. Much has been written about regional policy in Britain and Western Europe, the Thatcher years have developed into a major area of study in its own right, and even the SDA and its development activities have to some extent been the subject of academic analysis – and this project will review previous findings in order to develop its own distinctive argument. But at the same time the existing literature also constitutes a challenge because of the nature of the conceptual frameworks employed, often relying on relatively few parameters that easily become 'over-extended' when used to summarise complex social phenomena.

This chapter provides a critical overview of key arguments and concepts in the literature in three areas that are central to the themes of the book, namely the development of regional policy in Western Europe and Britain (Section 2.1), the politics and policies of Thatcherism (Section 2.2), and the history of the SDA (Section 2.3). The aim of the literature reviews is threefold. *Firstly*, to situate the analysis of the SDA firmly within the broader context of the development of regional policy in Western Europe, and in the specific context of British politics in the 1970s and 1980s. *Secondly*, to establish the current state of affairs with regard to the analysis of British regional policy in general and the SDA in particular in order to identify key themes and hypotheses about their development. And *thirdly*, to identify central features of the conceptual frameworks employed in the existing literature in order to assess their relevance from the perspective of the current project. On the basis of this the final section places the aims of the present study in relation to the existing literature and suggests ways in which the problems identified may be addressed.

2.1. Understanding Regional Policy

As regional policy has changed from the 1960s to the 1990s, so have the ways in which it has been studied in the academic literature. The purpose of this section is to outline the most significant policy trends in

Western Europe and Britain,[1] and to establish how these trends have been translated into different analytical traditions. A full discussion of the notion of regional policy, including a review of definitions employed in the existing literature, will be undertaken in Chapter 3 below, but for the purpose of the present section a preliminary delineation will suffice. Accordingly, in the following regional policy will refer to

> public policies aiming to influence the relative economic performance of regions *vis-à-vis* other regions

and two features in particular should be noted. *Firstly*, that the focus is on economic performance rather than e.g. wealth measured as disposable personal income, and thus interregional transfers through central government funding for regional and local government do not fall under this heading. *Secondly*, as an intention of differential impact is required, central government policies which happen to benefit firms in particular regions, e.g. support for particular forms of industrial research and development, are also excluded. Both types of activities are demonstrably important in terms of allocation of financial resources between regions,[2] but they are not considered in the following discussion because the focus of this study is on a particular form of territorially discriminatory economic policy rather than the overall spatial consequences of government activity.

Changing Policies

Working on the basis of this preliminary definition, the statement that regional policy in Britain and Western Europe has changed dramatically since 1945 is difficult to dispute. Although the precise timing of developments may differ between countries, three main periods can be identified on the basis of predominant actors and policies.[3]

The period until the early 1960s could be labelled the *initial phase* where the degree to which regional economic inequalities formed the basis of public policy varied greatly.[4] Some countries such as Britain and Italy introduced what by later standards turned out to be relatively

---

1 The terms "Britain" and "British" here are to be taken literally, i.e. *not* including Northern Ireland where separate programmes and institutions have existed (see e.g. Hart & Harrison 1992, Finnegan 1998).

2 In financial terms budgetary transfers to regional and local government and non-spatial economic policies with spatial consequences are very extensive indeed (see Wishlade *et al.* 1997, Martin 1993, Allen 1989).

3 European overviews are provided by Nicol & Yuill 1982, Bachtler 1997, Martin & Townroe 1992, Keating 1997, Stöhr 1989, Veggeland 1992. For introductions to the development of regional policy in Britain, see e.g. Armstrong & Taylor 1985 pp. 171-224, Prestwich & Taylor 1990 pp. 112-61.

4 'Initial' would seem to be less prophetic than Nicol & Yuill's 'dormant' (1982).

limited measures, while others like e.g. France, Germany and Denmark did very little despite having sizable interregional disparities.[5] Central government was responsible for regional policy and attempted to support weak regions through the strategic employment of 'hard' resources such as infrastructure, financial incentives for firms locating in designated problem areas, and spatial regulation of industrial location.

The second period, stretching from the early 1960s through to the late 1970s, could be termed the *national phase* because regional policy was near-exclusively dominated by central government measures similar to those pursued in the preceding decades. Financial and other 'hard' resources were employed to achieve interregional equality through redistribution of economic growth, especially through diversion of firms and investment from more prosperous parts of the country to designated problem regions with high levels of unemployment. Policy programmes were generally not selective, i.e. they did not focus on particular industries or types of firms, and they operated in a reactive manner with government offices considering applications from individual firms for e.g. financial support for particular investments. The main change in relation to the previous phase relates to the scale on which regional policy operated: in the second period central government involvement in addressing 'the regional problem' became a general feature in Western Europe and the level of expenditure rose significantly, although differences between countries still existed, and Britain, along with Italy, was a leader in terms of expenditure on regional measures.[6]

The third period, from the 1980s onwards, can be described as the *multi-level phase* in that both regional and European actors came to play important roles in regional development alongside central government,[7] and that an increasing number of policy programmes, not least those emanating from the European level, came to involve cooperation between several tiers of government. The regional programmes of central government were maintained in most countries,[8] but expenditure was reduced – in 1990/91 British real term expenditure constituted only 13%

5 See e.g. McCrone 1969, McCallum 1979, Vanhove & Klaassen 1987 pp. 263-81, Keating 1988a pp. 147-56, Halkier 2001.

6 In the period 1975-80 average net grant equivalent expenditure per head of national population (in fixed 1975 prices, pound sterling) varied from 14.0 (Italy) and 7.3 (the UK), via 2.8 (Germany) and 1.9 (France), to 0.6 (Denmark) (Yuill & Allen (eds.) 1982 p. 101).

7 The vertical expression 'multi-level' is preferred to 'multi-jurisdictional' (Armstrong 1986) because the text draws on the broader notion of 'multi-level governance' rather than the more narrowly economic 'fiscal federalism'.

8 Denmark is the exception in that all central government regional subsidies were terminated in 1991 (Halkier 2001).

of the maximum recorded in 1975/76[9] – and programmes became increasingly selective, not least with regard to the types of investment projects that were supported through financial incentives. Parallel to this, an explosive growth occurred in what became known as 'bottom-up' regional policy, initiatives specific to individual regions which often involved the setting up of separate development bodies,[10] and in Britain this not only involved local government initiatives 'from below' but also central government sponsorship of major RDAs in Scotland and Wales.[11] Regionally-based institutions would be able to target the specific needs of individual areas and operate in a more proactive manner in devising programmes and projects, and policies tended to focus mainly, but not exclusively, on attempts to strengthen the competitiveness of the region by supporting indigenous firms by means of 'soft' policy instruments like advisory services, although in many cases 'harder' forms of support, such as technological infrastructure or venture capital, were part of the armoury too. In parallel with this mushrooming of economic development initiatives 'from below', the European level also emerged as a major actor in regional policy.[12] Developing gradually from being mainly a system for reimbursing national expenditure on regional policy and supporting basic infrastructure projects, the European Structural Funds came to constitute a regional policy programme in its own right with a separate system of designated 'problem areas' and development programmes from 1988 onwards. Although 'hard' policy instruments such as infrastructure and investment subsidies continued to play a major role, support for 'softer' measures such as advisory services and network building gradually became increasingly important, and the introduction of regional development programmes as the vehicle through which support is channelled to individual regions could also be interpreted as similar to the specific targeting entailed in the bottom-up approach. Finally, it should be stressed that the three levels of regional policy have not only coexisted but also interacted in a variety of ways.[13] On the one hand through higher-tier regulation of lower-tier activities such as European rules about subsidies to individual firms in member states, or national control of the statutory powers of regional government. On the other hand through cooperation within particular policy

9 Calculated on the basis of Wren's grant-equivalent figures (1996b p. 328).

10 The two major surveys of RDAs in Western Europe are Yuill (ed.) 1982 and Halkier & Danson 1997. For case studies of individual agencies and countries, see e.g. Alden & Boland (eds.) 1996, Halkier, Danson & Damborg (eds.) 1998, Danson, Halkier & Cameron (eds.) 2000.

11 Lawless 1988, Cochrane & Clarke 1990, Danson *et al.* 1992, Tickell *et al.* 1995.

12 For overviews, see Wishlade 1996, Staeck 1996, Anderson 1990.

13 See Bachtler 1997, Wishlade 1996, Page & Goldsmith 1987.

programmes such as the European Structural Funds where the involvement of national and regional actors is required, and in cases where regional development bodies implement national policy programmes.

Since the 1960s regional policy in Western Europe has in other words changed profoundly, and the principal trends can be summarised as follows:

- Organisationally regional policy has ceased to be an area of public policy in which central government enjoys a monopoly, and instead three tiers of public authorities are now involved, acting alone or in conjunction.
- Development strategies have also become more complex, gradually supplementing the original emphasis on diversion of investment and jobs from more prosperous regions with attempts to strengthen the competitiveness of indigenous firms within the weaker regions.
- Policies no longer exclusively rely on 'hard' measures such as finance, infrastructure and regulation, allegedly entailing relatively strong incentives for private actors to comply with public priorities.[14] Instead 'softer' measures like advisory services and network building have become more important, and at the same time more selective and proactive approaches have been introduced.

As the national phase gave way to the multi-level phase, the field of regional policy did in other words become far more complex, both in terms of actors, development strategies and policies. Moreover, while British development were broadly in line with these Western European trends, some specific features are also worth noting: both the expansion and the subsequent contraction of central government programmes have been remarkable in terms of scale and pace, and at the same time differences exist between England and other parts of Britain with regard to bottom-up initiatives in regional development. Taken together, this suggests that the literature on British regional policy can be expected to combine general European features with traits reflecting specific historical circumstances.

Analytical Approaches to Regional Policy

Since the 1960s a growing body of literature on regional policy has emerged, and although diversity has been increasing since the 1980s, it is still possible to identify some important features that would seem to characterise significant parts of the existing contributions.

---

14 The relevance of the distinction between 'hard' and 'soft' measures will be discussed in Chapter 3.

Firstly, the organisational differentiation of regional policy has tended to be reproduced in the literature, with most writings focusing on either national, regional or European programmes. As the nature of policies have tended to differ between tiers of government – e.g. national programmes of financial subsidies versus regionally-based advisory services – this division of labour has resulted in three distinct sets of writings, each with their associated methodologies, sometimes cohabiting in e.g. the columns of the *Regional Studies* journal or edited volumes,[15] but still maintaining their separate characteristics. An important exception to this pattern of segmentation is a small, but important, body of literature attempting to account for the rise of bottom-up initiatives and contrast their key features with those of traditional top-down schemes of central government.

Secondly, an interest in the impact and evaluation of policies has generally been very conspicuous. As regional policy is a spatial economic policy, it is hardly surprising that this has led to a strong presence of studies emanating from the economic sciences,[16] macro- as well as micro-oriented, although it should be noted that this dominance is somewhat less pronounced in studies of European and regionally-based development initiatives where complex institutional set-ups have prompted interest from researchers within e.g. political science and organisational studies. In Britain Wyn Grant observed in 1982 that "there has been little interest in the systematic examination of the political forced surrounding the formation of regional policy",[17] and two decades later Wayne Parsons' *The Political Economy of British Regional Policy* (1988) is still the only book-length study available.[18]

Segmentation in combination with a predominance of studies oriented towards economic impact evaluation would in itself hardly seem to be conducive to the development of an integrated analytical framework capable of accounting for the origins and consequences of policy change, and this impression is reinforced by a closer look at the research traditions associated with the policies of each tier of government.

Most studies of central government regional policy have focused on establishing the effects on the regional economies. Both financial subsidies and the investments they promote lend themselves to quantification in monies or employees, and economists and quantitative methods have

15 See e.g.Maclennan & Parr (eds.) 1979, Townroe & Martin (eds.) 1992, Harrison & Hart (eds.) 1993.

16 Grahm 1988, Hogwood 1986 cf. the discussion below.

17 Grant 1982 p. 53.

18 Other contributions include Morgan 1985, Hudson 1989a Ch. 4, and, placing regional programmes in the wider context of industrial policy, Hogwood 1986, Grant 1982.

therefore played a prominent role in the study of policies conducted from the national level, developing increasingly sophisticated techniques in order to cope with the problems involved in estimating the aggregate economic impact on the regional level of particular measures.[19] In addition to this quantitative approach, qualitative dimensions of central government policies have also been studied: the increasingly complex and frequent policy changes have been recorded in detail,[20] and some work has been done on the interaction between public bodies and private firms.[21] All in all national-level regional policy has been analysed by means of concepts such as those listed in Table 2.1,[22] primarily focusing on quantitative variables concerning the amount of resources employed and the outcome in terms of jobs and investment, but also employing qualitative variables with regard to the selectivity of policy programmes and whether or not payment is automatic or subject to discretionary administrative assessment. Although the introduction of these qualitative variables made the study of national-level regional policy a rather complex task, it was still a widespread practice to compare different countries and periods by means of fairly simple dichotomies such as 'weak' and 'strong' policies, with the latter often perceived as being linked to more general approaches to economic policy such as 'interventionism' or 'Keynesianism',[23] and this produced some rather uncomfortable analytical dilemmas. If expenditure falls because of an increasingly selective focus on particular industries with good prospects in terms of job creation and competitiveness, has policy then become 'weaker' or 'stronger'? And what to make of an avowedly anti-interventionist government that institutes discretionary policy programmes in which civil servants impose their judgements about the market upon private firms?

---

19 See e.g. Moore & Rhodes 1973, Bölting 1976, Ashcroft 1980, Diamond & Spence 1983, Folmer 1986.

20 On a comparative Western European basis, *European Regional Incentives* (Yuill *et al.* 1981-99) remains the classic example of this approach, but it can also be found in the study of individual countries: see e.g. Gaardmand (1988) on Denmark, Neuperts (1986) on Germany, and McCrone (1969) on Britain.

21 E.g. Allen *et al.* 1988.

22 The table has been constructed on the basis of distinctions found in central contributions to the existing literature, i.e. McCrone 1969, Yuill *et al.* 1981ff, Armstrong & Taylor 1985, Vanhove & Klaassen 1987, Prestwich & Taylor 1990.

23 See e.g. Ashcroft 1980, Cambridge Economic Policy Group 1980, Grant 1982, Morgan 1985, Damesick & Woods 1987, Moore & Pierre 1988, Grahm 1988, Hansen *et al.* 1990, Prestwich & Taylor 1990, Martin 1985, 1989a, 1989b.

Table 2.1 Analysing national-level regional policy

Dimension	*Variables*
Policy instruments	* finance * information * regulation
Selectivity	* area * industry * project type
Implementation	* automatic * discretionary
Output	* expenditure * enforcement
Outcome	* investment * employment * competitiveness

In contrast to this relatively homogeneous picture of national-level regional policy, the smaller scale and much more heterogeneous nature of bottom-up initiatives have generally pointed towards the employment of more qualitative methods. Case studies or surveys via questionnaires have formed the foundation of much research in the area, and the study of regional policy therefore began to draw increasingly on approaches from the management, organisational and political sciences. After an initial period in which parts of the academic literature displayed a rather messianic tone, singing the praises of the emerging forms of bottom-up development initiatives,[24] much energy has been devoted to describing individual initiatives and appraising their impact on private firms.[25] Given the diverse nature of bottom-up regional policy and the variety of analytical approaches employed, it is perhaps unsurprising that no

---

[24] See e.g. Segal 1979, Veggeland 1983, Martin & Hodge 1983b, Rothwell & Zegveld 1985, Chisholm 1987, Ewers 1990, Christensen *et al.* 1990, Cooke & Morgan 1993, UNIDO 1997, Lambooy & Boschma 2001.

[25] International surveys such as Yuill (ed.) 1982 and Halkier & Danson 1997 are few and far between. Much more common are studies of bottom-up regional policy in individual countries such as Danson *et al.* 1992 (Britain), le Galès 1994 (France), Sturm 1991 (Germany), Neumann & Uterwedde 1986 Ch. 6 (Germany and France), Steiner & Jud 1998 (Austria), Sleegers 1998 (the Netherlands), Damborg & Halkier 1998 (Denmark), and along similar lines in the United States, Eisinger 1988. Studies of individual regions or development bodies abound, in England as contributions to the extensive literature on local economic policy (e.g. Campbell (ed.) 1990) and studies of individual RDAs in Wales and Scotland (e.g. Cooke 1980, Grassie 1983, Alexander 1985, Rees & Morgan 1991, Thomas & Drudy 1993, Griffiths 1996, Rees 1997, Hughes 1998, Morgan 1997 & 1998, cf. Section 2.3 on the SDA).

common conceptual core has emerged, although organisational features have been fairly prominent. In practice this does, however, also imply that a systematic analysis of the similarities and differences between various types of bottom-up initiatives, let alone between these and top-down forms of central government intervention, is not readily available, but although the more complex nature of bottom-up activities make them even less well suited to the use of dichotomous interpretative schemes, in Britain the ideological dimension has sometimes been foregrounded in order to position particular agencies or policies *vis-à-vis* the economic strategies and policies of the Conservative governments of the 1980s when e.g. English bottom-up initiatives have been grouped according to whether they entail "restructuring for labour or for capital".[26]

Just as the regional policies instigated from the European level could be said to combine elements from both the top-down and bottom-up approaches, so does the academic literature. On the one hand European programmes have, especially since the 1988 reform of the Structural Funds, been systematically evaluated, and this interest is, hardly surprising, also reflected in the academic literature.[27] On the other hand the complex institutional setting with its emphasis on partnership between European, national and regional actors has spawned a growing interest in organisational and political issues, highlighting the relationship between different tiers of government and public bodies currently involved in regional policy in Western Europe within a framework that has been dubbed 'multi-level governance',[28] with British contributions stressing particular issues such as e.g. the conflict between London and Brussels about "additionality" and the absence of a regional tier of government in England. Unfortunately, cross-fertilisation between the two lines of inquiry on the supra-national level would still seem to be fairly limited, at least in terms of developing an integrated conceptual framework for the study of regional policy.

Interpreting the Development of Regional Policy in Britain

The difficulties encountered when trying to account for policy change in the absence of a suitable analytical framework can be illustrated by taking a closer look at prevailing interpretations of the development of regional policy in Britain. The segmented character of the literature means that we should in effect be looking for three parallel

26 Totterdill 1989.

27 E.g. Armstrong 1993, Charles & Howells 1993, McAleavey 1993 & 1995, Bachtler & Turok (eds.) 1997.

28 See e.g. Marks *et al.* 1996, Hooghe (ed.) 1996, Heinelt & Smith (eds.) 1996.

stories, but as the development of European regional policies in Britain has generally been cast in a supporting role as an external influence on other policies, we are left with only two sets of interpretations to account for, focusing on the national and regional levels respectively.

With regard to the top-down policies operated from London in the name of inter-regional equality, two very different strands in the literature can be identified on the basis of their interpretation of the objectives of government policy. Most authors have analysed regional policy in terms of its stated or imputed *socio-economic* aims of limiting inter-regional inequalities, either in terms of unemployment or economic competitiveness.[29] Some observers have, however, emphasised the *political* aims of regional policy as a way of managing social processes and conflict regarding e.g. unemployment, class, or territory.[30] Although these two types of objectives may have coexisted historically[31] – improving regional economic indicators may bring political advantages – stressing one or the other would have rather different implications in terms of how the development of regional policy is likely to be interpreted.

Focusing on the political objectives implies that policies will be judged by standards of political management, i.e. containing discontent on a level that does not threaten basic economic and political institutions. Such criteria are not necessarily linked in any straight-forward manner to changes in economic disparities nor indeed to particular policy instruments or degrees of resource commitment: if high levels of unemployment become less politically sensitive, then the need for government to be seen to be doing something about inter-regional inequality is less urgent. Along these lines Parsons argues that "regional policy has […] been shaped by the demands of political expediency" in order to "maintain the territorial legitimacy of government" by "managing high levels of unemployment".[32] This leads to the conclusion that also – or perhaps especially – in the 1980s where regional policy became much more selective and expenditure decreased significantly

---

29 See e.g. McCrone 1969 Ch. 1 and 8, Ashcroft 1978, Martin & Hodge 1983a, Armstrong & Taylor 1985 Ch. 9, Vanhove & Klaassen 1987 pp. 274ff, Martin & Tyler 1992.

30 The core argument of Parsons 1988, also found in Morgan 1985. Dunford *et al.* (1981) see regional policy as management of class relations in that it provides support for employment and investment without interfering in the decisions of and bargaining within individual firms.

31 The role of political motives is also recognised by e.g. McCrone 1969 pp. 30f, Regional Studies Association 1983, Diamond & Spence 1983 pp. 8f, Young & Hood 1984 p. 30, OECD 1994.

32 1988 pp. viii-ix.

> from the political point of view [...] there can be little doubt but that the system of regional aid proved highly successful in containing the problems associated with unemployment and economic decline [...] Politically, the fact that the periphery looked to the centre for answers and aid greatly reinforced the legitimacy and power of the state.[33]

The story of top-down regional policy in Britain tend to look rather different when viewed from the perspective of its objectives with regard to inter-regional equality. Here policies can be judged on their impact on particular indicators and hence success could be expected to depend on the resources devoted to the policy area in a relatively direct way. If inter-regional differences persist, decreasing government expenditure as experienced in Britain in the 1980s can be seen as a lack of commitment to the fundamental goals of inter-regional equality, and it is therefore possible to interpret the long-term development of British regional policy in terms of a rise-and-fall scenario in which the "retreat"[34] from or "rejection of"[35] regional policy by central government leading to its "weakening",[36] "demise",[37] "dismantling"[38] and "phasing out".[39] Because of the rather direct link between public activity and measures of success, understanding how and why central government policies have changed becomes an important task from this perspective, and in the following existing interpretations will be discussed under the headings of ideological changes, changing political priorities, and policy dysfunction, the main sources of change identified in the literature.

Given that many of the changes took place in the 1980s and were presided over by a Conservative government committed to 'rolling back the frontiers of the state' and 'freeing the market', it is perhaps not surprising that the 'decline' of central government regional policies could be interpreted as the result of the dominance of *liberal ideologies* at the heart of the British state. Pursuing inter-regional equality by offering financial subsidies to firms locating in the assisted areas could be seen as part of the old postwar consensus in several ways: although like Keynesian macro-economic management it did not involve government direction of individual firms but relied on indirect measures such as subsidies that firms were free to ignore, regional policy did entail a commitment to spatial equality that could be seen as a supple-

---

33 Parsons 1988 pp. 198f.

34 Damesick 1987.

35 Anderson 1990a.

36 Damesick & Woods 1987.

37 Gibbs 1989.

38 Martin 1986 p. 269.

39 Balchin 1990 p. 67.

ment to the welfare benefits supporting individual citizens.[40] When a new government arrived in 1979 with a *laissez-faire*[41] or "market-oriented"[42] approach, this could not be expected to be compatible with a 'strong' regional policy, and the Conservatives were therefore believed to have maintained a limited regional policy for purely pragmatic reasons: as a symbol of political commitment to less prosperous regions and the territorial unity of the state,[43] and, not least, because nationally designated assisted areas had been made a prerequisite for obtaining support from the European Structural Funds.[44] The critical importance of ideology has, however, been disputed by others,[45] arguing that as the downgrading of regional measures began well before the first Thatcher government was elected in 1979 and similar trends can be found in European countries with a less radical ideological climate, factors other than rampant liberalism must have been at work.

This leads us on to the question of *changing political priorities*, a type of explanation that focus on the relative priority of regional issues *vis-à-vis* other central government policies. Three different policy areas in particular have been seen as contributing to the marginalisation of traditional forms of top-down regional intervention. *Firstly*, the introduction of new forms of macro-economic management inspired by monetarism lead to financial austerity and cutbacks in public sector programmes, hitting not only regional policy but also other areas.[46] *Secondly*, the weakening international position of British manufacturing made national competitiveness rather than the location of industry in particular regions the overriding priority, and hence various forms of national industrial policy became a political priority for central government.[47] And *thirdly*, problems in Britain's inner cities in terms of unemployment, deprivation and social unrest gave another form of spatially targeted intervention a new urgency, and urban policy eventually outgrew regional policy in terms of expenditure in the late 1980s.[48] More-

40 See Grant & Wilks 1983, Martin & Hodge 1983a, Martin 1986 p. 269, Keating 1988a pp. 147 & 167, Martin 1989a p. 28, Balchin 1990 pp. 65ff.

41 Balchin 1990 p. 99, Gudgin 1995 p. 61.

42 See Martin 1985 and 1989a, Townroe 1986.

43 Martin 1985, Jones 1986, Damesick & Woods 1987, cf. Morgan 1985.

44 Morgan 1985, Martin 1985 and 1986, Jones 1986, Damesick & Woods 1987, Parsons 1988 p. 199, Prestwich & Taylor 1990 p. 153.

45 Johnes 1987, Taylor 1992 p. 292, Barberis & May 1993 pp. 150ff.

46 Diamond & Spence 1983 p. 105, Regional Studies Association 1983 p. 9, Martin 1985 & 1986, Hudson 1989a Ch. 3, Gibbs 1989, Prestwich & Taylor 1990 p. 150.

47 McCallum 1979, Morgan & Sayer 1983, Ashcroft 1983, Diamond & Spence 1983 p. 19, Regional Studies Association 1983 pp. 17ff, Prestwich & Taylor 1990 p. 231.

48 Damesick 1987, Martin & Townroe 1992, Martin 1993 p. 271, Wren 1996a pp. 115ff.

over, in all three areas significant changes had taken place well before the advent of the first Thatcher government – monetarism and financial austerity had been introduced by the Labour government of James Callaghan[49] and the increased importance of industrial and urban intervention also predates 1979[50] – and thus perceiving changing political priorities as the basis for the 'demise' of regional policy in effect questions the overriding importance of party politics and ideology.

Explanations grouped under the heading of *policy dysfunction* argue that the changes which top-down regional intervention underwent from the mid-1970s can be traced back either to growing doubts about the effectiveness and relevance of traditional policies. *Firstly*, despite regional policy measures having been in place since the mid-1960s, unemployment generally remained at a markedly higher level in the designated assisted areas, and thus the effectiveness of the policy in achieving what was assumed to be one of its main goals was open to doubt.[51] *Secondly*, the cost-effectiveness of the existing policies as means of creating jobs in the assisted areas became a political issue from the early 1970s. Investment grants were automatically paid out to firms locating in the designated areas, also when these firms would have located there anyway for reasons of e.g. logistics, and thus support for e.g. capital-intensive oil-related facilities in Scotland could be used to raise doubts about the effectiveness and efficiency of the existing regional policies of central government.[52] *Thirdly*, the economic recessions that hit Britain in the 1970s and early 1980s did not only erode the very mechanism through which central government policy operated – low levels of investment meant that few firms and projects were available for 'diversion' to assisted areas – but also undermined political support for regional policy because diversion of investment and jobs became a much more sensitive issue.[53] And *finally*, it has been argued that the relevance of traditional regional diversionary policies had been fundamentally challenged by new economic trends, especially a shift away from mass-production towards new forms of flexible specialisation. Not only did the shift away from mass production undercut the supply of 'divertible' investment and link the fate of individual regions to the competitiveness of indigenous firms, but it also produced new spatial patterns in which

---

49 Martin 1986 p. 239.

50 McCallum 1979, Parsons 1988 p. 185, cf. Section 4.3.

51 Regional Studies Association 1983 p. 2, Parsons 1988 p. 194, Martin & Townroe 1992 p. 20, Gudgin 1995 p. 55.

52 Ashcroft 1982, Regional Studies Association 1983 p. 11, Jones 1986, Martin 1986 p. 271, Damesick 1987, Gudgin 1995 pp. 56f.

53 See e.g. McCallum 1979 pp. 34ff, Alexander 1981, Jones 1986, Anderson 1990b, Gudgin 1995 pp. 54f, Taylor 1992.

difference *within* regions were more significant than *between* regions, and thus the trend towards flexible specialisation would seem to undermine both the economic and political rationales for an inter-regional form of spatial economic policy.[54] Even this brief overview underlines that the apparently 'technical' preoccupations of the explanations grouped under the heading of policy dysfunction point towards political considerations about the desirability of particular goals and the likelihood of specific policy measures being able to bring about change.

Turning now to the interpretations of the rise of bottom-up forms of economic development policies, a similar pattern, albeit perhaps on a less grandiose scale, would seem to emerge. While some authors have traced this trend back to the same structural economic changes that undermined the working of top-down policies,[55] most writings have tended to interpret the development of bottom-up regional policy in relation to their sponsoring level of government. With regard to the non-English RDAs sponsored by central government, an important issue in the literature concerns the extent to which the regional agencies were transformed or adapted themselves to the economic priorities of the Thatcher governments in the 1980s:[56] while some focus on the continuities in the development of the RDAs,[57] others have argued that the agencies did move away from their interventionist past towards a much more market-oriented approach.[58] With regard to regional development initiatives by local authorities in England, their development in the 1980s has also to a large extent been interpreted in the light of central government policy.[59] On the one hand bottom-up activities were initiated in response to developments on the national level: to compensate for the increasingly selective nature of top-down regional policy, or to demonstrate practical alternatives to the economic policies of Thatcherism. On the other hand the way in which they were organised gradually came to resemble central government urban policy with a strong emphasis on formalised partnership between public and private actors. All in all it would appear that central government, directly and indirectly, is being seen as a major influence, and hence developments on the national level

54 See Martin & Hodge 1983b, STUC 1989 pp. 40f, Martin 1989a, Gibbs 1989, Morgan 1996, cf. Firn 1982 pp. 9ff..

55 Martin & Hodge 1983b, Martin 1989a, Moore & Booth 1986a pp. 62ff, Cooke 1990.

56 For a discussion of interpretations of the development of the SDA, see Section 2.3.

57 See e.g. Grassie 1983, McCrone & Randall 1985, Moore & Booth 1989 pp. 113ff, Morgan 1994.

58 Danson *et al.* 1989a, 1989b, 1990, and 1992; Griffiths 1996.

59 Goodwin & Duncan 1986, Moore & Pierre 1988, Totterdill 1989, Barnekov *et al.* 1989 pp. 206ff, Cochrane 1990, Pickvance 1990.

and the way they interact with sub-national actors became an important part of the overall picture of bottom-up regional policy in Britain.

Two conclusions suggest themselves on the basis of the preceding discussion. In relation to the concrete explanations of the development of British regional policy from the 1970s through to the 1990s, the role of the Thatcher government with its liberal rhetoric is clearly a contested one, with many seeing it as a major factor in its own right while others explicitly or implicitly dispute its role by stressing the continuities from the mid-1970s onwards, the contradictions inherent in the policies of the 1980s, or the importance of other explanatory factors such as the emergence of a new (post-fordist) paradigm of economic development. And in terms of conceptual frameworks, the explanations proffered for the change of direction taking place in British regional policy between the 1960s and the 1980s would again seem to rely predominantly on factors that are external to the analytical dimensions employed when analysing individual forms of regional policy, e.g. party politics and ideology, relations between different tiers of government or areas of public policy, and the impact of structural economic changes.

Interpreting Paradigmatic Change in Regional Policy

Although the complex realities of multi-level regional policies, in Britain and elsewhere, have hitherto primarily been translated into segmented programme-oriented forms of research rather than integrated attempts to gauge the overall development trends within a particular region or country, the existing academic literature does, however, also contain elements that can engender some degree of optimism with regard to the prospects of overcoming well-entrenched limitations. Ironically, particularly useful in this respect is the tendency to pinpoint the perceived limits of other, 'competing', forms of regional policy, especially between the two forms of regional policy that have been seen as being furthest removed from one another, namely the regionally-based bottom-up initiatives and the traditional top-down forms of regional policy pursued by central government at the national level.

Table 2.2 Regional policy paradigms – A preliminary comparison

Key features	*Top-down*	*Bottom-up*
Level of operation	National	Regional
Underlying value	Interregional solidarity	Interregional competition
Strategy	Redistribution of growth	Indigenous growth
Policy instruments	'Hard'	Mostly 'soft'
Mode of operation	Non-selective Reactive	Selective Proactive

From a *top-down perspective* the stressing of the need for individual regions to improve their competitive position *vis-à-vis* other regions by bottom-up policy would seem to fly in the face of the basic value underlying traditional central government policies, namely interregional solidarity, and at the same time the impact of bottom-up initiatives was likely to be limited by its reliance on internal and primarily 'soft' policy resources, and, of course, potentially cancelled out by the efforts of competing regions.[60] Conversely, from a *bottom-up perspective* the results of top-down policies have often been seen as questionable, because diversion of investment from more prosperous areas often resulted in transfer of low-grade assembly operations that would be vulnerable in times of economic hardship. Moreover, the use of non-selective policy instruments also meant that central government programmes ultimately relied exclusively on private sector investment projects and hence were insensitive to specific needs of individual regions.[61]

Especially proponents of bottom-up measures have attempted to spell out the difference between themselves and the traditional policies of central government,[62] and when these efforts are combined with the key characteristics outlined above, the contours of two competing paradigms in regional policy would seem to emerge, each promoting the position of some regions in relation to other regions, but doing so in rather different ways, as illustrated by Table 2.2. Regional development can be pursued either within a national framework based on interregional solidarity or through competition between individual regions, and historically these two approaches have come to be associated with particular forms of public policy, i.e. either national redistributive policies based on a non-selective and reactive application of 'hard' resources or regional indigenous-oriented strategies with a more selective and proactive application of 'soft' policy instruments. This pattern is of course not the only one imaginable or indeed practicable – central government may support indigenous development and RDAs may compete to attract footloose investment from outside – but the contention here is that these particular combinations of organisation, objectives and policies have nonetheless come to been seen as paradigmatic, i.e. as fundamentally different ways of addressing the political issue of economic inequality between regions.

---

60 See Armstrong 1986, Keating & Hooghe 1994, Martin & Townroe 1992, Bachtler 1997, Dunford 1998.

61 See e.g. Martin & Hodge 1983a, Rothwell & Zegveld 1985 Ch. 7, Stöhr 1986, Albrechts & Swyngedouw 1989.

62 Elaborated on the basis of Albrechts & Swyngedouw 1989, Stöhr 1989, Martin & Townroe 1992, Hallin & Malmberg 1996, cf. Halkier & Danson 1997.

It is, however, also interesting to note that despite the segmented pattern of much of the existing academic literature on regional policy, the fact that the top-down and bottom-up approaches can actually be subjected to comparison with regard to key features strongly suggests that it should be possible to develop a comprehensive conceptual framework capable of accounting for material differences between and within different policy paradigms. Having struck these encouraging notes, it is therefore well worth pausing to take a closer look at the limitations of the 'dual paradigm' interpretation of regional policy in order to assess the scale of the reconstruction work required to arrive at a conceptual framework suitable for the purpose of the present project. One way of doing this is to demonstrate the difficulties encountered when applying the limited range of dichotomous concepts in Table 2.2 to the current multi-level situation. *Firstly*, it is clear that regional policy in Western Europe in the 1980s and 1990s is characterised by the coexistence of different forms of regional policy. Although the relative importance of the traditional top-down policies of central government has decreased, no clear-cut transition from top-down to bottom-up dominance has occurred, and expenditure on national-level regional policy still constitutes a significant policy programme in most Western European states. As the policies pursued from the European level combine features from both paradigms – heavy reliance on financial grants and infrastructure in combination with regional programming and increasing importance of 'soft' policy resources – the two paradigms do clearly not reflect a historical succession of different forms of regional policy but must, at best, be seen as 'ideal types' to which individual countries, regions and policies may conform to a greater or lesser extent at a given point in time. *Secondly*, the limited range of variables entailed in Table 2.2 is also highlighted when we look at the explanations proffered for the reduced role of central government polices and the concurrent rise of European and regionally-based development programmes that have produced the current multi-level situation. While the rise of European policies can be seen as the combined result of inter-governmental bargaining in connection with successive enlargements of the Community and an attempt by the European Commission to enlarge its direct role in policy-making and implementation,[63] the reduction in national policies has been explained with reference to changing economic circumstances – recession causing fiscal austerity and limiting the amount of footloose investment available for diversion to problem regions[64] – and/or as a

63 See Hooghe 1996.

64 See Alonso 1989, Vanhove & Klaassen 1987, Rothwell & Zegveld 1985, OECD 1989, Gudgin 1995, Ewers & Wettmann 1980, Stöhr 1989, Albrechts & Swyngedouw 1989, cf. the discussion of the British case above.

result of political changes in terms of specific priorities or overarching ideologies.[65] Similarly the rapid rise of bottom-up regional policy has been associated with the development of new economic structures that necessitate more targeted forms of regional intervention in order to be effective.[66] Add to this the likeliness that developments within the three levels of government have mutually influenced one another – the increasing European role making it easier for national governments to reduce their activities, and subnational mushrooming being at least partly prompted by developments at the national and European levels – and it is clear that the changing face of regional policy has primarily been explained in terms of factors that are *not* included in the variables delineating the two policy paradigms, thereby underlining the limited potential of the approach.

The present situation with regard to the academic study of regional policy can in other words be summarised as follows. The prevailing traditions of academic enquiry have primarily focused on establishing the economic effects of particular types of policy programmes, and the result has been a segmented literature focusing on relatively few dimensions while concentrating on specific forms of regional policy. Although there are undoubtedly excellent historical explanations – such as the availability of external funds for policy evaluation – for this pattern, it is nonetheless not particularly helpful from a broader analytical perspective. *Firstly*, a comparison between different types of policy measures is hampered by the absence of a comprehensive conceptual framework that can account in a systematic fashion for basic features across policy paradigms. *Secondly*, the focus on economic impact means that significant aspects of policy-making such as methods of implementation and the institutional environment are often overlooked or given only cursory attention. *Thirdly*, and partly a product of the first two features, the study of the *politics* of regional policy – the reasons why some courses of public action are preferred to others – has either been neglected, remained steeped within simplistic dichotomies like liberalism-versus-intervention, or – at best – been undertaken in splendid isolation from the nature of the actual policy programmes. This is rather unfortunate because the general literature on public policy suggests that decisions about the design and implementation of public policies are not exclusively – or perhaps even predominantly – based on an assessment of the prospects of successful achievement of stated goals,[67] and hence the

65 Alonso 1989, Martin 1989a, Jones 1986, cf. the discussion of the British case above.

66 Tracing changes back to a transition from fordism to post-fordism are Albrechts & Swyngedouw (1989) and Batt 1996, but similar lines of argument are pursued by Martin (1989a) and Alonso (1989), cf. the discussion of the British case above.

67 See e.g. Hogwood & Gunn 1986, Ham & Hill 1984.

analysis must also be able to account for the role of e.g. the ideological connotations of particular policy instruments, the perceived need to accommodate particular organised interest groups, or preferences with regard to the organisational set-up for implementation.

2.2. Understanding Thatcherism and Postwar Consensus

British political history in the 1980s has attracted intense academic interest across a wide range of disciplines in the social and cultural sciences. For several decades the study of Thatcherism has been a growth industry in its own right, and although the expansion of this dedicated body of literature slowed somewhat after the turn of the century,[68] the prediction that the 1980s "will continue for the foreseeable future to provide a rich source of data to be used to address some of the key issues in political science and sociology"[69] would seem not only to stand the test of time but also to be relevant in other disciplines as well. Margaret Thatcher, the politician giving rise to the only prime-ministerial 'ism' in the 20th century,[70] reputedly boasted that "I have changed everything",[71] and while many have been sceptical about the accuracy of this claim, more would seem to accept that, like it or not, at least she had a jolly good try at breaking away from the politics and policies of the 'postwar consensus' which had been pursued by Labour and Conservative governments alike from the 1940s onwards.

This section first gives a brief historical outline of British politics from the Second World War to the early 1990s, focusing particularly on the nature of the original cross-party consensus, the attempts in the 1960s and 1970s to address Britain's increasingly difficult economic position from positions within the consensual framework, and the, supposedly different, approach heralded by the election of the first Thatcher government in May 1979. Following this, the nature of the analytical strategies brought to bear on Thatcherism is discussed in order to position the current text in relation to existing paradigms and suggest some implications for the conceptual and methodological approach to the study.

68 A search for titles containing the sequence of letters 'thatcher' in the *ISI Web of Science* on-line database produced 410 hits published in the 1980s, 520 in the 1990s, and only 101 in the four years from 2000 to 2003.

69 Marsh 1995 p. 595.

70 Although often claimed to have been the first and only one, Victorians had used the expression 'Gladstonianism' (Green 2002 p. 214), and in the 1950s the economic policy consensus between the major parties was dubbed 'Butskellism' by *The Economist* journal by combining the names of the Conservative Chancellor Butler and his Labour counterpart Gaitskell (Niss 1991 p. 18).

71 Pugh 1994 p. 302.

Postwar Consensus and the 'British Disease'

In most interpretations of British history, the Second World War is considered to be a watershed. In its wake the Labour Government of Clement Atlee launched a comprehensive set of reforms, building on preparations undertaken by the war-time coalition government and pertaining to the UK as a whole, which formed the basis of public policy for decades to come.[72] Comprehensive welfare measures were introduced which established a safety net covering British citizens 'from cradle to grave' via public involvement in e.g. education, housing, and health care, and through the institution of national schemes regarding pension and financial support for those unemployed or unable to work. In economic policy Keynesian demand management aimed to facilitate high and stable levels of employment, something that was not only an important goal in its own right but also served to limit the cost of the new comprehensive welfare state which provided standardised services and universal benefits. Moreover, the economic role of the state expanded through nationalisation of traditional industries (coal, steel), services (Bank of England, railways), and utilities (electricity, gas) that were considered to be of vital importance to the performance of modern industries and the economic well-being of the nation at large. In relation to the outside world the ambition to retain a significant British role in global affairs was intact, although it required close cooperation with the US and allies in Europe and was no longer envisioned as an endeavour to be achieved solely through Empire and Commonwealth. For several reasons this combination of domestic and international priorities later became known as the 'postwar consensus' for two reasons: they were adhered to by both the two major parties when in government, thus sending the strongest possible signal of cross-party agreement, and the importance of regular consultations with especially trade unions and employers associations was also recognised, indicating the ambition of maintaining a broad consensus in society about the aims and methods of public policy.

In many ways the spirit of the postwar consensus was summed up neatly by the campaign slogan used by the Conservative Prime Minister Harold Macmillan for the 1959 general election, "You never had it so good", which not only hinted at the undisputable fact that British citizens enjoyed unparalleled levels of prosperity but also demonstrated that the Conservatives as the party of government since 1951 was happy to associate itself with popular affluence as a central political ambition.

---

[72] For overviews of the politics and policies of the postwar consensus, see Middlemas 1989a, 1989b, 1991; Hall 1986 Ch. 2-4, Leys 1989, Overbeek 1990 Ch. 4-5, Grant 1993, cf. Halkier 1991a.

This positive complacency was, however, overshadowed by growing concerns about Britain's position in the world at large: not only was the Empire gradually being dismantled, but difficulties in keeping up with other industrialised countries in terms of economic growth gave rise to the notion of a 'British disease' which could explain the relative 'decline' experienced *vis-à-vis* the US and large parts of continental Europe.[73] In the long run it proved difficult to make Keynesian demand management work effectively in the British context where the co-existence of a relatively weak industrial base and the City of London as a global financial centre made policies oscillate between expansionary and deflationary measures, something which discouraged long-term productive investment and thus weakened the industrial base even further.[74] It is therefore hardly surprising that British economic policy in the 1960s and 1970s became dominated by a search for alternative policy instruments which could address the perceived causes of slow growth and thereby possibly reverse the country's deteriorating international position.

On several occasions, notably the mid-1960s under Harold Wilson's Labour government, and from 1972 to 1979 under the Conservative Edward Heath and the subsequent Labour governments, a strong interest was taken in the continental European experience with industrial policy, i.e. public measures designed to promote the development of specific firms or industries.[75] The underlying assumption here would seem to be that the economic decline was primarily caused by the failure of industrial managers to modernise British firms in terms of structure, organisation and technology, and hence tripartite fora were established to improve cooperation between key economic actors, specialist state bodies set up to promote industrial reorganisation and future-oriented investments in new technology, and grant schemes instituted to support the modernisation of traditional industries. Still, by international standards British efforts were, however, fairly limited, frequent changes in support schemes undermined their potential attractiveness to private firms, and

---

73 For central contributions to and overviews of the literature on Britain's decline, see Blackaby (ed.) 1979, Williams *et al.* 1983, Pollard 1984, Wiener 1985, Coates & Hillard (eds.) 1986, Elbaum & Lazonick (eds.) 1986, Anderson 1987, Sked 1987, Sevaldsen 1988, Newton & Porter 1988, Gamble 1990, Overbeek 1990, Halkier 1991c, Dintenfass 1992, Rubinstein 1994, Clarke & Trebilcock (eds.) 1997, English & Kenny (eds.) 1999.

74 For incisive accounts of the peculiarities of British macro-economic management in general and the stop-go cycle phenomenon in particular, see Pollard 1983 pp. 408-30, 1984 pp. 47ff; cf. Ingham 1984.

75 On industrial policy and decline, see Young & Lowe 1974, Stout 1979, Freeman 1979, Mottershead 1978, Grant 1982, Davenport 1983, Wilks 1984, Hall 1986, Martin 1989c, Kramer 1989, Middlemas 1989 & 1991, Coates 1994 Ch. 5-6.

forward-looking initiatives were dwarfed by government expenditure on subsidies for ailing firms and nationalised industries and grandiose prestige- or defence-oriented projects, and thus these policies could hardly be expected to fundamentally change long-standing patterns of behaviour within British industrial firms.

An alternative set of anti-decline policies focused on the immediate environment of industrial managers, i.e. the role of the workforce and the organisations representing their interests. Long-term investment in order to improve competitiveness would be hampered if e.g. introduction of new technology was resisted by trade unions in order to preserve existing jobs, and in the British case matters were complicated by a fragmented organisational pattern that gave rise to recurring 'boundary disputes' between different trade unions about the right of their members to perform particular functions within a firm, and, indeed, by the long-standing adherence to the principle of 'free collective bargaining' which exclusively made industrial relations a matter for management and trade unions to be handled primarily at the local level.[76] Several political attempts were made to reform the way in which industrial relations were conducted, aiming to give government a greater role and increase the authority of national-level bodies such as the Trade Unions Congress (TUC) and the Confederation of British Industry (CBI) so that central government would have reliable partners with which durable long-term agreements could be reached and implemented. In practice the success of government initiatives was, however, limited because legislation was undermined by resistance from especially trade unions, and thus neither Labour nor the Conservatives managed to break away from the hands-off approach to industrial affairs that was an essential part of the postwar consensus, and thus if the organisation of industrial relations had indeed been an important factor contributing to Britain's economic decline, this impediment to economic modernisation remained very much in place in the late 1970s.

The third type of anti-decline policy pursued revolved around the assumption that the relatively slow modernisation had been caused by an under-exposure of British firms to competitive markets, having depended for long primarily on government-sheltered domestic and Imperial markets.[77] Access to competitive international markets had been sought by Conservative as well as Labour governments from the early

---

[76] On trade unions, economic decline and industrial relations reform, see Jessop 1980 pp. 44ff, Hall 1986 pp. 44f, Hodgson 1986, Coates 1994 Ch. 4, Thomsen 1996 Ch. 4.

[77] On market access and economic decline, see Jessop 1980 pp. 70ff, Pollard 1983 pp. 352f, 366ff, Spence 1985, Hall 1986 pp. 40ff, Dunford & Perrons 1986 pp. 75f, Sked 1987 pp. 18f, Rowthorn & Wells 1988 p. 187, Nicholls 1988, Grant 1993 Ch. 2, cf. Bacon & Eltis 1976.

1960s through attempts to join the apparently successful continental Europeans in the Community, but French vetos delayed entry until 1973, and thus when British firms were finally exposed to the more competitive European markets, the international economic climate had changed dramatically in the wake of the break-down of the Bretton Woods exchange-rate arrangements and the first oil price shock. At the same time a 'domestic' version of market liberalisation as a means to improve competitiveness had been launched by the Heath government in 1970, but its determination to withdraw subsidies from uncompetitive 'lame duck' firms only lasted until 1972 when rising levels of unemployment and industrial unrest prompted a policy shift that undisputably made the mid-1970s one of the high points of industrial policy in Britain.

Whether through government grants and corporatist institution-building, more predictable industrial relations or increased market competition, long-term policies aimed at creating improved conditions for growth by making it more attractive for industrial managers to modernise production would seem to make little progress, and therefore it is hardly surprising that more short-term oriented measures were tried out. From the 1960s onwards British governments, Conservative as well as Labour, tried to improve international competitiveness by keeping prices and wages under control,[78] and although these policies were overwhelmingly based on voluntary agreements, wages did in fact grow significantly slower in Britain than in continental Europe in the 1960s and 1970s,[79] helping to improve the attractiveness of British goods *vis-à-vis* important competitors.

The 1970s saw the combined presence of unemployment and inflation throughout Western Europe, but in Britain the international crisis unfolded in a relatively weak economy presided over by governments in a rather tenuous position.[80] The long-standing inability to improve the performance of the economy limited the room for manoeuvre for government because financial resources were relatively scarce, and the lack of progress not only undermined the political credibility of political parties which regularly engaged in policies that appeared to be rather different from their electoral platform, from the much-publicised U-turn of the Conservative Heath government in 1972 to the no less remarkable strategic permutations of the 1974-79 Labour governments. When key

78 On incomes policy and economic decline, see Glyn & Sutcliffe 1986, Grant 1993 Ch. 3, cf. Bacon & Eltis 1976, Warwick 1985 pp. 108f, Hodgson 1986 pp. 326ff.

79 Nolan 1989 p. 112, cf. Pollard 1983 pp. 323ff.

80 For studies of the 1970s in general and the Labour governments in particular, see Middlemas 1991 Ch. 2-5, Hall 1992, Mullard 1992 Ch. 10, Grant 1993 Ch. 3, Evans 1999 Ch. 1.

policies were regularly challenged by e.g. striking workers, the authority of government at large was potentially being eroded, and difficulties escalated during the second half of the 1970s where the Prime Ministers Wilson and Callaghan were confronted with political challenges that were difficult to accommodate within the framework of the postwar consensus. *Firstly*, in terms of economic policy the international crisis continued to put pressure on the national economy and public finances to such an extent that emergency credit facilities had to be obtained from the IMF, and these entailed strict conditions with regard to public expenditure and inflation control that effectively reinforced the already developing preoccupation of the Labour government with monetary targets and fundamentally questioned the capacity of the British government to conduct an independent economic policy. *Secondly*, the relevance of the British state was also questioned from within by Scottish and Welsh nationalism. The surge in electoral support for the Scottish National Party (SNP) – polling 30.4% of the Scottish vote at the October 1974 general election left Labour with a lead of less than 6% north of the border[81] – lead the traditionally staunchly unionist Labour party to propose devolved assemblies in Edinburgh and Cardiff, but even this rather limited measure of home rule had to be forced through against much internal opposition both in England and, indeed, the Scottish and Welsh branches of the Labour party, and eventually voters in Scotland and Wales remained unconvinced about the merits of the devolution proposals and rejected them in referenda in March 1979. While the unity of the political system remained unaltered, national solidarity on the UK level, a central assumption underpinning the operation of the welfare state as part of the postwar consensus, could in other words no longer be taken for granted. And *finally*, after several years of voluntary wage restraint in an inflationary environment, trade unions and their members rebelled on a grand scale during the 1978-79 'winter of discontent', thereby again questioning the ability of the state to implement policies against opposition from powerful social interests. While the postwar consensus may have lived on in the political rhetoric of the Labour governments of the late 1970s, its actions called some of its key underlying assumptions into question – international independence, national solidarity, consensual governance – and thereby contributed to the 're-politicisation' of British politics. Although Labour excelled in the discipline of governing without a parliamentary majority, the government's ability to strike deals with key actors – TUC, IMF, internal opponents to devolution – was not matched by a similar degree of influence on civil society or, indeed, on the response of its political opponents to the challenges of the 1970s. Having had its economic

[81] Kellas 1989 p. 106.

policies undermined by striking workers and its devolution proposal rejected, the attempt of the Labour government to fashion what amounted to a new social and territorial compromise had effectively been blocked, and Prime Minister Callaghan was forced to call a general election for May 3rd 1979.

Thatcherism: Politics and Policies

The 1979 general election was won by the Conservative party under the leadership of Margaret Thatcher, as were the subsequent general elections in 1983 and 1987, and the combination of prime ministerial longevity and a confrontational political style coupled with a gift for down-to-earth populist rhetoric probably explains the emergence and durability of 'Thatcherism' as a term of political abuse or praise.[82] Some have questioned the analytical relevance of the term on the ground that it overstates the coherence of a complex phenomenon,[83] but this text follows Andrew Gamble's argument that "the term denotes a phenomenon for investigation, not a known entity",[84] especially when allowing for the complication that the same term leads two intertwined lives, in political discourse and academic analysis.

Thatcher was elected leader of the Conservative party in the wake of Edward Heath's second general election defeat in 1974, and as leader of Her Majesty's Official Opposition she began to formulate a general vision for Britain that consciously positioned itself in opposition to key elements of the postwar consensus, revolving around the two key themes of 'freeing the economy' and 'strengthening the state'.[85] Based on the belief that market forces were a superior means of coordinating economic activity, government intervention in 'the mixed economy' was seen as inherently distortive rather than potentially helpful, and reducing the size of the public sector and the extent of government regulation therefore became goals in their own right. Moreover, lowering taxation and widening property ownership would restore incentives to work and invest, rekindle an 'entrepreneurial spirit', and ultimately help to 'make Britain great again'. Likewise, the influence of organisations like trade unions which obstructed the workings of the free market had to be limited, and the central goal of economic policy became to control inflation which undermined markets by destabilising the value of money

---

82 The term is believed first to have been used by leftist critics but was quickly adopted by Conservatives and Thatcher herself (Jessop *et al.* 1988 pp. 21ff).

83 Hall 1983, Gamble 1988 pp. 20ff, Jessop *et al.* 1988 pp. 6ff, Phillips 1996.

84 Gamble 1988 p. 22.

85 On principles of Thatcherism, see Hall 1983, Gamble 1983, 1988 pp. 83-95; Kavanagh 1987 Ch. 1, Martin 1992b, Phillips 1996, Evans 1997 Ch. 1-2, Green 2002 p. 216.

and hampering exchange. In order to achieve a 'free economy' along these lines, a 'strong state' was needed, in the sense that it was necessary to recreate the authority of government which had been compromised in the 1970s not only by attacks from e.g. striking miners but also through the principle of trying to build consensual long-term relationships with vested interests.

All in all the general political views of the new Conservative leader were not only clearly formulated in opposition to the postwar consensus but also claimed to provide an alternative market-oriented strategy for the quest to restore Britain's international position. Wanting to break away from the policies pursued by the Conservative leadership for around three decades did, however, not take Margaret Thatcher into a political territory unknown to the party:[86] the emphasis on authority, law and order, and hostility towards organised labour had existed as more than a marginal undercurrent amongst its membership also after the Second World War, and especially from the 1960s onwards liberal economic thinking had gradually become more prominent. While Thatcher clearly wanted to distance herself from the One Nation Toryism of her immediate successors, she did so by reformulating existing traditions within the party and using them as a platform for addressing central concerns of the late 1970s, but despite having been elected by a majority of Conservative members of parliament, most of the Shadow Cabinet still supported many aspects of the postwar consensus,[87] as did, presumably, the departments of central government after decades of Keynesian policy-making. Thus, Thatcher and her closest colleagues instead drew extensively on external sources of policy advice even after having gained power and became intellectually associated with a diverse range of 'new right' theoretical traditions such as monetarism, Hayek's Austrian economics, supply-side economics, and the public-choice tradition in political science.[88]

The 1979 Conservative manifesto gave few details about specific policies but focused on two issues that the Labour government of James Callaghan had confronted without much success, namely getting inflation under control and, after a decade of industrial disputes culminating in the 1978-79 'winter of discontent', to bring government authority to bear on socially and economically disruptive behaviour by trade unions.[89] Although especially the first of these two areas remained central

---

86 Kavanagh 1987 Ch. 7, Gamble 1988 Ch. 2, Middlemas 1991Ch. 6, Evans 1999 Ch. 1-2, Green 2002 Ch. 8.

87 Gamble 1983, 1988 pp. 85f; Evans 1999 pp. 40-47, cf. Morris 1991.

88 Kavanagh 1987 Ch. 1, Gamble 1988 Ch. 2.

89 Kavanagh 1987 pp. 205ff, Riddell 1991 p. 8.

to the Thatcher government not just through the first term but also through the second and third ones, additional priorities gradually emerged, and in the following the main thrust of government initiatives until the early 1990s will be outlined in the following under the Gamblesque headings 'free economy' and 'strong state'.[90]

In terms of freeing the economy from the burdens and distortions brought about by the public sector, the Thatcher government moved forward in many different directions, albeit some areas much progress was only made from the mid-1980s onwards. From the beginning the fight against inflation clearly took priority, resulting in the adoption of deflationary budgets in the midst of deep recession in the early 1980s which reinforced the steely public image of Thatcher as the Iron Lady, originally a jibe by the Soviet press. While inflation control remained a priority, the means to achieve this did, however, gradually shift away from the initial monetarist-style reliance on money supply targets and towards the external exchange-rate discipline through shadowing the Deutschmark before eventually joining the European Exchange Rate Mechanism in 1990.[91] In contrast to this, the addressing of supply-side issues was rather uneven:

- reduction of the overall level of taxation was hampered in the early recessionary years by accelerating expenditure on unemployment benefit, although a shift from direct to indirect taxation eased the burden away on high incomes,
- especially from the mid-1980s attempts were made to reduce government 'red tape' regulations in order that enterprise could flourish, and despite Thatcher's well-publicised scepticism about European integration, her government strongly supported the creation of the Single market, and
- financial resources were gradually shifted away from direct grants to individual firms and came instead to focus on developing urban localities and the skills of individual unemployed.[92]

The two most high-profile ways in which the Thatcher government attempted to 'free' markets were, however, industrial relations reform and de-nationalisation. On the one hand various trade union practices

[90] Cf. the oft-quoted title of Gamble 1988. For general overviews of the policies of Thatcherism, see Hall & Jacques (eds.) 1983, Kavanagh 1987, Gamble 1988 Ch. 4, Savage & Robins (eds.) 1990, Marsh & Rhodes (eds.) 1992, Cloke (ed.) 1992, Wilson 1992, Evans 1997, Evans 1999 Ch. 3-5.

[91] Hall 1986 Ch. 5, Gamble 1988 pp. 98ff, 113ff; Thompson 1990 Ch. 3, Jackson 1992, Mullard 1992 Ch. 11, Grant 1993 pp. 50-63, Thomsen 1996 Ch. 5.

[92] Wilks 1985, Thompson 1990 Ch. 5, Middlemas 1991 Ch. 10, Sharp & Walker 1991, Martin 1992b, Barberis & May 1993 Ch. 8-9, Thomsen 1996 Ch. 7, Wren 1996a Ch. 7-8, Lee 1996, Theakston 1996, Cortell 1997.

were outlawed and immense government resources were invested in defeating the 1984-85 miners strike against pit closures.[93] On the other hand an attempt to redraw the boundary between the public and the private sector began in the early 1980s by giving tenants in local-authority public housings the 'right to buy' their home at a heavily discounted price, but from the mid-1980s many nationalised companies were transferred back to private ownership, often amidst a blitz of publicity and discounted share offers that should help transform Britain into a 'property-owning democracy' by furthering 'popular capitalism'.[94]

With regard to 'state strengthening', the dispute with Argentina over the Falkland Islands in the early 1980s presented the Conservative government with a winnable war, and the associated national fervour was an important turning point in securing the long-term popularity of Thatcher with a sizeable part of the electorate.[95] While the measures to curb excessive trade union powers were a high-profile way of detaching government from the influence of interest organisations, also employers' associations found their access to the corridors of power restricted, and in general the 1980s was a decade characterised by the increasing importance of the key departments of central government in London.[96] Local government implemented many national welfare policies and administered a sizeable part of total public expenditure, and hence making subnational actors comply with the strategies of central government with regard to individual services and overall levels of expenditure was essential for a government intent on fiscal retrenchment and reform of public services. Moreover, substantial Conservative majorities in the Westminster parliament meant that local government became a party-political battleground:[97] many Labour-controlled authorities in urban areas saw themselves as the last bulwark against the Thatcherite juggernaut, and seen from Downing Street conflicts about e.g. local government expenditure or parental choice in education took on the wider symbolic significance of eradicating 'municipal socialism'.

Interestingly, the fall of Margaret Thatcher as prime minister was not brought about by the parliamentary opposition, at least not in a direct way, and it reflected issues relating to both the freeing of the economy and the strengthening of the state. Although under Thatcher the Conser-

93 Scamble 1986, Moran 1988, Gamble 1988 pp. 103ff, 115ff; Marsh 1995, Thomsen 1996 Ch. 6.

94 Fine 1989, Veljanovski 1990, Marsh 1995.

95 See e.g. Hobsbawm 1983.

96 Kavanagh 1987 Ch. 10, Middlemas 1991, Marsh & Rhodes 1992, Holliday 1993, cf. Thomsen 1996.

97 Stoker 1990, Holliday 1992, Evans 1999 Ch. 5, cf. Rhodes 1988 pp. 180-208.

vatives never attracted much more than 40% of the popular vote at general elections, the defection of Labour's social-democratic wing and the emergence of a new centrist 'third force' in British politics meant that the opposition was divided, and making Labour 'electable' again after having fought the 1983 election on a leftish platform – famously dubbed 'the longest suicide note in history' – was a long process that only gained momentum with the centrist 1988 policy review.[98] Instead Thatcher was forced to stand down by her own MPs who were concerned about her ability to win a fourth successive general election. Her uncompromising style in relation to unpopular policies like the new flat-rate local government 'poll tax' was considered a major electoral liability, and the continued presence of leading Conservatives of a more 'caring', 'Europhile' and/or 'interventionist' leaning meant that dissatisfaction could crystallise around a number of different issues.[99] John Major succeeded Margaret Thatcher as party leader and prime minister in November 1990, and while some of the most unpopular policies, notably the 'poll tax', were discontinued and a more conciliatory tone adopted in dealings with European partners, early 1990s policy initiatives were still concerned with much the same issues, i.e. keeping inflation under control while progressing reform of welfare services in order to make them more accountable to individual citizens.[100] In other words, Thatcherism would seem to live on, even after the departure of its controversial figurehead.

Interpreting Thatcherism and the Postwar Consensus

While the previous pages have attempted to reconstruct key features of the general thinking and policy initiatives of the Conservative governments of Margaret Thatcher, the interpretation of the way in which the 1980s related to previous decades in British history is the subject of ongoing debate, revolving primarily around the relationship between Thatcherism and the postwar consensus. Official Conservative statements spoke of heroic ambitions or triumphant accomplishments – with Thatcher's claim that "I have changed everything" being the most extreme example – and within the academic literature responses to such claims range from heated denial via scepticism to tacit or explicit acceptance of 1979 as a watershed in contemporary British history. These positions are also in evidence in the study of politics and policy, and it is possible to identify four basic positions by combining the focus of

98 See e.g. Dunleavy 1993.

99 For a rounded account, see Evans 1999 pp. 127-34.

100 See e.g. Ludlam & Smith (eds.) 1995.

attention – general politics or implemented policies – with the degree of change identified, as suggested by Table 2.3.

Table 2.3 Interpreting Thatcherism and the postwar consensus

		Focus	
		Politics	*Policy*
Change	*Extensive*	Strategic revolution	Gradual revolution
	Limited	Ideational heritage	Implementation failure

The *strategic-revolution* perspective could be seen as the explicit or implicit starting point of all four positions in the sense that it claims that the ideas entailed in Thatcherism were radically different from those underlying the postwar consensus. The seminal 1983 book entitled *The Politics of Thatcherism* set the tone by denoting Thatcherism as 'authoritarian populism', a new hegemonic discourse aiming to replace the existing social democratic consensus by a combination of free-market liberalism and organic patriotic Toryism formulated in the populist language of 'compulsive moralism'.[101] Along similar lines others have focused on ideological aspects, arguing that the general notions within Thatcherism about ideal public-private and state-citizen relationships were fundamentally at odds with the emphasis on collectivism, egalitarianism and statism inherent in the postwar consensus.[102] The combination of seemingly contingent elements – the principles of free markets, state authority and Victorian morality may be at odds with regard to social phenomena like sex, drugs, and rock'n'roll – has prompted a search for the unifying principle of Thatcherism, and two types of answers have been provided to this question. Bob Jessop and Andrew Gamble have argued that Thatcherism constituted an attempt to refashion social relations in Britain so that an internationally oriented 'accumulation strategy' built around the City of London was supported by a 'two-nation' hegemonic strategy through a 'liberal-authoritarian

[101] Hall & Jacques 1983, Hall 1983.

[102] Much work along these lines has been inspired by traditions in discourse analysis, e.g. Keat & Abercrombie (eds.) 1991, Fairclough 1992 pp. 187ff, Phillips 1996, cf. Hedetoft & Niss 1991.

state'.[103] While the unity of the "free economy, strong state" paradigm[104] is identified by looking for the long-term beneficiaries of Thatcherism – certainly *not* industry, the unemployed or peripheral regions – others have explored its underlying assumptions from a history-of-ideas perspective. Shirley Robin Letwin argued strongly for the importance of personal 'vigorous virtues' – being "upright, self-sufficient, energetic, adventurous, [...], loyal to friends" – as the moral basis for individuals, families and the nation, and, indeed, as a principle of political action which was fundamentally sceptical about the role of government in society,[105] and in parallel with this E. H. H. Green interpreted Thatcherism as the ultimate embodiment of a general Conservative trust in the 'agencies of civil society' which had been subdued during the pragmatic statist years of the postwar consensus.[106] Despite revolving around the claim that the politics of Thatcherism constituted a radical break with the past, the strategic-revolution perspective still highlights the complex nature of the alleged ideational change, and that unity in the form of over-arching strategies or underlying assumptions cannot be taken for granted but has to be established through analytical efforts, possibly running the risk of overstating the coherence of Thatcherism as a long-term 'strategic project'.

In many ways the *ideational-heritage* perspective can be seen as an attempt to place the alleged newness of the politics of Thatcherism in a longer historical perspective by focusing on aspects that existed prior to the 1970s. The least radical versions of this approach focus on the development of particular elements of Thatcherite discourse, demonstrating the long-standing existence of an economically liberal tendency within the party or the traditional commitment to law and order,[107] thereby making the Conservatism of the 1980s appear rather unsurprising and/or emphasising the internal tensions in the party during the postwar decades. A more wide-ranging perspective on continuity is the idea advanced by Jim Bulpitt that Thatcherism is a continuation of traditional Conservative 'statescraft', i.e. a political strategy in which winning control of centre of the political system by seeking electoral support wherever possible and trying to insulate this centre from external and domestic pressures by concentrating on restricted areas of 'high

[103] The most prominent examples of this approach are found in the work of Bob Jessop (Jessop *et al.* 1988 cf. 1990) and the more understated but succinct writings of Andrew Gamble (1983, 1988).

[104] The title of Gamble 1988.

[105] Letwin 1992, quote from p. 33.

[106] Green 2002 Ch. 9.

[107] E.g. Green 2002 Ch. 9, Gamble 1988 Ch. 2.

politics'.[108] Although the Conservatives had deviated from this approach by letting central government become increasingly entangled in corporatist networks in the 1960s and 1970s, subscribing to the original postwar consensus with its emphasis on national economic and welfare policies could still be interpreted in terms of 'statescraft' in response to popular hardship during the 1930s and the Second World War, and the perception of the party as ruthlessly pragmatic centralists defending what is perceived as UK national interests has important implications because it introduces a potential for great flexibility of policy-making. If Bulpitt is right, then the translation of ideological discourse and underlying assumptions into policy initiatives cannot be taken for granted but must be established through empirical analysis.

Turning now to policy change, it is difficult to find analysts who claim that everything changed in the wake of 1979, except in some political tracts by Thatcherite acolytes and their opponents. Instead the dividing line in the literature runs between those who interpret the 1980s as a decade of 'gradual revolution' in which, slowly but surely, the Conservative government reconstructs Britain in the image of the general ideas identified by the strategic-revolution paradigm. A *gradual-revolution* perspective on Thatcherism would thus chart the gradual shift of political attention from macro-economics and industrial relations in the early 1980s toward the issues of privatisation and reform of the welfare state in the late 1980s,[109] often being primarily concerned with the fact that policies and institutions considered important during the postwar consensus were now being questioned, reshaped or replaced with new policies operating in a different organisational set-up. Conversely, a more sceptical approach is in evidence in the *implementation-failure* perspective which tends to concentrate on the relationship between ideological principles and/or stated aims on the one hand, and implemented policies and their impact on the other. Unsurprisingly, this position has often been adopted by academic specialists who have seen 'their' field being subjected to interpretations of the 1980s which too readily seem to accept the radical-change rhetoric of political discourse,[110] but also more wide-ranging accounts have been produced from this perspective by political scientists[111] as well as historians.[112]

[108] E.g. Bulpitt 1983, cf. Gamble 1988 pp. 167ff, Kavanagh 1987 Ch. 10.

[109] This interpretation is prominent e.g. Aughey 1983, Hall & Jacques (eds.) 1983 Section II, Martin 1988, Riddell 1991, cf. Jessop 1989, 1995.

[110] Examples include Wilks 1985 & 1987, Green (ed.) 1989, Thompson 1990, Savage & Robins (eds.) 1990, Mullard 1992 Ch. 10, Martin 1992b, Cloke (ed.) 1992, Holliday 1992, Thomsen 1996, and, indeed, specialist contributions to Marsh & Rhodes (eds.) 1992.

[111] E.g. Rhodes & Marsh 1992b, Marsh & Rhodes 1992, Hogwood 1992, Hood 1994, Rose & Davies 1994, Marsh 1995, Walsh 2000, cf. Bevir & Rhodes 1998.

The overall impression is clearly one of very uneven progress, with reforms being attempted in some areas but delayed or eschewed in areas deemed to be politically dangerous, and resulting in outcomes that involve relatively limited change, partly, according to Marsh & Rhodes,[113] because the centralised style of governing which consciously tried to minimise consultation with interest groups and implementing organisations made central government less well-informed and created more problems of compliance than in the much-maligned 1960s and 1970s.

The review of interpretations of Thatcherism and its relationship to the postwar consensus would seem to suggest that each of the positions listed in Table 2.3 to some extent will be useful sources of inspiration for the current study. While the overall perspective adopted is clearly parallel to the implementation-failure position in that it is sceptical about the nature, extent and origins of policy change within the SDA, the study will also have to take into account the interplay between, on the one hand political discourse and underlying assumptions, and on the other hand the 'statescraft' of pragmatic politics and the ability of government to control implementation and influence outcomes in order to bring about a gradual revolution. In short, a conceptual framework is called for that can account in a consistent way for the development of politics and policy in their institutional and discursive setting.

2.3. Understanding the SDA

The following pages give a brief outline of central features in the development of the SDA by way of background, and then move on to consider central contributions to the existing literature on the agency in order to identify its key characteristics and the main interpretative schemes entailed. Together with the previous literature reviews, this should enable us in the ensuing section to identify the road that must be travelled in order to improve our understanding of the Agency's development and its position in the broader context of British politics and changing paradigms of regional policy.

Developing Agency in Scotland

In the following a brief outline of central features of the SDA and its development is given on the basis of the existing literature.[114] The

[112] E.g. Kavanagh 1987, Wilson 1992, Evans 1997, and, last but certainly not least, Evans 1999.

[113] Marsh & Rhodes 1992.

[114] For overviews of the history of the SDA, see Hood 1991a, Halkier 1992, Danson *et al.* 1993.

importance of these features and not least the causal relationship between them has been interpreted in rather different ways, and hence this *ex ante* overview takes the form of an enumeration that should make it easier to follow the ensuing review of these interpretations rather than attempt to produce a quick-and-dirty draft history of the Agency.

The Scottish Development Agency was set up in 1975 by the then Labour government as a statutory body funded by public money, sponsored by the Scottish Office as the territorial department of central government for Scotland, and situated at arm's-length outwith the departmental system. While the notion of an RDA had for some years enjoyed support from many Scottish actors, the eventual adoption of a proposal that gave low-land Scotland preferential treatment compared to equally crisis-ridden regions in the north of England has often been seen as an attempt to counter the rise of political nationalism in the shape of the Scottish National Party (SNP).[115] Moreover, like the National Enterprise Board (NEB) on the UK level, it would have the power to invest public money in private firms, and thus the Agency could easily be presented by the Conservative opposition as a vehicle of back-door nationalisation or an attempt to prop up lame ducks with public money.[116] From the outset the new body was thus placed firmly in the midst of political controversy along territorial and ideological lines.

The SDA was headed by an independent board appointed by the Secretary of State for Scotland in order to protect the new body from political interference and bureaucratic second-guessing in its day-to-day activities. In its early years the Agency's policy profile was relatively limited, reflecting functions inherited from existing organisations incorporated into the new body, but it gradually broadened over the years. Its policies combined traditional activities such as building of advance factories with programmes that were more innovative in the context of regional policy such as public investments in private firms, regionally-based advisory services, attraction of inward investment and development initiatives for particular sectors of the Scottish economy.

Central government sponsorship of the SDA was undertaken by the Scottish Office, and the first high-profile change brought about was to give the Agency responsibility for a major urban renewal project in Glasgow's East End in 1976.[117] Although the SDA had been set up by a Labour government and the Conservative opposition had denounced it as a vehicle of back-door nationalisation and interventionism, the in-

115 McCrone & Randall 1985 p. 234, Keating & Boyle 1986, Hood 1991a p. 4, Midwinter *et al.* 1991 pp. 63f, Halkier 1994.

116 Keating & Boyle 1986 pp. 22ff, Hood 1991a p. 5, Halkier 1994.

117 Midwinter *et al.* 1991 p. 64, Danson *et al.* 1990 p. 173, Hood 1991a pp. 9ff.

coming Thatcher government decided to maintain the organisation. New guidelines did, however, make direct investment in private firms a measure of last resort and demanded strict commercial discipline, and furthermore the Agency's overseas promotion of Scotland as an location for inward investment became a joint effort with the Scottish Office. In February 1987 a government review concluded that the SDA operated in accordance with the general principles of economic policy adhered to by the Conservative government, and thus the activities of the Agency appeared to have been accepted as a valid contribution to regional policy in Scotland.[118]

Not long after the publication of the Scottish Office report, the SDA did, however, become subject of another review, this time of a much more overtly political nature, when Bill Hughes, chairman of the Confederation of British Industry (CBI) in Scotland, proposed a merger of the Agency with the Scottish arm of the Training Agency. Circumventing the Scottish Office and obtaining the public support of the Prime Minister in September 1988, the proposal for creation of a two-tier network of RDAs consisting of a strategic national core and a network of Local Enterprise Companies swiftly became official government policy. Although the functional strategies behind the merger – combining development of firms and human resources and decentralising policy delivery – could have obvious advantages, the initiative was generally seen as having a strong party-political dimension, attempting to exorcise the ghost of Labour interventionism by refashioning Scottish regional policy in a more palatable form.[119] On the first of April 1991 the SDA had ceased to exist and the merged entity, Scottish Enterprise, began to operate.

Analytical Approaches and the Interpretation of the SDA

The treatment of the SDA in the regional policy literature fits neatly into the general pattern of fragmentation identified earlier in this chapter. While studies of the Agency generally focus on its policies and only place these in the context of broader British trends on a very general level, writings focusing on traditional central government programmes have very little to say about the organisation, merely recording its existence[120] or describing it as "an *ad hoc* agency"[121] which represented

[118] Hood 1991a p. 12, Danson *et al.* 1992 p. 299, Halkier 1992 pp. 14ff.

[119] Moore 1989; Hood 1991a pp. 18ff, 1991b; Danson *et al.* 1993 pp. 176ff, MacLeod 1998a.

[120] E.g. Prestwich & Taylor 1990 Ch. 7, Temple 1994 Ch. 8.

[121] Armstrong & Taylor 1985 p. 250 – an interesting choice of words given the constant flux in central government policies recorded elsewhere (e.g. Armstrong & Taylor 1985 Ch. 9, cf. Section 4.3).

"a move from the passive to the active mode for government involvement".[122] Although a new departure, the Agency is in other words presented as a somewhat marginal phenomenon,[123] and thus for treatment of a less cursory nature one has to consult writings that focus specifically on the SDA and its policies.

The literature on the Agency is by no means extensive, and given the multi-functional nature of the organisation it is hardly surprising that much work focuses on individual policy areas. Some of these have clearly received more attention than others, with the so-called area initiatives attempting to promote the development of particular localities by means of especially industrial property and environmental improvement being fairly well researched, often in connection with broader studies of central government urban policy or local economic policies.[124] In contrast to this, policies oriented directly towards private firms such as industrial investment, advisory services, attraction of inward investment and sectoral initiatives have been the subject of only a limited number of in-depth academic studies, mostly seeing these activities as a regional form of industrial policy.[125] A number of texts do, however, attempt to provide an overview of the development of the SDA, and as they entail rather different interpretative schemes that can also be found in more specialist writings, these broader studies provide a useful starting point for the discussion. It is possible to distinguish between two main perspectives according to the nature and source of change in the development of the Agency.

Some authors have interpreted the history of the SDA in a way which can be summarised under the heading of *external revolution*, stressing that profound changes took place from 1975 to 1991 and that these were primarily the result of external political pressure. This approach, resembling the gradual-revolution interpretation of Thatcherism cf. Section 2.2 above, is typified by the work of Mike Danson, Greg Lloyd and David Newlands who have argued that the Agency of the 1980s was fundamentally different from that of the 1970s and that the

---

122 McCallum 1979 p. 37.

123 Anthologies on British regional policy tend to fare better by including separate contributions on RDAs and other sub-national economic development initiatives, e.g. Townroe & Martin (eds.) 1992, Harrison & Hart (eds.) 1993.

124 Important contributions include Wannop 1984, Gulliver 1984, Keating & Boyle 1986, Moore & Booth 1986d, Donnison & Middleton (eds.) 1987, McCrone 1991a.

125 The sectoral initiatives of the Agency are central in both Moore & Booth 1989 and Hood & Young (eds.) 1984, and the latter also touches upon issues related to industrial investment, inward investment attraction and small firm development.

root cause of this was the advent of the Thatcher government in 1979.[126] While in the early years the Agency pursued not only economic but also social goals within an interventionist strategy, the 1980s came to be dominated by purely economic aims and strategies that worked with rather than against the market.[127] This change at the strategic level was also reflected in the policy instruments employed: in the early years the Agency operated "primarily as an investment bank",[128] but in the 1980s other activities that relied more on the positive response of private actors came to the fore, i.e. attraction of inward investment, advisory services for indigenous firms, and the reliance on partnership with the private sector in policy implementation generally increased.[129] These changes in the strategies and policies of the SDA were instigated by the advent of the Thatcher government: new guidelines marginalised the role of industrial investments and made the Agency focus single-mindedly on economic development goals,[130] and the threat of outright termination continued to loom in the background and prompted the Agency to undertake further adjustments of its policies in order to "out-Thatcher Thatcher".[131] The mid-1980s review could therefore confirm that the organisation now operated according to the "modified market" approach of the Conservative government according to which policies should only be instituted where markets had failed.[132] But despite compliance with regard to strategies and policies the SDA had, however, never been fully accepted politically, and this ultimately paved the way for the adoption of the Scottish Enterprise proposal.[133] In the end not even radical

126 Danson *et al.* 1988, 1989a, 1989b, 1989c, 1990a, 1990b, 1992, 1993, cf. Danson 1997. This line of argument can also be found in other overview articles by e.g. MacLeod (1996, 1998a) and Hood (1991a). While the former shares the reservations of Danson and his colleagues about the impact of Thatcherism on the SDA, Hood sees the 1980s as a "flourishing period" or "golden years" (1991a p. 12), although at the same time some degree of continuity between the 1970s and 1980s is also hinted at (p. 12). The most recent article by Danson (1999) is a rather more cautious version of the external-revolution interpretation.

127 Danson *et al.* 1988 pp. 1f, 1989b pp. 71f, 1989c pp. 4f, 1990 pp. 172ff, 1992 pp. 299f. See also Hood 1991a pp. 12ff, Swales 1983, Rich 1983, Midwinter *et al.* 1991 pp. 187ff, Barnekov *et al.* 1989 Ch. 7.

128 Danson 1980 pp. 12f, Danson *et al.* 1990 p. 173, 1993 p. 169. The expression refers not only to the SDA's direct investments in private firms, but also to its role as provider of industrial property.

129 Danson *et al.* 1990 pp. 174ff, cf. Barnekov *et al.* 1989 Ch. 7.

130 Danson *et al.* 1988 p. 9, 1989b pp. 71f, 1990 pp. 168f, 174, 1992 p. 299. See also Hood 1991a p. 12.

131 Danson *et al.* 1989a p. 562, cf. 1989b p. 71, 1993 p. 162. See also Rich 1983 p. 273, Hood 1991a.

132 Danson *et al.* 1990 p. 174, 1992 p. 299. See also Hood 1991a pp. 15f.

133 Danson *et al.* 1989a pp. 357f, 1989b p. 72, 1990 pp. 176ff, 1993 pp. 171f.

changes in response to government priorities was enough to save the Agency as a separate and integrated RDA for lowland Scotland.

An alternative perspective on the development of the SDA can be encapsulated under the heading of *internal evolution*. While perhaps less oriented towards reconstructing the history of the organisation as such, the work of Chris Moore and Simon Booth on a wide range of Agency activities[134] represents a very different interpretation that emphasises the gradual nature of change and the continuities between the early years and the 1980s, much like the implementation-failure perspective on Thatcherism discussed in Section 2.2 above. Individual policy areas – and indeed the overall policy profile of the Agency – changed through a predominantly incremental process in which new initiatives were introduced and subsequently modified on the basis of the experience acquired. Compared to the external-revolution perspective, the early years are seen as less 'social' or 'interventionist' in terms of strategies, and conversely it is stressed that the 1980s saw the continuation and development of activities that are clearly not in accordance with the liberal economic philosophy espoused by the Thatcher government.[135] Instead policy development reflected specific needs within the Scottish economy in e.g. particular industries and localities, and as these problems were of a long-standing nature, the activities of the Agency acquired a high degree of continuity.[136] Although the political environment was clearly less well-disposed after the change of government in 1979, the SDA and its policies survived, presumably because of the Conservatives being a minority party in a region suffering great economic difficulties is seen as having had a moderating influence.[137] It is therefore seen as significant that the Scottish Enterprise initiative only became government policy after a 'political maverick' bypassed the 'usual channels' of the Scottish political system and ensured the public backing from the British level by the Prime Minister herself.[138]

---

[134] The two major works are Moore & Booth 1989 and chapters by the same authors in Lever & Moore (eds.) 1986 (1986a, 1986b, 1986c and 1986d); short versions can be found in Moore 1994 and 1995.

[135] Moore & Booth 1989 pp. 11, 142f, Ch. 7, 1986b, Moore 1995 pp. 237f, Booth & Pitt 1983. See also Firn 1982 pp. 14f, Keating & Midwinter 1983 pp. 36ff, Young & Hood 1984 p. 47, McCrone & Randall 1985, Turok 1987 pp. 240ff, Parsons 1988 p. 183, MacLeod & Jones 1999.

[136] Booth & Pitt 1983 p. 12, 17ff; Moore 1994 pp. 53ff, 1995 pp. 234ff. See also Grant 1982 p. 120, Midwinter *et al.* 1991 p. 192.

[137] Moore & Booth 1989 p. 69, 1986c p. 117, Booth & Pitt 1983 pp. 17ff. See also Keating & Boyle 1986 pp. 24f, Kellas 1989 p. 225.

[138] Moore 1989 pp. 237ff.

These two interpretations of the development of the SDA would certainly seem difficult to reconcile because their basic contentions about "what happened and why" contradict one another, and this impression is reinforced by the different conceptual frameworks employed: the external-revolution analysis tends to be cast in terms of traditional typologies of economic policy, contrasting interventionism with a 'modified market' approach, while the internal-evolution perspective has primarily been couched in terms inspired by traditions within political science, especially those concerned with corporatism and institutionalism. On closer inspection it is, however, interesting to note that the two perspectives also share a number of important features. *Firstly*, both of them would seem to base their analysis on selected aspects of the Agency's activities: the external-revolution perspective focuses particularly on changing corporate strategies and policies related to 'hard resources' such as industrial investment and area initiatives based on factory building and environmental improvement, while the internal-evolution perspective has been developed on the basis of a study of the SDA's sectoral policies and organisational development.[139] This underlying selectivity certainly suggests that a degree of caution is necessary: the various policies of a multi-functional body like the Agency may have developed in different ways, and statements of corporate strategy, including those emanating from the sponsor department, are not necessarily an accurate reflection of policy development in practice, and thus extrapolating findings from selected areas of activity is hardly the best way of approaching the study of the Agency as such. *Secondly*, in practice the attempts to explain the predominance of change or continuity come close to being mono-causal: in the external-revolution scenario the Agency's 'own' initiatives are interpreted as adjustments to the ideological pressures of Thatcherism, and from the internal-evolution perspective SDA policies come to reflect the ownership structure and market position of particular sectors in an almost functionalist way, despite the importance of ideological factors being recognised in principle.[140] But new policies might be introduced in response to changing economic circumstances, and even in the Thatcher years other political considerations than ideological ones are likely to have influenced government policy,[141] and thus explanations of change and continuity in the

139 It should, however, be noted that with regard to area initiatives, essentially spatially focused employment of the SDA's powers in factory building and environmental renewal, Moore & Booth (1986d) recognises a significant degree of change towards a less social and more economic approach and cite both external economic and political pressures and internal learning processes as explanations.

140 Moore & Booth 1989 p. 69.

141 Cf. the discussion in Section 2.2.

SDA could therefore certainly benefit from a more systematic exploration of a wider range of options.

All in all this would seem to suggest that an understanding of the development of the SDA in the context of broader trends in regional and other public policies will require more than just a critical reexamination of the existing literature:

- more empirical work is needed to make up for the uneven coverage of different aspects of the Agency's activities undertaken on the basis of a unitary conceptual framework in order to enable meaningful comparisons, and
- in order to illuminate the factors that may have contributed to their development, the history of the SDA must be explored in its British and European contexts, especially the advent of the Thatcher government and the general shift towards bottom-up regional policy.

In short, there would seem to be ample room for a book-length study of an organisation that has often been extolled as a paradigmatic example of the RDA approach to regional policy.

2.4. Three Steps to Heaven

Having reviewed three fields of academic endeavour relevant to the study of the development of the SDA in its wider British and European context, some important features have emerged.

On the one hand the body of literature concerning regional policy has been shown to focus on economic impact rather than organisational and political aspects, and to reflect institutional divisions between top-down and bottom-up policies. This has resulted in a fragmented situation where comparison between different policies is impeded by the absence of a conceptual framework able to account for more than one policy paradigm. On the other hand studies of Thatcherism fall into two main groups, some emphasising the New Conservatism of the 1980s as a radical ideological departure from the postwar consensus that was gradually translated into radical policy change, and others focusing on the limited extent of policy changes and their impact on British society due to the enduring pragmatism of Conservative 'statecraft'.

The review of the literature on the SDA demonstrated that these general features are also in evidence in this particular field of study. On the one hand, like other forms of bottom-up regional policy the Agency's activities have mostly been treated as a field of investigation in its own right or discussed from the perspective of non-regional forms of public intervention like urban or industrial policy. On the other hand the main interpretations of the development of the Agency – external revolution and internal evolution – would seem to correspond to different approaches to the study of Thatcherism, like those focusing on ideological change and policy continuities respectively. The existing writings on the SDA are in other words 'typical' in that they have key characteristics in common with academic studies of similar phenomena, and as such they also share the problems associated with these. *Firstly*, the absence of a comprehensive conceptual framework makes it difficult to compare different forms of policies for regional development; this is a problem both with regard to the Agency itself when the existing literature has tended to focus on particular aspects of its activities, and for attempts to relate the Scottish experience to wider policy developments in Britain and Europe. *Secondly*, the complex and evolving nature of Thatcherism greatly complicates attempts to establish the relationship between external and internal sources of change in the history of the SDA, and hence the conceptual framework should also be capable of interpreting its wider political, economic and social context.

All in all this strongly suggests that a study of the history of the SDA as a bottom-up development body in the context of British regional policy will have to involve three major elements:

- comprehensive empirical research is needed, both with regard to individual policy activities and corporate aspects of the Agency, in order to overcome the uneven and fragmented nature of the existing literature and allow for the fact that a multi-functional development body may not have developed in a coherent way,
- if the sources of continuity and change are to be illuminated, the analysis of the SDA must be situated firmly in its British political context, including the effects of Thatcherism on national-level regional policy, and related to broader Western European trends in regional policy, and finally
- undertaking these tasks requires the development of a conceptual framework capable of identifying key features of and distinctions between different forms of regional policy, and of taking the political, economic and social environment in which regional policy unfolds into account in a systematic and coherent manner.

The text will therefore proceed by discussing general and conceptual issues in order to establish a suitable platform for the ensuing empirical analysis.

CHAPTER 3

Towards an Institutionalist Approach to Regional Policy

As underlined by the preceding chapter, undertaking an in-depth study of the history of the SDA set within the wider UK context requires the fashioning of an analytical framework capable of identifying key features of both traditional and new forms of regional policy while at the same time illuminating the role of politics, institutions and discourse. What is needed is a coherent set of concepts eschewing reductionist simplicities and being sufficiently multi-dimensional to account for the complexities of empirical analysis, but still capable of identifying significant differences between types of policies by highlighting central features of material consequence.

Before embarking on the more detailed work of developing appropriate concepts to be employed in empirical analysis it is, however, necessary to reconsider a number of fundamental issues in order to ensure the coherence of the framework, and therefore this chapter proceeds in four steps. Starting with more abstract matters, Section 3.1 introduces the basic theoretical foundations of the text through a discussion of the role of institutions, organisations, agency, and discourse. Establishing a coherent institutionalist perspective on the relationship between public and private organisations in the policy process should not only provide coherence within the present text but will, hopefully, also help to establish an analytical framework relevant to the study of other areas of public policy. In order to delimit the field of inquiry of the book, Section 3.2 considers definitions of regional policy in order to position it in relation to other forms of public policy with a spatial and/or economic dimension. Following this, Section 3.3 outlines an institutionalist approach to the study of public policy and identifies central dimensions of the analysis of regional policy, drawing inspiration from a discussion of literatures inside and outside regional studies. On the basis of this, finally, a comprehensive analytical framework is formulated in Section 3.4 which should be able to make sense of the development of the SDA in the context of British politics and wider European trends.

3.1. Institutionalist Foundations

The study of regional policy essentially concerns a particular form of interaction between public bodies and private sector firms, and as the relationship between public and private has been one of the central problems in 20th-century social science – and the major political dividing line in Western politics – no shortage of writings can be claimed. What we are looking for is, however, a body of thought not just drawing attention to the obvious differences between these two elements of society, but also one that is capable of analysing them within a unitary framework and establishing their similarities and modes of interaction. It is hoped that an institutionalist approach will be able to ensure this.

Proclaiming an adherence to 'new institutionalism' does, however, not amount to much, because this label has been used by distinct theoretical traditions – within e.g. political science, organisational studies and economics – formulated in opposition to dominant theoretical trends of the 1960s and 1970s in general and their assumptions about the relationship between structures and agency in particular. In order to position this book in relation to the new institutionalisms, the thinking of prominent contributions will be traced in the following, but first it is worth stressing that this author also shares the realist assumptions inherent in most institutionalist traditions[1] and remains sceptical about the tenability or usefulness of many constructivist arguments. As a major point in the following will be to integrate discourse – a field of enquiry often associated with constructivism – into an institutionalist and realist approach, it will therefore be useful to make some general remarks about the basic assumptions informing the current study. *Firstly*, in terms of ontology it is paramount to distinguish between natural and social phenomena: while it is difficult to accept the former as being 'constructed' except in the traditional and rather weak sense that conceptual frameworks make a difference also in the natural sciences,[2] the latter clearly involve a 'subjective ontology' in the sense that social agency involves intention and interpretation of human behaviour – and hence inter-subjective recognition – in order to exist.[3] It is, however, questionable whether using the term constructivism in this connection is particularly helpful because social phenomena have not necessarily been brought about through the deliberate actions of collective or individual human actors, and thus a term with strong intentional and

---

1 See e.g. North 1990a, March & Olsen 1984 (cf. Pedersen 1990 pp. 100ff), Ostrom 1986 (cf. Bogason 1994). Some institutionalist traditions within organisational studies operate on the basis of a constructivist epistemology (Esmark 1998).

2 Wenneberg 2000 pp. 122f, 160ff.

3 Collin 2002, cf. Wenneberg 2000 pp. 102ff, Fairclough 1992 pp. 62ff.

voluntarist connotations is expected to cover also evolving social institutions[4] – something which is hardly desirable, except, of course, from a deconstructivist perspective with the limited ambition of demonstrating the long-term 'contingency' of all things social. *Secondly*, from an epistemological perspective a 'strong' version of constructivism[5] – where in the famous words of Derrida "there is nothing outside of the text"[6] – implies that knowledge is contingent because the external world does not have a distinct essence and meaning can only be constructed through language. This anti-essentialist line of reasoning does, however, rest on the simplistic assumption that the only alternative to essentialism is radical contingency, it entails a contradictory claim about absolute relativism,[7] and it would also, ironically, seem to undermine itself by essentialising language as an absolute barrier between social actors and the external world which appears to be impervious to e.g. physical experience.[8] It is therefore hardly surprising that in practice also self-proclaimed anti-essentialists have come to doubt the philosophical underpinnings of 'strong' constructivism because 'weaker' versions have emerged which, while insisting that perceptions of the external world will always be framed by a particular language, maintain that it is possible to devise criteria for choosing between different interpretations.[9] In terms of epistemology such a 'weak' constructivism does, however, become rather difficult to distinguish from more traditional approaches to conceptualisation which recognise the impact of conceptual frameworks and analytical approaches.[10]

Instead of attempting to carve out a tenable position on the basis of the least problematic constructivist traditions, the present study draws its inspiration from what has become known as 'critical realism'. This

---

4 Wenneberg 2000 p. 133.

5 The distinction between 'strong' and 'weak' constructivisms has been taken from Bredsdorff 2002, but can also be found in Wenneberg 2000.

6 Quoted from Anderson 1983 p. 42.

7 Collin 2002.

8 Kjørup 2001 p. 149, Bredsdorff 2002 pp. 55ff, Collin 2000 & 2002, Wenneberg 2000 Ch. 11, cf. the comments on the 'empty realism' of Laclau & Mouffe in Halkier 2003.

9 Wenneberg 2000 pp. 150ff, Bredsdorff 2002 pp. 62ff. Examples of this would seem to include Jørgensen & Phillips (1999 pp. 17f), and Foucault (cf. Ifversen 2000b).

10 Bredsdorff 2002 p. 63, Neumann 1999 pp. 163ff, Wenneberg 2000 pp. 34ff, 107ff, Collin 2002, Halkier 1990a, cf. a range of prominent introductions to 'the historians craft' such as Bloch 1954 pp. 64ff, Dahl 1980 Ch. 1, Clausen 1963 p. 58, Carr 1987 Ch. 1, and even the polemical constructivist Jenkins 1991.

approach, associated with the work of Roy Bhaskar and Andrew Sayer, entails four propositions,[11] namely

- the external world exists independent of conceptualisation or observation,
- some structural aspects of the external world are not immediately discernible,
- social structures depend on human perception of their own practices, and
- scientific endeavour can identify structures of the external world, including those relating to human interaction in society.

From this perspective knowledge is *not* completely determined by external realities or their conceptualisation because it is the product of more or less deliberate social activity such as work, science, and/or communication. Knowledge is in other words not random because it is related to an outside world with specific properties, but at the same time it is also fallible because it depends on the conceptual framework employed and the limits of empirical experience. A cause for concern could be the emphasis on 'structural explanations' as opposed to 'empirical generalisations', but as critical realists also insist that alternative conceptual frameworks and interpretations can be subjected to comparative scrutiny,[12] the privileged position accorded to 'structural explanations' can be dispensed with while still maintaining that there is more to society than meets the eye.

New Institutionalisms – An Overview

Within the theoretical umbrella that has become known as 'new institutionalism', this 'more' has been conceptualised in rather different ways, and this section gives a brief critical survey of key contributions before turning to the task of formulating the position on the basis of which the current text will proceed.

In political science the term 'new institutionalism' has been closely associated with the writings of James March and Johan P. Olsen[13] who in their 1984 manifesto *Organisational Factors in Political Life* insisted that political institutions cannot be reduced to "arenas for contending

11 Sayer 1984, cf. Benton & Craib 2001 Ch. 8, Ougaard 2000, Wad 2000, Collin 2000, Thomsen 1991a.

12 Wad 2000, Collin 2000, Ougaard 2000.

13 Notably March & Olsen 1984, 1989. The relationship between the 'new' and earlier generations of institutionalists in political science seems to be more indirect than is the case in economics (see Thelen & Steinmo 1992 p. 3, Rhodes 1995), and the work of March & Olsen would certainly seem to draw as much upon organisational theory as political science (cf. Esmark 1998).

social forces" because they are also "collections of standard operating procedures and structures that define and defend interests".[14] From this perspective institutions are seen as "political actors in their own right", coherent, autonomous, and capable of moulding political preferences in society at large,[15] but despite the prominent status this essay has achieved,[16] its key concepts remain ambiguous. Given their claim that organisational factors structure political institutions, it would have been good to know whether organisations and institutions belong to the same level of analysis and whether their relationship is interpreted as a case of one-way determination,[17] and thus March & Olsen would hardly appear to be the obvious starting point for an attempt to develop an analytical framework extending beyond the sphere of politics.

The so-called historical institutionalism elaborates on some of the themes pursued by March & Olsen, drawing on the study of history and organisations and establishing itself in opposition to functionalist and rationalist traditions.[18] In an approach primarily concerned with comparative empirical studies of the role 'intermediate level organisations' play in policy-making,[19] Peter Hall has presented the most elaborate statement of the underlying assumptions, emphasising "the institutional relationships, both formal and conventional, that bind the components of the state together and structure its relations with society".[20] By establishing particular relationships between actors, institutions influence "both the degree of power that any one set of actors has over policy outcomes" and "an actor's definition of his own interests",[21] and thus historical institutionalists see actors not as rational 'maximisers' but rather as path-dependent 'satisficers'.[22] By having institutions define the interests of actors, Hall runs the same risk as March & Olsen, namely to erect structures that recreate their own conditions of existence and hence

---

14 March & Olsen 1984 p. 738.

15 March & Olsen 1984 p. 739.

16 See e.g. Bogason 1989, Thomsen 1994, Lane 1995, and, acerbic but acute, Rhodes 1995.

17 Reservations remain about the implications of March & Olsen seeing institutional rules as limits to political behaviour, seemingly replicating the problems of structuralism in being unable to explain social change because of the capacity of the structures to reproduce themselves (Thomsen 1994, Esmark 1998).

18 Hall 1986 pp. 5-15, cf. Rhodes 1994 p. 54.

19 Thelen & Steinmo 1992 pp. 10-13, cf. Hall 1986 p. 19, Esmark 1998 pp. 8-10. The fruitfulness of the approach can be seen in the many empirical studies emanating from it, e.g. Hall 1986, Elbaum & Lazonick (eds.) 1986, Steinmo *et al.* (eds.) 1992, Anderson 1992, Moore 1994.

20 Hall 1986 p. 19.

21 Hall 1986 p. 19.

22 Thelen & Steinmo 1992 pp. 7-10.

should exist in perpetuity,[23] but in practice this problem would seem to be resolved by the co-existence of several institutions that provide dynamic tensions in societies.[24] Hall later came to take a strong interest in the development of new policy ideas, thereby transgressing the limitations of the original approach,[25] but perhaps also suggesting a need to take the more general formulations back to the drawing board again.

Within organisational studies the new institutionalism comprises a diverse set of contributions, many of which share key characteristics with the two traditions discussed above. Defining institutions as conventions that take on "a rule-like status in social thought and action", they are seen as cultural phenomena rather than the result of conscious design, and norms and discourse become integrated parts of institutions.[26] Also this approach runs the risk of portraying institutions as static structures, and like in the case of historical institutionalism change has to be introduced through external factors.[27] An important advantage would, however, seem to be that within organisational theory, perhaps precisely because of its subject-matter, institutions and organisations are more often seen as two distinct phenomena, with abstract institutional rules providing the framework in which individual organisations operate.[28]

In what has become known as the new institutional economics[29] a prominent figure is Douglas C. North whose work has mainly concentrated on patterns of economic growth and performance. According to North "institutions are the rules of a game in a society [...], they structure incentives in human exchange, whether political, social, or economic".[30] Institutions may have been consciously created and formalised like a constitution, or they may exist as conventional codes of social behaviour, but either way they "define and limit the set of choices"[31] available to social actors and thereby reduce the fundamental uncertainty involved in having to negotiate a complex environment by means of a limited capacity to process information.[32] In contrast to this, organisations – e.g. firms, political parties, or churches – are defined as

---

23 Cf. Esmark 1998.

24 Cf. Thomsen 1994.

25 E.g. Hall 1992, 1993; Hall (ed.) 1989.

26 DiMaggio & Powell 1991 pp. 9ff.

27 DiMaggio & Powell 1991 pp. 27ff, cf. Esmark 1998 pp. 11-14.

28 Jepperson 1991 cf. Esmark 1998.

29 In contrast to the 'old' pre-Keynesian institutionalism (Petr 1984, Hodgson 1989 Ch. 1).

30 North 1990a p. 3.

31 North 1990a p. 4.

32 North 1990a Ch. 3.

"groups of individuals bound by some common purpose",[33] and according to North the interaction between organisations and institutions is a crucial feature of social and historical development. Organisations are created to take advantage of opportunities that exist because of the institutional make-up of society, and at the same time organisations are "major agents of institutional change",[34] either through unforseen consequences of attempts to accomplish their own objectives or, more seldom, in the form of conscious attempts to change the rules of the game.[35]

Although the rational-choice inspired institutionalism of North is not unproblematic with regard to the manner in which it has been applied to the analysis of economic development,[36] some more general advantages would seem to be in evidence. *Firstly*, the distinction between institutions and organisations establishes two different levels of analysis, related yet separate, and thereby makes it possible to distinguish between underlying rules and specific historical processes. *Secondly*, although individual and collective actors operate on a terrain structured by the incentives inherent in existing institutions, they still have a reasonable degree of freedom with regard to choosing their strategies,[37] and thus while institutions do structure social intercourse by making some options more attractive than others, they do not reduce actors to 'structural dopes'. And *thirdly*, the possibility of actors challenging existing institutions or creating new ones not only makes it possible to perceive the interaction between institutional structures and social agency as a two-way process, but also allows for institutional transformations to be the result of an intentional process. The most conspicuous problems in North's position concerns the static nature of institutions where change seems to be external (induced by actors),[38] and that the reduction of uncertainty as the main *raison d'être* of institutions tends to overshadow the potentially unequal distribution of incentives inherent in a particular institution, thereby underplaying the conflictual nature of society.[39] Given the importance generally attached by North to institu-

---

33 North 1990a p. 5.

34 North 1990a p. 5.

35 North 1990a Ch. 9.

36 Although North stresses that in principle institutions do not exist because they are optimal from the perspective of economic efficiency (North 1990a pp. 7f), this assumption would still seem to be implicitly present in some of his studies of economic history (Villumsen 1994).

37 Cf. Thelen & Steinmo 1992 pp. 7f.

38 Esmark 1998 pp. 4-8.

39 Cf. North's attempt (1990b) to develop a theory of political institutions on the basis of market analogies.

tional structuring of social incentives, this would, however, seem to be more a question of over-extending a specific institution (the market)[40] than a fundamental flaw in the basic conceptualisation, and hence North would still seem to provide a suitable starting point for an institutional approach to the study of social phenomena, also outwith his original point of departure in economics and economic history.

Institutions, Actors, Organisations, and Discourse

Despite significant differences between the new institutionalisms, it is also clear that much common ground can be found in the attempt to avoid structural determinism and atomistic voluntarism on the basis of a realist perspective. These features are shared with other contributions to the broader discussion on structures and agency, and additional inspiration has therefore been found in the structuration theory of Anthony Giddens,[41] the 'strategic-relational' state theory of Bob Jessop,[42] and the 'institutional history' of Niels Åkerstrøm Andersen.[43] In the following the attempt to formulate a general foundation for an institutionalist approach to the study of regional policy is undertaken by defining and elaborating upon four key concepts, namely institutions, actors, organisations, and discourse. Given the very different ways in which these terms have been used within the various institutionalisms, this task of clarification is clearly essential.[44]

Institutions are seen as sets of rules structuring social relations by defining options and distributing the incentives associated with particular courses of action.[45] This means that institutions can be both limiting

40 Cf. Villumsen 1994.

41 Giddens' position has been repeatedly stated since the 1980s but *The Constitution of Society* (1984) remains a *locus classicus*.

42 Clearly stated in Jessop 1990 Ch. 9.

43 Andersen 1994, Andersen & Kjær 1996.

44 The lack of internal coherence within the 'new institutionalism' is underlined by the conceptual confusion. Structural properties are referred to as both institutions (North) and organisational features (March & Olsen), specific historical entities are referred to as both organisations (North) and institutions (March & Olsen), and some use the term 'organisation' "as a virtual synonym for 'institution'" (Hall 1986 p. 19).

45 The resemblance with North's definition is obvious, although the use of the expression 'social relations' rather than 'social exchange' is intended to avoid specific connotations to market-related institutions and instead highlight the pervasive nature of institutional structuring of social interaction. North's position has, not surprisingly, been echoed by other institutional economists (e.g. Johnson & Lundvall 1989, Johnson 1992, and Hodgson 1989) and similar statements can be found in the writings of strategic-relationist state theorists (Thomsen 1991b pp. 156ff, Hay 1995 pp. 199ff). Giddens' definition of social structures as "rules and resources" (1984 p. 17) is in fact closely related because he appears to see resources as a special type of rules, cf. the discussion below.

and enabling in that rules may either prohibit, permit or require certain acts.[46] Institutions may exist either as more or less informal norms and ideals, or as highly formalised written procedures embodied in particular organisations vested with the power to enforce them,[47] but either way they are inherently social phenomena that can only be reproduced through the continuous agency of the actors operating on the basis of a particular set of rules.[48] While a general rationale for social institutions would seem to be the need for routinisation in order to cope with information and decision overload,[49] it is also clear that institutions may entail sets of actors amongst whom resources are not distributed equally – resources themselves, of course, being a set of social rules denoting 'what counts' in particular situations. Institutions may in other words create an uneven 'playing field' by limiting the options available to actors and privileging those in possession of certain resources, and in this sense it will be possible to speak of 'institutional influence', i.e. effects of social institutions upon the strategies or resources of actors that have not been brought about directly by any actor.

Actors are defined as specific historical entities, individual or collective, with a capacity for agency: being capable of having acted differently.[50] Although the behaviour of actors is embedded in institutions, their agency through choice is still intact because institutions structure the environment of actors by defining options rather than determine their behaviour directly.[51] From this perspective actors have the capacity to produce effects upon other social actors, operating through strategic employment of resources within the rules of a particular social institution. At the same time it is, however, important to stress that the strategies of actors will be limited by the resources available to them and their

46 The insistence on structures being not just negatively limiting but also positively enabling from an actors perspective is widespread (e.g. Giddens 1984 pp. 17ff, Hodgson 1989 p. 132, Jepperson 1991, Thomsen 1994 p. 13, Hay 1995 p. 200), and the more precise distinction between prohibition etc. is inspired by Bloomington public-choice theorist Elinor Ostrom (1986 pp. 5f, cf. Bogason 1994).

47 The role of informal or 'cultural' norms is generally recognised (North 1990a pp. 4ff, Ch. 5; Hodgson 1989 pp. 123-34, Hall 1986 p. 19), but it is essential to make an analytical distinction between the abstract institutional rules and the historical organisations that either uphold specific rules or operate in accordance with them.

48 Giddens argues this point forcefully as part of his structuration theory (1984 pp. 25ff), but it would also seem to be fundamental to strategic-relationist state theory (Hay 1995 pp. 199ff).

49 Hodgson 1989 p. 128.

50 Giddens 1984 p. 9.

51 Ostrom 1986 pp. 5ff. As social relations are a defining characteristic of society (cf. Giddens 1984 Ch. 1), it would take a hermit (or a suicide) to 'opt out' of them altogether.

cognitive maps of the environment in which they operate. Actors are not omniscient but guided by 'bounded rationality'.[52] They pursue their objectives on the basis of a perception of their environment influenced by

- their vantage point (institutional position),
- prevailing discursive interpretations (non-formalised institutions), and
- their capacity for gathering and processing information (resources available).

Although actors may attempt to improve their position *vis-à-vis* other actors or indeed to deliberately attempt to modify or eliminate particular institutions, the likelihood of constant or radical challenges to existing social relationships would therefore seem to be fairly limited.

Organisations are defined as collective actors, namely "groups of individuals bound by some common purpose",[53] and like individual actors they operate in a strategic manner in relationship to a structured environment. Organisations may owe their existence to a variety of reasons: many will have been set up with a view to exploit opportunities or defend interests generated by existing institutions,[54] but other organisations function as embodiment of a particular institution with the purpose of making other actors act in accordance with a specific set of rules. Like individual actors, organisations may pursue a range of different objectives, including maintaining their position *vis-à-vis* other actors with regard to e.g. resources, but the existence of internal social relations makes the co-existence in organisations of parallel strategies, official or otherwise, even more probable.

Discourse is defined as ways of using language that ascribe meaning to the world,[55] and as such it entails assumptions about the state of the world and the roles of actors within it. A particular discourse may privilege certain forms of behaviour and thus function as an informal institution – which may be formalised through the setting up of organisations which promote corresponding forms of behaviour, although of

52 Originally the sociologist Henry Simon's expression (Hodgson 1989 pp. 79ff), but also central in the institutionalist critique of neoclassical economics.

53 North 1990a p. 5.

54 North 1990a Ch. 9.

55 A fairly 'narrow' definition inspired by Fairclough (1992 Ch. 3), a discourse analyst that sees language as a social practice which is *both* shaped by pre-existing social structures *and* at the same time "constituting and constructing the world in meaning" (1992 p. 63). Although he is subsumed under the constructivist umbrella by Jørgensen and Phillips (1999 p. 13), his position would seem to be difficult to distinguish from the position of critical realists, cf. Halkier 2003 and the discussion above.

course such a development cannot be taken for granted but will depend on concrete circumstances, i.e. actor strategies and existing institutions.[56] Like institutions in general, many different and often conflicting forms of discourse can be present in a society at a given point in time, and thus the make-up of what could be called the 'discursive terrain' cannot be taken for granted – e.g. reduced to reflections of economic interests or the views of dominant elites[57] – but must be established through empirical analysis. Any organisation will through its activities embody more or less compatible values and ideals, either deliberately or simply by making particular forms of behaviour appear normal or otherwise attractive,[58] and thereby contribute to the maintaining of associated forms of discourse.

The relationship of the definitions of the four key concepts to the existing literature has been spelt out in the footnotes to the preceding paragraphs, but before moving on to the task of applying these concepts to areas relevant from the perspective of regional policy, it will perhaps be helpful to indicate what appears to be the advantages of the position outlined above. *Firstly*, the conceptualisation of the relationship between social institutions and social actors clearly allows for mutual influence. While institutions structure the environment in which actors operate, agency will, intentionally or otherwise, reproduce, modify or discontinue particular rules, and the way in which both institutions and actors exercise influence hinges on their capacity to affect the options available to social actors through rules and resources. The fundamental objective of establishing institutional structuring and historical agency as two separate, yet related, levels of enquiry has in other words been achieved. *Secondly*, denoting general rules and specific collective actors by the Northian pair institution-organisation rather than the organisational-feature/institution coupling of March & Olsen[59] underlines the potentially systemic nature of sets of rules entailed in an institution:

---

56 Niels Åkerstrøm Andersen's 'institutional history' contains similar ideas, albeit couched in a different terminology (Andersen 1994, 1995 pp. 15ff; Andersen & Kjær 1996, cf. the discussion above), as does the work of Norman Fairclough (1992 pp. 65f, cf. Jørgensen & Phillips 1999 p. 74).

57 This was the point of departure for Laclau & Mouffe's critique (1985) of class reductionism in traditional Marxism.

58 A similar argument has been forcefully made by Laclau & Mouffe (1985 pp. 107f, cf. Jørgensen & Phillips 1999 pp. 46ff).

59 Stressing the systemic nature is actually closer to the thinking of the strategic-relational tradition (Jessop *et al.*) because institutions in North's writings tends to be relatively simple and static constructs (cf. Esmark 1998). The same effect could also have been achieved by using the word structures to denote sets of rules, but despite – or perhaps because – the efforts of Anthony Giddens to rid this concept of its structuralist connotations, this option was considered to be less attractive.

rules may embody particular development tendencies or interact in complex patterns.[60] *Thirdly*, it is worth recalling that the reemergence of the spectre of reductionism in the guise of structuralism is made even more difficult by the co-existence of several institutions in a particular historical setting, making it possible for actors to 'escape' by moving from one set of social relations to another. *Fourthly*, it is possible to understand social transformation as an open-ended structured process because change can either result from the strategic behaviour of individual actors and organisations, be propelled by tendencies and tensions on the institutional level, or be brought about through the complex interaction of developments on these two levels of analysis. And *fifthly*, because formal and informal institutions are both seen as rules structuring social relations, the relationship between discourses giving privilege to particular forms of behaviour and organisations embodying or operating on the basis of particular rules is potentially a close one. The main focus is, however, primarily on the 'extremes' of the discursive spectrum, i.e. the overall discursive terrain and the assumptions of individual actors, while the 'intermediate' category, that of particular discourses, is seen as analytical shorthands for commonly held and/or promoted sets of beliefs at a particular point in time rather than well-defined and systematically structured entities. Moreover, it is also important to stress that the agency of organisations or individual actors also involve non-discursive aspects, physical or material, so that what is being said and what is being done may potentially diverge and, indeed, be interpreted differently by actors with different vantage points. While this makes it possible to integrate communicative actions and symbolic meanings as an important element of the overall analytical approach, discourse analysis does *not* become *the* overarching perspective of the present project.[61]

Combining inspiration from new institutional economics, the strategic-relationist school, and institutional history, it is hoped that this platform is not only suitable for its immediate purpose but could perhaps also prove useful in the context of other empirical research projects by charting a road that could be seen as occupying the middle-ground between the large-scale conceptual engineering projects of Jessop and Giddens on the one hand and the more minimalist approach of Douglas North on the other.

---

60 Something that has come to the fore in the growing literature on social 'systems' (Luhmann, Willke), cf. Esmark 1998, Andersen 1999 Ch. 5. For a discussion of this from a macro-historical perspective, see Halkier 1990b.

61 By collapsing all social phenomena into discourse, constructivist versions of discourse analysis such as e.g. Laclau & Mouffe (1985) would appear to run the risk of underplaying the potential tension between different aspects of social practice, especially between constructed meaning and physical experience (Jessop 1991 p. xixf). For a discussion of discourse analysis and institutionalism, see Halkier 2003.

3.2. What's in a Name? Defining Regional Policy

In order to develop a conceptual framework and pursue an empirical study, an operational definition of regional policy is needed in order to delimit the field of inquiry. An immediate and crucial problem is how to define regions in relation to other geographical entities. Even when focusing on sub-national 'micro regions' rather than supra-national 'macro regions', each academic discipline has its own criteria according to which regions can be separated using e.g. political, economic or cultural parameters. Some of these approaches may produce more orderly maps than others, but attempting to merge these criteria into a multi-dimensional definition of "a region" that can be applied across e.g. Europe is likely to lead to disappointment, simply because cultural and economic phenomena do not tend to co-variate systematically within administrative boundaries.[62] Instead the term 'region' will be used in a much more limited fashion, namely to denote an intermediate spatial level, situated between the nation state above and the local level inhabited by individual cities and communities below,[63] and this multi-dimensional 'intermediateness' introduces a degree of fluidity into the delimitation of regions. On the one hand areas sharing economic characteristics like e.g. declining manufacturing industries may straddle across several administrative regions, and thus the designation of assisted areas for central government regional policy could in effect institute a new set of regional divisions on top of the existing administrative ones and lead to the coexistence of several forms of intermediate tiers. On the other hand it is also important to point out that although regions in the guise of e.g. administrative organisations or economic structures are in most cases fairly stable institutions in the short term, they can be subject to change for a variety of economic and political reasons. From a historical perspective regions are often not only "transitory phenomena",[64] but also open to attempts of redefinition, and thus the spatial target of regional policy would seem to be if not moving, then at least a movable one.

Regional policy has been delimited in different ways in the existing literature. Some have proposed more general definitions, trying to capture particular features that characterise regional policies,[65] thereby running the risk of excluding public activities that are generally considered to be part of 'regional policy as we know it', but many have adopted a more pragmatic approach of simply including activities described by public authorities as being regional policy and hence being

62 Keating 1997, MacLeod 2001, Sagan & Halkier (eds.) 2005.

63 This solution is parallel to the arguments of Sharpe (1993) and Hogwood (1996).

64 Wannop 1995 p. xx.

65 Cf. the discussion below.

able to reflect changing forms of public intervention.[66] As the present study is concerned with different forms of public policy commonly included under the 'regional' heading rather than with a comprehensive study of the spatial consequences of public activities, it would be preferable to have a broad definition that is able to comprise common forms of regional policy, but still capable of setting this area apart from other areas of activity in terms of the rationale for public intervention. Such a definition will seem to rationalise commonly held views about what constitutes regional policy in a particular historical and geographical setting, i.e. Western Europe in the second half of the 20th century, and thus the borders of regional policy are by no means closed once and for all.

A very broad definition has been employed by Niles Hansen and his collaborators in their comparative study of regional policy in industrialised and industrialising countries, namely "any and all conscious and deliberate actions on the part of government to alter the spatial distribution of economic and social phenomena".[67] While stressing the intention to change existing spatial patterns would seem to exclude a-spatial policies with spatial side effects, the authors' claim that the definition includes also trade, monetary and fiscal policy[68] would seem to raise doubts about the importance of this criterion. At the same time including social phenomena as a policy target would also seem to make the definition very extensive indeed, comprising practically all public activities that are spatial in nature, e.g. any form of infrastructure investment. As the present project focuses on a more narrow range of policies traditionally considered to be regional, a less inclusive definition would seem to be preferable, and this requirement is fulfilled by Gavin McCrone who in his classic book on regional policy in Britain concentrated on "the development of policy to influence the location of economic activity".[69] This could in many ways be seen as a more satisfactory starting point because of its economic focus, but still the use of the word 'location' would seem to reflect its origins in redistributive forms of top-down regional policy. The formulation of Folmer, "those acts of the central, regional or local governments which are consciously aimed at influencing the economic situation in one or more regions",[70] avoids this particular problem by including several tiers of regional policy-making and

66 Examples of this include Armstrong & Taylor 1985, Jones 1986, Vanhove & Klaassen 1987 pp. 263ff, Taylor 1992, Bachtler 1997.

67 Hansen *et al.* 1990 p. 2.

68 Hansen *et al.* 1990 p. 2.

69 McCrone 1969 p. 22. For parallel approaches, see Ashcroft 1982 p. 51, Dunford *et al.* 1981 p. 396.

70 1986 p. 14 (italics original).

insisting on the importance of deliberate intention, but the wording could actually include even macro-economic policies affecting "more regions" (namely all of them) because no reference is made to the relative position of regions *vis-à-vis* one another.[71]

While some of the existing delimitations are not far off the mark, there would still seem to be room for improvement, and accordingly the present study defines regional policy as

> public policies designed to influence the relative economic performance of one or more regions *vis-à-vis* other regions by establishing options and incentives that discriminate between firms according to spatial criteria.

This formulation implies a series of choices which in combination would seem to live up to the original ambition, namely to include activities normally referred to as regional policies (cf. the discussion in Chapter 2) while excluding other forms of public policy that are not deliberately discriminatory on an intermediate spatial level. More specifically, this would seem to be the case because

- no reference is made to a particular *level of government* as instigator of development policies, and therefore both European, national, regional, and even sub-regional initiatives are included,
- the *spatial level* on which policies operate is merely specified as intermediate, i.e. sub-national and supra-local, and thus it allows for a wide range of spatial delimitations, including joint efforts by groupings of local authorities, but excludes local economic policies and urban policy,
- *spatial selectivity* is a requirement, but the criteria for selection is not specified, and thus the definition covers both nationally designated assisted areas and self-designation by individual regions, opening up the possibility of regional policy being pursued not only for or by 'problem regions' but also by relative well-off regions,
- the insistence on *intentionality* excludes government interventions that happen to have an uneven spatial impact such as e.g. macro-economic policies or defence procurement,
- the focus on *economic performance* excludes policies targeting other forms of spatial inequality, e.g. welfare policies focusing on the social situation of individual citizens or block grants to local authorities,[72]

---

71 Something that is prominent in the definitions proposed by e.g. Neuperts (1986 p. 65) and Ashcroft (1982 p. 51).

72 The definition in other words corresponds to what in the Scandinavian literature has often been referred to as the 'narrow' regional policy (Oscarsson 1988).

- the *objectives* remain relatively vague and therefore potentially conflicting aims can be included, e.g. top-down policies designed to promote inter-regional equality with regard to employment and bottom-up initiatives aimed at improving the competitiveness of indigenous firms, and finally
- the nature of *policy measures* is also left undefined, allowing for the employment of the widest possible range of policy instruments – from regulation and infrastructure provision to financial assistance and advisory services – which could target either individual firms, providing collective facilities for groups of firms, or changing the nature of the environment in which economic activities take place.

The definition in other words leaves plenty of scope for variation and controversy within the broadly defined policy area itself, while at the same time also establishing external borders that are open to historical reinterpretation, e.g. with regard to the nature of economic objectives, and thus an appropriate balance would seem to have been struck between clarity and flexibility.

3.3. Developing an Analytical Framework

Public Policy – An Institutionalist Perspective

Before turning to the task of developing a conceptual framework for the study of regional policy, it is necessary to consider public policy as a social phenomena in general, and hence this section will attempt to establish an overview of policy processes by focusing on the actors involved in and the dimensions entailed in their interaction.

The study of public policy has traditionally been dominated by two competing approaches.[73] On the one hand a top-down perspective has proceeded on the basis of an ideal-type 'perfect administration' where perceived problems are systematically tackled from the setting of objectives via implementation to evaluation.[74] On the other hand bottom-up oriented policy research has emphasised the lack of central control with implementation and the importance for the outcome of 'front-line bureaucrats' and their interaction with the actors targeted by public intervention.[75] Although "the different approaches have comparative

[73] For overviews, see Ham & Hill 1984, Hogwood & Gunn 1986, Winter 1994, Parsons 1995, cf. Halkier 1996.

[74] See Hogwood & Gunn 1986 pp. 207ff, Ham & Hill 1984 pp. 98f, Sabatier 1993 pp. 267ff.

[75] See Elmore 1979, Lipsky 1980, Hjern & Porter 1993, Bogason 2000, cf. Ham & Hill 1984 pp. 136-42, Sabatier 1993 pp. 276-80, Winter 1994 pp. 78-86, Parsons 1995 pp. 467ff.

advantages as explanations in different contexts",[76] the scope for combining elements from the two traditions has already been demonstrated by e.g. Ham & Hill, Sabatier, Parsons and Winter, and it should therefore be possible to develop a more comprehensive approach to the study of public policy.

Figure 3.1 Organisations and the policy process – A basic model

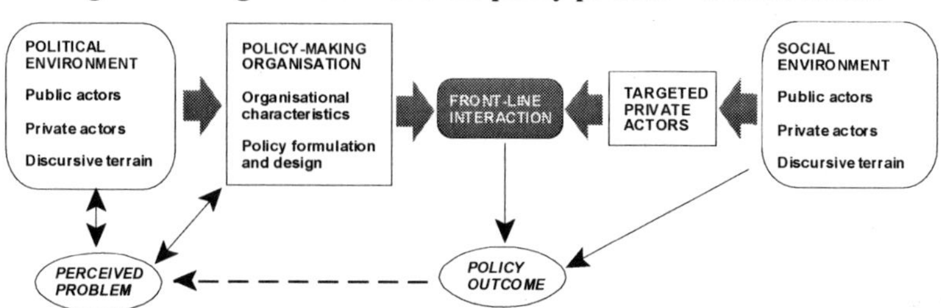

From an institutionalist perspective, the obvious starting point for the analysis of policy processes is the public and private organisations involved and the rules governing their interaction. As illustrated by Figure 3.1,[77] at the heart of the process we find the relationship between the policy-making public agency and the private actors targeted by policy initiatives, because the basic rule underlying public policy is that government can legitimately institute options and incentives in order to influence the behaviour of private actors in accordance with political priorities. Furthermore, the interactions of these key actors are of course embedded in a wider political and social environment where other actors and institutions influence the options available to both public policy-makers and private targets, and thus despite the apparent focus on actors and organisations the institutional level of analysis remains central, as will become evident when each of the elements in the basic model are subjected to closer scrutiny in the following section. Moreover, it is important to underline the complex nature of the policy process: policy formulation and design are limited by the general conditions of incomplete information,[78] while policy-makers *may* be able to control policy output – money, staff or other resources committed to a particular purpose – the ultimate outcome of public intervention also depends on the response from the private actors targeted and will be influenced by

76 Sabatier paraphrased by Parsons (1995 p. 489).

77 The graphics probably come closer to a revised top-down model like the syntheses proposed by Winter (1994 pp. 58ff) and Jenkins (1993 p. 40).

78 Milgrom & Roberts 1992 pp. 28ff, Andersen & Davis 1986, Hodgson 1989 p. 204, North 1990a p. 27.

what Anderson has called 'problem logic',[79] i.e. actors and institutions dominating a particular policy area. From this perspective many things *may* go when it comes to policy-making, but not anything, because public policy is not only about symbolic gestures or the entrenched powers of organisations but also related to 'real world problems', albeit in a rather tenuous way.

This "useful heuristic map"[80] can, through inspiration from inter-organisational approaches to the study of public policy and network theory,[81] be translated into three levels of analysis which taken together can be used to uncover key elements of the policy process, namely

- *organisational dimensions* concerning aspects internal to each of the organisations interacting,
- *relational dimensions* the way in which two organisations interact with one another, and
- *multi-organisational dimensions* characterising the way in which the relational dimensions are inscribed into a larger social context.

It is important to stress that these levels are intimately linked analytical perspectives on the same phenomena, and that neither of them are meaningful without the other two: no organisation exists in a vacuum and their individual features are only of interest when seen in relation to other organisations as a potential starting point for interaction, and at the same time both organisations and their interactions are inevitably set in a larger social context dominated by particular organisations, institutions and forms of discourse. For each level of analysis a set of dimensions have been identified on the basis of existing literature which can then be pursued in the development of a conceptual framework for the study of concrete forms of social interaction, i.e. in this case regional policy.

Looking first at the *organisational* level of analysis, three dimensions appear to be crucial for understanding the position of an individual organisation *vis-à-vis* its surroundings: the resources at its disposal, the assumptions it holds about itself and the environment in which it operates, and the strategies pursued. It must be stressed that all of these are of course relational in the sense that their importance stem from how they compare to corresponding features of other organisations and may form the basis for future agency, and the organisational dimensions also clearly reflect more general institutional circumstances such as rules concerning 'what counts' as resources or the make-up of the discursive

79 Anderson 1992 pp. 55ff, cf. Hogwood & Gunn 1986 pp. 16ff.

80 Jenkins 1993 p. 42.

81 Thorelli 1986, Jordan & Schubert 1992, van Waarden 1992, Hanf & O'Toole 1992, Rhodes & Marsh 1992a, Östholl, Svensson & Halkier 2002.

terrain which is likely to influence the assumptions held by individual organisations. Notwithstanding this, the organisational dimensions are, however, still the starting point from which any organisation proceeds and thus crucial to establish. From an institutionalist perspective commanding *resources* is intimately linked to the capacity of an organisation to influence other actors.[82] Both public and private organisations are generally characterised by having the authority to perform certain tasks in a particular geographical setting, and hence they face two basic challenges, namely to perform particular functions and maintain their position *vis-à-vis* other actors with similar functions. Both these tasks involve strategic deployment of resources – authority, finance, organisation and information[83] – and therefore establishing the position with regard to resources of each of the interacting bodies is an important task.[84] At the same time an organisation embodies explicit or implicit assumptions about the external world that are likely to influence its relations with other organisations.[85] Inspired by the work of Niels Åkerstrøm Andersen these assumptions have been grouped into three *assumptional orders* containing a series of dichotomies which, like those of one of his main inspirations Reinhart Koselleck, function as lines of orientation along which a particular actor ex- or implicitly situates itself *vis-à-vis* the surrounding world in terms of space, time and agency:[86]

- *Topographical assumptions*, a 'social mapping' that entails two dichotomies, *inside/outside* (us/them) and *up/down* (what defines social hierarchies).
- *Temporal assumptions*, a 'historical mapping' that entails two dichotomies, *past/future* (experience/expectation) that defines directions of change, and *cause/consequence* that orders objects (physical or social) according to the way in which they are seen to impact on one another.

---

82 E.g. Rhodes 1988, Thorelli 1986, Christensen *et al.* 1990, Gustaffson & Seemann 1985.

83 Cf. the discussion below.

84 A distinction between various types of resources is undertaken in connection with the discussion of policy instruments below.

85 The importance of e.g. values and ideologies in policy-making and inter-organisational conflicts is widely recognised (see Parsons 1995 pp. 519ff, Hogwood & Gunn 1986 Ch. 7, Rhodes 1988 pp. 93f).

86 For a discussion of the work of Koselleck and Andersen, see Halkier 2003. The description of the individual dichotomies is inspired not only by Andersen (1994) but in most cases also by Koselleck (1990), and like Koselleck before him, the dimensions of Andersen's narrative orders vary between texts: the original outline of his 'institutional history' approach (1994) contained five aspects, while Andersen and Kjær (1996 pp. 10f) and Andersen 1995 operate on the basis of a simplified three-aspect scheme.

- *Operational assumptions*, an 'agency mapping' that entails two dichotomies, *subject/object* (agent of change, object to be manipulated) and *assistant/adversary* (support/resistance).

All three assumptional orders and their dichotomies combine cognitive and normative aspects: while they involve perceptions of social realities within the external world – defines collectivities, designates roles, distinguishes between now and before – they also ascribe values in that they designate good and bad, friends and foes, and desirable directions of change.[87] The assumptions of a particular organisation can be more or less distinct, more or less explicit, more or less in line with prevailing assumptions in its environment, and more or less coherent, not just internally but also in relation to the actual strategies pursued by the organisation, but still a central analytical task will be to establish the specific ways in which a particular organisation positions itself within the general assumptional orders, i.e. what criteria are used to distinguish between e.g. inside and outside. Finally the *strategies* of an organisation, the more specific guidelines according to which the organisation intends to proceed, need to be taken into consideration, and here two sub-dimensions can be identified.[88] On the one hand the functional strategies relating to the pursuit of the substantial aims of the organisation such as product development in private firms or the policy programmes of public bodies. On the other hand the positional strategies *vis-à-vis* other organisations which operate in the same domain and therefore could be construed as either competitors or allies for e.g. bureaucratic, party-political or territorial reasons. In both cases strategies are statements of intended future of action which may or may not correspond to the courses of action eventually taken.

Turning now to the *relational dimensions* concerning the interaction between individual organisations, two areas are of crucial importance, namely the rules guiding the relationship and the eventual outcomes produced. *Interaction rules* define the options and incentives available to the parties involved,[89] e.g. access to particular forms of expertise from private consultants or the availability of public support for particular types of investments. Clearly policy instruments are central interactional rules in public policy, but the broader term is used in order to indicate that the dimension also covers e.g. relations between public bodies or

---

87 Andersen to some extent collapses these two aspects when the inside/outside dichotomy is immediately translated into positive/negative, something that makes dimensions such as subject/object and assistant/adversary superfluous because these roles have been assigned once and for all.

88 E.g. Rhodes 1988, van Waarden 1992, Christensen *et al.* 1990, Bogason 1992 Ch. 1, Gustaffson & Seemann 1985.

89 Rhodes 1988 pp. 91f, cf. van Waarden 1992 pp. 39ff, Parsons 1995 p. 306.

amongst private firms which may affect policy outcomes. The *outcome* can either be an immediate exchange of resources – e.g. sharing of information between firms in a network, financial transfers to private firms complying with certain conditions, or a market transaction – and/or an undertaking to carry out particular activities in the future, perhaps enshrined in organisational change in one or both of the parties involved. The outcome will obviously reflect not only the rules according to which the interaction takes place but also the resource interdependencies between the actors involved, the degree to which they share particular assumptions, and the strategies they pursue, i.e. both institutional and actor-oriented elements, and thus a prerequisite for studying the relational dimensions is knowledge produced by analysis on the organisational level. Moreover, it should also be noted that this will make it possible to account for the role of communicative actions and symbolics within public policy, not just in the form of difference between what organisations say and do, but also how shared or diverging assumptions influence their interaction, and the possible impact of changes in the overall discursive terrain.

Because individual interactions between organisations do not take place in a vacuum but are embedded in broader organisational and institutional settings, it is also necessary to consider what has been dubbed the *multi-organisational dimensions* of inter-organisational relations. Three dimensions are regarded as important, namely the actors involved in a particular set of relations, the way in which their activities are coordinated, and the discursive terrain in which the interaction takes place. With regard to *actor configuration*, establishing the degree of homogeneity between the actors currently involved and the possible access for new actors to a particular type of interaction is important, e.g. a small number of private organisations enjoying a monopoly of access to particular public resources versus an open market with low entry costs, because it will influence the ability of individual actors to exert influence.[90] The *coordination rules* which govern a particular set of interactions – mainly through the institutions of market, hierarchy, or network as modes of social coordination[91] – is significant because it indicates whether one organisation will be able to exert either formal authority or to bring other forms of influence to bear on the rest of the actors within a particular policy network or area of economic activity.[92]

90 See Jordan & Schubert 1992, Rhodes & Marsh 1992a, Thorelli 1986, Gustaffson & Seemann 1985, van Waarden 1992, cf. Moore & Booth 1989 pp. 3ff, 143ff.

91 Halkier & Damborg 1997, cf. the discussion below.

92 The importance of asymmetrical resource dependencies within networks is of course widely recognised, cf. Thorelli 1986, Christensen *et al.* 1990, van Waarden 1992, Rhodes & Marsh 1992a.

Not least in the context of policy-making organisations, this ultimately ties in with the way in which they are placed within the overall structure of the state, and indeed positioned in relation to the *discursive terrain*, i.e. the informal and often conflicting and explicitly contested social institutions constituted by prevailing topographical, temporal, and operational assumptions of relevance to their particular policy area. These assumptions may be nodal points specific to a particular form of public policy or may be part of a more extensive political ideology, i.e. a stable set of assumptions concerning e.g. the relationship between public and private or the territorial unity of the state, and from the perspective of public policy-making a crucial analytical endeavour will be to establish nodal points and relations of dominance between different perspectives because these may propel e.g. the activities of an implementing organisation in particular directions.

Table 3.1 An institutionalist approach to politics and policy – Analytical dimensions

Level of analysis	*Dimensions*	*Sub-dimensions*
Organisational	Resources	Authority Finance Organisation Information
	Assumptional orders	Topographical Temporal Operational
	Strategies	Functional Positional
Relational	Interaction rules	Options Incentives
	Outcomes	Resource exchange Organisational change
Multi-organisational	Coordination rules	Market Hierarchy Network
	Actor configuration	Homogeneity Access
	Discursive terrain	Specific nodal points Ideology

All in all the attempt to arrive at an institutionalist approach to the study of public policy has been brought to a fruitful conclusion in the sense that debates within the existing literature has been elaborated upon and restated in a systematic fashion that, as illustrated by Table 3.1, will allow a linking of the seemingly trivial details of policy design and political discourse with 'big issues' such as social institutions.

The Implementing Organisation

Reflecting the approach summarised in Table 3.1, three dimensions are central to the study of individual organisations, namely resources, assumptional orders and strategies, and in the following the way in which these dimensions can be analysed will be discussed in turn.

Resources

Although the existing literature on regional and related policies is clearly aware of the importance of resources in the policy process, the interest seems to focus primarily on the level of financial resources available, and much can be gained by taking other policy-relevant resources into account in a systematic manner. Inspired by the work of Christopher Hood,[93] four basic policy resources can be identified, as summarised in Table 3.2, which should be taken into account when attempting to establish the position of an implementing organisation, or indeed other organisations involved in regional and similar policies. Three aspects would seem to be particularly interesting, namely the stock, generation and depletion of resources, and here the four basic resources pose rather different challenges.[94]

Table 3.2 Basic resources in the making of regional policy

Authority	Legal and/or legitimate position to define rules
Information	Capacity to mobilise and process cognitive data
Finance	Stocks of moneys or other means of general exchange
Organisation	Capacity to coordinate physical and human assets

93 Hood 1983. Broadly similar views can be found in Windhoff-Héritier (1987 pp. 35-41) and Rhodes (1988 pp. 80f), cf. the discussion in Halkier 1996 pp. 57ff.

94 This does not imply that they are the only policy resources imaginable, merely that they have been deemed to be central from the perspective of public policy and inter-organisational relations in general and for the purpose of studying regional policy in particular, and when studying other policy areas such as e.g. caring or education other forms of resources that cannot be reduced to combinations of authority, information, finance and organisation may be relevant.

From an analytical perspective *authority* could be seen as a relatively accessible resource in the sense that basic features of the remit of a particular body will often have been put down in writing.[95] Such documents are, however, seldom exhaustive, partly in order to maintain a degree of flexibility and partly because of the difficulties involved in describing complex relations in minute detail, and thus informal forms of authority also need to be taken into account, i.e. types of actions on part of an organisation that have come to be acceptable to other parties involved. Although formal authority is generally vested in a particular organisation by some other organisation, authority may be augmented rather than depleted through use.[96]

Information is another type of resource that is potentially enhanced rather than depleted through use.[97] Establishing the informational resources of a particular implementing organisation, i.e. its capacity to mobilise and process cognitive data, could lead into very detailed and time-consuming empirical studies, not least if the relevance of the information to the policy area in question were to be assessed. Instead more indirect approximations can be used such as e.g. the educational backgrounds of staff, the presence of internal think tanks, interaction with other actors, or outputs in the form of publications or advisory services.

From an analytical perspective *finance* as a resource has the advantage of being quantifiable, and since it can be converted into other, more specific, forms of resources such as information or organisation more easily than the other way around, it is tempting to see it as a central resource due to its flexibility. Contrary to authority and information, financial resources are, however, "immanently depletable" in that "they are used up as they are used",[98] and although knowing the overall size of the budget of an implementing organisation is of course relevant, it is often even more interesting to know the extent to which the organisation has the authority to generate resources and/or shift them between different activities.[99]

Finally, *organisation* is also a depletable resource in the sense that any particular body can only take on a limited array of tasks at the same time.[100] Appraising the position of an implementing body with regard to organisational resources may start from the number of staff employed, but will also involve the internal structures of the organisation, e.g.

---

95 Hood 1983 Ch. 4.

96 Hood 1983 p. 144.

97 Hood 1983 p. 144.

98 Hood 1983 pp. 144f.

99 Cf. Dunleavy 1989.

100 Hood 1983 pp. 144f.

functional divisions and the degree of decentralisation. Although these features can often be readily established, they are, however, still indirect ways of measuring capacity for direct action.

All in all the situation with regard to resources as an analytical dimension would seem to be complex, not only because four different types of resources need to be taken into account, but also because some of the variety sources through which the position of a particular organisation can be illuminated.

Assumptional Orders

The next dimensions to consider on the level of the individual organisation are the assumptions made about the world and its own role in the great scheme of things. From a rationalist perspective these assumptions can be seen as the basis on which an organisation devices strategies for employing its resources, but whether or not e.g. 'mission statements' have a bearing on strategic pronouncements and actual interactions with other organisations, the implicit nature of many assumptions means that in terms of analysis, underlying assumptions often have to be identified *through* readings of strategy documents because they have not been made explicit in the form of e.g.'corporate values'.

Table 3.3 Assumptional orders in regional policy

Assumptional orders	*Regional policy dichotomies*	*Central aspects*
Regional relations (topographical)	Territorial positioning (in/out)	* spatial targeting * central-regional relations * inter-regional relations
	Social configuration (up/down)	* underlying values
Regional problems & solutions (temporal)	Inter-regional disparities (past/future)	* issue definition * future promises
	Remedial procedures (cause/consequence)	* perceived causalities
Agents of regional development (operational)	Primary actors (subject/object)	* policy instigators * policy targets
	Secondary actors (assistant/adversary)	* supportive organisations * obstructive organisations

Earlier in this chapter, regional policy was defined as "public policies designed to influence the relative economic performance of one or more regions *vis-à-vis* other regions by establishing options and incentives that discriminate between firms according to spatial criteria", and hence the assumptional orders and dichotomies introduced in Section 3.3 can be translated into a series of specific assumptions which taken together can be said to characterise a particular approach to regional policy, focusing on

- the position and central characteristics of the region designated for support,
- the nature of the regional problem and possible ways of improving the situation, and
- the distribution of roles between social actors in the policy process.

Table 3.3 summarises the assumptional orders associated with regional policy, and the following paragraphs expand on the key dichotomies entailed on the basis of the existing literature on regional policy.

Given the spatial nature of regional policy, it is hardly surprising that topographical assumptions are important, and the heading *regional relations* covers first the horizontal issues concerning the position of the region *vis-à-vis* the external world, and then the vertical issues of social mapping with regard to the central features of society that can justify the translation of inter-regional disparities into public intervention.

As regional policy is inherently discriminatory along spatial lines, the in/out dichotomy is obvious central. Basically, the territory to which a policy applies must be defined, and this will position the included territory in relation to central government and other, excluded, regions – and on all these counts the scope for variation is clearly significant.[101] The authority to instigate spatially discriminatory policies can be exercised both internally, when regions designate themselves, and externally, when a region is designated for support by a national or supranational tier of government. This immediately introduces not only the next dimension, namely assumptions about the relationship between the region and central government but also underlines the importance of inter-regional relations because the designation of a particular territory may either be part of a systematic institutionalisation on a wider territorial scale or be a unique expression of political support, two options that position the designated region in very different ways *vis-à-vis* other regions within the national territory. But these points are still variations on a particular theme, namely that regional policy cannot avoid entailing assumptions about the relation between the designated region and its

[101] See e.g. Wishlade 1999, 2003.

environment based on the way in which a particular policy programme institutes discrimination through exclusion and inclusion of economic actors along spatial lines.

In addition to these horizontal topographical assumptions, regional policy will also, at least implicitly, entail assumptions about the social configuration of the region (and/or the entire nation), because in order for regional disparities to be translated into regional problems, value judgements must be made as to why such differences are unacceptable and hence a legitimate object of public policy. This in turn refers back to the central aspects of society which the public initiative is expected to sustain, and within the literature on regional policy a wide range of 'underlying values' have been identified, cf. the adjoining Table 3.4. These values are not necessarily compatible: competitiveness may undermine employment, growth may threaten equality, and structural change be inimical to social and political stability. Although inter-regional disparities are the specific focus point of this particular form of public intervention, the underlying reasons for being concerned are neither necessarily nor primarily of an economic nature.

Table 3.4 Values underlying regional policy

Inter-regional equality
Regional competitiveness
Regional self-reliance
National growth
Socio-economic stability
Political stability

Based on: McCrone 1969, Bölting 1976, NordREFO 1978, Cambridge Economic Policy Group 1980, Nicol & Yuill 1982, Ashcroft 1982, Diamond & Spence 1983, Begg 1983, Regional Studies Association 1983, Chisholm 1984a, Neuperts 1986, Folmer 1986, Oscarsson 1988, Parsons 1988, Hudson 1989a, Scouller 1989, Gordon 1990, Hansen *et al.* 1990, Keating 1992, OECD 1994.

Like other forms of government activity, the temporal assumptions involved in regional policy are, at least on a general level, rather predictable:[102] an undesirable past/present is contrasted with the prospects of a more desirable future, and causal relations are suggested that will enable policy-makers to take the region in this direction. In practice assumptions about both the nature of the regional problem and its possible causes will of course be of a much more specific nature, as illustrated by Table 3.5 which provides an overview of perceptions of inter-regional disparities and their causes based on existing contributions to the literature on regional development. *Firstly*, it is interesting to note that infor-

102 For a parallel argument, see Therkelsen & Halkier 2004.

mation on the phenomena used to identify the bleakness of the past/present is more likely to be readily available because it can be produced as regional break-downs of statistics that are central to macro-economic policy, but at the same time it does not give many clues as to what the origins of underperformance might be, and hence the indicators can be associated with many different underlying causal relations which may have brought about what is considered to be an undesirable state of affairs. *Secondly*, with the possible exception of external ownership, all the specific problems perceived as being regional focus on features of the relatively weak regions themselves, and hence strategies developed on the basis of such perceptions could well, for better or worse, come to focus primarily on the problem regions themselves rather than their relationship with more well-off parts of the country. And *thirdly*, the diversity of the list would appear to present policy-makers with inherent dilemmas because some of the ways in which the regional issue is defined may be difficult to reconcile, at least from a short-term perspective.[103] For example, making firms within the regions more competitive by introducing more efficient technologies may lead to redundancies, and improving the sectoral structure by attracting investment from outside will strengthen the externally owned sector of the regional economy. All in all the multiple variations and permutations with regard to inter-regional disparities do not lend themselves to a limited set of fixed variables, and thus when analysing a particular implementing body the task will be to identify the specific explicit and implicit temporal assumptions about the character and origins of the regional problem on the basis of the statements and activities of the organisation.

Table 3.5 Perceptions of inter-regional disparities

Issue definitions	Unemployment
	Slow growth
	Income disparities
	Population decline
Causes of disparity	Distance from markets
	Outdated sectoral structure
	External ownership
	Uncompetitive firms
	Inadequate infrastructure
	Inadequate human resources

Based on: McCrone 1969, Bölting 1976, Ashcroft 1980, Cambridge Economic Policy Group 1980, Nicol & Yuill 1982, Diamond & Spence 1983, Regional Studies Association 1983, Vanhove & Klaassen 1987, OECD 1989, Beckman & Carling 1989, Martin 1989a, Scouller 1989, Funck 1990, Keating 1995, Yuill *et al.* 1999.

103 For succinct statements of this dilemma, see Townroe 1986, NordREFO 1978 p. 74, Oscarsson 1988.

Also the operational assumption within regional policy would seem to present a fairly predictable picture, at least on a general level of analysis.[104] When regional policy is defined as public activities aimed at addressing problems in society by attempting to influence the behaviour of private actors, then a public agency will take the role of the subject, the object to be manipulated will be the private firms targeted by public policy, while a variety of public and private actors will appear as supporters or obstructors of change. But also among the primary actors the scope for variation is considerable: different public agencies may want to be involved (or try to disengage themselves from) particular regional problems, and the firms targeted by policy measures may be a more or less select group. Moreover, alternative assumptions about the operational distribution of roles may come about when policies operate in an indirect manner, e.g. by making it more attractive for local government to erect subsidised factory space, and the dominance of liberal ideological thinking may make it convenient to stress the active role of the targeted firms and thus make them appear as subjects rather than objects of policy while the implementing body takes on a more limited assisting capacity. In short, while ultimately we are likely to find a public agency as instigator of change and private actors as the ones whose behaviour is supposed to change, the distribution of roles, not to mention the way in which policy is being presented in statements of strategy, may to varying degrees diverge from the basic pattern.

All in all a limited set of assumptional orders have been identified which, taken together, can be used to position organisations, and indeed individual development initiatives, in relation to other forms of regional policy on the basis of the assumptions about their position in terms of space, time and agency. Although sometimes inferred and ambiguous – and when stated in no uncertain terms not necessarily in accordance with strategic statements or inter-organisational interaction – these dimensions provide important input into the study of the politics of regional policy because perceived underlying assumptions may inform the views of other actors in the policy process. In short, action may sometimes speak louder than words, but often reading between the lines is equally important, both for practitioners in and analysts of the policy process.

Strategies

An organisation involved in implementing regional policy will need to engage in strategic behaviour along two principal lines: operational development strategies in relation to those private firms it seeks to

[104] For a parallel argument, see Therkelsen & Halkier 2004.

influence, and positioning strategies in relation to other public organisations involved in the same policy area.

Looking first at the broad guidelines for the deployment of resources in pursuance of the functional goal of regional development, one could hope that these operational strategies would reflect the perceived nature of the regional problem. The existing literature on regional policy suggests that such rationalist expectations are in fact not frustrated, as operational strategies have included attempts to:[105]

- increase the number of jobs and the levels of income and growth via support for private firms in problem regions and diversion of economic activity from more prosperous areas,
- modify the sectoral structure of the regional economy through promotion of modern industries,
- stimulate local firms in order to strengthen the indigenous sector of the economy,
- upgrade existing infrastructure and skills within the region, and
- support the modernisation of individual firms in order to make them more competitive.

On a general level there would indeed seem to be a high degree of correspondence between assumptions about the nature and causes of regional problems (Table 3.5) and the operational strategies listed above, but this does of course by no means guarantee a similar correspondence within individual policy-making organisations. Most regional problems can be tackled in more than one way, and as 'problem logic' is less than determinate, in practice 'other considerations' – e.g. existing competences within the organisation or the potential for building 'political capital' – are also likely to influence the choice of strategies.

Turning now to the question of positioning strategies *vis-à-vis* other organisations, the most important inter-organisational relations for a body implementing regional policy are likely to be with its political sponsors[106] and with other development bodies operating in the same region. As long as regional policy was exclusively the domain of central government departments, the relevance of these issues was limited and they were therefore not given much attention until the 1980s where of a host of new actors on the European and regional levels appeared.[107] The analysis of positioning strategies has generally been based on various theories of networks, intergovernmental relations and multi-level governance, and as can be seen from Table 3.6 a range of strategic options

[105] See references for Table 3.5.

[106] Discussed separately in the following section.

[107] Cf. Section 2.1.

have been identified. Although the strategies enumerated do not constitute an exhaustive list but are merely some commonly used terms attempting to capture particular ways in which resources can be employed to maintain or change relations with other organisations, it will be noted that different positioning strategies would seem to reflect different ideals with regard to how coordination of activities between organisations should be achieved, i.e. through hierarchical means, via network-type collaboration or market-style patterns of un-commitment. As the realisation of these strategies requires deployment of particular resources, not all of them will be open to any organisation, and the scope for choice that exists for individual organisations will of course also be circumscribed by the resources and strategies of other actors.

While implementing organisations often make public statements about their strategic intentions with regard to their functional objectives, this is less likely to be the case with regard to positioning strategies *vis-à-vis* other public actors in regional policy, but in both cases the possibility of discrepancies between words and deeds must be taken into account, and hence the analysis should not confine itself to official statements of intent, but must also consider actions in relation to other actors.

Table 3.6 Positioning strategies

Positioning strategies
Incorporation
Control
Delegation
Recognition
Cooperation
Imitation
Circumvention
Competition

Based on: Rhodes 1988 pp. 92f, van Waarden 1992, Christensen *et al.* 1990, Gustaffson & Seemann 1985, Cromie & Birley 1994, Cameron & Danson 2000.

The preceding pages have developed an approach to studying the front-line public body vested with the task of implementing regional development initiation, and from the discussion above it is clear that very different forms of regional policy could be accounted for within this framework, and hence the scope for variation and conflict has been maintained. Moreover, the analysis of many of the variables will to a large extent have to rely on interpretations of words and actions, and thus the need for the application of qualitative methods in the study of regional policy is clearly evident. Although the dimensions are all measured on the level of the individual organisation, the underlying approach ensures that they are capable of situating a particular body in the context of its organisational and institutional environment, and therefore the following sections will be able to draw upon many of the points made above.

Political Sponsoring

Bodies implementing regional policy have an ongoing relationship with their political sponsors, and this can be studied along the same lines as other inter-organisational relations. Political sponsorship involves authorising another body to perform certain functions, and it thus entails a hierarchical relationship between the two organisations. The degree to which political sponsors attempt to exercise authority over policy implementation does, however, vary greatly, and front-line bureaucracies often control resources that are significant for successful implementation, and it is therefore important to look not only at the interaction rules as laid down in legislation and operational guidelines, but also to consider the eventual outcome of the interaction between the implementing body and its political sponsors.

Within the literature on regional policy the issue of political sponsoring has mainly been associated with bottom-up forms of regional policy, presumably because a fairly unambiguous situation prevailed with regard to the responsibility for policies and their implementation when mainstream departments of central government administered traditional subsidy schemes. With the introduction of a new array of implementing bodies outwith the mainstream apparatus of government, the nature of sponsoring relations came much more to the fore, but mostly in a form that either noted the issue or questioned the position of particular bodies claiming to be in a semi-autonomous position. The following discussion will therefore primarily build on previous efforts to conceptualise different forms of sponsorship relations[108] in combination with literature of a more general nature.

When studying the rules governing political sponsorship, the relation of authority between the implementing body and its political sponsors is the obvious starting point. From an institutionalist perspective, two distinctions would seem to be central:

- whether the object of sponsorship is the implementing organisation as such or just a particular policy activity, and
- the extent to which authority is granted to the implementing body.

Looking first at circumstances in which the sponsor relation pertains to the *implementing organisation* as such, the relationship between the two

[108] Halkier 1992 pp. 5f, Halkier & Danson 1997 pp. 244ff, cf. Halkier & Damborg 2000.

actors is clearly a hierarchical one because the implementing body does not have an existence independently of its political sponsors. In this situation the general incentive for the implementing body to comply with the rules established by its sponsors is a very strong one indeed, namely to ensure its continued existence.[109] Still the degree of authority vested by political sponsors in the implementing body – bureaucratic autonomy for short – may vary significantly, and in the literature on public administration a distinction has been made between departmental and arm's-length (non-departmental) types of relationships. In a *departmental* relationship, elected executives such as ministers are responsible for any decision made by the implementing organisation and can ultimately be held accountable for the activities of their department by e.g. parliament. Day-to-day business is, however, largely conducted according to guidelines delegating authority, and the bureaucracy may even enjoy some scope for initiative in the development of policy programmes.[110] Contrary to this, an *arm's-length* relationship[111] describes a situation where political sponsors have established a separate organisation outside the normal structures of the government apparatus and vested it with a relatively large measure of authority, both with regard to policy development and implementation. Here the involvement of the political sponsors is limited to a fairly general level, i.e. allocation of resources and broad policy guidelines, and the possibilities of direct parliamentary scrutiny is much more limited.[112] An arm's-length relationship in other words places both strategic initiative and important discretionary powers with front-line bureaucracy, albeit subject to political scrutiny. It is important to note that 'departmental' and 'arm's-length' relations are ideal types on a continuum of bureaucratic autonomy,[113] and thus it would be possible to have departmental relations with

109 Beesley & White 1973 p. 65, Thiel 2001 Ch. 1.

110 Greenwood & Wilson 1989 pp. 21f, Hood & Schuppert 1988 pp. 5f, Wilding 1982, Thiel 2001 Ch. 1.

111 A term originally used to describe the relationship between nationalised industries in Britain and their sponsoring department in central government (see e.g. Greenwood & Wilson 1989 pp. 31ff, Fine & Harris 1985 pp. 153-65, Williams *et al.* 1983 pp. 104-9, Pollard 1983 pp. 310ff, Scouller 1987 pp. 6.31-46).

112 Beesley & White 1973 p. 65, Hood & Schuppert 1988 pp. 6f, Wilding 1982 p. 40, Greenwood & Wilson 1989 Ch. 11, Thiel 2001 Ch. 1. By focusing on the relationships of sponsorship, the not always fruitful search for a definition that can separate out a well-defined group of organisations such as e.g. quangos (Quasi Non-Governmental Organisations) can be avoided (see Greenwood & Wilson 1989 Ch. 11, Barker (ed.) 1982, Hood 1986, Modeen & Rosas (eds.) 1988, Hood & Schuppert (eds.) 1988, Hogwood 1995).

113 Greve *et al.* 1999, Thiel 2001 Ch. 1.

more or less direct involvement of elected politicians,[114] and arm's-length agencies being kept under different degrees of control by their sponsors, ranging from detailed scrutiny of implementation to the 'technocratic dream scenario' of being given money and told to 'get on with it'.[115] Determining 'the length of the arm' is, however, not a straight-forward task, as, on the one hand, formal rules of authority may in themselves be difficult to interpret. Many British implementing bodies positioned at arm's-length from central government can for instance be issued with specific instructions to perform particular tasks, but the fact that these powers are only very rarely used could give the impression that in practice the arm is a fairly long one – unless its function is simply to be a general deterrent and last-resort measure while pressures normally are brought to bear in much more discreet ways.[116] On the other hand even in cases where "the arm of sponsorship" is a relatively short one, an implementing organisation may have other resources that could strengthen its position *vis-à-vis* its political 'masters'.[117] Indeed Rhodes in his extensive study into sub-central government in Britain found that "the capacity of (arm's-length agencies) for independent action is rooted in their command of organisational and informational resources",[118] and thus by playing a central role in the development and implementation of policy, such an agency may even be able to accumulate informal political authority that can further reinforce an arm's-length position, especially if its activities are perceived as being important by a wider constituency.

Instead of using directly subordinate organisations such as government departments or arm's-length agencies, political sponsors may decide that the best way to pursue particular policy objectives is to leave implementation to an organisation sponsored by another political authority. When e.g. a central government scheme in regional policy is being administered by local government bodies, political sponsorship relates to a *particular policy* rather than the implementing organisation as a whole. In this case the relationship between the policy sponsors and the implementing body can best be described as a network-type rela-

---

114 The presence of a separate advisory council of e.g. private sector representatives could strengthen the position of implementing organisations *vis-à-vis* its political sponsors, and having an arm's-length agency sponsored jointly by a large number of co-sponsors may have the same effect (Halkier & Danson 1997 pp. 246f).

115 Having stressed the wide range of meanings attached to the two ideal types, the two additional 'extreme' ideal types proposed earlier (Halkier 1992 pp. 5f) can be dispensed with.

116 See Johnson 1982.

117 Rhodes 1988 pp. 170ff, Hood 1986 pp. 195-98, Christensen & Christiansen 1992, Ham & Hill 1984 Ch. 9.

118 Rhodes 1988 pp. 180.

tion[119] because of the long-term nature of the relationship and the absence of formal sub-ordination between the parties involved, and the wider resources and strategies of the actors therefore take on particular significance: how easy would it be for the political sponsors to substitute the implementing organisation? How important are the externally sponsored activities to the implementing body? And what are the long-term positioning strategies of the parties involved?

When analysing the relationship between an implementing body and its political sponsors – be it departmental, arm's-length or networked – it is therefore crucial to consider not only the rules of the game but also the outcome, i.e. how the overall patterns of resource dependencies and the strategies of the actors have been translated into outcomes: Who influences what decisions in practice, and how are relations of authority related to other forms of resource exchange such as flows of finance and information or long-term organisational change? Other things being equal, the less bureaucratic autonomy, the greater the likelihood of e.g. party-political preferences influencing policy development and implementation – but other things are seldom equal because formal relations of authority are mitigated by other resource dependencies and the strategic ingenuity of actors involved, and thus the question of 'the length of the arm' can itself take on political significance. What might have looked as a fairly straight-forward task in terms of analysis, has acquired a high degree of complexity and turned out to involve both analysis on the level of individual organisations – assessing the resources and strategies of sponsors and sponsored – and inter-organisational relations including both the rules governing the sponsorship relation and the outcome of the interaction.

The Political Environment

The political environment of an implementing body comprises phenomena related to the political processes surrounding the making of public policy, i.e. the main institutions influencing the policy process, the configuration of actors in the form of public and private organisations that may provide input into the process, and prevailing forms of political discourse. The following sections will consider these three aspects in turn.

Coordination Rules and Actor Configurations

Even with the increasing importance of the supra-national level for EU member states since the mid-1980s, the national political system clearly remained the single most important institution from the perspec-

[119] Halkier & Damborg 1997, cf. the discussion below.

tive of coordinating activities associated with regional policy in the period under scrutiny. While the political system establishes the relative permanent institutional basis on which public and private actors interact, the specific configuration of actors, their resources and strategies, will decide how interaction unfolds historically, and hence both institutional and actor-oriented aspects must be accounted for in empirical studies. In the context of regional policy, by definition a spatially discriminatory form of economic policy, especially two aspects of the political system would appear to be of particular importance. On the one hand rules affecting the competition between political parties, especially the extent to which a territorial dimension is integrated into the political process via e.g. electoral procedures, because if the electoral process is organised in a territorially fragmented way where geographical constituencies are crucial, then local interests are more likely to become politicised than if the elected representatives are less tied to particular localities. On the other hand the territorial structures of the public sector will also have an impact because the spatial hierarchy of government tiers will influence the possibilities of articulated e.g. regional grievances, and specifically the extent to which sub-national governance structures apply across the entire territory of the state or involve differential treatment of e.g. individual regions will create a more or less even playing field for sub-national actors pursuing a regional development agenda.[120]

Turning now to the configuration of actors within regional policy, three groups would seem to be of particular importance, namely public bodies, private sector organisations, and political parties, and the following pages will deal with each of these in some detail.

The activities of other *public organisations* can influence the activities of an implementing body both directly and indirectly. On the one hand a number of different public organisations will be involved in most policy areas, each with different roles and perspectives, and these activities tend to be coordinated through what has become known as policy networks, directly influencing on the activities of individual implementing organisations. On the other hand the presence of other public bodies also have an indirect influence in that their policies affect the environment in which an organisation is operating by instituting alternative options and incentives.

The general notion of *policy networks* can readily be applied to regional policy, not only in the multi-level phase where a large number of actors emerge on the scene, but also in the preceding national phase where the way in which sub-national actors are involved can vary

120 See e.g. Loughlin & Peters 1997, Keating 1997, le Galès 1998.

greatly.[121] What needs to be considered is, in other words, which organisations are the key actors within the network,[122] and what roles they play, especially with regard to ensuring the overall coordination of activities within a particular area of public policy.

The existence of indirect *policy impacts* in the sense that the activities of one organisation may affect those of other organisations is undoubtedly one of the very reasons for the existence of policy networks, but also the activities of public bodies outside the immediate network may have a bearing on an implementing organisation by creating alternative incentives that make private firms ignore its best laid plans. Given that regional policy has had both economic and social rationales, a fairly broad range of policies may have an indirect impact on regional development activities, and this has indeed been widely recognised within the existing literature. *Firstly*, regional policies operating in other geographical contexts may have a significant impact because they influence the competitive environment, both with regard to attraction of investment from external sources and with regard to support for indigenous firms. This horizontal impact does not only apply to regions within the same national context but also has an international dimension, especially with regard to attraction of footloose investors.[123] *Secondly*, other forms of spatially targeted development measures such as urban policies or the economic policies of local government may also be an influence,[124] e.g. by giving preference to particular geographical areas or working with individual firms in ways that differ from regional priorities and strategies. *Thirdly*, the macro-economic policies of central government may have a major impact on the regional economy, and e.g. national incentives schemes relying on diversion of investment would be hampered by dampening of economic growth through tight fiscal and monetary policies.[125] *Fourthly*, other economic policies of central gov-

121 Although notions of interaction between mutually dependent actors are common in especially writings on bottom-up regional policy, explicit adoption of some form of policy-network perspective is less common. For examples of this, see Anderson 1990 & 1992, Hogwood 1977a & 1987, Ackermann & Steinmann 1982, Cooke & Morgan 1993, Balme & Bonnet 1994, Heinelt & Smith (eds.) 1996, Östholl, Svensson & Halkier 2002.

122 The question of delimitation and classification of individual policy networks is notoriously difficult (Jordan & Schubert 1992, Lane 1995 Ch. 4, Parsons 1995 pp. 191ff), but not necessarily a particularly fruitful line of enquiry.

123 An official rationale for EU regulation of national regional policy is to avoid "'competitive outbidding' between nations and regions" (Wishlade 1998 p. 573), while the competition between regions via non-financial means is a more discreet, but nonetheless real, phenomenon, cf. e.g. Brown & Raines 1999, Armstrong & Twomey 1993, Keating 1995, Damborg & Halkier 1998.

124 See e.g. Barnekov *et al.* 1989, Christensen 1990, Keating 1993, Pickvance 1990.

125 See e.g. Firn & Maclennan 1979, Martin & Hodge 1983a, Prestwich & Taylor 1990.

ernment attempting to improve the performance of particular groups of firms – e.g industrial policies, technology policy, or promotion of small firms – can have spatial effects that may well run counter to regional policies aiming for inter-regional equality priorities because firms in relatively prosperous areas can be better positioned to take advantage of such schemes.[126] And *finally*, limited welfare provisions for individual unemployed may prompt migration and run counter to regional policy attempts to halt population decline in problem regions. In each of these areas a lack of 'strategic fit' between regional and other policies can clearly affect the performance of the former, and thus such effects – or even perceived differences between different policies with regard to e.g. goals and underlying values – may lead to strategic adjustment by one or more of the organisations involved.

All in all not only the position of the implementing organisation in the internal organisation of the political process in the form of policy network, but also the relationship with the policy output of other public bodies are clearly important parts of the political environment which need to be taken into account.

The role of *private sector organisations* has generally attracted a good deal of attention as part of the attempt to contextualise the policy process, and as the essence of regional policy is to make private firms behave in ways they would not otherwise have done, one could expect a high degree of interest in these matters from interest organisations representing employers and employees. The issue of spatial discrimination can, however, be highly contentious because interest organisations operating on the national level may find it difficult to endorse unequal treatment of their members,[127] and on the regional level the enthusiasm of individual firms and managers for promotion of particular localities depends amongst other things upon economic interests and intangible things such as place attachment.[128] The general thrust of the analysis of the role of organised interest has shifted several times, from the early pluralist perspective via the corporatism discussion to the current somewhat more flexible emphasis on policy networks and governance,[129] and similar tendencies can also be found in the literature on regional and related policies, with network-oriented perspectives becoming increas-

126 See e.g. Cameron 1979, Armstrong & Taylor 1985 pp. 211-19, Amin & Pywell 1989, Anderson 1992 pp. 68ff, Begg 1993, Harrison & Mason 1993, Charles & Howells 1993.

127 Kenworthy 1990, Sidenius 1984.

128 See the literature on formalised public-private collaboration cited below, cf. Kooistra 1998.

129 Jordan & Schubert 1992, Parsons 1995 pp. 257ff, Thomsen 1996 Ch. 1, Rhodes 1988, 1996.

ingly significant in the 1990s.[130] Numerous typologies of public-private cooperation have been developed, e.g. growth coalitions,[131] urban regime theory,[132] development coalitions,[133] and public-private partnerships,[134] and many move along lines that are parallel with the approach to public policy developed in Section 3.1, emphasising the importance of resource inter-dependencies and the strategies of actors.

From this perspective two scenarios would seem to present themselves: private actors may either be directly involved with the implementing organisation, or play a more or less formalised indirect role within the broader policy network. *Indirect involvement* of private actors in the broader policy network can be assessed in the same way as that of public actors, and again a key issue will be the strategies and resources employed and the extent to which the eventual outcome is seen to have been influenced. *Direct involvement* of private actors with the implementing body can be studied as other inter-organisational relations in terms of rules and outcomes, and of particular interest are of course what private actors are involved, which aspects of the policy-making process they are involved in, what the consequences of their involvement appears to be both in terms of the policies pursued and the response of private firms to particular policy initiatives, and how their presence affects the relationship between the implementing body and its political sponsors.

Although the centrality of *political parties* in the development of regional policy, and indeed public policy in general has been disputed,[135] party politics are still involved in two important aspects of the political process, namely representation of constituencies within civil society and executive roles within the government administrative system, and thus parties cannot be ignored within an approach to public policy,[136] although their influence needs to be established empirically. Accordingly political parties will be seen as a particular form of organisation with functional strategies pertaining to the overall direction of public policy

130 See e.g. Ackermann & Steinmann 1982 and Sturm 1991. Some authors combine aspects of corporatism and network perspectives (Moore & Booth 1989, Olsson 1995, Totterdill 1989), while others argue that their analysis aims at situating their case studies on a corporatism-pluralism continuum (Anderson 1992 pp. 25ff, Moore & Pierre 1988).

131 Molotch 1976, cf. Logan & Molotch 1987.

132 Lauria (ed.) 1997, Wong 2000.

133 Keating 1991 Ch. 7 and 1993, cf. Morgan 1998.

134 See e.g. Moore & Pierre 1988, Eisinger 1988, Harding 1990, Keating 1995, Batt 1996, Östhol, Svensson & Halkier 2002.

135 See e.g. Rose & Davies 1994, Hogwood 1992, Hood 1994, cf. Chapter 2.

136 Rhodes 1988 pp. 83f, Hall 1986 pp. 91f, cf. Anderson 1992.

and the state as such, and positioning strategies centred on maintaining and strengthening political support in order to gain direct access to executive roles within the political system.[137]

A key resource for political parties is therefore authority in the form of political legitimacy generated through success in the electoral process, and discursive interventions in support of their strategic projects are therefore an important way to further party interest. For political parties spatial considerations are, however, both crucial and cumbersome at the same time. Although most parties in Western Europe are national in the sense that they are committed to maintaining current state boundaries and promoting particular interests and ideas throughout the national territory, the spatial dimension of politics does not disappear. Regionalist or separatist parties may have a presence, electoral support for national parties may be unevenly distributed geographically, and subnational parts of the party organisation may pursue particular regional interests,[138] and thus national parties may adopt strategies that discriminate between different areas by means of e.g. regional policies in order to maintain influence. Moreover, these interactions take place in an environment structured by political institutions that have spatial dimensions and consequences, e.g. the role of territory in the electoral process or the relative influence of different tiers of government, which may be conducive or otherwise to the inclusion of regional problems on the political agenda.

With regard to political parties and regional policy it is in other words important not to see parties solely in terms of their official discursive interventions, but also to take both institutional features and organisational positioning into account.

The Discursive Terrain

An important part of the political environment in which regional policy is conducted is the discursive terrain, i.e. the prevailing assumptions that as informal institutions make certain claims seem 'natural' and other claims inherently dubious. Establishing or maintaining the dominance of assumptions that are conducive to a particular political strategy is important because successful discursive interventions can contribute to the entrenchment of certain assumptions and values that will also influence future decision-making. Like modification of formalised political institutions, 'ideational' battles are a form of long-term strate-

[137] A formulation combining Jessop's notion of strategic agency and a more traditional perspective of competition between parties (Hogwood 1977a pp. 13-16, Hague *et al.* 1992 pp. 234ff, Lane & Ersson 1994 pp. 102f).

[138] Keating 1997, Rhodes 1988 pp. 208ff, Townroe 1986 p. 368.

gic behaviour by political actors, and thus implementing organisations are at the same time both actors on the discursive terrain and objects of such interventions in that they may embody, espouse or have ascribed to them certain values and assumptions. As 'embodiment' can be more or less ambiguous – a particular policy may be compatible with different underlying assumptions – the room for discursive association is often considerable, and presenting activities to the public in an acceptable 'discursive wrapping' may thus become a strategic tool in its own right.

We saw earlier that the core assumptions underlying regional policy were primarily, and unsurprisingly, concerned with the nature of the regional problem, the position and character of the region in relation to the external world, and the distribution of roles in the process of attempting to address perceived weaknesses. Given that policy initiatives and networks concerned with regional development have existed for decades, this is likely to lead to the emergence of specific nodal points, i.e. concepts or lines of argument around which the debate on regional policy revolves at a particular point in time, e.g. the 'growth poles' in the 1960s, 'indigenous potential' in the 1990s, or the recent interest in 'partnership' as an organisational vehicle for regional development.[139] At the same time it is also evident that regional policy as an attempt to influence relations between regions within the national space through policies aiming to change the behaviour of private economic actors is intrinsically linked with two issues that in various ideological guises have been key themes in Western European politics since the 19th century, namely the relationship between government and private economic actors on the one hand, and the relationship between national and sub-national tiers of government on the other.

Table 3.7 Government and private economic actors in political discourse

Planning
Interventionism
Market failure
Laissez-faire

Differences with regard to the relationship between government and private economic actors have to a large extent been the organising principle behind the traditional ideological distinction between left and right in politics,[140] and although no generally agreed nomenclature

[139] Armstrong & Taylor 1985 pp. 76ff, 207ff; Östhol, Svensson & Halkier 2002, cf. Chapter 2.

[140] See e.g. Bogason 1992 Ch. 1, Grant 1993 Ch. 1.

prevails, it is possible to identify a number of ideal-type perceptions of the desirable nature of this relationship, all of which have been associated with regional policy. While the desirability of economic growth would seem to be a shared underlying value, the means by which this is to be achieved vary according to what is assumed to be the comparative advantages of markets and public agencies in realising growth, cf. the adjoining Table 3.7. At one end of the scale *planning* assumes the superiority of government agencies because they can take coordinated decisions on the basis of comprehensive information whereas the horizon of private firms operating on the market is limited by short-term self-interest and a comparatively weak informational basis.[141] Presenting policies in terms of *interventionism* could be seen as a more pragmatic version of this approach, relinquishing the claim of extensive public control of the economy, but in effect still applying the same basic thinking on a more selective basis in e.g. particular industries.[142] The antithesis of planning is the *laissez-faire* approach that sees the market as the most efficient allocator of scarce resources and insists that public interference will only keep the spontaneous workings of the economy from delivering its benefits and should therefore be kept at an absolute minimum.[143] The corresponding 'pragmatic' *market failure* approach adheres, in principle, to the fundamental virtues of the market as extolled by the *laissez-faire* position, but at the same time it is recognised that markets do not always produce economically or socially desirable results, and *if* this market failure occurs due to e.g. imperfect information, problems may be addressed through targeted public intervention.[144] From an institutionalist perspective it is immediately obvious that both the two 'extreme' positions, planning and *laissez-faire*, would seem to be based on the unrealistic notion that their privileged agents of economic development operate on the basis of complete and perfect information, but this has of course not prevented any of them from functioning as the general discursive setting for political parties or individual

[141] Hansen *et al.* 1990, Wigley & Lipman 1992, Lane 1995 Ch. 1, cf. Grant 1982 pp. 12ff, Kristiansen 1987 pp. 52ff.

[142] Parsons 1986, Ewers & Wettmann 1980, Imberg & Northcott 1981, Alexander 1981, Grant 1982, Keating & Midwinter 1983 pp. 171ff, Hindley 1983, Turok 1987, Moore 1988, Hodgson 1989, Prestwich & Taylor 1990, Danson *et al.* 1990, Barberis & May 1993. Dirigism has sometimes been used as a roughly equivalent term, e.g. Sidenius 1989, Morgan 1994.

[143] Parsons 1988, Lane 1995, Grant & Wilks 1983, Hansen *et al.* 1990, Balchin 1990, Prestwich & Taylor 1990, Gudgin 1995. Liberalism or a 'market-based' approach have often been used as parallel terms, e.g. Morgan 1994, Kristiansen 1987, Sidenius 1989, Martin 1989a, Christensen 1990, Keating 1993, Morgan 1994.

[144] E.g. Sawyer 1992. Alternative terms are 'social market strategy' (Grant 1982 pp. 12ff, Hudson 1989a Ch. 4) or 'modified market approach' (Danson *et al.* 1989b & 1990, Moore & Pierre 1988, Cochrane 1990).

policy programmes. Moreover, the difference between 'selective intervention' and the addressing of selected 'market failures' would seem to be difficult to define in a clear-cut way,[145] and thus the four ideal types are probably best thought of as being part of a continuum where the importance of public and private actors in economic development gradually shifts as we move from planning-type assumptions via interventionism and market failure towards a *laissez-faire* perspective. These ambiguities underline the dual nature of assumptions about the ideal roles of public and private actors as being informal institutions, functioning both as a guide to future action and as a form of 'wrapping' that aims to situate particular actions within the framework of a socially or politically acceptable discourse.

The other central set of assumptions with which regional policy is associated relates to the spatial organisation of the political system, i.e. the degree of decentralisation within a particular nation state and the position of regions relative to one another. Perhaps because of its metapolitical nature – focusing on the distribution of authority rather than particular policies – its presence on the political agenda has been more uneven,[146] but it is still clearly possible to identify different assumptions about the desirable relationship between the national and regional levels – which may of course be in accordance with actual state organisation to a greater or lesser extent. From a *vertical* perspective, the central issue is the distribution of authority between central and sub-national levels of government.[147] In a *unitary* political system authority is primarily seen as being vested at the national level with parliament and central government, while in a *decentralised* context a more or less extensive transfer of authority to sub-national actors has taken place, either politically through the setting up of sub-national legislative assemblies or administratively through spatial deconcentration within functional departments of central government or the presence of multifunctional territorial departments within central government responsible for a geographically limited part of the country. Historically this may reflect a combination of pressures from regional actors and the perceived functional or political advantages of off-loading particular functions to subnational tiers of government.[148] Either way, the constitutional preeminence of the national level is being maintained and the division of labour between different tiers of government can therefore at least in

---

145 Taylor & Wren (1997) actually equates the two, but this pragmatic solution would seem to obscure the different ideological connotations suggested by assumptions about the (desirable) primacy of public and private actors respectively.

146 See e.g. Keating 1988a, 1997, 1998; Sharpe 1993, Anderson 1994.

147 See Rhodes 1988, Loughlin & Peters 1997, Wiehler & Stumm 1995, Anderson 1992.

148 Keating 1998, Wright 1998, Sharpe 1993, Marks 1996.

principle be unilaterally modified by central government.[149] Contrary to this a *federal* setting not only involves more extensive transfer of authority but also enshrines this constitutionally and hence makes it more difficult to encroach upon the position of sub-national actors. Parallel to these vertical inter-tier relations, assumptions are also being made along *horizontal* lines with regard to the relationship between different regions within the national political system where the ideal can either be *equal* treatment, i.e. every region being vested with the same degree of authority, or *preferential* treatment, i.e. different regions enjoying different degrees of authority or access to resources through what could be termed 'preferential regionalism'. In this context regional policy initiatives can easily be interpreted as being part of vertical or horizontal positioning strategies, with particular measures becoming controversial simply because they contravene existing divisions of authority, like the new bottom-up regional policies which have often been associated with a general strengthening of sub-national actors *vis-à-vis* central government.[150]

Organisations implementing regional policy do in other words find themselves in a political environment in which certain assumptions tend to prevail because these are sustained by and help sustain dominant political actors. The presence of such informal institutions as part of the historical setting has important consequences for the conduct of regional policy because they will be taken into account by the organisations involved. On the one hand a sponsoring body or a political party may attempt to associate itself with an implementing organisation by claiming it for particular values in order to e.g. influence its activities or make it part of an acceptable political discourse – something that is of course particularly likely to happen because the assumptions underlying regional policy touch upon central issues in Western European politics. On the other hand the implementing organisation itself may use discursive means as part of its positioning strategy, i.e. by making sure that it publicly espouses values that are in accordance with dominant assumptions, not least those of its political sponsors. Although many regional policy activities could be said to be ambiguous with regard to their underlying values, there are limits to this flexibility, and therefore it becomes important to establish the relationship between the assumptions embodied in the policies of an implementing organisation and the way these activities are interpreted publicly by the organisation itself and by other political actors, because differences may suggest tensions and possible future directions of change.

---

149 Hogwood 1982 pp. 34ff.

150 See e.g. Keating 1992, 1997.

The preceding pages have demonstrated the importance of the political environment in which regional policy is situated, both in terms of direct interaction between the implementing body and public or private organisations, and the more indirect, but nonetheless very real, influence of prevailing political discourses and institutions. But even the impact of this impressive array of external pressures still depends on the concrete interaction between actors in concrete historical circumstances: resource interdependencies and conflicting political assumptions and strategies may well leave considerable room for manoeuvre, and thus the extent to which the activities of an implementing organisation reflect values and strategies that prevail in its political environment needs to be investigated empirically and cannot be taken for granted.

Policy, Implementation and Private Firms

Having explored the political aspects at some length, attention is now turned to the other key inter-organisational relationship within regional policy: the direct interaction between the implementing body and private firms. The importance of the response from actors targeted by public policy has been widely recognised not only within the general literature on policy analysis,[151] but also in writings on regional and related policies,[152] and the discussion proceeds in two steps, looking first at the relationship with individual firms – under the headings of policy instruments, modes of implementation and outcomes – and then in the following section situating these interactions in the broader context of the economic environment.

Policy Instruments and Modes of Implementation

As options and incentives designed to influence the behaviour of private actors, policy instruments constitute the general rules that govern the interaction between an implementing body and private firms, and the choice of instruments and the way in which these are put to use can therefore be expected, to some extent at least, to reflect considerations about what measures would bring about an appropriate response by a sufficient number of private firms in order to meet overall policy objectives.

151 See e.g. Christensen & Christiansen 1992 Ch. 1, Christiansen 1993 p. 23, Winter 1994 Ch. 4.

152 See e.g. OECD 1977 pp. 26ff, Ackermann & Steinmann 1982, Pinder 1982, Swales 1993, Henning 1983, Heclo 1986, Grahm 1988, Beckman & Carling 1989, Hudson 1994 – cf. the extensive literature on evaluation which could be said to owe its existence to the uncertainty created by regional policy depending on the response of private actors so that the outcome of public intervention cannot be taken for granted.

Table 3.8 Classifications of regional policy measures

Authors	*Key variables*	*Authors*	*Key variables*
Vanhove & Klaassen	Infrastructure Financial incentives Discouragement measures Public office dispersal Public purchasing RDAs	McCrone	Controls Grants Inducements Public investments
		Coffey & Polese	Social animation Information Finance
Demko & Fuchs	Infrastructure Monetary incentives State ownership Controls and disincentives Migration/training policies	Young & Lowe	Economic regulation Legal regulation Exhortation Advice and services Financial inducements

Sources: Vanhove & Klaassen 1987, Demko & Fuchs 1984, McCrone 1969, Coffey & Polese 1985, Young & Lowe 1974.

In the regional policy literature, the existence of different forms of policy instruments is generally acknowledged, as illustrated by the classification schemes summarised in Table 3.8. These classifications mostly focus on the different resources being employed and often imply a contrast between inherently 'strong' (controls) and 'weak' (advice) instruments,[153] and thus the notion of policy instruments found in the literature on regional policy is broadly in line with general trends in the specialist literature on policy instruments,[154] with major weaknesses being that typologies either appear to be closely tied to specific historical settings and/or implying a fixed hierarchy between instruments based on the degree of constraint they entail *vis-à-vis* the private sector actors targeted.

Table 3.9 Policy instruments as options of interaction

Mandatory	Compulsory behaviour
Conditional	Access to resources on certain terms
Voluntary	Resources made available unconditionally

153 See e.g. McCrone 1969 Ch. 8, Allen 1979, Hogwood 1983, Richardson 1984, Folmer 1986, Beckman & Carling 1989, Alonso 1989, Kern 1990, Allen & Yuill 1990, Halkier & Danson 1997, Yuill *et al.* 1998.

154 For a critical survey, see Halkier 1996 pp. 51-57.

From an institutionalist perspective the starting point when analysing policy instruments must be their position within the overall policy process where they can be defined as a particular set of rules instigated by policy-makers in order to influence private actors in certain directions by defining options and incentives. Depending on the overall position and strategies of the public and private organisations involved, policies may or may not succeed in bringing out the desired response from the private actors targeted, but the role of policy instruments remains the same: changing the rules in order to influence the game. From an institutionalist perspective rules are seen as defining options and incentives,[155] and an understanding of policy instruments will therefore have to take into account *both* the nature of the resources (incentives) involved *and* the options of interaction regarding their use.[156] Concerning the latter, a two-tier analytical scheme is proposed, distinguishing between on the one hand the basic options of interaction between public and private actors, and on the other hand the secondary rules governing the specific modes in which the policy resources and basic options are employed. Turning first to the *basic options of interaction*, the concepts summarised in Table 3.9 can be said to be inspired by Adrienne Windhoff-Héritier,[157] and in sharp contrast to the resource dimension the three basic options of interaction clearly institute a hierarchy of constraint between policy instruments. Once mandatory rules have been instituted, they can effectively be imposed upon individual citizens and organisations in the sense that attempts to avoid them will normally carry legal sanctions, while private actors can choose either to accept or decline the terms on offer when policy instruments are conditional, and no strings are attached to the resources on offer through voluntary policy instruments, thereby giving ingenious private actors the possibility to use these resources for purposes other than those originally envisaged by policy-makers.

By combining the four basic policy resources (Table 3.3) and the three basic interaction options (Table 3.9), we arrive at twelve combinations of one option and one incentive, all of which have been part and parcel of welfare policies in Western Europe in the second half of the 20th century.[158] Moreover, most of these have also formed part of re-

155 Cf. the discussion in Section 3.1.

156 By far the most consistent attempt to integrate resources and options of interaction into a unified framework has been found in the work of Adrienne Windhoff-Héritier (1987).

157 Alterations include adoption of concepts with a more, for want of a better word, legal ambience, and a regrouping of concepts in order to remove subdivisions that would appear to reflect resource- rather than rules-oriented criteria, cf. the discussion in Halkier 1996.

158 See Halkier 1996 pp. 60f.

gional policy, as illustrated by Table 3.10,[159] although in practice the use of several forms of resources at once often occurs in the form of giving private actors access to informational or organisational resources on subsidised conditions, i.e. by renting factory space or obtaining advice at prices considerably lower than the cost of obtaining similar services through market transactions.

Table 3.10 Policy instruments as rules in regional policy – Examples

	Authority	*Information*	*Finance*	*Organisation*
Mandatory	location control	(primary education)	taxation	(policing)
Conditional	quality assurance	specialist advice	grants	joint marketing
Voluntary	(summer time)	general advice	(social security)	motorway

Note: examples of general welfare measures in brackets.

The way in which particular policy instruments are employed can also make a difference, both with regard to making firms respond and securing outcomes that are in accordance with public priorities. These secondary rules, or *modes of implementation*, are specific policy designs that define the roles of public and private bodies in the interaction on a more detailed level.[160] As the present study is primarily concerned with the relative influence of the actors involved in the implementation process, the following modes, reflecting key stages of the process through which public resources are used to influence private actors, would seem to be particularly relevant:

- *selectivity* – the targeting of specific forms of economic activity for support,
- *project generation* – the origins of individual projects considered, and
- *project appraisal* – the way in which decisions on individual projects are taken.

---

159 The table is by no means exhaustive and merely provides some typical examples, drawing on the brief overview of the development of regional policy in Chapter 2. In practice the position with regard in particular the options dimension may vary according to specific circumstances, cf. the detailed discussion of the policy instruments of the SDA below.

160 The existing literature on policy instruments has pointed to a number of design features that could form part of a mode of implementation analysis, see e.g. Hood 1983, Salamon & Lund 1989, Linder & Peters 1989, Lindblom 1977 p. 38, Bogason 1985, Hucke 1983, Halkier & Danson 1997.

Turning first to *selectivity*, a focus on particular firms or groups of firms sets not just regional policy but industrial policy in general apart from macro-economic policies aimed at the national economy as a whole.[161] By publically designating particular forms of economic activity for support and excluding others, selectivity can be seen as an expression of political priorities that involves potentially controversial choices.[162] Any form of regional policy is by definition *spatially* selective, promoting the development of one region *vis-à-vis* other regions, but the nature of political controversy is likely to differ between various forms of regional policy. Top-down policies operated from the national or European level institute divisions between designated and undesignated areas *within* the polity, and thus the process of area designation could in itself be contentious. Conversely, bottom-up policies applying to a particular region are only selective in the sense that the same support measures are not available elsewhere, and in this case politicisation is more likely to take place after new measures have been introduced by e.g. neighbouring regions.

Table 3.11 Selectivity in regional policy

Dimension	*Criteria (examples)*
Spatial	Top-down designation Bottom-up self-designation
Sectoral	Macro sectors (manufacturing, services) Industries (electronics, textiles)
Firm type	Size (SMEs, large firms) Ownership (indigenous, external, multinational) Supply-chain position (e.g. subcontractor)
Project type	Expenditure items (wages, property, equipment) Firm strategies (expansion, modernisation, relocation)

Based on: Allen *et al.* 1988 and Yuill et al. 1981ff.

With regard to non-spatial criteria in Western European regional policy, selectivity has historically included and excluded firms on many grounds, and the most common ones are listed in Table 3.11. The reasons for introducing different forms of selectivity range from targeting of specific regional weaknesses via restriction of public expenditure to the perceived political 'sexiness' of e.g. high-tech industries, but from the perspective of private economic actors the result is the same: some

161 Young & Lowe 1974 pp. 16ff, Wren 1996a pp. 5ff.

162 Wilks 1984, Zukin 1985, Townroe 1986, Eisinger 1988 Ch. 1, Armstrong & Taylor 1985 pp. 208f.

forms of activity are included and some are excluded from support, and thus selectivity is one way of introducing specific public priorities into regional policy. Selectivity along *sectoral* lines is one of the most common forms: in the case of policy instruments based primarily on authority or financial resources it has often involved an explicit exclusion of groups of firms, e.g. in the service sector, but it can also be found in connection with e.g. advisory services or collective marketing efforts designed to promote particular industries. Selectivity according to the *type of firm* has also been fairly common, both as formal exclusion of e.g. firms above/below a certain size from subsidy schemes, and by informal means through provision of advisory services that are mainly of interest to e.g. either subcontractors or relatively large firms. Finally restrictions on *project types* have historically been important in connection with financial subsidies both in terms of restrictions on the items that can be supported by grants, and by linking support to e.g. expansionary firm strategies through the setting of targets for job creation. Project-type selectivity can, however, also be found in connection with other policy instruments: a public testing facility for new products and processes would for instance seem to be relevant mainly for firms aiming to modernise their production facilities. With this range of selectivity criteria, the scope for variation is very wide indeed, ranging from the minimalist option of purely spatial criteria to combinations of several sets of criteria that make only very specific forms of economic activity eligible for support.

As a mode of implementation *project generation* is important because the location of the initiative may have implications for the degree of public influence on the nature of a particular project. In the literature on regional and related policies the distinction between reactive and proactive forms of public policy is widespread,[163] with the former denoting situations in which private actors propose a project which is then considered for support by the implementing body, and the latter referring to situations where public bodies are actively encouraging individual private actors to pursue particular types of projects for which public support may be available. The basic difference between a proactive and a reactive approach relates to the extent to which firms are prompted to do things they would not otherwise have done, and here a reactive approach relying on projects initiated by the firms themselves would seem to be more likely to end up supporting activities that are in line with the current strategies of private actors. Contrary to this a proactive

[163] E.g. OECD 1977, Ewers & Wettmann 1980, Imberg & Northcott 1981, Grant & Wilks 1983, Martin & Hodge 1983b, Rothwell & Zegveld 1985, Moore & Booth 1986c, Beckman & Carling 1989, Totterdill 1989, Martin 1990, Rees & Morgan 1991, Hogwood 1992 Ch. 9, Olsson 1995.

approach shifts the initiative to the implementing organisation, and thus a potential would seem to exist for policies to stimulate strategic changes within firms. It would, however, still be misleading to equate proactive strategies with promotion of innovative projects: on the one hand proactivity could for instance simply be a way of raising the profile of an implementing organisation *vis-à-vis* other public bodies through aggressive marketing of policies, and on the other hand qualitative change could also be brought about by e.g. a reactive scheme combining a highly selective approach with adequate incentives.

The final mode of implementation concerns the way in which individual projects are *appraised* by the implementing organisation before it is decided to provide access to public resources. In the existing regional policy literature the basic distinction being made with regard to appraisal is between automatic and discretionary forms, where the former involves overt conditions and fixed rates of support and the latter gives the implementing body discretion with regard to whether a project should be supported, and if so, to what extent.[164] Although this distinction has been used mainly in connection with financial subsidies, more or less formalised screening of individual projects also takes place in connection with other forms of support, and thus the notion of discretion would seem to have a more general relevance.[165] Policy programmes operating in an automatic manner presuppose that eligible forms of economic activity can be specified to such an extent that firms can predict whether their projects are likely to be eligible or not; in this case public priorities are transmitted primarily through what in effect amounts to selectivity criteria setting out in advance what will and will not be supported. In contrast to this, discretionary programmes only need to have such criteria defined for internal use[166] or can rely on the tacit knowledge and concrete judgements of experienced staff, and thus for relatively complex criteria for public support – relating to e.g. technological change or the profitability of individual projects – a discretionary appraisal process would seem to be an obvious choice. The borderline between automatic and discretionary modes of implementation is, however, not entirely clear-cut because an implementing organisation may announce broad criteria that projects will have to comply with – e.g. long-term profitability or creation of a minimum number of jobs – and then subject the documentation submitted by firms in support

164 For a succinct definition, see Allen *et al.* 1988 p. 19.

165 See e.g. Hughes 1982.

166 The link between discretion and the difficulty of specifying criteria of support is well made by Ham & Hill (1984 Ch. 9).

of their claims to detailed scrutiny and negotiation.[167] Overt, but vaguely defined criteria would in other words seem to introduce a considerable element of discretion into the appraisal process, and it is therefore probably productive to think of the automatic/discretionary distinction in terms of a continuum.

The implications of automatic and discretionary modes of implementation for the actors involved in regional policy have been discussed extensively in connection with financial subsidies. Automatic grants, historically the form that originally prevailed across most of Western Europe, have been seen to have the advantage of being predictable and hence possible to integrate into the investment assessment procedures of private firms while at the same time being relatively easy to administer for both the implementing body and private sector applicants. Automatic programmes with limited selectivity did, however, also come to be seen as unduly expensive, financially as well as politically, because large capital-intensive investments with no real choice of location had to be supported if they happened to be situated in designated problem areas, resulting in 'windfall gains' that bolstered the profitability of e.g. large multi-national companies without additional gains for the regional economy in terms of productive capacity or employment.[168] In contrast to this, discretionary grants have often been claimed to increase public influence on the investment decisions of private firms because discretion allows the implementing organisation to give priority to 'good' projects or to 'improve' proposals put forward by private firms,[169] something that should also help to increase the precision of policy measures and economise with depletable policy resources.[170] Discretionary programmes do, however, clearly entail more administration for both the public and private actors involved, something that larger firms with more organisational resources may be better equipped to deal with,[171] and as the outcome of an application for discretionary support is fundamentally uncertain, it cannot be incorporated into the formal investment appraisal procedures of firms and could therefore constitute a windfall gain because only projects that are viable on their own terms would

---

167 Conversely, the existence of internal administrative guidelines for awarding discretionary regional grants may in effect make them semi-automatic (Allen *et al.* 1988 pp. 21f).

168 See e.g. Maclennan & Parr 1979, Imberg & Northcott 1981, Richardson 1984, Neuperts 1986, Damesick & Woods 1987, Allen *et al.* 1988 & 1989, Balchin 1990, Harris 1993.

169 See e.g. Wilks 1987, Parsons 1988 p. 201, Balchin 1990 Ch. 3, Yuill *et al.* 1992 p. 12.

170 See e.g. Vanhove & Klaassen 1987, Allen *et al.* 1989, Beckman & Carling 1989.

171 Walker & Krist 1980, Allen *et al.* 1988, Swales 1989, Allen & Yuill 1990.

have been put forward.[172] Although the latter argument may overestimate the importance of both formal investment appraisal procedures in private firms and windfall gains as an economic problem,[173] it does underline the difficulties associated with establishing the impact of discretionary modes of implementation. While purely automatic schemes, at least in the absence of stringent selectivity criteria, would mainly seem to accelerate existing investment projects and increase the overall volume of investment in assisted regions, a discretionary approach would seem to be more likely to produce individual projects of a 'path-breaking' nature through public targeting of support and/or dialogue with applicants,[174] although this is of course critically dependant on the organisational and informational resources of the implementing body and subject to the general conditions of incomplete information. But at the same time bringing out projects of a qualitatively new nature may be more or less necessary, depending on the nature of the regional problem: Significant change does not necessarily require the abrupt introduction of completely new patterns of behaviour at the level of individual firms, but may also be brought about by a cumulative effort, supporting trends already in existence as an undercurrent in the regional economy.

All in all policy instruments and the ways in which they are put to use through modes of implementation constitute the rules on the basis of which the interaction between the implementing organisation and private firms takes place. The eventual outcome does, however, still depend on the response of private actors who will judge the options and incentives put in place by public policy on the basis of their own objectives and strategies, a perspective that may well differ significantly from that of the sponsors and implementors of regional policy.

Outcomes

The outcomes of regional policy can be studied from a number of perspectives, each of which could potentially influence the perception of a particular form of public intervention as more or less successful.

---

[172] This has been argued strongly by Kevin Allen and his colleagues (Allen *et al.* 1986 & 1988, cf. Ashcroft & Love 1988, Scharpf 1983), but other studies find less of a difference between the way firms cope with discretionary and automatic grants (McGreevy & Thomson 1983).

[173] Walker & Krist (1980) argue that even by improving the profitability of one round of investment by a particular firm, the next round is being brought forward.

[174] The use of the notion of path-breaking here refers exclusively to individual projects and does not in any way imply that path-supportive policies are inherently inferior in bringing about change in the regional economy, as sometimes implied in the literature (e.g. Sidenius 1984 pp. 123ff, Goodwin & Duncan 1986 p. 18).

Because the majority of policy schemes involves interaction with individual firms that can be quantified and the level of response to non-mandatory policy instruments cannot be predicted with any great certainty, measuring the *output* of an implementing organisation in terms of e.g. grant expenditure, firms advised or factory floor-space constructed becomes an important task in its own right. Output measures produce more or less extensive information about the resource exchanges taking place between public and private actors, and while such figures are of course an uncertain guide to what impact a particular policy makes on both individual firms and the regional economy as a whole, they easily feed into the political process as symbols of public (in)action.

Table 3.12 Firm-level impact of regional policy

Type of firm	*Nature of change*
New	Indigenous creation Incoming location
Existing	Liquidation Change of ownership Preservation Expansion Modernisation

With policy objectives focusing on improving the relative position of particular regions by influencing the behaviour of individual firms, the *firm-level impact* of regional policy is not only a central form of outcome but also one that to a certain extent can be inferred from e.g. the way particular policy instruments are employed: promotion of inward investment is likely to attract new firms from outwith the region, while provision of advice on new technology in a particular industry is likely to result in modernisation of existing indigenous firms. As illustrated by Table 3.12,[175] two distinctions can be made here, namely between the types of firms involved and the nature of change taking place. These firm-level outcomes are interesting not just because of their relevance in the context of various policy strategies, but also because they entail different socio-economic perspectives for private actors and policy-

[175] Although recognised on a general and strategic level in the form of choices between strategies based on indigenous and incoming growth (see e.g. Ashcroft 1978, McCallum 1979, Martin & Hodge 1983a, Vanhove & Klaassen 1987, Eisinger 1988, Hart & Harrison 1990, Halkier 1992), attempts to account for the different possibilities in a systematic way are rare. Table 3.12 is an expanded and reworked version of the typology employed by Yuill *et al.* (1981ff) in their annual surveys of central government regional policy in Western Europe.

makers that may be translated into political repercussions even as policies are in the process of being devised. The implications of the advent of *new firms*, i.e. creation of additional productive capacity in the shape of new separate entities, is likely to vary according to their origin. While both indigenous start-ups and inward investment from outwith the region will provide additional employment, their importance for existing firms depends on their respective markets and strategies: some may welcome the advent of new potential subcontractors or customers, while others could construe newcomers as unwelcome competitors. At the same time the political pay-off from strategies relying on incoming rather than indigenous new firms would also seem to differ: while the former represents a 'lumpy' form of economic growth which is highly visible and produces excellent photo opportunities – "500 new jobs, minister pictured opening new factory" – the latter will result in a low-profile long-term trickle of smaller start-ups. *Liquidation* of existing firms inevitably has direct and immediate adverse consequences for the workforce and mostly occurs as a result of deliberate non-intervention, e.g. a government decision *not* to support firms in crisis; head-on decisions to eliminate productive capacity has only been a realistic policy option with regard to industries in public ownership. *Change of ownership* in existing firms, ranging from nationalisation or privatisation via more or less friendly mergers and take-overs to management buy-outs, may lead to anything from liquidation to modernisation, depending on how and why the present owners are displaced, and hence the political ramifications are likely to be equally varied, especially because some forms of change of ownership have historically been inscribed into particular forms of public political discourse, e.g. for and against public or foreign ownership of productive facilities. *Preservation* is defined as a situation where the activities of an existing firm are maintained in terms of output and employment, despite the immediate pressures of the market. *Expansion* of an existing firm is probably the least controversial option available: increasing the volume of production without revolutionising organisation or technology could potentially increase employment, output and profits at the same time, and hence keep both management and employees of the supported firm happy, if not necessarily their competitors. Contrary to this, *modernisation* involves significant organisational or technological changes and could therefore be a mixed blessing for all parties involved: certainly more demanding in terms of management resources and a potential threat to shop-floor employment in the short run as productivity increases or resources are transferred to R&D activities. The impact of regional policy on private firms does in other words vary greatly, and this has important implications not just for the aggregate socio-economic effect of public intervention, but also, potentially, for the politics of regional policy because of the sometimes very visible impact on particular firms and groups of employees that a political party may want to position itself in relation to.

Ultimately, the *aggregate impact* on the regional economy and its position *vis-à-vis* other regions is of interest as a measure of the effectiveness of regional policy, and especially from the 1980s onwards a growing body of writings has been devoted to establishing the effects of various forms of public initiatives. Given the broad range of social, economic and political goals associated with regional policy and the coexistence of many forms of public intervention in any given region, evaluation is certainly no mean task, and most studies have tended to concentrate on the explicit or implicit economic goals of regional policy. For the purposes of this project, the (varying) reliability of the evaluations themselves is, however, less interesting than the way such studies are being used in political discourse to underpin different perceptions and claims about the success or otherwise of particular policies and organisations.

All in all it is clear that understanding the implementation of regional policy from an institutionalist perspective will involve drawing on a wide range of methodologies, primarily qualitative but also quantitative, in order to account for the interplay between actors, strategies and institutions. The interaction between the implementing body and individual firms leaves plenty of room for the former to institute different options and incentives reflecting public priorities with regard to regional development: a wide range of policy instruments are available, and the various modes of implementation allow support to be targeted by selectively focusing on a particular group of firms or by screening private sector projects according to more or less transparent appraisal criteria. Whatever their lineage, the resulting policies still constitute a new set of rules that may alter the behaviour of private firms by making some options more attractive than others, but the eventual outcome of public intervention still depends on the way in which private actors respond to particular policies, and this, in turn, will have been influenced not only by the resources and strategies of the individual firm, but also by the wider economic environment in which they operate.

The Economic Environment

The economic environment of regional policy is in a sense even more difficult to delimit than the political environment. Whereas the latter refers to a fairly well-defined and relatively stable set of organisations, most of which operate within the context of one and the same territorial state, the economic environment is a much more fluid phenomenon. Private economic actors within a region may operate on international markets, have formalised links with firms in other countries or simply be affected by the changing fortunes of the global markets, and thus focusing on a fixed spatial entity like an administrative region or the national economy would not be particularly helpful. At the same time both the structure of the regional economy and the way its firms are positioned in the international division of labour are subject to change, and we thus find ourselves with the seemingly unenviable task of trying to take a closer look at something that is spatially indeterminate and temporally variable. The problem of fluidity encountered when attempting to arrive at a definition of the economic environment of an organisation implementing regional policies could, however, be circumvented by focusing instead on the way the private firms targeted by public intervention are positioned in the economic division of labour.[176] In a sense, of course, this results in a situation where an implementing organisation has as many 'economic environments' as there are firms within 'its' region, but as the position of individual firms in the regional, national and global economy is likely to influence the resources at its disposal and the strategies it pursues, the approach would still seem to make sense, although it may well produce a relative complex account of the economic environment in which regional policy operates. The text therefore proceeds in two steps: first different ways in which regional firms are integrated into the wider division of labour is considered, and then role of the discursive terrain is briefly revisited in relation to private actors and their strategies.

Coordination Rules and Actor Configurations

Over the last century a prominent feature of the social sciences has been the attempt to distinguish between major areas of social activity that proceeds on the basis of different 'standard operation procedures' and associated rationalities,[177] and an important advantage of an institutional approach would seem to be that it is not assuming that e.g. political and economic organisations necessarily operate on the basis of two sets of distinct institutions but instead on the basis of potentially overlapping 'modes of social coordination', i.e. social institutions on the basis of which actors interact in order to achieve particular outcomes. Three forms of coordination have generally been singled out as particularly relevant for the study of economic and political phenomena in Western societies in the second half of the 20th century: markets, hierar-

---

176 A parallel approach to the study of the economic environment of regional policy can be found in the work of Moore & Booth in which characteristics of industrial sectors are associated with particular forms of regional policy (1989 pp. 82ff, cf. Halkier 1992 pp. 3ff), and the following pages essentially rework this idea within an institutionalist framework.

177 Giddens 1984 p. 31, Bogason 1989 pp. 221ff.

chies and networks.[178] While there is little disagreement about defining market relations as discrete exchange between formally independent actors through the price mechanism and hierarchy as a long-term relationship involving formal subordination of one actor to the commands of another,[179] the notion of networks as a specific mode of social coordination has given rise to a number of more or less satisfactory definitions.[180] Inspired by the work of Hans Thorelli,[181] this text defines modes of social coordination on the basis of the specific combination of the formal relationship and resource dependencies between individual actors, the duration of individual interactions, the key medium through which coordination is achieved,[182] and the outcome produced by the process of coordination. As indicated by Table 3.13, this results in a clear-cut profile also for network relations which are seen to involve two or more formally independent actors establishing a long-term relationship on the basis of mutual dependence upon resources controlled by other participants which operates through bargaining about and exchange of specific combinations of resources. It is, moreover, important to note that as the three modes have similarities with regard to several aspects, the potential for relations between actors moving from one form of coordination to another would seem to be a very real one, and that particular areas of social activity may involve different modes of coordination at the same time. As argued above, government is not just about hierarchies when many public agencies collaborate in a policy area, and similarly the economic sphere of social activity – the production and exchange of goods and services[183] – is not just about market exchanges. A firm *may* opt to use market exchange to achieve its goals, but depending on transaction-specific circumstances other options could be attractive alternatives, e.g. internalisation of sub-contractors through acquisition (hierarchy) or customised production for long-term private

---

178 Focusing on these three modes of social coordination does not imply that these are the ones thinkable or currently existing – like e.g. democracy as a means of decision-making (see e.g. Lindblom 1977, Bogason 1989, Heinelt & Smith 1996) – but merely reflects pragmatic considerations of direct relevance from a regional policy perspective.

179 Levacic 1993, Mitchell 1993, Rhodes 1997 Ch. 3.

180 Halkier & Damborg 1997 pp. 7ff.

181 1986.

182 It should be noted that the 'medium' category denotes the central way in which coordination is achieved in each of the modes, and that this does not preclude e.g. that a degree of trust may exist between individuals within a hierarchy, that command-like forms of communication may occur as part of enforcing market obligations, or that transfer of resources between networking partners can involve a form of pricing.

183 Hodgson 1989 pp. 15ff, cf. Johnson & Lundvall 1989, Andersen & Davis 1986, Gee 1991.

or public clients (networking) – and although such courses of action are likely to involve money changing hands at some point, these relationships between private actors should not be confused with market exchange proper.[184] Finally it must be stressed that the economic environment in turn is affected by government and its activities, right from the very basic upholding of property rights and markets as social institutions,[185] via an array of public policies, to the emergence of a qualitatively new form of economic governance and a gradual blurring of the distinction between the public and private sectors[186] and the growing importance of intricate inter-organisational networks which cut across the public-private divide.[187] All in all this underlines the importance of being able to draw on a coherent analytical framework that is not tied to particular spheres of social activity, and in the following each of the three modes of coordination will be considered in turn, especially with regard to their implications for regional development strategies.

Table 3.13 Modes of social coordination compared

Mode of coordination	*Formal relationship between actors*	*Resource dependence between actors*	*Duration of inter-action*	*Medium of coordination*	*Outcome*
Market	Independence	Mutual	Short-term	Prices	Exchange
Hierarchy	Subordination	One-sided	Long-term	Command	Obedience
Network	Independence	Mutual	Long-term	Trust	Cooperation

Hierarchical relations are part and parcel of the economic environment in the sense that firms are hierarchical organisations and different *patterns of ownership* are likely to influence the ways private actors respond to public policies. While the relative advantages of particular ownership structures are the subject of, sometimes fierce, debate which essentially concern the perceived nature of 'the regional problem',[188] ownership structures define options and incentives for private actors that public policy-making has to take into account. Three features are generally identified in the literature on regional policy as being of particular importance, namely the extent to which ownership is

184 Hodgson 1989 Ch. 8, cf. Johnson & Lundvall 1989 pp. 92ff.

185 North 1990a Ch. 6 & 7, Hodgson 1989 pp. 182-94.

186 See Nielsen & Pedersen 1988, Berrefjord *et al.* 1989, Pedersen *et al.* 1992.

187 E.g. Jordan & Schubert 1992, Hanf & O'Toole 1992, Rhodes & Marsh 1992a, Christiansen 1993, cf. the discussion in Section 3.3.

188 Prominent issues have included the relationship between external and indigenous ownership, small and larger firms, and the role of public ownership.

- *fragmented or concentrated* – a limited number of key players provide a very different target for public policy than many smaller ones,[189] both when attempting to deal with firms on an individual and a collective basis;
- *external or indigenous* – regardless of what other advantages external ownership may have in terms of e.g. access to additional resources, it will to some extent limit the decision-making capacity of branch plant managers by integrating them into the overall strategies of multi-regional or global firms;[190]
- *private or public* – while the *raison d'être* of privately owned firms is to achieve and maintain profitability, the goals of public firms have historically been of a much more complex nature, in theory more amenable to goals like e.g. employment retention but in practice often integrated into multi-regional (i.e. externally owned) corporate structures and/or controlled by public bodies whose primary remit lies outwith regional development.[191]

Market relations are important to the extent that firms within the region (or indirectly their external owners) buy their inputs, including labour, and sell their products in competitive markets. The *competitive position* of individual firms has been analysed from different perspectives,[192] including

- *the industrial sector* in which it operates, giving a broad indication of the general ways its field of economic activity is developing,
- *product characteristics*, e.g. intermediate products or original equipment,
- *market location*, i.e. local, regional, national or international,
- *market shares*, ranging from monopoly to being a minor player, and
- *entry barriers* which may vary considerably according to the level of financial, organisational and informational resources needed to establish a new venture in a particular area of economic activity.

Again aspects that will influence the way in which a firm is likely to respond to public policies: both in terms of resources and strategies there are obvious differences between e.g. a major regional player in a stagnating industrial sector and a minor subcontractor in a rapidly expanding international field of economic activity.

---

189 Moore & Booth 1989 pp. 82ff.

190 Hood & Young 1976, Massey 1984 Ch. 3, Rothwell & Zegveld 1985 Ch. 7, Moore & Booth 1989 pp. 82ff, Amin & Thrift 1994, cf. Morgan 1995.

191 See e.g. Hudson 1989a Ch. 8, Moore & Booth 1989 pp. 63f, Anderson 1992 p. 69.

192 Cf. Cooke & Morgan 1998 Ch. 2.

Network relations, long-term cooperation between independent actors on the basis of mutual resource dependencies, is the third way in which firms interact with other economic actors. Network relations in the form of *inter-firm collaboration* have often been associated with spatial agglomerations in the form of 'industrial districts', but this type of economic interaction is not necessarily confined to a particular region.[193] Similarly, it would be unwise to associate networking with a capacity to adopt in a flexible manner to new challenges: a situation of mutual resource dependency could also produce an impasse where no new strategies can be agreed upon and no individual actor carries enough weight to take the lead,[194] and thus once more the specific circumstances need to be established. From the perspective of regional policy, a feature of particular interest is the extent to which formalised associations or networks along e.g. sectoral lines exist within the region, because the presence of this form of collective organisational resource changes the options and incentives faced by an implementing body: coordination between private actors could improve their bargaining position, but may also present a potential vehicle for collective solutions to particular development problems.[195]

Clearly the overall importance of each of the three modes of coordination will differ between regions and change over time, and it is therefore paramount to take the various modes of social coordination and their key aspects into account when analysing the economic environment in which regional policy operates. The way in which firms in a region interact with other economic actors clearly influences the resources at their disposal and the strategies they pursue – and hence makes some regional policy strategies more feasible than others.

The Discursive Terrain

Although the imperative to stay profitable and competitive looms large in the horizon of managers of private firms, it would be unhelpful to see their strategic choices as reflecting the economic interests of the firm pure and simple. On the one hand the interaction between private firms and public bodies takes place in a discursive terrain dominated by particular assumptions about the world: actors may have varying degrees of attachment to particular nodal points and ideological assumptions, but still their presence could affect the ways private actors come to view particular organisations and public policies and hence possibly their willingness to modify their economic activities in response to

---

193 Amin & Thrift 1994, cf. Morgan 1994.

194 For a sobering perspective on networking industrial districts, see Bellini & Pasquini 1998.

195 Moore & Booth 1989 pp. 82ff.

political initiatives. On the other hand it is important to remember that private economic actors are potentially more 'footloose', i.e. less tied to a particular geographical location, than public organisations tend to be: while the latter operate in a defined geographical setting and hence by definition have a 'territorial remit', the attraction of or attachment to a particular region for private economic actors cannot be taken for granted,[196] something that has prompted extensive use of images and place promotion as part of regional policy strategies.[197]

It could be argued that private economic actors are less likely to subscribe to strong views about the nature of the *regional problem*, being more concerned about the profitability of their own activities, but still such assumptions may be of importance, for instance if a particular policy initiative is seen to be in line with ("lack of external finance") or contradicting ("incompetent management") the assumptions held by private economic actors, or if a generally poor image of a region undermines its desirability as location for economic activity for external investors. With regard to the ideal form of *economic governance*, liberal views extolling the benefits of a 'market economy' and the autonomy private firms might be expected to prevail, but historically the need for varying degrees of public sector involvement has also been accepted,[198] and thus establishing the nature of the prevailing assumptions and their possible impact on the response of private actors to regional policy measures remains an empirical task. In terms of assumptions of a *spatial* nature, two aspects would seem to be relevant because they could potentially influence the disposition of individual firms. On the one hand the vertical relationship between tiers of government, because the degree of attachment of private sector managers to the national and regional levels respectively may influence the perception of development initiatives emanating from particular tiers of governance. On the other hand assumptions about the horizontal relationship between regions are important because the extent to which discrimination along spatial lines is seen as acceptable may make it more or less desirable to be associated with particular forms of regional policy.

From the perspective of regional policy the relevance of the make-up of the discursive terrain is in other words not limited to the political environment, but should also be considered in connection with the interaction between an implementing organisation and private economic actors within and outwith the region.

---

196 See e.g. Burgess 1982, Kooistra 1998.

197 Cf. Therkelsen & Halkier 2004.

198 See e.g. Pedersen *et al.* 1992, Wilks 1984 Ch. 1, Berrefjord *et al.* 1989, Christiansen 1993 Ch. 2.

Table 3.14 Regional policy – An institutionalist perspective

Areas of analysis	*Dimensions*	*Sub-dimensions*
Political environment	Discursive terrain	Public-private assumptions Central-regional assumptions Nature of regional problem
	Coordination rules	Political system Policy networks
	Actor configuration	Political parties Private bodies Public bodies
Political sponsoring	Interaction rules	Organisational sponsorship Policy sponsorship
	Outcomes	Resource exchange Organisational change
Implementing (and other) organisations	Resources	Authority Information Finance Organisation
	Assumptional orders	Regional relations Regional problems/solutions Agents of regional change
	Strategies	Development strategies Organisational positioning
Policy implementation	Interaction rules	Policy instruments Modes of implementation
	Outcomes	Output Firm-level impact Aggregate impact
Economic environment	Actor configuration	Existing firms Entry barriers
	Coordination rules	Ownership structure Competition Inter-firm collaboration
	Discursive terrain	Public-private assumptions Central-regional assumptions Nature of regional problem

Regional Policy – An Analytical Framework

The development of a conceptual framework was based on the notion that regional policy is essentially about making private economic actors change their pattern of behaviour in accordance with public spatial priorities. A framework would therefore have to take two key aspects into account, namely

- the political process through which aims and means emerge, and
- the implementation process in which private actors interact with public bodies on the basis of rules instituted by the latter.

Working on the basis of an institutionalist approach to the study of public policy, the framework takes as its point of departure the implementing organisation, focuses on its key inter-organisational relationships and situates these interactions in their broader political and economic context. This identifies five areas of analysis that must be considered, and in each of these the discussion of analytical dimensions incorporated issues highlighted by the existing literature on regional policy as well as more general points emanating from traditions within institutionalism, policy analysis and inter-organisational studies, ultimately resulting in the conceptual framework summarised in Table 3.14.

From the perspective of the current project the main advantages of the framework would seem to be that it manages to overcome the principal limitations identified in the analytical approaches of the existing regional policy literature, including the side-lining of political and organisational issues, and the reliance on quantitative or dichotomous conceptual schemes. Instead a comprehensive approach has emerged, capable of accounting for different forms of regional policy and providing a multi-dimensional picture of the interaction between public and private actors that does not imply that e.g. individual actors will always be consistent in terms of what they do or, indeed, that e.g. policies will be consistent with regard to underlying assumptions, causal mechanisms and policy instruments. While this is likely to require employment of a broad range of methods and eventually produce a relatively complex picture, building the framework around core ideas drawn from institutionalism and organisational studies should result in a reasonable degree of consistency. Moreover, the systematic use of basic concepts – institutions, actors, resources, assumptions and strategies in particular – should also ensure that the proposed framework is not just capable of illuminating permanent features of regional policy, but also able to identify and account for historical patterns of change. If these claims are supported by the experience of employing the framework in the empirical analysis of the development of the SDA in the context of British regional policy, then the approach might even serve as a possible source of inspiration in connection with studies of other policy areas, notably, of course, those involving attempts to change patterns of behaviour amongst private firms according to public priorities.

3.4. Paradigms Revisited

Having developed the analytical framework, this section immediately puts it to work by applying it to two key paradigms in regional policy. Reconsidering these 'ideal types', summing up features that tend to characterise different forms of regional policy in the existing literature, should provide a reinterpretation on the basis of the new framework of the broader historical context of regional policy in Western Europe and thereby function as a first stepping stone between the conceptual discussions in this chapter and the ensuing empirical analysis of the Scottish case in its British and European context. The two paradigms in question are, not surprisingly, the top-down approach of central government, the dominant form of regional policy in Western Europe till the late 1970s, and the RDA approach, a particular version of bottom-up regional policy of which the SDA has been seen as a prominent example. The paradigms should, however, not be seen as normative ideals, prescribing how policy ought to have been,[199] but rather as empirical generalisations that can be used to highlight areas in which individual nations and regions diverge from the predominant pattern of regional policy in Western Europe.

A preliminary outline of the top-down and bottom-up paradigms was given in Table 2.2, contrasting central government policies attempting to redistribute economic growth through the reactive use of financial grants with bottom-up attempts to improve the competitiveness of indigenous firms through selective and proactive deployment of a wide range of policy instruments. The main difference between the latter and the RDA approach is one of organisation, namely that the implementing organisation is a semi-autonomous public body rather than a government department.[200] The RDA approach in other words sees the meeting of two trends, the growth of bottom-up initiatives in regional development and a more general trend towards implementation 'by proxy'. The increasing importance of various forms of what has sometimes been dubbed 'quangos' (quasi non-governmental bodies) in public policy has been re-

199 The tongue-in-cheek use of the term 'model agency' as short-hand for the characteristics of RDAs in earlier work on bottom-up regional policy in Western Europe (Halkier & Danson 1997) could give this impression, although the actual use of the model as a vehicle for identifying different types of regionally-based development bodies was analytical rather than prescriptive.

200 Cf. the discussion in Halkier & Danson (1997) which defined an RDA as "a regionally based, publicly financed institution outside the mainstream of central and local government administration designed to promote indigenous economic development through an integrated use of predominantly 'soft' policy instruments", in parallel with the definition suggested by Velasco for DG XVI of the European Commission (Velasco 1991 p. 12).

corded throughout Western Europe and elsewhere,[201] and while the forces behind this development differ between countries and policy areas, the origins of quangos in the field of regional policy have been traced back to two types of considerations in particular. A frequently cited feature of RDAs is the *insulation* from political pressures achieved by vesting implementation in a semi-autonomous body; this should have the advantage of keeping party-political or local territorial interests at bay and thus promote long-term development strategies at the expense of short-term gesturing.[202] From an *organisational* perspective a position outside the mainstream government apparatus is believed to make it possible to recruit specialist staff and adopt a more flexible approach to policy implementation and collaboration with private actors, something which should further the development of strategies tailored to address the specific weaknesses of particular regions, allow a more integrated and proactive approach use of public resources, and also generate a business-like image more attractive to private actors than the traditional civil-service image of traditional government departments.[203]

By combining these organisational features with the general characteristics of the bottom-up approach outlined in Chapter 2, it is possible to arrive at a profile of the RDA approach that can be compared to that of the traditional top-down approach of central government on the basis of the new analytical framework. As can be seen from Table 3.15, the comparison comprises all five areas of analysis identified as relevant from the perspective of regional policy in Table 3.14, although some dimensions and sub-dimensions have been omitted because they have not been recognised in the existing literature as exhibiting regular features in either approach.

---

201 See Hood 1978 & 1986, Barker (ed.) 1982, Curnow & Saunders (eds.) 1983, Modeen & Rosas (eds.) 1988, Hood & Schuppert (eds.) 1988, Ridley & Wilson (eds.) 1995, Thiel 2001.

202 See Stephen 1975, Barberis & May 1993, EURADA 1995, Keating 1998, cf. Harden 1988.

203 See McCrone 1969 p. 205, Regional Studies Association 1983, Vanhove & Klaassen 1987 Ch. 7, Velasco 1991, Danson *et al.* 1992, EURADA 1995, Olsson 1995, Batt 1996, Halkier & Danson 1997.

Table 3.15 Regional policy paradigms compared

Key analytical dimensions	*Main features*	
	Top-down approach	*RDA approach*
Political environment	National policy networks Political parties	Regional/national policy networks Insulation from party politics
Political sponsorship	Central government Departmental relation	Regional/central government Arm's-length relation
Implementing organisation * key resources * underlying values * development strategy * organisational positioning	Government department National authority Finance Interregional equality Diversion of investment Policy fragmentation	Semi-autonomous body Regional authority Specific information Regional competitiveness Indigenous growth Policy integration
Policy instruments	Conditional finance Mandatory authority Voluntary/conditional org.	Voluntary/condit. information Conditional finance Conditional organisation
Modes of implementation	National spatial selectivity Reactive Automatic/discretionary	Non-spatial selectivity Proactive Discretionary
Firm-level impact	Incoming location Indigenous expansion	Indigenous creation Indigenous modernisation
Economic environment	Hierarchy/markets	Markets/networks

The first conclusion that can be drawn from Table 3.15 is that each of the paradigms are complex forms of public intervention, involving choices made on the basis of limited information, potentially conflicting interests both amongst public actors and in their interaction with private firms, and uncertain outcomes depending on the wider economic environment. Success in regional policy, by whatever standard, can be difficult to measure, let alone achieve, and it would therefore be unhelpful to describe top-down policies as being *per se* 'stronger' than those associated with the RDA approach – or *vice versa*. The resourcefulness of central government with regard to authority and finance has to be balanced against the informational and organisational resources of RDAs, and the use of mandatory authority within the top-down approach must be balanced against the potential for promoting public priorities through selectivity, proactivity and discretion. Both policies depend on the response of private actors, and according to specific historical circumstances some forms of public policy may be better equipped than other forms to elicit strategic changes amongst the firms targeted.

Comparing the two approaches to regional policy, the extent to which they display material differences across the whole range of analytical dimensions is difficult to overlook:

- the *political environment* and *sponsorship relations* differ both with regard to the spatial dimension and the degree to which elected politicians are involved in regional policy,
- the *implementing organisations* operate in different territorial domains on the basis of different resources, pursuing objectives and strategies that are not necessarily compatible,
- *policy implementation* includes some overlap with regard to individual policy instruments but important contrasts with regard to capacity for integrating different instruments, modes of implementation and the firms targeted by public policy, suggesting rather different forms of interaction between public and private actors, and
- the *economic environment* in which each of the two approaches is likely to make a sizeable impact would also seem to differ, with top-down policies to a large extent depending on the presence of private firms establishing branch plants and RDAs needing indigenous firms willing to improve their market position and network relations.

A transition from one paradigm to another would in other words certainly make a difference to regional policy, and in situations where the two paradigms coexist, cohabitation may result in national and regional actors adopting positioning strategies ranging from cooperation to competition. On this background it is hardly surprising that the two approaches have been seen as alternative paradigms in regional policy, two forms of public intervention that are both concerned with remedying spatial inequalities but which involve different actors, different spatial levels of operation, different underlying assumptions, different strategies, and different forms of interaction between public and private bodies – and thus ultimately different consequences for the politics of regional policy.

This brief sketch would seem to suggest that the analytical framework may well become a useful tool. By examining regional policy from a new and hitherto neglected perspective, an adapted version of an institutionalist approach to policy studies, it is possible to focus more precisely on key issues and, most importantly, bring a much wider set of dimensions into the analysis in a systematic fashion. On this positive note attention can now be turned to matters Scottish.

CHAPTER 4

Contexts: Peculiarities of the Scottish

In order to understand the background for decision to establish a regional development agency in Scotland in the mid-1970s and the environment in which the new body would subsequently operate, it is necessary to explore three contexts in particular, namely the prevailing patterns of political governance, the nature of the economic problems faced, and, last but not least, earlier policy initiatives aiming to promote the economic development of the region. This chapter will therefore examine the 'peculiarities of the Scottish'[1] in the period until 1975 with regard to political governance (Section 4.1), economic development (Section 4.2), and regional policy measures (Section 4.3). While the chapter is primarily based on the existing specialist literature, the text will not only reinterpret developments on the basis of the conceptual framework developed in the previous chapters, but also to some extent draw on interpretation of documentary sources, especially with regard to the development of British regional policy.

4.1. Within the Union State: Governing Scotland in Britain

The state structure of 20th-century Britain has sometimes been seen as archaic due to e.g. the existence of a hereditary chamber of parliament, and certainly neither the continued coexistence of multiple nationalities nor the 'asymmetrical' spatial organisation of the British state would seem to sit easily with ideals about the nation state as a homogenous social and political system.[2] The position of Scotland is an excellent example of this lack of institutional 'neatness', and the present section first charts the development of the core institutions and actors in what James Kellas has called 'the Scottish political system',[3] and then examines the development of political discourses in relation to this, including 'national' issues about the possible reconfiguring of the relation between Scotland and the rest of the United Kingdom.

---

1 Apologies to E. P. Thompson (1978).

2 From a historical perspective see e.g. Perry Anderson (1964, 1987), Tom Nairn (1977, 1981), Hugh Kearney (1989, 1991).

3 Kellas 1989 pp. 16ff, cf. the discussion below.

From State Union to Administrative Decentralisation

After centuries of uneasy coexistence in the British Isles, the two major powers England and Scotland came to be ruled by the same (Scottish) monarch in 1603, and after another century of political upheaval – the inappropriately named 'English Civil Wars' followed by what became trumpeted as the 'Glorious Revolution'[4] – the United Kingdom of Great Britain was established in 1707 when the parliaments of England and Scotland voted for themselves to be replaced by a joint parliament. This created a unified political superstructure but at the same time explicitly safeguarded the continued existence of separate Scottish institutions in central spheres of social life such as law, religion, education and local government, and thus the new state would seem to entail the promise of more peaceful coexistence on the British Isles by institutionalising difference within a shared political framework that would allow both parties to enjoy the benefits of a large unified market with a growing number of overseas colonies.[5] Unlike in e.g. France no attempt was made to homogenise what Rokkan and Unwin have dubbed 'the union state',[6] and thus also in the 19th century Roman-type Scottish law continued to differ from the English common-law tradition, the Church of Scotland remained Presbyterian in contrast to the Anglican Church of England, and distinct Scottish traditions in education and local governance were maintained.[7]

In the Victorian age the growing role of local public bodies in providing e.g. infrastructure lead central government to establish numerous public boards to oversee developments, and reflecting its distinct form of local government these boards were generally specific to Scotland. In 1885 the Scottish Office was established as a decentralised territorial department of central government, designed to serve several purposes:[8] from an administrative perspective it would facilitate coordination of activities hitherto governed by free-standing boards, and from a political perspective it could help diffuse demands for more parliamentary attention to Scottish matters and for Scottish 'home rule' in line with what was being sought for Ireland. The creation of the Scottish Office was a concession by the then Conservative government to repeated demands

4 While the former was clearly more than an internal English affair (Kearney 1991 p. 64), the latter was essentially a reassertion of the collective parliamentary strength of the nobility *vis-à-vis* the monarchy (Kearney 1989 p. 128, cf. Jensen 1991).

5 Kellas 1989 Ch. 2, Midwinter *et al.* 1991 Ch. 1, Paterson 1994 Ch. 3, Brown *et al.* 1996 Ch. 1.

6 Rokkan & Unwin 1983 pp. 181ff, cf. Østergård 1986.

7 Keating 1996 Ch. 2.

8 Kellas 1989 Ch. 3, Midwinter *et al.* 1991 Ch. 3, Paterson 1994 Ch. 1, Keating 1988a pp. 86ff.

from Scottish MPs, and while expectations in Scotland according to Prime Minister Salisbury were "approaching to the Archangelic", the first, clearly reluctant, incumbent of the position as Scottish Secretary saw the new office as "quite unnecessary".[9] For many decades the responsibilities of the Scottish Office grew only slowly, and it was not until the interwar period that the current shape of the organisation began to emerge: politically the Secretary of State for Scotland was strengthened by becoming a cabinet-level position within the British government, administratively the process of incorporating free-standing boards was completed, and for practical and symbolic reasons the Scottish Office was given a major physical presence in Edinburgh.[10]

The development of the welfare state after 1945 resulted in further expansion of the public sector in Scotland, but while some of the new social programmes did become the responsibility of the Scottish Office, especially via its supervision of local government, others were administered centrally (e.g. macro-economic policy) or through deconcentrated offices of London-based central government departments (e.g. social security), and thus while the number of persons employed directly by the Scottish Office grew significantly, they were still outnumbered by civil servants working in Scotland for other central government departments.[11] One of the most important additions to the responsibilities of the Secretary of State for Scotland was an increasing involvement in economic matters, not only in terms of general oversight,[12] but also more specifically through attempts to coordinate existing activities with a view to promoting economic growth and lobbying in favour of location of major industrial projects in Scotland,[13] and eventually the Scottish Office assumed responsibility for some aspects of UK regional and industrial policy in Scotland which would otherwise have been administered either directly from London or via the deconcentrated DTI office in Glasgow.[14]

By the mid-1970s Scottish political actors were therefore involved in three different types of policy-making processes which can be separated

---

9 Correspondence in August 1885 quoted by a 20th century successor to the office, William Ross (1978 p. 4).

10 Ross 1981 pp. 2ff, Kellas 1989 Ch. 3, Midwinter *et al.* 1991 Ch. 3.

11 Hood *et al.* 1985, Kellas 1989 pp. 62ff, Midwinter *et al.* 1991 pp. 61ff.

12 MacDonald & Redpath 1980, McCrone 1992 Ch. 6. An example of this was the annual publication by the Scottish Office of *Industry and Employment in Scotland* in the period 1947-63.

13 Cf. Section 4.3 and Chapter 5.

14 Cf. Section 4.3.

on the basis of the political authority according to which action was taken:[15]

- *British policy arenas* such as foreign policy where the specific Scottish input depended on the performance of the Secretary of State in the UK cabinet,
- *Multi-level policy arenas* where authority was either ambiguous or shared between London and Edinburgh, e.g. when UK-wide schemes were being implemented in more or less specific ways by the Scottish Office, and
- *Scottish policy arenas* which involved specific legislation or initiatives but still were subject to the collective political responsibility within the UK government.

All in all this pattern of policy-making authority clearly ensured that, one way or the other, a specific Scottish input into policies pertaining not just to Scotland itself but also to Britain in general, much in contrast to the regions of England and, until the creation of the Welsh Office in 1964, Wales. On the other hand it is, however, also evident that a British input can be found even in policy arenas labelled as 'Scottish' because the Secretary of State for Scotland was a member of the British cabinet and thus his politics reflected those of the British parliament and *not* a body elected exclusively in Scotland. This is of course particularly conspicuous in periods when the party-political preferences have differed between Scotland and England, but also in situations where the same party dominated North and South of the Border, the room for manoeuvre at the Scottish Office was limited by the perceived need for policy coordination at the UK level and political concerns about English reactions to what might be construed as undue favouritism towards Scotland.

The role of the Secretary of State for Scotland has therefore been described as a dual one, representing *both* specific Scottish interest in London *and* the interests of the British government in Scotland,[16] and within the existing literature it has been debated which of these functions have generally tended to prevail. Some have maintained that Scottish autonomy has been limited in the sense that all major UK policy programmes have been implemented in Scotland,[17] but others have argued that Scottish actors have exercised a considerable degree of influence with regard to the way in which policies have been implemented and even suggested the existence of a distinct Scottish 'political system' or a 'semi-state' enjoying a degree of autonomy little different

---

15 Inspired by Hogwood 1977b, Ross 1981, Midwinter *et al.* 1991 Ch. 4, pp. 200ff; Brown *et al.* 1996 Ch. 5.

16 E.g. Kellas 1989 Ch. 5.

17 Midwinter *et al.* 1991 Ch. 3, Keating 1996 p. 168.

from small sovereign nation states like e.g. Denmark.[18] From the perspective of the current project it will be more fruitful to interpret the Scottish political set-up on the basis of the approach outlined in the preceding chapters. From this perspective the position of the Scottish Office can be seen to be at the intersection of several policy networks coexisting within an asymmetrical unitary political framework which allows for varying degrees of territorial autonomy in different areas of public policy,[19] and while Figure 4.1 is an attempt to capture this complexity in just two dimensions, some of its central features will be elaborated upon in the following.

Figure 4.1 The political environment of the Scottish Office

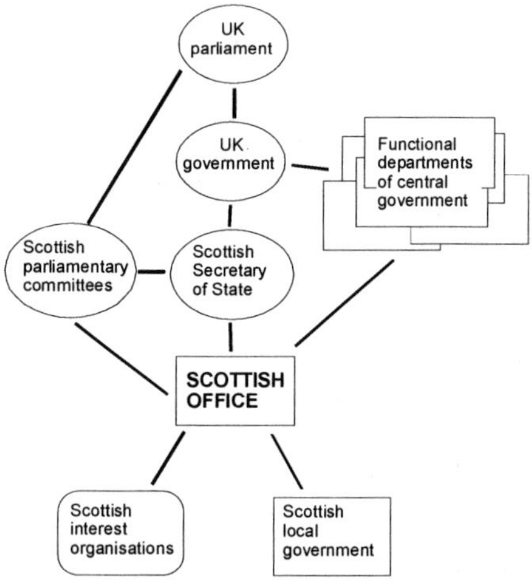

Inspired by Midwinter *et al.* 1991 p. 62.

Because the Scottish Office was a territorial department of central government, it was an integrated part of a series of UK-wide *functionally defined policy networks* in which the large, functionally defined departments in London – e.g. the Department for Trade and Industry or the Department for Education – were the leading actors.[20] The weight of input from the Scottish Office into the policy process was likely to vary between policy areas not only according to the degree of formal authority vested in the territorial department – responsibility for implementation or the existence of separate Scottish legislation could clearly make a

[18] Kellas 1989 Ch. 1 & 14, Paterson 1994 Ch. 1, Brown *et al.* 1996 Ch. 1 & 3.

[19] For perspectives parallel to this, see Hogwood 1982, Moore & Booth 1989 pp. 15ff, Brown *et al.* 1996 Ch. 3.

[20] Cf. Hogwood 1982, Kellas 1989 Ch. 12, Brown *et al.* 1996 Ch. 3.

difference – but also depending on the perceived need for a coordinated British approach in view of the political salience of particular issues and the extent to which a specific Scottish approach would run counter to strategies and assumptions of the governing party.

At the same time the Scottish Office was also part of a specific *territorial policy network* which brought it into frequent contact with interest organisations and local government in Scotland through consultations, advisory boards, supervision etc. As the system of local government differed from that in other parts of Britain and most non-monetary welfare services were handled at the sub-national level, many professional and interest organisations were separate Scottish bodies, or Scottish branches of British associations with a high local profile and political independence.[21] Territorial policy networks in Britain have been described as being generally

> integrated stable networks with continuity of membership, […] a high degree of vertical interdependence […] [and] a high degree of horizontal interdependence rooted in shared territorial interests[22]

and Scotland in particular has been characterised as a 'village', an informal network of the great and the good which vigorously pursued what was perceived to be Scottish interests.[23] While such a notion probably inflates the importance of the territorial policy network, it still highlights an important feature that set Scotland apart from England or the English regions, namely the existence of a relatively small and tight-knit policy community centred around a decentralised territorial department of central government, something which could not only further the emergence of consensual 'Scottish views' on various issues but also points towards a variety of ways – from broad-based public campaigning to discreet lobbying through the Scottish Office and the Secretary of State – in which these views could be promoted.

The double role of the Scottish Office as the focal point of a territorial policy network and an actor in UK-wide functional policy networks was mirrored at the parliamentary level through the existence of what can be interpreted as a Scottish sub-system within the British parliament at Westminster. On the one hand separate parliamentary committees existed to handle Scottish issues:[24] two Scottish standing committees

---

21 Kellas 1989 Ch. 10, Moore & Booth 1989 pp. 29ff, Midwinter *et al.* 1991 pp. 75ff. Examples of the former are the Scottish Trades Union Congress and the tripartite Scottish Council Development & Industry, while the Scottish branch of the employer's association Confederation of British Industry is an example of the latter.

22 Rhodes 1988 p. 284.

23 E.g. Moore & Booth 1989 pp. 85f, Midwinter *et al.* 1991 pp. 84ff, MacLeod 1996.

24 Kellas 1989 Ch. 5, Midwinter *et al.* 1991 pp. 64-70.

processed Scottish legislation, a Scottish select committee occasionally scrutinised the work of the Scottish Office and its associated bodies, and the Scottish Grand Committee – comprising all MPs elected for Scottish constituencies – debated more general political issues. On the other hand the Secretary of State was assisted by a team of junior ministers with responsibility for particular policy areas like industry or education, who in turn were being shadowed by members of the opposition parties.[25] Compared to the administrative level, the parliamentary set-up did, however, reflect the unitary nature of the British political system because Scottish legislation was voted on by all MPs and not just those representing Scottish constituencies, and the Scottish Secretary of State and his ministers were appointed by the UK Prime Minister and hence reflect British political views which were not necessarily dominant in Scotland. This distribution of formal political authority made it possible for a government to pursue policies in Scotland which had been rejected at the ballot box by a majority of Scottish voters, but the temptation to do this is likely to have been tempered by the importance of the Scottish vote in maintaining the government in power[26] and the extent to which government policies challenged consensual 'Scottish views'.

Scottish Agendas

Given the uniqueness of the Scottish institutional environment compared to other parts of Britain, it is hardly surprising that political actors and forms of discourse in Scotland have also deviated from the British pattern, and, indeed, increasingly so from the late 1950s onwards. In addition to issues which were high on the agenda all over Britain such as social welfare and economic growth,[27] Scottish politics was also characterised by the presence of a territorial issue, namely the position of Scotland *vis-à-vis* the rest of the UK.

The notion that 'Scotland' was not just the northernmost parts of Great Britain but a distinct territorial entity can be traced back to the conscious creation of the UK with Scotland as one of its predecessor states, i.e. to the historical experience of Scottish statehood, but its persistent prominence in political discourse has undoubtedly been facilitated by the continued presence of a broad array of social institutions that were distinctly Scottish in the sense that they only existed north of the Border or were organised differently from similar activities in England. This ensured that the notion of Scotland as a distinct territo-

25 Kellas 1989 Ch. 2, Midwinter *et al.* 1991 pp. 58ff.

26 The preferential treatment of the former in terms of the average constituency size of course increased Scotland's importance in party-political terms (Kellas 1989 Ch. 5).

27 Cf. Section 2.2.

rial unit would potentially appeal to both elite and popular groups because religion, law, education and local political governance were spheres of social activity in which most people were engaged on a regular basis.[28] Moreover, with the creation of a growing number of separate political institutions, from the Victorian administrative boards to the Scottish Office, the territorial positioning of Scotland unequivocally acquired a political dimension:[29] it became a natural 'self' in political discourse, delimited from external 'others' like the rest of the UK and hence possible to associate with distinct problems and remedial policies.

The presence of this territorial 'self' in Scottish political discourse could, at least from the 19th century onwards when the idea of the nation permeated politics in Europe,[30] be interpreted as a form of national identity in the sense that it involved individuals invoking a particular 'imagined community', but this did not result in (or from) the emergence of a separatist nationalist movement, and even calls for some form of 'home rule' were fairly low-key compared to e.g. the Irish campaign. This has often been seen as reflecting the new possibilities created by the success of industry and empire within the Victorian union state,[31] but also in the first half of the 20th century the British level retained its crucial role by successfully organising the military defence against continental aggressors and then after three decades of popular hardship instituting a comprehensive set of social welfare measures as a 'just reward'.[32] In parallel with the presence of distinct Scottish social and political institutions within the union state, British institutions also clearly had a tangible presence in many spheres of social activity north of the Border, and Scotland could therefore be construed as two different territorial 'selves' at the same time: a separate entity but also part of something bigger. Consequently it is hardly surprising that a pattern of 'dual identity' should have emerged with varying degrees of identifica-

---

28 These distinct Scottish institutions have often been referred to collectively as 'civil society' (e.g. Midwinter *et al.* 1991 Ch. 1, McCrone 1992 Ch. 1, Paterson 1994 Ch. 3, Brown *et al.* 1996 Ch. 1), but subsuming especially local government under this heading would seem to stretch an already wide-ranging concept (cf. Nielsen 1994), and hence the more inclusive expression 'social institutions' is preferred.

29 Midwinter *et al.* 1991 Ch. 1, McCrone 1992 Ch. 1, Paterson 1994 Ch. 4, Brown *et al.* 1996 Ch. 2, MacLeod & Jones 1999 pp. 586f, cf. Kellas 1989 Ch. 7, Macdonald & Thomas 1997.

30 See e.g. Hobsbawn 1990 pp. 104f, Thomsen 2001 Ch. 4.

31 Paterson 1994 Ch. 4.

32 Cf. Section 2.2.

tion with Scotland and Britain coexist, not just within the Scottish electorate at large but also at the level of the individual.[33]

Figure 4.2 General election results in Scotland 1945-74

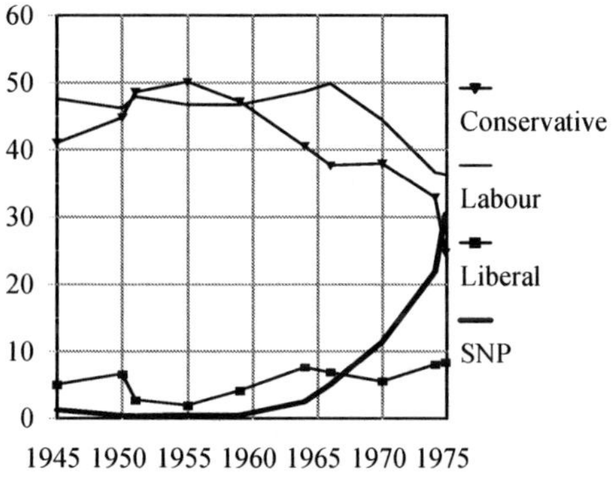

Per cent of votes.
Source: Brown *et al.* 1996 p. 146.

While the growth of the Scottish Office had encouraged preferential regionalism, i.e. demands for special measures to remedy the allegedly unique difficulties facing Scotland put forward by political actors supporting the union state,[34] the emergence of separatist nationalism as an organised political force was slow and, as illustrated by Figure 4.2, only really took off in electoral terms after the Scottish National Party (SNP) won a by-election in Hamilton in 1967. But although the dominant position of the two major British parties came increasingly under pressure from the mid-1960s onwards, countering the challenge of a new territorial politics pursued in the name of 'Scotland-as-nation' proved to be more than a little difficult. In terms of political principles the most significant difference between the SNP and the other parties clearly did concern territorial politics: the SNP favoured independence (or at least a significant degree of home rule),[35] the small liberal party mostly argued for Scottish devolution as part of a more general scheme to decentralise the governance of Britain,[36] and the dominant British parties continued to support the economic and social advantages of the unitary union state,

---

33 Paterson 1994 pp. 65f, Brown *et al.* 1996 pp. 197ff, cf. Hedetoft 1995 pp. 162f, 442ff.

34 Paterson 1994 pp. 112ff, cf. Chapter 5.

35 On the development of the SNP's territorial politics, see Brown *et al.* 1996 p. 140, Hutchison 1999 p. 124, and, for a detailed inside account, the autobiography of SNP Chair William Wolfe (1973).

36 Midwinter *et al.* 1991 pp. 37f, cf. Hutchinson 1999 pp. 117ff.

although both Labour and the Conservatives were internally divided and contained minorities favouring some form of constitutional realignment.[37] At the same time it was obvious that the economic and social policies proposed by the SNP did not deviate from the existing postwar consensus, but after a succession of British governments had failed to address effectively the problems of industrial decline and interregional inequalities,[38] the SNP could claim that an independent Scotland would have the financial means to pursue such policies much more vigorously on the basis of income generated by the recently discovered North Sea oil.[39] The extensive use made of the slogan "It's Scotland's oil" by the SNP in the 1974 election campaigns[40] would seem to suggest that an important part of the party's appeal was the tying together of a separatist nationalist form of territorial politics with the prospects of being able to address the longstanding weakness of the Scottish economy, but as some opinion polls showed that even among SNP voters only a minority supported Scottish independence,[41] it is hardly surprising that the pragmatic economic aspect of nationalism led both Labour and the Conservatives to step up their preferential regionalist efforts to secure positive economic discrimination in favour of Scotland within the union state.[42]

This did, however, not mean that the challenge of the new territorial politics of the SNP could be ignored in the long run, and thus in the wake of the parliamentary break-through of the SNP at the Hamilton by-election both the two main parties reconsidered in very public ways the constitutional position of Scotland within the UK, although the proposals eventually put forward were of a fairly limited nature and the commitment of the parties to constitutional reform remained open to doubt due to frequent policy changes and internal divisions. The major party of the right had officially called itself "Conservative and Unionist" until 1965,[43] and its post-Hamilton study group headed by former Prime Minister Sir Alec Douglas-Home produced a proposal for a directly elected body to handle the committee stages of Scottish legislation, but

37 Keating & Bleiman 1979 pp. 59-74, Midwinter *et al.* 1991 Ch. 2, Harvie 1994 Ch. 6, Brown *et al.* 1996 Ch. 6, Hutchinson 1999 Ch. 4.

38 Cf. Section 4.3.

39 Hutchinson 1999 pp. 122ff, cf. Kellas 1989 p. 138.

40 See SNP 1974a. The issue of oil did, however, appear in SNP materials already in the mid-1960s when explorative drilling began in the North Sea (e.g. SNP 1965d, cf. William Wolfe 18.7.90).

41 E.g. the October 1974 System Three Scotland Survey quoted by Kellas (1989 p. 148).

42 Cf. Chapter 5. The role of economic issues in the rise of the SNP has been much debated, e.g. Webb & Hall 1978, Rose 1982, Paterson 1994 Ch. 6, Brown *et al.* 1996 Ch. 6, Hutchinson 1999 p. 121.

43 Midwinter *et al.* 1991 pp. 21f.

while this duly appeared in the Conservation manifesto for the 1970 general election,[44] it was never implemented by the Heath government – and creation of some form of Scottish assembly only returned to the Conservative manifesto for the second general election in 1974 in the wake of a strong SNP showing in the February general election.[45] Labour, on the other hand, had historically been associated with the Scottish Home Rule cause, but while many Labour candidates in Scotland had campaigned for some form of devolution in 1945, no legislation was introduced and the policy was officially dropped in 1958.[46] After the SNP victory in Hamilton the Labour government established a Royal Commission in 1968, but although submissions to the Commission showed widespread support in Scotland for an elected Scottish Assembly, the 1970 election Labour manifesto firmly "reject[ed] separatism and also any separate legislative assembly" in order to preserve the integrity of the UK.[47] Although the SNP had clearly overtaken the Liberals in terms of votes at the 1970 general election, nationalist support was spread fairly evenly across Scotland and hence the British electoral system left the SNP with only one MP at Westminster, and while this perhaps explains the absence of any reference to devolution in the February 1974 Labour manifesto,[48] it also made it easy to construe the inclusion – after well-publicised internal conflicts – of devolution as a manifesto commitment for the October 1974 general election as a somewhat opportunistic attempt to counter the electoral success of the SNP by signalling willingness to reconsider the constitutional position of Scotland within the British union state.

4.2. Empire and Beyond: Scotland in the British Economy

While the differences between Scotland and e.g. the northern regions of England are conspicuous with regard to political governance, from an economic perspective the British localities which had been at the forefront of the industrial revolution clearly had many things in common. When Scotland joined England and Wales in forming the United Kingdom of Great Britain, this involved a customs and monetary union that not only created a unified internal market but also combined the colonial possessions of the constituent parts of the new union state, and for nearly three centuries economic development in Scotland has therefore been intrinsically linked with the shifting fortunes of the much larger

44 Reproduced in Craig 1990 p. 128.

45 Webb & Hall 1978, Warner 1988 pp. 213ff, Kellas 1989 pp. 145ff.

46 Webb & Hall 1978, Keating & Bleiman 1979, Kellas 1989 p. 145.

47 Reproduced in Craig 1990, quote from p. 148.

48 Reproduced in Craig 1990 pp. 186ff.

country south of the border. This section first surveys key aspects of Scottish economic history in their British context and then takes a closer look at some of the postwar characteristics that have sustained Scotland's status as a 'problem region'.

As the delimitation of regions by means of economic criteria is notoriously difficult,[49] it is worth noting that in this text expressions like 'the Scottish economy' or 'the regional economy' are often used instead of the more cumbersome 'economic activities taking place in Scotland', but this does *not* imply any assumptions about the distinctiveness of the Scottish situation compared to other British regions or, indeed, the relationship between economic actors within and outwith Scotland. Effectively delimiting the Scottish economy on the basis of geo-political criteria is, however, clearly compatible with institutionalist-type perspectives in which no correspondence between the spatial organisation of different types of social activities is assumed,[50] and even economists have come to accept – albeit sometimes rather reluctantly and for pragmatic reasons like data availability or discursive salience[51] – the usefulness of such an approach.

Markets and Industries till the Mid-1970s

In the 19th century Scotland was integrated into the British economy via its position in the division of labour evolving in response to the expanding markets of the growing overseas empire, and the Victorian version of the Scottish economy was dominated by three types of activities, namely heavy industries centred around shipbuilding in and around Glasgow, other first-generation manufacturing such as textile production spread across central Scotland, and internationally-oriented financial services primarily located in Edinburgh.

Central Scotland was among the first British areas to become industrialised, drawing on a fortuitous combination of natural and human resources: coal and iron ore in the ground, access to sheltered deep-water facilities, and a ready supply of workers from the agrarian hinterlands north and south of the River Clyde.[52] At the centre of the regional economy was an "extraordinary integrated complex" which combined "coal, engineering, shipbuilding and metal manufacture and processing"

---

[49] Cf. Section 3.2.

[50] Cf. e.g. Moore 1988a pp. 2f.

[51] See e.g. Lythe & Majmudar 1982, Ingham & Love (eds.) 1983.

[52] For general introductions to Scottish economic history, see Lee 1995, Payne 1996, Smith 1985a. The early decades are covered in intelligent detail by Campbell 1980, the postwar period by contributions to Saville (ed.) 1985 and Hood & Young (eds.) 1984.

and "sent products to all parts of the globe".[53] Shipbuilding is widely regarded as the driving force in what would nowadays have been characterised as a cluster, and the highly specialised region came to be a major player in important niches of the imperial economy, with Scottish yards accounting for no less than 18% of all new ships launched throughout the world in 1913.[54] The patterns of ownership and the organisation of production did, however, continue to reflect the early expansionary phases of industrialisation,[55] with many small family-owned firms collaborating in what appears to have been long-term network relations mediated through e.g. interlocking directorships,[56] shipyards located in the middle of heavily urbanised areas which imposed severe physical constraints on modernisation and resulted in a production process in which workers with highly specialised skills played a central role.[57] The combination of these characteristics obviously limited the room for manoeuvre – concentration of ownership was difficult to translate into more efficient forms of production – and with "an industrial structure [...] dependent on competitive success in a relatively narrow range of activity",[58] even in its heyday the core of Scottish manufacturing looked potentially vulnerable to externally induced upheavals in demand, market access, or production technologies. Even so, the traditional heavy industries still had a sizeable presence in the late 1950s because as the demand for their products had been maintained by imperial preference and wartime destruction of innumerable vessels,[59] but from the 1960s onwards competitive pressures grew after the dismantling of the British Empire and the exhaustion of local coal supplies. The following decades witnessed what was effectively an orderly rundown – through nationalisation and other forms of government intervention – of what had been the manufacturing core of the regional economy so that by the mid-1970s the cluster that at the beginning of the 20th century had secured employment for one third of a million workers had all but disintegrated.[60]

---

[53] Devine 1996 p. 2.

[54] Payne 1996 p. 116, Lee 1995 p. 32.

[55] Lee 1995 pp. 84ff.

[56] Cf. Scott & Hughes 1980.

[57] Lee 1995 pp. 84ff.

[58] Lee 1995 p. 47. The use of terms like structural 'dislocation' (STUC 1989 p. 43) would seem to suggest that the sectoral make-up *per se* was the problem, thereby backgrounding the organisational and social factors that complicated adaption to external pressures.

[59] Payne 1985, Lee 1995 Ch. 4.

[60] Payne 1985, 1996 p. 22; Lee 1995 Ch. 4.

In contrast to this, the second area in which Scottish firms excelled in the Victorian age, financial services, adapted much more successfully to the changing environment.[61] Although originally associated closely with both local industry and investments in the expanding British Empire, the regional financial cluster – comprising three note-issuing clearing banks, major insurance companies and managers of investment funds – maintained its dynamism also in the 20th century. Keeping themselves at arm's-length of vulnerable industrial activities while to some extent being sheltered by the distinct regulatory environment in Scotland meant that Scottish banks and insurance companies could readily shift their activities to new markets in Britain and beyond. Edinburgh was therefore able to remain the second financial centre of the UK, after the City of London but well ahead of other regions, and thus by the mid-1970s financial services came close to outnumbering the combined employment of traditional heavy industries like shipbuilding and metal manufacturing in terms of employment.[62]

In the 19th century 'lighter' forms of manufacturing, i.e. textiles and consumer products, played a less prominent role in the regional economy, and although around 25% of the Scottish labour force was engaged in textiles and clothing, this was still somewhat below the UK average.[63] While a reduced role for the economic activities that had dominated Britain during the first waves of industrialisation could perhaps be expected as overseas competitors began catch up, the relatively limited extent to which new industries emerged in Scotland is more remarkable: production of consumer goods remained at a low level even after the advent of the 'affluent' 1960s, allegedly because of the smallness of the regional market and a weak tradition for marketing, a skill which had not been nurtured by the old network-based cluster economy.[64] Instead the new industries which became increasingly important in Scotland – first electronics and then from the late 1960s oil-related activities – were largely incoming in the sense that economic activity was dominated by firms from outwith the region, and at least till the mid-1970s regionally-based firms were primarily providers of more basic products and services in these expanding areas of the economy.[65]

All in all Scotland would seem to encapsulate many of the features more generally associated with the development of the British economy,

---

61 Draper *et al.* 1988 Ch. 1, Payne 1996 pp. 23f.

62 Calculated on the basis of Bain & Reid 1984 Table 11.2, Young & Reeves 1984 Table 5.2.

63 Campbell 1980 Table 5.

64 Lee 1995 pp. 59ff, Payne 1996 p. 20.

65 Lee 1995 Ch. 5, Payne 1996 p. 34.

i.e. the long-term importance of patterns reflecting early industrialisation and reliance on sheltered imperial markets, the position of heavy manufacturing sustained through warfare and the continued presence of British troops throughout the world, the new consumer-goods industries relatively weakly developed, an increasing share of modern manufacturing controlled by foreign multinational companies, and an internationally-oriented financial sector with relatively limited direct involvement in domestic manufacturing. On an aggregate level the combined long-term effects of these characteristics have often been discussed under the heading of 'industrial decline' because eventually many of the countries industrialising in the wake of the first industrial revolution not only caught up with but also overtook the UK in terms of prosperity because the relatively slow growth of the British economy,[66] and from the perspective of the present study the importance of Britain's relative industrial decline is twofold. *Firstly*, it strongly suggests that the relatively poor performance of the Scottish economy may have reflected problems akin to those at work at the British level, and that the problems experienced in Scotland could therefore be difficult to address at the regional level because they concerned UK-wide institutions such as free collective bargaining or dominant assumptions about the proper relationship between financial institutions and manufacturing firms. And *secondly*, the relatively slow growth of the aggregate wealth of the British nation will also have circumscribed the ability of central government to remedy interregional disparities brought about by e.g. the dismantling of traditional forms of industrial activity. As summed up by Neil Buxton, "growth was generally sluggish relative to that of the rest of the UK, and the UK itself, in turn, performing badly in relation to her industrial competitors abroad",[67] and hence Scotland ceased to be a prosperous part of the 'workshop of the world' and became instead a 'problem region' in a slow-growing economy at the north-western periphery of Europe.

Figures of Decline

The gradual slide of the Scottish economy away from its Victorian prominence and prosperity became increasingly conspicuous in the decades after the Second World War, and the economic position of Scotland continued to be intrinsically linked to its role in the spatial division of labour within the British economy.

Several features of Scottish manufacturing have repeatedly been identified as weaknesses underlying the problems experienced by indi-

[66] Cf. the discussion in Section 2.2.

[67] Buxton 1985 p. 47.

vidual firms and the regional economy as a whole, and for obvious reasons over-reliance on the 'traditional' industries has been high on the agenda as a tangible way on pinpointing material disadvantages *vis-à-vis* other parts of the UK. 'Old' forms of economic activity continued to account for relatively large shares of total employment compared to the British average while 'new' forms of manufacturing were 'under-represented', and at the same time the former were the areas in which activity contracted sharply while the latter expanded: between 1953 and 1978 nearly half the Scottish employment in shipbuilding, metal manufacture and textiles disappeared – all in all more than 100,000 jobs – and this could not be offset by rapid growth in some of the new industries because this proceeded from a very low starting point.[68] Variations on the theme of 'over-reliance' on 'old' industries was also found in other regions which had flourished in the 19th century, especially Wales and the northern parts of England,[69] and in policy terms this pattern could be used as a platform for attempting to 'modernise' the regional economy by attempting to increase the role of 'new' industries.

Figure 4.3 Location of ultimate ownership of Scottish manufacturing plants in 1973

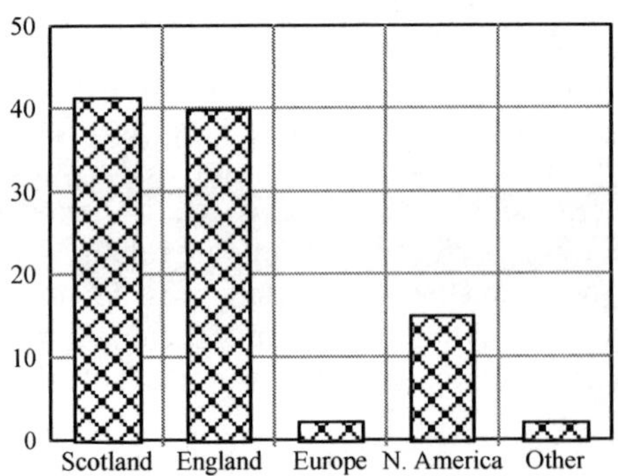

Share of employment.
Source: Firn 1975 p. 402.

Ownership was another feature which became associated with the weakening of Scottish manufacturing. While the maritime heavy engineering cluster and most other Victorian industries had been predominantly indigenous in the sense that they were owned, financed and managed locally, by the early 1970s direct control of manufacturing had moved firmly outwith the region even on the basis of a relatively con-

68 Calculated on the basis of Randall 1985 p. 248.

69 Massey 1984 pp. 128ff, 1986 pp. 33f; Martin 1986 pp. 87f.

servative measure, as illustrated by Figure 4.3.[70] This was true not only of traditional industries, many of which had been nationalised and hence brought under British control, but also of many modern industries such as e.g. electronics which had grown significantly in the postwar decades due to the arrival of US-based multinational firms.[71] While nationalisation in many respects served as a public attempt to ensure an orderly demise of traditional industries in Wales, Scotland and the north of England,[72] incoming firms like IBM brought modernisation through new products, technologies and management methods but at the same time also served as tangible reminders of the weakening of indigenous entrepreneurialism. Like other British regions which had prospered as part of the Victorian imperial economy, the industrialised central parts of Scotland did not perform particularly well with regard to formation and growth of new firms, something which helped to increase the importance of incoming firms as a source of economic change.[73]

Similar changes in patterns of ownership and the importance of particular industries can be observed in other British regions, albeit of course varying according to the nature of the economic inheritance from the glory days of the 19th century,[74] and the long-standing importance of local industrial clusters ensured that firm-level developments were subject to much public attention at the sub-national level. In the postwar decades aggregate indicators of interregional disparities increasingly came to the fore, and in terms of wealth the British spatial hierarchy remained stable in the first three postwar decades, with the old industrial regions clearly below and South East England clearly above the national average,[75] resulting in major population movements away from the peripheral northern and western areas towards more prosperous localities, with Scotland as the leading exporter of manpower.[76] Despite the large number of Scots who 'got on their bikes', the level of unemployment did, however, remain significantly above the UK average, and while registered unemployment was of course low in the late 1950s and

[70] See e.g. the detailed comments by Scott *et al.* 1980.

[71] Scott & Hughes 1980 Ch. 3.

[72] Hudson 1989b pp. 114ff, MacLeod 1998b p. 856.

[73] Firn & Swales 1978, Hood 1984 Table 3.2, Mason 1987 pp. 138ff, Hart *et al.* 1993 pp. 185f.

[74] See e.g. Hudson 1989a on the North East of England, and on Wales Cooke & Rees 1981, Morgan 1994 .

[75] McCrone 1969 p. 20, Prestwich & Taylor 1990 p. 171.

[76] Prestwich & Taylor 1990 p. 174. Migration nearly cancelled out the natural population increase in the 1950s and early 1960s.

early 1960s compared to the crisis-ridden 1930s, the Scottish level remained around twice the British average.[77]

All in all the various indicators illustrating Scotland's changing position in the British economy would seem to warrant two conclusions. *Firstly*, during the 20th century the primary way in which Scottish economic actors were integrated into the British economy changed dramatically: while in the Victorian age integration was largely ensured by means of the mechanism of the market – Scottish access to selling and buying throughout the union state and its Empire – by the mid-1970s core parts of Scottish manufacturing had now become integrated in corporate hierarchies headquartered outwith the region or, indeed, the UK. *Secondly*, while the economic development of Scotland reflected specific features, the region could still readily be construed as part of a larger group of 'problem regions' which – especially when seen through the lens of aggregate measures like unemployment – displayed broadly similar characteristics and consistently compared unfavourably with the UK average.

The 'figures of decline' would in other words seem to suggest two central dilemmas which makers of regional policy would be facing: on the one hand the question of the spatial level of public intervention, i.e. the balance between UK-wide and regionally-based measures, and on the other hand the question of the firm-level targeting of policy measures, i.e. whether to attempt to revive indigenous Scottish enterprise or compete for the attention of external, English or overseas, investors. But of course the position eventually adopted would not only reflect the economic attractiveness or feasibility of the various options but also the political preferences of policy-makers, and thus although Scotland in economic terms was broadly similar to other British 'problem regions', the politics of regional policy could still entail a distinctly Scottish dimension.

4.3. Uneven Developments: British Regional Policy before 1975

This section presents a reinterpretation of the development of regional policy in Britain prior to the setting up of the SDA, with the aim of clarifying the policy context into which the Agency was introduced and establishing the foundation for the subsequent analysis of the environment in which the organisation operated. A brief outline of the basic features of the history of British regional policy bears strong resemblance to the general trends in Western Europe, but on closer inspection

77 Calculated on the basis of Lythe & Majmudar 1982 p. 11.

it becomes clear that "the British way" has important characteristics that may influence the ways in which it is interpreted in the academic literature.[78] The text falls into three parts: first the initial and national phases of British regional policy are analysed, and then the specific Scottish experience is considered.

The Initial Phase, 1928-58

Regional policy in Britain can be traced back to central government responses to spatial concentration of unemployment in the wake of the economic crisis from the late 1920s onwards. In 1928 the *Industrial Transference Board* was established, providing unemployed miners and other workers with training and economic support in order to enable them to move and find employment in other parts of the country, and by offering conditional financial and informational incentives to individual unemployed, the relocation of around 20,000 persons was assisted every year in the 1930s.[79] From the mid-1930s, at the height of national economic hardship, the focus of regional policy was extended significantly through attempts to promote economic development in crisis-ridden parts of the country. The foundations for the new approach was laid in 1934 through the *Special Areas (Development and Improvement) Act*, designating four "special areas" eligible for support, appointing two Commissioners responsible for implementation, one for England and Wales and one for Scotland, and allocating a sum of 2 million pounds. Initially the Commissioners' main areas of activity came to be environmental improvement, e.g. sewerage schemes and clearing of derelict industrial sites,[80] and provision of industrial floor space through the setting up of industrial trading estate companies for each of the four Special Areas. But later in the 1930s the possibility of small-scale financial support for firms located in the Special Areas was instituted, and thus by the end of the 1930s some of what became the core elements of central government regional policy had been introduced – nationally designated areas in which financial subsidies and provision of factories were available – although the prominence of industrial transference ensured that the private actors targeted by public policy were not just firms but also to a significant extent individual workers. A variety of very different underlying assumptions could be associated with

78 For introductions to the development of regional policy in Britain, see e.g. Armstrong & Taylor 1985 pp. 171-224, Prestwich & Taylor 1990 pp. 112-61, and, for the early years, McCrone 1969, McCallum 1979.

79 McCallum 1979 pp. 3f, Prestwich & Taylor 1990 p. 115, Parsons 1988 pp. 6f, McCrone 1969 p. 92.

80 90% of the expenditure committed by the Commissioner for Scotland in his first year of activity related to sewerage schemes (McCrone 1969 p. 95).

prewar regional policies, ranging from inter-regional equality with regard to unemployment to concerns about social and political stability in a decade of great tensions in Europe at large,[81] but even at the highest level of government regional policy was apparently seen as having primarily a political rationale, as shown by the following comment, presumably uttered in private, attributed to the Chancellor of the Exchequer Neville Chamberlain in 1936:

> [...] the Commissioner has made certain proposals and politically it might be helpful to try. If they failed, as I think they would, we should have done little harm, but we should have met the reproach that we neither accept others' suggestions nor produce any of our own.[82]

Still, the introduction of the Special Areas and the associated policies constituted a significant break with the traditional liberal 'orthodoxy' espoused by the Treasury,[83] and thus regional issues became linked with emerging changes in the general discourse about economic policy and the role of the public sector *vis-à-vis* private economic actors.

The first major statement about regional problems and policies in Britain, the report of the Barlow Commission on the *Geographical Distribution of the Industrial Population*, was published in January 1940 and served as a central point of reference for deliberations about regional issues for decades,[84] but in terms of policy development other factors appear to have been even more influential, i.e. war-time experience with government regulation of the spatial distribution of economic activity and the emerging consensus about a new, Keynesian, approach to economic policy which saw "maintenance of a high and stable level of employment" as a primary aim of government.[85] The *Distribution of Industry Act* introduced in 1945 by the new Labour government essentially modified ad-hoc prewar measures with regard to the geographical coverage of the assisted areas, financial support for individual firms was consolidated into one programme administered by the Treasury, and the Board of Trade, an economic department of government, took over the administration of factory building and environmental improvement from the Development Commissioners.[86] The final element of postwar re-

---

81 Cf. Parsons 1988 Ch. 1.

82 Wren 1996a p. 24.

83 Parsons 1988 pp. 1f, 18f.

84 Parsons 1988 p. 67.

85 McCrone (1969 p. 106) quoting the introduction to the 1944 white paper on employment, cf. McCallum 1979 p. 7, Armstrong & Taylor 1985, Parsons 1988 p. 65.

86 McCrone 1969 pp. 107ff, Slowe 1981 Ch. 2, King 1986, Prestwich & Taylor 1990 pp. 121ff, Wren 1996a pp. 31ff. While the assisted areas in the 1930s comprised 8.5% of the working population, the figure rose to 20% after the extensions in 1948 (Prestwich & Taylor 1990 p. 122)

gional policy, location control, only emerged as a separate policy instrument following the 1947 *Town and Country Planning Act* which stipulated that industrial projects involving more than 5000 square feet – equivalent to a one storey building of little more than 22 by 22 metres – should apply for an Industrial Development Certificate (IDC) which could be withheld by the Board of Trade in congested or economically 'over-heated' parts of the country, thereby making it more attractive for firms to expand or relocate in e.g. development areas.[87] Central government in other words was no longer merely the sponsor but also the chief implementor of policy, public intervention now exclusively targeted private firms, and thus an important strategic shift had taken place in that policies were now designed to, in the words of President of the Board of Trade Stafford Cripps, "take the work to the people and not the people to the work".[88] Still, although the average level of annual expenditure more than trebled compared with the 1930s,[89] direct financial support to individual firms continued to play a minor role while factory building accounted for nearly 90% of the regional policy budget.[90]

While the period from 1946 to 1958 was marked by continuity in terms of the policy instruments available, the extent to which they were employed varied, and the immediate postwar years have often been contrasted with the 'policy-off period' of the 1950s,[91] but while the level of policy output in the 1950s was clearly lower than in the years immediately after the war due to government fiscal restraint and dramatically reduced levels of unemployment, regional policy was certainly not 'off' compared to the 1930s: average annual expenditure on regional policy in the period 1949-58 was around 2.5 times that of the pre-war period,[92] and, more importantly, central government still had a wide range of policy instruments at its disposal, should the economic and political environment change.

---

87 McCrone 1969 p. 111, McCallum 1979 p. 7, Prestwich & Taylor 1990 p. 124.

88 Quoted by Slowe 1981 p. 16.

89 Annual average expenditure 1928-1938 was less than 2 million pounds while the equivalent figure for the period 1946-1958 was more than 6 million pounds (calculated on the basis of Prestwich & Taylor 1990 p. 119, McCrone 1969 p. 114).

90 Calculated on the basis of McCrone 1969 p. 114.

91 E.g. Armstrong & Taylor 1985 p. 173, Wren 1996a p. 41, and Parsons 1988 pp. 103ff. The use of the 1950s as the 'policy off' period with which later decades were compared has been a central feature of many attempts to measure the effects of British regional policy (Armstrong & Taylor 1985 pp. 278ff).

92 Calculated on the basis of Prestwich & Taylor 1990 p. 119, McCrone 1969 p. 114.

The National Phase, 1958-75

The national phase of regional policy in Britain in which central government was the undisputed force behind spatial economic intervention began in the late 1950s and lasted into the mid-1970s where new actors emerged both on the regional and European level. The areas eligible for support were gradually increased until they comprised 44% of the working population in 1977, IDCs were again used to dampen expansion in prosperous parts of the country, albeit to somewhat varying degrees, and as the range of financial instruments was extended, annual expenditure rose from 4 million in 1960/61 to 1010 million in 1975/76.[93] From a comparative Western European perspective, the British level of expenditure per head of national population was only exceeded by Italy and Ireland,[94] and thus regional policy was not only a growing area of activity but also in relative terms clearly a priority for central government.

With central government as the sole sponsor of regional policy, one could hope that the organisational set-up would be reasonably constant, but while government departments continued to play a dominant role in policy implementation, important qualifications apply on a more detailed level of analysis. Within central government responsibility for regional policy was subject to frequent changes, the administration of some programmes gradually became more deconcentrated with regional offices of central government departments becoming involved in the administration of various forms of grant support to individual firms, alongside a succession of advisory committees that eventually also acquired a regional dimension.[95] Moreover, non-departmental bodies continued to be responsible for the implementation of policy programmes involving primarily organisational or informational resources, with the two most prominent examples being factory building which continued to be administered by Industrial Estates Management Corporations,[96] and the use of regional development agencies to address development problems in non-industrial areas, i.e. the Highlands and Islands Development Board (HIDB) in the remote and sparsely populated parts of Scotland and the Development Commission operating in rural parts of Britain.[97]

With regard to spatial selectivity, the basic distinction was maintained between growth areas subject to location control and assisted

---

[93] Wren's grant-equivalent figures in 1980 prices (1996b p. 328).

[94] Yuill *et al.* 1982 p. 101.

[95] Hogwood 1977a p. 9, 1982 p. 39; McCallum 1979, Field & Hills 1976 pp. 2-14.

[96] McCrone 1969 p. 141, Prestwich & Taylor 1990 pp. 128f.

[97] Cf. the discussion below.

areas in which various forms of support were available to firms, and as these two concerns were not necessarily complementary, a third type of areas, subject to neither location control nor supportive measures, emerged as a residual category. Especially the decade from the mid-1960s onwards was clearly one in which the assisted areas rapidly expanded beyond the traditional problem regions in the northern and western parts of the country, and eventually by the mid-1970s more than 40% of the British population lived within a multi-tiered system of assisted areas. In the national phase of regional policy the role of direct financial support for individual firms increased significantly, and a variety of subsidy schemes was introduced, modified, and replaced from the late 1950s to the mid-1970s: discretionary forms of support continued throughout the period but automatic forms of financial assistance were increasingly made available through a series of schemes which provided additional funds for various aspects of manufacturing activity on the basis of relatively simple claims procedures based on overt conditions of eligibility and fixed rates of support. The operation of the subsidy schemes were invariably reactive, relying on firms to generate activities that could be considered by public organisations for support,[98] and non-spatial selectivity continued on the basis of very general criteria: both existing and incoming firms were eligible for support, capital as well as labour-related activities were assisted, and towards the end of the period the sectoral focus was broadened through the inclusion of certain service sector activities, albeit only in discretionary schemes. The cumulative effect of these developments was a massive increase in the financial incentives available in the assisted regions, and the adjoining Figure 4.4 presents the development of grant equivalent expenditure in constant prices.[99] It is evident that the main period of growth was the four financial years from 1965/66 to 1968/69 which saw the value of financial incentives to firms increase by nearly twenty times, but the figure also provides an insight into the distribution of these incentives between different types of policy instruments:

- labour subsidies provided nearly half of all financial incentives available in the assisted areas,

[98] McCrone 1969 pp. 131-38, cf. Field & Hills 1976.

[99] Grant equivalent expenditure is a synthetic measure that attempts to calculate the financial value of different policy programmes to private firms by estimating the incentive involved rather than relying on direct government outlays which would e.g. overestimate the relative importance of expenditure on loans (see Wren 1996b). For 1958/59 and 1959/60 no grant equivalent figures are available, and instead 'raw' expenditure figures in current prices are used. Given the limited level of subsidy involved in loans, still the predominant form of direct financial support, this would seem to counteract the lack of deflation.

- while investment-related support was increasing rather steadily, tax allowances were only significant in two short periods, and
- in financial terms discretionary grants, traditionally the main form of financial support, came to play a very minor role from the mid-1960s onwards, accounting for only 3% of the incentives available.[100]

Figure 4.4 Financial regional policy instruments, 1958-75

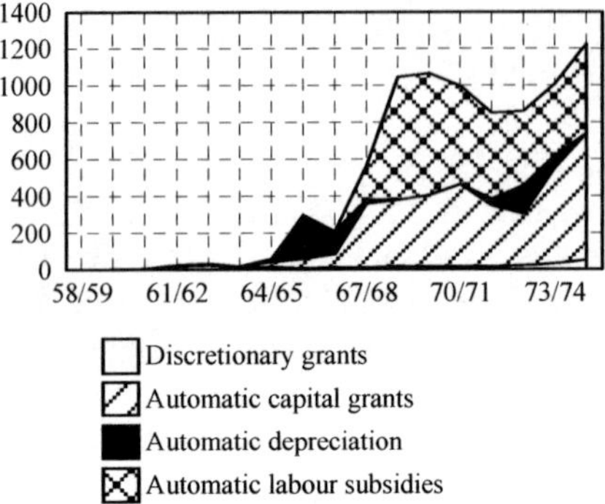

Grant equivalent expenditure (£m 1985/86).
Source: McCrone 1969 p. 114, Wren 1996b p. 328.

From having played a limited role in the initial phase of British regional policy, the rapid expansion of financial incentives had made them a very visible part of the overall picture by the mid-1970s.

The second main policy instruments carried forward from the 1950s was the use of IDCs as a means of limiting development in congested regions and instead encouraging firms to expand in the assisted areas. As illustrated by Figure 4.5, the threshold under which a permit was required changed regularly from the mid-1960s onwards, but the discretionary way that this policy instrument was used meant that there was not necessarily a straight-forward relationship between the formal limits and the ways in which these were implemented. While location control as a negative incentive for firms to relocate to assisted areas was certainly implemented much more stringent than in the mid-1950s,[101] its relative importance *vis-à-vis* financial incentives would, however, seem to decrease in the 2nd half of the 1960s when IDC limits increased and refusal-rates dropped while at the same time financial support for regional development measures increased dramatically.

[100] Calculated on the basis of Wren 1996b p. 328.

[101] Moore *et al.* 1986 p. 28.

The third main pillar of postwar regional policy, provision of subsidised factory space in the assisted areas, was also maintained in the 1960s and beyond, with the basic set-up remaining the same: Industrial Estate Management Corporations built and managed industrial property which was let to private firms on terms which involved a degree of subsidy, sometimes in the form of rent-free periods.[102] The resumption in 1959 of factory building in advance of specific demand was one of the measures that marked the beginning of a new phase in regional policy, and only in the late 1960s was a 'rolling' approach adopted which linked the construction of new advance factories to the take-up of existing units in an attempt to keep the empty capacity relatively constant.[103] In terms of selectivity it is also worth noting that it was only in 1974 that increased emphasis was given to the construction of smaller factory units,[104] indicating that this form of support had formerly primarily targeted larger firms.

Figure 4.5 IDC limits and refusals, 1958-75

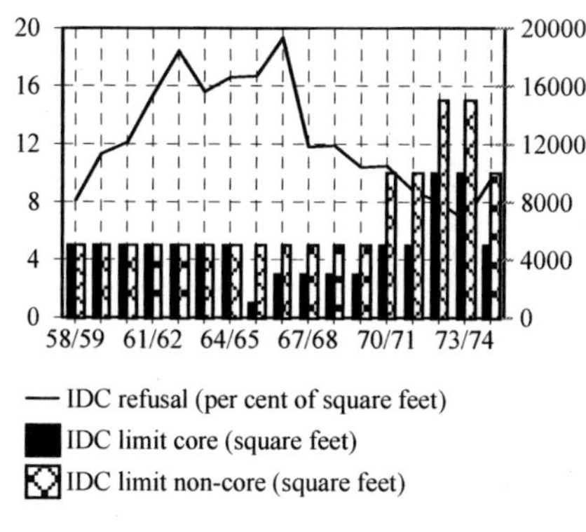

Per cent / square feet.
Core defined as South East and Midlands 1965-72,
as the South East from 1972 onwards.
Source: Moore *et al.* 1986 p. 28.

The final element of regional policy carried over from the 1950s, support for the clearing away of left-overs from earlier waves of industrialisation – e.g. disused factories and mining sites – was also retained, mainly in the form of grant support for local authorities engaging in the clearing of derelict land. Also this form of activity increased over the

102 McCrone 1969 pp. 141f, Wren 1996a p. 48, Prestwich & Taylor 1990 pp. 128f.

103 Moore *et al.* 1986 pp. 26f.

104 Prestwich & Taylor 1990 p. 147.

years, especially in the early 1970s where support was made available in the newly designated Intermediate and Derelict Land Clearance Areas.[105]

Regional policy had originally focused primarily on urban areas hit by industrial decline, but during the national phase the rather different problems of 'under-industrialised' localities gradually came to be seen as a 'regional problem' in its own right and hence a legitimate object of public policy. Having served as a multi-purpose quango since 1909, the *Development Commission* gradually concentrated on support for economic as well as social development in rural parts of Britain.[106] Operating on what its chairman described as "a shoe-string budget",[107] the Commission provided financial support and advice via *CoSIRA*, the Council for Small Industries in Rural Areas, and its Scottish counterpart *SICRAS*, the Small Industries Council for Rural Areas in Scotland, and from the late 1960s the Commission engaged in a programme of building small advance factories in selected localities.

Having charted developments from the late 1950s onwards, it is necessary to consider the assumptions underlying public intervention because, despite its subsequent reputation as a 'golden age', British regional policy in the 1960s and 1970s was essentially an experimental activity – or in the less diplomatic words a committee of the House of Commons "empiricism run mad, a game of hit-and-miss"[108] – where few attempts had been made to evaluate the effectiveness and efficiency of regional policy measures.[109] Although official government statements of the aims and operational strategies of British regional policy were held in vague and general terms,[110] the policy profile outlined above in combination with the existing literature would still seem to give fairly clear-cut pointers. With area designation based on inter-regional differences in unemployment, it is hardly surprising that inter-regional equality is the most commonly cited rationale, but especially in the climate of economic growth in the 1960s other notions were also common, i.e. making effective use of idle resources and relieving inflationary pressures by diverting economic activity away from the congested areas in the South of England, or attempting to stimulate 'growth points' in the assisted areas by a combination of regional policy and planning meas-

[105] McCrone 1969 p. 114, *Local Employment Act Annual Reports* 1969ff, *Industry Act Annual Reports* 1973ff.

[106] For overviews of the early years of the Development Commission, see Rogers 1999 Ch. 1-3, Tricker & Martin 1984, cf. Chapman 1975.

[107] Chapman 1975 p. 5.

[108] HC Paper 85, 1973, quoted by McCallum 1979 p. 24.

[109] Ashcroft 1978, Moore *et al.* 1986 Ch. 2, Prestwich & Taylor 1990 pp. 220f.

[110] Ashcroft 1978 pp. 1f, Regional Studies Association 1983 p. 3, Diamond & Spence 1983 Ch. 2, Armstrong & Taylor 1985 pp. 176ff.

ures.[111] The regional problem was generally perceived as a legacy of the concentration in particular localities of traditional, and now declining, industries like coal, steel and textiles, and thus diversification through the introduction of 'modern' industries came to be seen as essential for the purposes of regional policy.[112] By instituting incentives that made it more attractive to expand economic activities in the assisted areas – and especially in the 1960s more difficult to do so in prosperous localities – the general trust of regional policy in Britain would seem to have been a strategy that redistributed economic activity across the face of the country from non-assisted to assisted areas:[113] taking "work to the workers" rather than "workers to the work". This redistributive approach could easily be construed as being based on the direct attraction of jobs, investments and firms to the assisted areas while not paying "sufficient attention to encouraging growth from within the assisted areas",[114] but the policy profile in place from the mid-1960s onwards would, however, seem to suggest that this interpretation is a misrepresentation. With automatic financial incentives available to indigenous as well as incoming firms, redistribution will have taken place both directly via new production capacity being transferred to the assisted areas from outside and indirectly through the cost advantage of indigenous firms *vis-à-vis* competitors in non-assisted areas,[115] and from the perspective of indigenous firms British regional policy will in a sense have been "money for nothing" because the specific combination of policy instruments (financial) and modes of implementation (unselective, reactive, automatic) did not put a premium on e.g. investment in new technology or expansion of R&D activities but left it up to private firms to decide their individual strategies with regard to investment. Sectoral modernisation was therefore probably more likely to be brought about by incoming firms from prosperous regions while the decline of traditional indigenous firms was primarily being financially 'cushioned' via general investment and

---

111 McCrone 1969 p. 21, Ashcroft 1978 p. 2, Regional Studies Association 1983 p. 3, Armstrong & Taylor 1985 pp. 174ff, Parsons 1988 p. 169, Larsson 1988, Hudson 1989a Ch. 3, Scott 1996.

112 Regional Studies Association 1983 pp. 1f, Martin & Hodge 1983a p. 135, cf. McCrone 1969 Ch. 1.

113 Regional Studies Association 1983 p. 3, Diamond & Spence 1983 p. 18, McCallum 1979 p. 36, Gudgin 1995.

114 Armstrong & Taylor 1985 p. 210. Examples of this 'anti-indigenous' interpretation of the national phase of British regional policy include McCallum 1979 pp. 36f, Firn 1980 p. 253 cf. 1982 pp. 9ff, Martin & Hodge 1983b, Diamond & Spence 1983 p. 18, Randall 1985 p. 246, Martin 1989a.

115 A points also made by Armstrong & Taylor 1985 p. 210, cf. Ashcroft 1978, Armstrong & Taylor 1985 Ch. 14, cf. Moore *et al.* 1986.

labour subsidies,[116] and thus the same regional policy measures could in other words produce two rather different, but parallel and complementary, outcomes at the level of individual firms – unless, of course, indigenous firms should decide of their own accord to pursue modernisations strategies on the back of regional subsidies. As argued by Martin & Hodge, regional policy was essentially a form of "spatial Keynesianism":[117] central government provided a framework that should bring about a more equitable distribution of employment opportunities through modification of the price of capital and labour in designated 'problem regions', and the extent to which this resulted in the modernisation of the overall sectoral structure or individual firms within the problem regions depended on the response of private actors to the incentives instituted by public policy.

In the national phase regional policy was part of the cross-party consensus of the postwar settlement on which the welfare state and Keynesian economic policies were based, and the vastly increased levels of resources committed to this purpose throughout the period would certainly seem to reflect a high degree of adherence from both the main political parties to goals of inter-regional equality. Despite the existing consensus, it is, however, also important to note the differences between and within the two parties. With its core voters more likely to be affected by industrial decline and unemployment, Labour had every reason to portray itself as 'the party of the regions', and this impression was strengthened by the left wing of the party arguing that 'stronger' measures were required, e.g. by combining economic development policies with physical planning and new forms of public ownership.[118] The Conservative party, on the other hand, had presided over the 'policy-off' 1950s and wanted to do away with labour subsidies,[119] something which could readily be construed as a lack of commitment to the goal of inter-regional equality. Moreover, the liberal wing of the Conservative party was fundamentally sceptical about what was seen as excessive intervention in the workings of the free market, and while the short-lived flirtation of the Heath government with liberal economic policies in the early 1970s was hastily abandoned in the face of mounting unemployment and industrial action,[120] the (in)famous U-turn of

---

116 For a parallel argument, see Haughton & Lawless 1992 pp. 3ff.

117 1983a p. 133, cf. Martin 1989c, Martin & Tyler 1992 p. 144. Parallel arguments can be found in Dunford *et al.* 1981, Keating 1988a Ch. 6, 1997 p. 18.

118 Parsons 1988 Ch. 7 and p. 183, cf. Chapter 5.

119 The 1970-74 Heath government repeatedly announced its intention to bring forward the discontinuation of the Regional Employment Premium (Wren 1996a pp. 52ff, 74-78).

120 Parsons 1988 pp. 179ff, Young & Lowe 1974 pp. 122-59.

1972 could still be interpreted as a sign of the shallowness of Conservative commitment to goals of inter-regional equality. While spatial Keynesianism was still practised in the mid-1970s, the dominant policy paradigm seemed to have come under increasing pressure due to changes in its political and economic environment:[121] internal tensions within the two major political parties with regard to economic policies highlighted the increasingly conflictual discursive terrain in which regional policy operated, and thus the existing consensus could be severely tested when economic crisis made high unemployment a national phenomenon and thereby questioned the political and economic viability of continuing an extensive programme of inter-regional redistribution of industrial production.

Regional Policy in Scotland before 1975

Before the SDA was established in 1975, regional policy in Scotland was, with one important exception, the application of nation-wide British policies in a Scottish context. On the basis of the outline of the Scottish political environment in Section 4.1, this context could, however, in some cases make a difference, and thus in the following the development of regional policy in Scotland is outlined in order to document its essentially British character and identify developments that were distinctively Scottish.

The original British conception of problem regions had a strong Scottish dimension: Scotland had its own Development Commissioner and large parts of the industrial heartlands of the Clyde Valley (with Glasgow excluded like other urban centres) were designated as Special Areas. Postwar legislation extended regional policy coverage and included Dundee and Inverness in the new Development Areas,[122] and in 1946 the *Clyde Valley Regional Plan*,[123] a major exercise in physical and economic planning initiated by the Scottish Office, provided a general framework for regional development which combined dispersal of population and economic activity away from congested metropolitan areas like Glasgow with the setting up of three so-called New Towns, government-sponsored quangos with responsibility for physical and economic development of a green-field locality.[124] Building of factories remained the most important resource employed by government in the assisted areas during most of the initial phase of British regional policy, and by the mid-1950s industrial units located in Scotland accounted for

---

121 McCallum 1979, Parsons 1988 p. 186, Regional Studies Association 1983 pp. 7ff.

122 McCrone 1969 pp. 92ff, 107ff; Prestwich & Taylor 1990 pp. 116, 121ff.

123 Abercrombie & Matthew 1949.

124 See Smith 1985b, Randall 1985.

35% of all Board of Trade funded floor space, confirming that the region had indeed been a major recipient of regional assistance.[125]

Also in Scotland the national phase involved a significant increase in regional policy activities. The main policy programmes were administered by central government departments, either directly from London or via the regional offices of the Board of Trade (and its successor departments), and as a major development area, Scotland was catered for throughout by separate offices, although the level of discretion exercised by these deconcentrated outposts of central government was limited.[126] With regard to area designation, developments in Scotland reflect the general British pattern of expansion – by the early 1970s Scotland as a whole had become an assisted area[127] – and the extended area designation was accompanied by an increasing flow of financial resources. Available evidence suggests that British regional policies were implemented in much the same way in Scotland as in Wales and the English regions: expenditure per employee in the assisted areas moved along broadly the same lines,[128] and the Scottish share of regional policy expenditure remained remarkably stable in time and across different financial incentives programmes, amounting to between 35 and 40%, probably due to the predominance of automatic forms of assistance.[129] In other policy areas separate Scottish institutions existed such as the Scottish Industrial Estates Management Corporation, building and managing factories under central government guidelines applying throughout Britain, and SICRAS, the Scottish implementing organisation of the Development Commission, but neither of these appears to have been granted significant degrees of discretion by their political sponsors at the British level. Also with regard to factory building and clearing of derelict land the Scottish share of total expenditure amounted to 40 and 30% respectively over the period as a whole,[130] although the position with regard to both activities varied over time with Scotland taking early leads in the 1960s.

While deconcentration of regional policy administration applied to development areas throughout Britain, the Scottish Office continued to provide an alternative focus point in regional development within the region, despite not being directly involved in the administration of

125 Calculated on the basis of McCrone 1969 p. 113.

126 Hogwood 1982 p. 39, Field & Hills 1976 pp. 2-14.

127 Prestwich & Taylor 1990 p. 127, *Department of Employment Gazette* November 1974 pp. 1021ff.

128 Hogwood 1977a Table A.2.

129 Calculated on the basis of Hogwood 1977a Tables A.3, A.4 & A.6.

130 Calculated on the basis of figures given in *Local Employment Act Annual Reports* (1960-72) and *Industry Act Annual Reports* (1972-75).

mainstream policy programmes. As a territorial minister, the Secretary of State for Scotland had a general political responsibility for the economic well-being of the region, but from the 1960s onwards this was increasingly being translated into concrete initiatives: the Scottish Development Department was established in 1962, a separate Scottish Economic Planning Department was formed within the Scottish Office in 1973,[131] and continuing the tradition established by the Clyde Valley Regional Plan, several planning exercises were undertaken in which the Scottish Office played a pivotal role.[132] Moreover, in 1965 the Scottish Office became sponsor of Britain's first major RDA when the then Labour government decided to set up a separate agency, the *Highlands and Islands Development Board* (HIDB) to tackle the problems associated with depopulation and remoteness in an area largely bypassed by industrial development.[133] The remit of the Board included promotion of both economic and social development in this part of Scotland through a wide range of policies, including provision of financial support, industrial property (via the Scottish Industrial Estate Management Corporation), and advisory services. In fact the HIDB Act stipulated that it "could engage in such other activity as the Board may deem expedient for the introduction, operation or development, whether by the Board or other persons, of industrial, commercial and other enterprises in the Highlands and Islands".[134] These very broad powers were, however, subject to the sponsorship of the Scottish Office, providing general guidelines and scrutinising individual activities, but the upward trend in public funding evident in Figure 4.6 suggests that both Conservative and Labour governments must have appreciated the Board and its activities. The rise and rise of the HIDB was clearly a new departure in British regional policy, not because it was a non-departmental body – so were e.g. the 1930s Special Areas Development Commissioners and the postwar industrial estate management corporations – but because it brought together for the first time a wide range of powers within an organisation with responsibility for one cohesive region – unlike the spatially fragmented parts of rural England where the Development Commission operated – and hence would allow for an integrated approach to regional development where different policy instruments were

131 See MacDonald & Redpath 1980, McCrone 1985 p. 203.

132 The most important ones being Scottish Development Department 1963, Scottish Office 1966, and West Central Scotland Plan Team 1974a, cf. McCrone 1969 pp. 125f, 232ff; Firn 1985 pp. 131ff, Wannop 1995 pp. 124ff.

133 On the early years of the HIDB, see Grassie 1983, and Alexander 1981 & 1985. Lee (1995 pp. 155ff) sees the North of Scotland Hydro Electric Board as a specialised forerunner to the HIDB.

134 HIDB Act 1965 Section 7.

combined according to the specific circumstances of particular localities and firms.

Figure 4.6 HIDB funding 1965-75

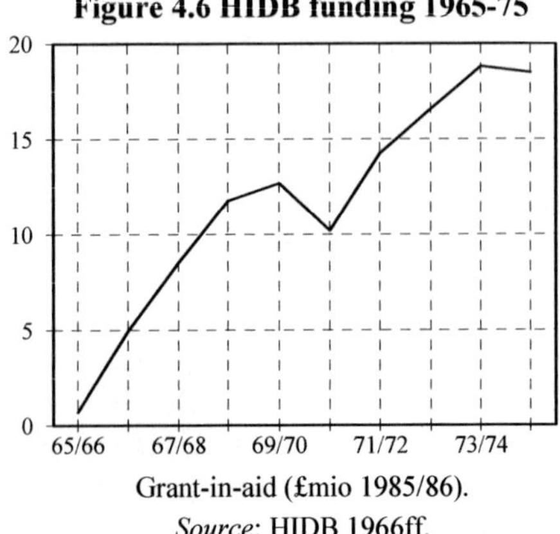

Grant-in-aid (£mio 1985/86).
Source: HIDB 1966ff.

In organisational terms, the Scottish dimension of regional policy in other words consisted of three elements:

- separate implementing organisations with limited discretionary powers with regard to their particular regional policy programmes,
- the presence of the HIDB operating in parts of Scotland, giving preferential treatment to the Highlands and Islands and introducing a new integrated RDA-type approach to regional policy, and
- the Scottish Office as provider of political authority and coordination that differed from the much more fragmented practice in the English regions, served by regional offices of several departments of central government.[135]

Although regional policies operating in Scotland – with the exception of the HIDB – continued to be British, the gradual build-up of organisational capacity under the political authority of the Secretary of State for Scotland would seem to suggest that distinct Scottish solutions to the problems of regional development could be pursued, provided political authority was granted from London.

Like elsewhere in Britain the growth of regional policy in Scotland was essentially experimental, driven by political will rather than knowledge about the effects of the various measures. It was only towards the end of the period that evaluations began to emerge, and one of the earliest ones actually had a specific Scottish dimension, estimating that 70-80,000 additional jobs had been created within the region, mostly by

[135] Hogwood 1996 pp. 11ff, cf. Section 4.1.

incoming firms,[136] but as these efforts did, however, not result in a radical improvement of performance of the regional economy relative to more prosperous parts of Britain, the economic environment made it possible for issues of inter-regional inequality to remain on the political agenda. At the same time the existence of a distinct political system provided a high-profile platform for a range of actors with a vested interest in regional development issues:

- even with its fairly limited autonomy, the Scottish Office provided an organisational focal point and enhanced the authority of regional representatives *vis-à-vis* central government,
- recurring parliamentary debates on economic and industrial development in Scotland institutionalised the involvement of political parties, also when in opposition,
- the existence of a distinct trade union movement ensured that the employment issue was given voice, and
- the activities of the Scottish Council Development & Industry (SCDI) as a tripartite organisation concerned with economic development not only provided high-profile input into the discussion of regional development strategies,[137] but also contributed to the building of consensus between private and public actors within the region, perhaps not about strategies, but certainly about economic development as a legitimate area of public policy.

Until 1975 economic development in Scotland, like other assisted areas, was supported through regional policies that were essentially oriented towards increasing the number of jobs available within the region by instituting a series of options and incentives, designed to make it more attractive for private firms to expand or relocate by lowering the price of capital and labour within the region. Regional policy was administered as a series of separate programmes, often by different government departments or arm's-length bodies, and from the mid-1960s the policy profile had become increasingly dominated by financial incentives distributed in an automatic and non-selective manner. It would be unreasonable to describe these activities in Scotland as constituting a distinctly Scottish regional policy: policies did not originate within the region and were not tailored to addressing its specific weaknesses. Outside the Highlands and Islands, British regional policy was the only show in town, but as the politics of regional policy gradually acquired a distinct Scottish dimension, based on political institutions

---

[136] Moore & Rhodes 1974.

[137] Cf. Chapter 5.

and interest organisations which were unique to the region, this situation might not be a permanent one.

4.4. A Distinct Relational Context

The most important characteristic of the context in which the SDA came to operate was the dual spatial nature of the environment, Scottish and British at the same time, and the specific ways in which these two aspects were linked constituted what can be seen as 'peculiarities' that, while retaining a more or less pronounced British dimension in most social activities, set Scotland apart from Wales and the English regions.

From an economic perspective the creation of the union state in 1707 established a unified market, and the expanding British empire made it possible for Scottish entrepreneurs to operate on a truly international scale. Adaptation to this market resulted in early and rapid industrialisation, revolving around a closely networked cluster in central Scotland centred on shipbuilding and heavy engineering. The limited flexibility of this specialised industrial complex became a major problem after the importance of the Empire and the navy as 'captive customers' started to wane in the postwar decades, and production capacity was gradually reduced, often under public management after crises-ridden private firms had been nationalised. Like in other areas dominated by traditional manufacturing activities, the growth of new industries was relatively slow, and although Scotland was comparatively successful in attracting investment from foreign multinational companies in e.g. electronics, this did not create sufficient jobs to prevent large-scale out-migration and unemployment rates significantly higher than the British average. By the mid-1970s the economic position of Scotland had in other words changed profoundly from being a prosperous core region in the Victorian Empire to a situation characterised by industrial decline, massive redundancies, and an increasing degree of 'external control' of economic activity that integrated Scottish manufacturing in the British economy not through direct market access but as lower-tier parts of inter-regional and international corporate hierarchies. While each of the British regions that prospered industrially during the 19th century had distinct features, their shared economic and social characteristics in terms of poverty, outward migration and unemployment still made it possible to see them as variations of the general theme of 'problem regions'. Especially in the welfare-oriented postwar decades such localities were potential objects of public policy, something which could potentially institute an additional economic link between Scotland and the British level, this time around the transfer of resources in order to promote regional development.

While the union state had created a single market which integrated the economies of Scotland and its southern neighbours, the new British political entity had deliberately maintained distinct forms of governance north of the border, and the expansion of the public sector in the 20th century had resulted in a growing role for the Scottish Office as a territorial department of central government. This in itself was enough to ensure that territorial politics remained firmly on the political agenda because the presence of separate Scottish institutions and public organisations made it possible to interpret social phenomena – even if they also existed south of the border – as distinctly Scottish, but the salience of territorial issues was further enhanced by the strategies routinely pursued by key political actors. Both the longstanding preferential regionalism practised by the unionist parties and the advent of a vocal nationalist party which raised the question of constitutional change helped to ensure that the discursive terrain was heavily influenced by the notion of Scotland as a separate entity.

In the national phase of regional policy, this general political distinctiveness did, however, only make a limited difference. Policies were overwhelmingly British programmes implemented according to guidelines applying also in England and Wales, and thus from the perspective of private firms Scotland was little different from other assisted areas in terms of the regional incentives on offer, although most British policies were actually administered within the region, either by deconcentrated offices of e.g. the DTI, or by implementing organisations in e.g. factory building separate from but parallel to similar bodies in England and Wales. Nonetheless, in some policy areas Scottish initiative was in evidence, notably the HIDB and the major exercises in physical and economic planning sponsored by the Scottish Office, and, most importantly, the politics of regional policy had acquired a distinct Scottish dimension, revolving around the general responsibilities of the Secretary of State for Scotland and a number of organisations with a vested interest in development issues.

All in all the distinctiveness of the Scottish context in which the SDA came to operate from the mid-1970s onwards was in other words tied primarily to civil society and the political sphere through the existence of separate institutions, actors and agendas, although what some have called 'the Scottish political system' was of course firmly inscribed in relations of authority to the British political system through administrative and party-political links. But for several reasons this political distinctiveness had, however, become increasingly salient also from an economic perspective. After World War II the public sector began to play a much more prominent role in the Scottish economy through a wide array of policies ranging from Keynesian macro-economic management via nationalisations to regional development measures, and thus the question of whether public economic intervention should have a specific Scottish dimension could readily be posed. Moreover, the perceived severity of the regional industrial malaise made economic issues central to the positioning strategies of both unionist and nationalist parties, and thus economic and territorial policies became increasingly intertwined. By the mid-1970s the Scottish economy had in a very literal sense become a 'political economy', not just because of the pervasiveness of – and consensus about the desirability of – public intervention in the region but also because of the influence, directly or indirectly, of territorial politics revolving around the position of Scotland *vis-à-vis* the rest of Britain.

Part II

Corporate Developments

CHAPTER 5

A Regional Development Agency for Scotland

Before beginning to reconstruct the history of the SDA, it is worth recalling that the existing literature entails rather different interpretations, right from the deceptively simple question of how policies and strategies developed to the more complex issues of the origins of change and continuity. As a reminder of major themes in the development of the SDA, Table 5.1 summarises the arguments of the main traditions and indicates the way they position themselves in relation to interpretations of Thatcherism and the RDA approach to regional policy. Apart from highlighting the strong internal coherence of the opposing stories and their comprehensive nature – both of them making claims relating to a large part of the dimensions entailed in the analytical framework developed in Chapter 4 – the table does, however, also serve to underline their limitations. Firstly, in a multi-functional agency the development of various areas of activity could be uneven, either because some policy instruments were more controversial than others or the private actors targeted were more or less inclined to adapt their strategies in response to the incentives provided by different Agency policies. Secondly, the degree of change brought about by the Thatcher governments in the 1980s may have been overstated, in British regional policy as a whole or specifically with regard to the strategies pursued by Conservative ministers at the Scottish Office. Thirdly, party-political interests might prompt continuity rather than change and thereby undermine the patterns of causation implied by the two interpretations: both the dominant unionist parties had a long history of engaging in preferential regionalist measures that entailed positive discrimination in favour of Scotland, and thus territorial politics might be in conflict with e.g. political urges driven by the ideal nature of the relationship between public and private actors. In short, although the two interpretative schemes are useful sources of inspiration, it is also important to look beyond them, and as every story has a beginning which may have consequences in the longer term, the empirical analysis starts by tracing the origins of the SDA.

Table 5.1 The development of the SDA – Contrasting perspectives

External revolution	*Internal evolution*
The setting up of the SDA was a party-political move designed to increase the role of public intervention in the Scottish economy and at the same time counter the surge in voter support for Scottish nationalism	The SDA was established as a new approach to regional policy in Scotland because existing central government measures had failed to address the specific problems of the Scottish economy
Strategies and policies changed radically from the interventionist 1970s to the market-oriented 1980s	Strategies and policies evolved gradually with clear continuities between the 1970s and 1980s, also with regard to the role of commercial criteria in public intervention
The capacity of the SDA to influence private firms was eroded by these changes	The evolution of the Agency's policy profile did not significantly erode its capacity to influence private economic actors
Changes were parallel to those affecting central government regional policies which were drastically reduced in the 1980s	The history of the SDA is thus in stark contrast to central government regional policies which were drastically reduced in the 1980s
The parallel development between SDA and central government policies was possible because the Agency enjoyed only a limited degree of autonomy *vis-à-vis* its political sponsors	These divergent paths demonstrate the relatively high degree of autonomy enjoyed by the Agency *vis-à-vis* its political sponsors
The case of the SDA is a good example of a field of public policy being radically transformed by the Thatcher government	The case of the SDA provides an example of an area of public policy in which the advent of Thatcherism had only a limited impact
The case of the SDA demonstrates that as implementing organisations RDAs are no better at sheltering regional policies from political pressures than mainstream government departments	The case of the SDA constitutes a good example of the advantages of the RDA approach to regional policy in terms of sheltering implementing bodies from political pressures

Based on: Chapter 2 above.

Origins Contested

In 1975 the Labour government proposed legislation to set up a new statutory body to be called the Scottish Development Agency, charged with furthering economic development and environmental improvement in Scotland, funded by public money and with a board appointed by the Secretary of State for Scotland but situated outwith the departmental organisation of the Scottish Office.

What persuaded a British government to give Scotland preferential treatment compared to e.g. the equally crisis-ridden north of England? What made the Labour government opt for an institutional set-up that was a radical departure from the regional policies hitherto pursued and that, at least in theory, would diminish the direct influence of ministers and parliament by transferring responsibility for policy development and implementation to a semi-autonomous body? And how was this initiative interpreted by other actors in Scottish politics? Providing answers to questions about the origins and reception of the proposal for an RDA for lowland Scotland is important because it helps to set the scene for the ensuing analysis of the Agency and its activities by identifying the initial roles of the various actors and establish discursive patterns that are likely to have influenced subsequent political developments.

Within the existing academic literature and, indeed, among political actors and policy practitioners, the origins of the SDA has been interpreted in a number of alternative ways that, if substantiated, would seem to entail different expectations about the future development of the new body. Some have seen the organisation as an example of *policy innovation* in the sense that the organisation was created to supplement or replace traditional forms of regional policy because, unlike central government financial subsidies, an RDA would be able to address the specific problems underlying the slow growth of the Scottish economy by working on a long-term basis with individual firms, drawing on a wide range of policy instruments, and operating at arm's-length outside day-to-day political pressures. This perspective on the origins of the SDA, focusing on the perceived functional efficiency of an innovative approach to regional policy, is common amongst policy-makers at the Scottish Office and the Agency and others associated with the 'internal-evolution' perspective on the subsequent development of the Agency.[1] In contrast to this others have argued that the SDA was created as a

---

1 Randall 1980 pp. 114f, Firn 1982, Rich 1983, Wannop 1984 & 1995, McCrone & Randall 1985, Moore & Booth 1986b, Hart & Harrison 1990 pp. 198ff, Hood 1991a, Gavin McCrone 20.6.90, Sir Kenneth Alexander 25.7.90, Craig Campbell 16.7.90, John Firn 15.6.90, Lewis Robertson 20.7.90 (curiously, Wannop is quoted with approval by external-revolutionists Danson *et al.* 1988).

vehicle for interventionism, an integrated part of what became the 'industrial strategy' of the Labour government in the 1970s, designed to increase public sector influence within the Scottish economy and using direct investments in private firms as its main policy instrument. Explaining the origins of the Agency in terms of its expected impact on the relationship between public and private actors is particularly common among Conservative commentators (and their opponents on the left) and some texts adopting an 'external-revolution' perspective in which the change of government in 1979 is seen as a qualitative turning point in the history of the organisation.[2] Finally, some have emphasised the importance of *territorial politics*, essentially seeing the SDA as the economic element in the Labour party's attempt to fend off the electoral threat of the SNP, a much-needed high-profile measure that could be brought into being much faster than the concurrent proposal for setting up a devolved assembly in Scotland. While this perspective can be found as a secondary element also in accounts falling within the two traditions previously outlined, it is particularly common among SNP observers and academic analysts for whom the main analytical interest is governance rather than regional policy.[3]

This chapter will pursue the origins of the SDA in two steps, both of which are primarily based on a comprehensive analysis of the documentary sources available. First the ideational and political origins of the proposal for a Scottish regional development agency will be examined by investigating administrative precedence within government policy as well as demands and blueprints put forward by interest organisations and political parties in Scotland on the other (Section 5.1). This is followed by an analysis of the political debate on the legislation proposed by the Labour government (Section 5.2), and finally the possible implications of the way in which the Agency came into existence will be assessed (Section 5.3).

2 Danson 1980, Parsons 1988 p. 183, George Mathewson 9.7.90, George Robertson 30.10.90, Helen Liddell 21.6.90.

3 E.g. Cooke 1980, Hogwood 1982, Lochhead 1983 p. 27, Rich 1983 p. 272, Keating & Boyle 1986 Ch. 1, Moore 1988 pp. 3f, Keating 1988b pp. 47f, Barnekov *et al.* 1989 p. 198, Midwinter *et al.* 1991 pp. 63f, Middlemas 1991 p. 190, Wannop 1995 p. 39.

5.1. Ideational and Political Origins

Government Planning and Economic Development

The task of rebuilding Britain after World War II created a fruitful environment for exercises in public planning, and the notion of bringing physical and economic powers to bear on development problems in a coordinated manner became popular, also because the government-led war efforts had functioned as a large-scale demonstration project.[4] In Scotland the Secretary of State, Tom Johnston, initiated a series of regional planning exercises of which the *Clyde Valley Regional Plan*,[5] covering the Greater Glasgow conurbation, was the most ambitious one.[6] The report stressed the need for industrial diversification to counterbalance the decline of existing traditional Scottish industries and the need to steer economic activity into both areas of high unemployment and to new towns erected to alleviate congestion in the old industrial centre of Glasgow, and in order to achieve these aims a new 'Regional Authority' was proposed with responsibility for the establishment of New Towns, environmental improvement, and location of industry.[7] Despite these extensive powers, the new authority was envisaged as working closely with local authorities and relying on the continued commitment at the UK level to redistribution of industry, although it was also stressed that "Scottish enterprise should play a greater part in the development of new industries in Scotland and not leave the field so much to southern firms."[8] While the Clyde Valley Regional Plan was clearly innovative as a study of a congested and crisis-ridden regional economy, it is, however, also striking that despite the intention of integrating physical planning and economic development within one organisation, the document remained skewed towards the former: very detailed proposals were put forward with regard to ways of dealing with urban congestion, while concrete industrial development measures were largely confined to provision of 'hard infrastructure' in the form of New Towns and industrial estates.[9] Given this rather unbalanced profile, it is hardly surprising that the direct impact of the plan was primarily to provide further impetus to the setting up of New Towns in order to alleviate urban congestion in the Clyde Valley – initially vehemently

---

4 Smith 1985b pp. 18ff, Wannop 1995 pp. 6ff.

5 Abercrombie & Matthew 1949.

6 Cf. Wannop & Smith 1985 p. 241.

7 Abercrombie & Matthew 1949 p. 344.

8 Abercrombie & Matthew 1949 p. 112.

9 Abercrombie & Matthew 1949 p. 111, cf. Firn 1985 pp. 115f.

resisted by local government in the City of Glasgow[10] – while the limited industrial development momentum quickly disappeared.[11]

The second *locus classicus* within the Scottish economic planning tradition is the *Inquiry into the Scottish Economy 1960-1961*,[12] known as the Toothill report after the Chair of the SCDI Committee that oversaw its production for the Scottish Office, and approaching the regional problem from a predominantly economic perspective, perhaps as a response to the difficulties experienced in Scotland by the Conservative government in the 1959 general election.[13] The analysis singled out the over-dependence on declining traditional industries as the major weakness and saw sectoral diversification as the solution, especially through indigenous enterprise and "the creation of regional complexes of science-based firms", while the Scottish Office rather than a freestanding 'development corporation' was identified as the organisation which could best promote development along these lines.[14] From a regional policy perspective the new, and oft-quoted, ideas in the Toothill report were its dual emphasis on promotion of indigenous growth and the importance of spatial concentration of economic activities, but even though Toothill was clearly much more industrial than its predecessor, the emphasis on 'industrial complexes' could still be construed as an argument for a planning-led approach building on the experience of the New Town programme, i.e. a strategy which would not necessarily address the central issues confronting indigenous firms. And by designating the Scottish Office as its preferred vehicle of implementation, the report effectively reinforced the centrality of physical planning because this was an area in which the government department had already developed considerable expertise, not least in connection with the ongoing programme of New Town construction in Scotland.

The political sponsors of the Toothill report clearly agreed with many of its recommendations, because shortly after the role of physical planning within the Scottish Office was expanded through the creation of the Scottish Development Department and an official response was produced through the 1963 government White Paper *Central Scotland. A Programme for Development and Growth.*[15] Its central idea was to designate a limited number of localities as 'growth areas' where infra-

10 Randall 1980 pp. 107f, Smith 1985b pp. 26ff, Keating & Boyle 1986 pp. 31f, Wannop 1995 pp. 114ff.

11 Cf. Parson 1988 Ch. 3-4.

12 Toothill *et al.* 1961.

13 Cf. Hutchison 1999 p. 104.

14 Toothill *et al.* 1961 pp. 183-91 (quote p. 37).

15 Scottish Development Department 1963, cf. McCrone 1969 Ch. 5, Pottinger 1979 p. 158, Keating & Boyle 1986 pp. 31ff, Parsons 1988 pp. 114ff.

structure investment could be concentrated and special efforts would be made to eradicate environmental scars left behind by declining traditional industries.[16] In terms of physical planning the White Paper's high-profile insistence on giving priority to selected localities on the basis of their future potential rather than their present predicament did of course run counter to the traditional focus of regional policy on 'problem areas',[17] but in practice most of the growth areas designated were related to New Town developments, and thus the suspicion that the White Paper had primarily drawn together existing initiatives was bound to emerge.[18] Moreover, much like the Clyde Valley Regional Plan, the non-physical aspects of economic development were simply the application of existing UK regional subsidies, and thus the most important innovation in the 1963 White Paper was its symbolic endorsement of the quest for growth and sectoral diversification.

Shortly after, following the election of the first Labour government of Harold Wilson, an innovative approach to regional policy was introduced in the least industrialised parts of Scotland in the shape of the HIDB which, despite an initial penchant for attraction of large-scale industrial projects from outwith the region, relied primarily on support for local firms and did not become involved in physical planning.[19] While at the political level the Labour government initially argued that the remote and sparsely populated north-west of Scotland was an exception that deserved preferential treatment,[20] its existence is likely to have made the notion of implementing regional policy through an arm's-length body more palatable for civil service administrators,[21] and, possibly, in the longer term also for ministers at the Scottish Office.[22]

When work began on the last major pre-SDA planning exercise, the *West Central Scotland Plan*,[23] it was undertaken by staff seconded from

16 Scottish Development Department 1963 Ch. 1.

17 See Labour Party 1964 p. 88, *Glasgow Chamber of Commerce Journal* February 1964 p. 81.

18 See STUC 1964 p. 24.

19 Cf. Danson *et al.* 1993 p. 167.

20 Secretary of State Willy Ross in a letter dated 16.9.66 dismissing demands for lowland RDAs (reprinted in STUC 1967 pp. 8f), cf. *HCPD* vol. 708 col. 1095.

21 E.g. according to Sir Kenneth Alexander 25.7.90, Lewis Robertson 20.7.90, Sir Robin Duthie 12.7.90. Gavin McCrone, who joined the Scottish Office as a top civil servant in 1970, had in his previous life as an academic written appreciatively about the Italian experience with arm's-length 'state holding companies' as part of regional policy (1969 pp. 204f) and suggested that from a British perspective this was "a field which government would do well to examine more closely" (p. 206).

22 Cf. Gavin McCrone 20.6.90.

23 West Central Scotland Plan Team 1974a.

the Scottish Office,[24] but a large number of especially public but also private actors became involved in a steering committee which appointed external academic consultants to undertake the preparation of a sprawling set of reports.[25] Like its predecessors it maintained a dual focus on physical planning and economic development, and although its main emphasis was on the latter, it also proposed to discontinue preparations for the creation of a 5th New Town at Stonehouse and instead intensify urban renewal in Glasgow. With regard to economic development the report focused on what was seen as the underlying reasons for the poor performance of indigenous firms and industries, i.e. deficient managerial skills and poor industrial relations, something which would require new policy instruments in the form of subsidised advisory services and proactive financial support for new small enterprises and high-risk ventures, supplemented by measures to facilitate industrial reorganisation among existing firms. Unlike the Toothill report which simply stated its preference for the Scottish Office as the vehicle for implementation, the 1974 report discussed several options at some length before rejecting a central government department in favour of a regional development agency – Strathclyde Economic Development Corporation (SEDCOR) – on the grounds that general civil service qualifications were not sufficiently industrial for purposes like providing management advice and assessing the commercial prospects of new ventures for e.g. mainstream regional policy grants.[26]

Looking back at the series of planning exercises initiated by the Scottish Office, it is evident that the department showed a sustained interest in economic development and that the ensuing proposals, at least in their intention, went beyond the traditional central government approach of redistributing industrial activity by means of financial incentives and location control: the ambition of combining physical planning and economic development was conspicuous from the Clyde Valley Regional Plan through to the 1970s, and the strategic emphasis would seem to shift a more explicit focus on the weaknesses of indigenous firms within the regional economy, although neither Toothill nor the West Central Scotland Plan had outright dismissed the potential contribution of incoming firms or inward investment.[27] While SEDCOR, in many ways a low-land version of the HIDB complete with grant-giving powers, was clearly the most detailed blueprint for a new form of bottom-up regional policy, its chances of being implemented were at the

[24] Keating 1988b p. 26, Wannop 1995 pp. 124ff.

[25] Cameron & Mulvey (eds.) 1973, West Central Scotland Plan Team 1974b vol. 1-5, cf. Randall 1980 p. 114, Wannop 1995 pp. 124ff.

[26] West Central Scotland Plan Team 1974a pp. 62-72.

[27] Toothill *et al.* 1961 pp. 183ff, West Central Scotland Plan Team 1974a pp. 64-68.

very least open to doubt: several sources would seem to indicate that the involvement of the Scottish Office and other sponsors of the planning exercise had been fairly limited,[28] it was launched at politically turbulent point in time with a minority Labour government and impending local government reform,[29] it included politically sensitive assumptions about management failure and poor industrial relations, and it proposed an organisational break with civil service implementation of regional policy. In short, the possibility of the report being quietly shelved would seem to be a very real one indeed.

Organised Interests and Regional Institutional Engineering

For decades regional problems had primarily been identified with high levels of unemployment, and it is therefore hardly surprising that Scottish Trades Union Congress (STUC) campaigned actively for the setting up of various development bodies from the 1960s onwards,[30] and it is possible to discern three main phases in the development of official trade unionist thinking on how regional policy in Scotland could be strengthened through institutional engineering. Beginning in 1963, the main annual congress resolution on economic matters urged government to reconvene the Scottish Economic Conference, a tripartite body fallen into disuse under the Conservatives, and to do so specifically with a view to consider setting up a Scottish Development Authority "to take effective charge of the Scottish economy" and "recognise the peculiarly distinctive nature of the Scottish economy",[31] something which was dismissed by the Conservative Secretary for State of Scotland. After the HIDB had been created by the new Labour government, the main STUC demand was that three regional boards should be created in lowland Scotland so that the whole country would be covered,[32] but despite persistent campaigning by the STUC the Highlands and Islands were still seen as a unique case by the Labour Secretary of State for Scotland.[33] In the early 1970s under the Conservative Heath government a

---

28 This is suggested both by study director Urlan Wannop (1995 p. 124), and by the comment of a Scottish Office official that the report was "the work of academics" (*Glasgow Herald* 30.4.74).

29 Legislation to create a new two-tier local government structure with extensive regional authorities had been passed in 1973 and were due to take effect in 1975 (Kellas 1989 p. 163).

30 Midwinter *et al.* (1991 p. 63) claim that STUC argued for the setting up of a Special Development Agency for Scotland already in the 1930s.

31 Resolution reprinted and oral comments summarised in STUC 1963 pp. 195ff, reiterated the following year (STUC 1965 p. 4).

32 STUC 1967 p. 5, 1969 p. 5, 1970 p. 334.

33 STUC 1967 pp. 8f.

new round of demands for economic development institutions was launched, maintaining the proposal for three lowland boards but now also campaigning for two additional bodies: a "high-powered Development Agency",[34] originally described as the equivalent of a new town corporation at the Scottish level but later given a broader remit,[35] and a "state holding company" that could "supply capital to industry" and "ensure effective control by the public of industries benefiting from state funds".[36] Despite the limited detail given, it is still possible to identify a shift of priorities in the thinking of STUC: while the first proposals seemed to focus mainly on physical planning (and by implication perhaps on attraction of inward investment), by the mid-1970s the emphasis was now on tackling the financial problems of indigenous firms and redrawing the border between the public and private sectors in order to ensure investment in job creation and industrial modernisation.

In sharp contrast to the well-documented and energetic trade union activism with regard to regional policy innovation, the role of employers' associations seems to have been mainly a reactive one, possibly because Scottish-wide organisations found it difficult to promote new initiatives that could benefit some members more than others.[37] CBIS commented on government initiatives rather than proposing new ventures themselves, and proactivity seems to have been limited to the local level where e.g. Chambers of Commerce were involved in promotion of particular areas to potential foreign investors,[38] and thus the main private sector input into regional policy innovation in Scotland will therefore have been via the work of the tripartite SCDI. This organisation had a unique position in Scottish regional policy because whilst it was essentially a lobbying body on behalf of its members in the public and private sectors, at the same time it also became increasingly involved in policy implementation when the Scottish Office began to financially support its ongoing programme of promotion of Scotland as a location for inward investment,[39] and this dual role could well be part of the explanation for the diversity of ideas with which it contributed to the ongoing debate on regional policy innovation. In November 1973 an extensive report entitled *A Future for Scotland* was published,[40] reviewing the regional economy in the light of the prospects of political devolution and rec-

34 STUC 1971 p. 20.

35 E.g. STUC 1973 p. 277.

36 STUC 1973 p. 277, cf. Expenditure Committee 1972 p. 79, STUC 1974 p. 339.

37 Cf. Sidenius 1984.

38 E.g. the *Scotland West* promotional initiative of the Glasgow Chamber of Commerce (*Glasgow Chamber of Commerce Journal* October, November and December 1970).

39 SCDI 1971 p. 4, cf. *HCPD* 21.12.71 vol. 828 cols. 1564f.

40 SCDI 1973.

ommending that in addition to "an established body" (presumably the SCDI) assisting government as a provider of advisory services to industry,[41] a Scottish Special Development Corporation should be created, funded primarily from oil and gas royalties and acting as a "mobile New Town Development Corporation", with responsibility for major infrastructure developments and having "the powers of a holding company and be able to participate in industrial developments".[42] Seven months later a newly-formed SCDI research subsidiary published a report entitled *Economic Development and Devolution*, introduced as "a more detailed elaboration of some of the themes contained in *A Future for Scotland*" but also carefully placed at arm's-length as "the independent work and conclusions of the Research Institute Staff".[43] Stressing the need for a Scottish Development Corporation to employ "people with commercial and industrial rather than civil service experience in what are essentially commercial operations",[44] the main functions of the proposed body were envisaged as provision of finance for industrial growth and new ventures by acting as a risk-taking merchant bank, provision of advice to small firms, factory building, and distribution of regional grant aid as an agent of the Scottish Office.[45] While it could be argued that the two SCDI proposals were not necessarily incompatible, their strategic orientation would, however, still seem to differ: the (preferred) Scottish Special Development Corporation was largely oriented towards major physical projects such as airports or oil-related installations, while the much more firm-oriented Scottish Development Corporation was merely presented as a worthwhile contribution to the ongoing policy debate in which the SCDI would not necessarily invest much political capital.

Being Party to Scottish Economic Development

In the period under review, the SNP was a regional opposition party in the unitary British political system, and its policy development was therefore driven by the efforts of individual members and the limited resources available at party head quarters. From the mid-1960s onwards the SNP gradually developed a proposal for an Industrial Development Corporation,[46] and according to the manifesto for the 1966 general

41 SCDI 1973 p. 21.

42 SCDI 1973 p. 143.

43 SCRI 1974 (preface by the chairman).

44 SCRI 1974 p. 35.

45 SCRI 1974 pp. 37-45.

46 First mentioned in SNP documents in April 1965, it was originally associated primarily with cooperative forms of enterprise (SNP 1965a, 1965b, cf. 1965c), before oil revenues started to dominated SNP financial thinking, the new body was to be

election its purpose would be "to stimulate and assist new and developing Scottish-controlled industry" through provision of risk capital, technical and professional advice, and to instigate market research in order to "ascertain the most appropriate spheres of industrial development".[47] Although some organisational aspects varied over the years, the general shape and profile of the proposed body was maintained till the mid-1970s,[48] i.e. a strategy focusing on indigenous firms implemented through an arm's-length body covering the entire country and having a range of financial and informational policy instruments at its disposal.

The other non-governing party in Scottish politics, the Liberals, had most of their MPs elected in rural constituencies,[49] and this probably explains the party's proposal for a development board to operate in the Scottish Borders with powers similar to the organisation already operating in the Highlands and Islands.[50] By the mid-1970s the geographical horizon of the Liberal party had, however, expanded significantly as they argued for a "proper Scottish development authority" financed from oil revenues,[51] but this was no more than a brief oral exchange in the House of Commons and probably best serves to illustrate the extent to which a link between oil revenues and intensified economic development efforts in Scotland was being made at this point in time, not just by the SNP but also by other parties, and it was only after Labour had proposed the setting up of the SDA that the Liberals published a proposal for a Scottish Development Corporation, funded by oil revenues and engaged in infrastructure development and provision of "finance for new and commercial ventures in Scotland".[52]

As a regular party of government in Britain, the Conservatives were much better placed to take additional initiatives to promote economic development in Scotland, but until the mid-1970s this did not result in creation of new development bodies. Apart from ensuring that mainstream British-level policies were duly applied in the region,[53] the 'special measures' brought about by Conservative governments were

---

funded through the issue of a special government savings bond (e.g. 1966a p. 4, cf. 1966b, 1966c), and at one point in time the corporation was elevated to a Ministry of Development and Industry (SNP 1969 pp. 33f).

47 SNP 1966a pp. 3f.

48 Compare e.g. SNP 1968a and 1973a.

49 Hutchison 1999 pp. 117ff, Midwinter *et al.* 1991 p. 37.

50 Borders Development (Scotland) Bill 1969, *HCPD* 22.1.69 vol. 776 cols. 486ff.

51 *HCPD* 21.2.73 vol. 851 col. 445.

52 *Glasgow Herald* 3.10.74.

53 Cf. Section 4.3.

primarily what the STUC would have called 'direction of industry',[54] i.e. leaning heavily on private firms to make them invest in Scotland such as the steel mill at Ravenscraig, the pulp factory at Fort William, and car production at Bathgate and Linwood.[55] Tory politicians had been sceptical about some of the HIDB's allegedly 'marxist' powers,[56] and demands for additional development bodies were routinely dismissed because of worries about having 'too many organisations',[57] the weak political accountability of arm's-length agencies,[58] or – in response to STUC demands for a Scottish Development Authority in 1973 – because the Conservative government preferred to rely on the working of the markets in the context of existing regional policies.[59] On the background of this, the ability of the party to quickly embrace the notion of a semi-autonomous development body in 1973-74 would seem to be a good example of the pragmatism of traditional Toryism: little more than six months after the STUC proposal had been turned down, the official Conservative campaign guide tacitly endorsed the SCDI's proposal for a New-Town-style Scottish Special Development Corporation,[60] and in May 1974 the Conservative opposition leader Edward Heath announced the intention of establishing a Scottish Development Fund to tackle both the problems created by the expansion of oil-related economic activities and the decline of traditional industries, a commitment included in the party's manifesto for the October 1974 general elections.[61]

Having established the HIDB in the mid-1970s, one could perhaps expect the Labour party to be a major source of innovation in Scottish regional policy, but apart from allowing for the unique circumstances of the Highlands and Islands, the philosophy of the party was neatly summed up by the conclusion of the lengthy 1958 policy document *Let Scotland Prosper* which stated that "it is only through socialist planning on a United Kingdom scale [...] that Scotland's many special problems

54 Although the Conservative President of the Board of Trade preferred the expression 'steering', his answer to STUC's demand for more 'direction' clearly appears to have been quietly appreciated (minutes of meeting 31.8.62 reprinted in STUC 1963 p. 15).

55 STUC 1961 p. 15, Conservative Central Office 1964 p. 362, cf. Kellas 1989 p. 47, Lee 194 p. 184.

56 Cf. *HCPD* 16.3.65 vol. 708 col. 1089, 11.2.71 vol. 811 col. 791.

57 Letter from Scottish Office Minister George Younger to STUC reported in the papers for the meeting of the STUC Economic Committee 30.3.71.

58 Letter from government minister Christopher Chataway reproduced in the papers for the meeting of the STUC Economic Committee 2.8.72.

59 STUC 1974 p. 37.

60 Conservative Central Office 1974 (February) p. 463.

61 Conservative Party 1974 p. 330, Conservative Central Office 1974 (October) pp. 91f, cf. *Glasgow Herald* 20.5.74, Harvie 1994 p. 189.

can be tackled".[62] This unitarian approach to economic policy was regularly invoked throughout the 1960s where emphasis was given to extending traditional redistributive measures such as investment grants and location control, backed up by the occasional threat of direct public action "if private concerns are not prepared to go where they are needed".[63] The Labour manifesto for the 1966 general election proclaimed that "we see the economic well-being of Great Britain as indivisible",[64] but in the beginning of the 1970s when Labour was in opposition, increased priority was clearly given to support for the indigenous sector of the economy,[65] and a 1972 policy document – issued in the wake of the Conservative government's sudden re-embracing of regional and industrial policy – outlined the new approach of the party.[66] Although further strengthening of the traditional instruments of regional policy would be required, the document proposed the setting up of a State Holding Company, because underinvestment was seen as a major reason for the economic weakness of Britain in general and the regional economy in particular, and hence the new organisation would inject capital into industries while providing "accountability through equity holdings in exchange for the large amounts of public money now being given to private industry".[67] One year later the proposed organisation had not only acquired a name, the *National Enterprises Board*,[68] but also a good deal more detail regarding its rationale and function. Starting again from the assumption that direct influence through public ownership of shares was needed to ensure "some relocation of existing establishments which private enterprise is unlikely to accept",[69] a two-tier structure – a British organisation with a Scottish subsidiary – was envisaged, and the role of the latter was specified as[70]

- acquisition and relocation to Scotland of growing English company,
- restructuring industries suffering from "badly coordinated production units, inadequate investment or poor management" in order to preserve employment,
- proactively promoting new indigenous enterprises where markets exist, and

62 Labour Party 1958 p. 81.

63 Labour Party 1962 p. 8.

64 Labour Party 1966 p. 8.

65 E.g. *HCPD* 22.12.71 vol. 828 cols. 1555ff.

66 Labour Party Scottish Council 1972.

67 Labour Party Scottish Council 1972 pp. 6f.

68 Also the title of Labour Party Scottish Council 1973.

69 Labour Party Scottish Council 1973 p. 6.

70 Labour Party Scottish Council 1973 pp. 13f.

- acting as a "development bank [...] heavily engaged in the task of promoting small and medium sized Scottish enterprises", a task which would also include advice on management and technical matters.

In addition to this, the administration of discretionary investment grants in Scotland should also be transferred from central government to the Scottish subsidiary of the new organisation in order to avoid duplication.

The new investment-based Labour approach to regional policy which evolved in the early-1970s would seem to reflect a number of, sometimes conflicting, considerations. *Firstly*, the need for taking specific Scottish needs into account was frequently reiterated, but still most attention was given to UK-level measures, i.e. the ways in which the new British-level organisation would be able to relocate the economic activities of private firms by using the mechanism of public ownership. *Secondly*, while an important argument for the need for public ownership was to establish a relation of 'accountability' when private firms receive large amounts of public money, at the same time "the new body will have to be [...] substantially independent of day to day political and administrative interference".[71] On this reading the new approach of the Labour party could be interpreted as an attempt to maintain 'economic unitarianism' while at the same time being able to claim that specific Scottish problems could be addressed professionally within the new set-up, and thus compared to the party's position in the 1960s its two most innovative elements were the dual emphasis on public ownership as instrument of regional policy and the need for a technocratic body at arm's-length of government as the organisational framework for policy implementation.

Political and Ideational Origins: A Reassessment

In order to be able to assess the ideational and political origins of the SDA – why such a proposal at this particular point in time – it is necessary to take a closer look at both the sequence of events surrounding its conception and the policy profile proposed for the new organisation.

After the Conservatives had lost the February 1974 general election a minority Labour government came to power, but there was no immediate indication that an arm's-length economic development body would be established. The setting up of a Scottish RDA had not been foreshadowed in the Labour manifesto for the February 1974 general elections,[72] and in early April the returning Scottish Secretary of State Willy Ross appeared to be unaware of the West Central Scotland Plan which had

[71] Labour Party Scottish Council 1973 p. 20.

[72] Reprinted in Craig 1990 pp. 186ff.

been prepared while Labour was in opposition, but still expressed his private sympathy for the idea of "a new body in Scotland with finance".[73] Towards the end of May the editor of the *Glasgow Herald* could still only speculate about the possibility of Labour "stealing" the Conservative proposal for a Scottish Development Fund,[74] and then suddenly on July 11th 1974 the White Paper on *UK Offshore Oil and Gas Policy* announced that "in order to strengthen, in an effective and lasting way, the instruments available for promoting the development of Scotland's economy and undertaking major environmental projects" it had decided to legislate for the setting up of a Scottish Development Agency, financed by the UK Treasury and responsible to the Secretary of State for Scotland.[75] This commitment was reiterated in Labour's manifesto for the October 1974 general election which also stressed that this extra funding would "reflect revenue from offshore oil",[76] and after having won a narrow majority the Labour government produced a more detailed consultative document which outlined the functions proposed for the new arm's-length development body and invited comments from actors with a vested interest in the SDA's various areas of responsibility, i.e. STUC, CBIS, SCDI and the recently created local authorities.[77]

In circumstances of persistently high levels of unemployment and an extensive heritage of environmental dereliction – phenomena that for years had been seen as legitimate objects of public policy – it would have been difficult to argue that the problems targeted by the proposed functions were not worthy of attention, and, indeed, when the Labour government first announced its plans they were welcomed in principle by interest organisations from CBIS to STUC.[78] Still, the time and place chosen by the Labour government to announce the proposal meant that motives other than the functional one of finding effective ways to deal with social problems could easily be inferred. The February 1974 election had not only produced a minority Labour government but also turned the SNP into a major political force with 22% of the Scottish vote and 7 MPs following their evocative "It's Scotland's oil!" campaign. As the next general election was likely to follow shortly and the Conservatives had launched their Scottish Development Fund in a bid to win back voters attracted by the SNP's blend of nationalism and economic pragmatism, the need for Labour to take a high-profile initiative was obvi-

73 Report of meeting between STUC and Willy Ross 5.4.74 in the papers for the STUC Economic Committee meeting 28.5.74.

74 *Glasgow Herald* editorial 20.5.74.

75 Department of Energy 1974 p. 9.

76 Craig 1990 p. 246.

77 SEPD 1975 p. 1.

78 See *Glasgow Herald* 12.7.74.

ous.[79] Not having made any reference to plans for a regional development agency for the whole of Scotland in its February manifesto and then announcing a proposal with very limited detail in a white paper on oil in July was a sequence of events that could easily be construed as an attempt to gain party political advantage by launching a highly visible form of preferential regionalism that would demonstrate the advantages of remaining within the union state with Labour in power. In short, being launched in a way that made it look like an anti-SNP (and anti-Conservative) measure ensured that the SDA, whatever its functional merits, would be seen as part and parcel of Scottish party-political competition and, indeed, British territorial politics from the very beginning.

The way in which the new body would be interpreted would, however, not only depend on the circumstances under which the proposal emerged, but also proposed policy profile because this created expectations about its future activities and made it possible to make assumptions about the 'ideational fatherhood' of the new organisation by linking it with the many proposals for a semi-autonomous development organisation which had been put forward since the mid-1960s. According to the consultative document circulated by the Scottish Office in January 1975, the SDA would be a statutory arm's-length body charged with promoting industrial and environmental development, appointed by and acting "in accordance with policy determined by" the Secretary of State for Scotland. In terms of economic development this would include

- "undertaking joint commercial ventures with private companies and, where necessary, launching new industrial ventures to create employment",
- provision of managerial and financial advice to especially small firms,
- "generally promoting growth and modernisation in Scottish industry, including opportunities arising from North Sea oil",
- promotion of Scotland to foreign investors "in concert with existing development bodies", and
- providing and managing industrial property.[80]

The environmental functions were described in much greater detail and included on the one hand taking over from the Scottish Office the

[79] This was of course not lost on contemporary observers, see e.g. the *Glasgow Herald* editorial 20.5.74 and the acerbic SNP comments at the launch of the proposal (*Glasgow Herald* 12.7.74).

[80] SEPD 1975 pp. 1f.

overseeing of clearance and rehabilitation of derelict land, and on the other hand large-scale urban renewal projects in collaboration with local authorities, especially in old industrial areas in the West of Scotland.[81]

In order to establish the relationship between the policy profile proposed for the SDA and previous blueprints for semi-autonomous development bodies, Table 5.2 lists the policy functions mentioned in the consultative document and compares them with the proposals by interest organisations and political parties analysed in the previous section.[82] Looking first at the overall policy profile of the proposals it is clear that while the proposed SDA was clearly a multi-functional body, this set it apart from most other proposals (except the West Central Scotland Plan's SEDCOR blueprint) which tended to be more tightly focused around a limited number of functions oriented towards industrial firms, often focusing either on promotion of Scotland to external investors or optimising support for indigenous firms by giving the new organisation responsibility for mainstream discretionary regional policy grants. A closer look at the individual functions demonstrates broad consensus about two functions, namely provision of investment capital and advisory services for private firms, while the inclusion of environmental powers targeting industrial dereliction and urban renewal had been demanded only rarely. The details of the process that lead to the precise make-up of the proposal could well remain under the veil of civil service discretion, but observed from the outside the final result looks very much like a combination of the RDA thinking espoused by some of the more detailed 1970s blueprints, the notion of public investment which was generally supported in Scotland and popular within the Labour party, the environmental powers of the Scottish Office and the new urban renewal priority thrown in to add financial volume and maintain some degree of central government control with local government activities, backed up by the reassuring precedence of the HIDB which had demonstrated to Scottish Office ministers and administrators that an arm's-length body would still be governable. With all these rather parallel ideas floating around in Scottish political discourse in the mid-1970s, drawing them together in event of pressing political circumstances would seem to be an imminently doable task for experienced policy-makers.

---

81 SEPD 1975 pp. 2ff.

82 The Conservative and Liberal 1974 proposals for which little or no detailed information has been found have been omitted.

Table 5.2 Policy functions of proposed Scottish RDAs

Policy resource / function		*STUC 1963ff*	*HIDB 1964*	*SNP 1966ff*	*NEB 1973*	*SCDI 1973*	*SCRI 1974*	*WCSP 1974*	*SDA 1975*
Finance	Investment	•	•	•	•	•	•	•	•
	Grants		•		•		•	•	
Infor-mation	Promotion	•				•		•	•
	Advice		•	•	•		•	•	•
Organi-sation	Factories		•			•	•		•
	Dereliction					•		•	•
	Urban renewal							•	•
Authority	Planning	•				•			

At the same time the complex ideational origins of the SDA did, however, also create a situation where many organisations would be able to see themselves as having played a part in the process and therefore potentially could develop rather specific expectations about the way in which the new organisation should function. Urlan Wannop, the director of the West Central Scotland Plan study, could rightly claim that the proposed Scottish-wide body was effectively SEDCOR with a wider geographical remit,[83] and in political terms this meant that the local authorities who had co-sponsored the venture would have expectations with regard to urban renewal, and the consultative document did indeed give strong hints about the SDA becoming involved in a major project in the east end of Glasgow.[84] Despite having promptly welcomed the government proposal, the SCDI would have noted that the strategic thrust of the government proposal was closer to the indigenously-oriented blueprint of its research subsidiary than the tripartite body's own infrastructure oriented 'mobile new town' approach – and could therefore worry about being sidelined as a policy implementing organisation because business advice would be provided directly by the arm's-

83 Wannop 1995 p. 126.

84 SEPD 1975 p. 4.

length body rather than, as implied by its 1973 proposal, by the SCDI itself. The inclusion of the investment function was one of the major innovations in the government proposal, but although public investment in private firms in order to stimulate regional development was widely supported, different organisations viewed this function in rather different ways: the SCDI had stressed the need for providing risk-oriented venture capital through a public merchant bank, while the STUC and the Scottish Council of the Labour Party had campaigned for the setting up of a State Holding Company. The consultative government document alluded at the former but referred explicitly to the new Scottish body carrying out 'appropriate functions of the NEB' simultaneously being proposed for British level, thereby mirroring the two-tier nature entailed in the model supported by STUC and the Labour Party Scottish Council. From the perspective of the Labour government this could simply have been a way of appeasing English Labour MPs by being able to claim that the new Scottish body was part of an overall British scheme and not just a particularly expensive form of preferential regionalism pursued in order to stop the forward march of separatist nationalism north of the border, but still the NEB connection of the investment function claimed in the consultative document, coming as it did on the back of recent left-wing proposals, introduced an additional political dimension to the discussion about the SDA, namely the extent to which extension of public control over private firms would become a goal in its own right.

All in all the timing and nature of the SDA proposal ensured that the new organisation would not be debated solely on the basis of its functional merits as a means to further economic development in Scotland, but would also be questioned from at least three additional perspectives, namely

- in terms of British territorial politics because it could be seen as an anti-SNP move,
- in terms of internal Scottish territorial politics because it was place as a central coordinator of local authority activities in environmental improvement and urban renewal, and
- in terms of conflicting ideals about the proper role of the public sector *vis-à-vis* private economic actors.

If the SDA was originally designed through ingenious policy bricolage at the Scottish Office on the basis of a wide-ranging Scottish consensus about the desirability of a regional development agency, this obviously suggested that the new body would have a good many friends to defend it if necessary. But its wide-ranging remit also ensured that the Agency was immediately positioned in the midst of several fiercely contested discursive fields where even its supporters could, if disappointed with the Agency's strategies and results, easily be tempted to interfere with its operational freedom – and thereby undermine what was supposedly one of the arm's-length position of the organisation, namely the safeguarding of industrial development activities from bureaucratic and political second-guessing.

5.2. Discursive Preludes: The Politics of the 1975 SDA Act

The first indication about how the politics surrounding the new semi-autonomous development body would play out in practice can be established through an analysis of the debate about the legislation that created the SDA as a statutory body. Each of the perspectives suggested in the previous section were brought into play, but as will become evident in the following one of them came to dominate the political discourse surrounding the creation of the Agency.

Territorial Politics, Development Strategies and Accountability

Establishing the SDA required legislation by parliament in Westminster, and the SDA Bill would therefore be debated by MPs from all over the UK. As the proposal clearly entailed discrimination between different parts of Britain in the sense that new body was charged with furthering development in Scotland only while no similar body was planned for e.g. the equally crisis-ridden regions in the north of England, one could have expected that issues relating to territorial politics would have been played a prominent role because the new agency would affect the relationship between Scotland and the rest of the UK, but in practice the principle of preferential regionalism underlying the proposal was, however, not an issue that was taken up by parties and interest organisations. Despite querying some of the functions and organisational detail, the Conservative opposition supported the notion of an RDA for Scotland and did not vote against the bill at the final Third Reading in the House of Commons, and while the Labour minister Bruce Millan derided them as "extremely reluctant converts",[85] the underlying position was probably best summed up by senior Tory backbencher Lord James Douglas-Hamilton who reiterated the commitment of the Conservatives to a Scottish Development Fund and insisted that "[t]he debate today concerns not whether there should be an agency but what the powers of that agency should be".[86] The only recurring issues that could be interpreted in terms of territorial politics were on the one hand the question of funding – SNP[87] and STUC[88] argued for more money to be allocated

[85] *HCPD* 21.10.75 vol. 898 col. 411.

[86] *HCPD* 25.6.75 vol. 894 col. 551, cf. col. 564.

[87] *HCPD* 25.6.75 vol. 894 col. 522, cf. *Glasgow Herald* 1.2.75, Isobel Lindsay 6.6.90.

[88] STUC 1975 p. 79, cf. *Glasgow Herald* 1.2.75, 28.2.75, cf. Douglas Harrison 19.7.90.

to the new body while the Conservatives made some play of what they regarded as a low level of additional funding[89] – and on the other hand whether the new organisation should be allowed to establish offices abroad, something which the SNP supported[90] while Conservatives wanted a coordinated UK approach in order to avoid wasteful competition.[91] All in all this discursive pattern clearly suggests that in terms of territorial politics the debate took place on the consensual premises that a Scottish RDA was agreeable as a measure of preferential regionalism, but perhaps the most important contributions were the ones that were notable by their absence, namely Labour voices from especially problem regions in the north of England who could have questioned the lack of territorial evenhandedness exercised by government. A likely explanation for this display of party discipline would seem to be recognition of the need to deal with the SNP threat to the large contingent of Labour MPs elected in Scotland and assurances from the Labour government that the NEB proposed for UK would entail a strong regional dimension in the English Assisted Areas,[92] and thus the SDA begun its life without having being publically construed as an object of envy by other regions in the UK.

Also the question of the organisation and overall development strategies of the SDA were subjects of relatively limited political attention. Apart from simply listing the functions and applauding the integration of economic and environmental functions, government ministers particularly stressed the need for the new organisation to adopt a proactive approach,[93] something which both SNP[94] and STUC[95] clearly supported but the Conservatives were sceptical about, especially in the context of industrial investments. Another major innovation from a regional policy perspective, the inclusion of environmental powers, drew few comments: STUC feared that they might distract the Agency's attention from what the organisation saw as the main tasks of investing in indigenous firms and attracting inward investment,[96] and the Conservatives seemed to quietly approve of the SDA as a central agency coordinating

89 *HCPD* 25.6.75 vol. 894 col. 487, 573.

90 *HCPD* 21.10.75 vol. 898 col. col 325ff.

91 *HCPD* 25.6.75 vol. 894 col. 542, 21.10.75 vol. 898 col. 337.

92 The main point of the final government intervention in the Second Reading debate on the Industry Bill containing the NEB proposal (*HCPD* 18.2.75 vol. 886 cols. 1239ff).

93 *HCPD* 25.6.75 vol. 894 col. 581.

94 *HCPD* 25.6.75 vol. 894 cols. 522ff.

95 1975 Congress resolution on the Scottish economy (STUC 1975 pp. 426ff), Helen Liddell 21.6.90.

96 STUC 1975 pp. 82ff, Douglas Harrison 19.7.90.

local government.[97] Actually more attention was given to organisational issues, especially the question of the accountability of the new organisation: despite not being averse to the principle of arm's-length agencies within regional policy,[98] both STUC and the Conservatives demanded more direct influence on the activities of the organisation in the name of the trade unions[99] and the Westminster parliament respectively,[100] but government ministers insisted on maintaining a sponsorship relation revolving around the Secretary of State who would appoint board members rather than accept interest organisation nominees and who would, like other ministers, be subject to parliamentary scrutiny.[101]

A combination of several factors would seem to explain the fairly limited debate about the policy profile and arm's-length approach proposed for the new organisation. On the one hand most of the Agency's functions were either perceived as fairly uncontroversial 'technical' operations (clearing of derelict land, environmental renewal) or would be inherited from existing arm's-length bodies (factory building, small firm support) which for decades had operated throughout Britain as supplementary regional policy measures and enjoying political support from all major parties. On the other hand one particular function, industrial investments, seemed to eclipse everything else in terms of the political energy devoted to debating its merits or otherwise because it could be associated with other policy controversies that had dominated British in the 1970s and reached a climax in 1975.

Investing in Industry: Shifting the Public-Private Divide?

By far the most controversial element in the SDA proposal was the industrial investment function that gave the arm's-length body the power to invest in existing firms through loans and equity and proactively create new ventures where economic opportunities had been identified. Both sides of industry, STUC and CBIS, articulated strong views about the function, and in Westminster the Conservative opposition managed through its majority in the House of Lords to remove the investment powers in what amounted to a high-profile symbolic gesture. The Labour government immediately declared its attention of restoring the "vital powers which are fundamental to the entire concept of the agency", but while the Conservatives voted against the bill at the end of the Second Reading in the House of Commons, the bill, albeit in a slightly amended

97 *HCPD* 25.6.75 vol. 894 col. 479.

98 Cf. the proposals analysed above.

99 STUC 1975 p. 79.

100 First Scottish Standing Committee 1975 p. 139.

101 First Scottish Standing Committee 1975 p. 141.

form, went unopposed through its Third Reading with the investment powers restored despite continued Conservative misgivings.

The amendments proposed by the opposition through the legislative process persistently focused on the question of public enterprise and how the SDA's investment powers would affect the relationship between public and private actors in the regional economy, but as the debate progressed, a change of emphasis can be detected away from the initial focus on the principle of public enterprise towards more technical issues relating to the management of Agency invested firms. The Labour government consultative document had originally presented the SDA as having NEB-type functions,[102] and it could therefore readily be assumed that the new Scottish body would, like its British counterpart, see extension of public ownership as something which would be desirable in its own right. The Scottish employers association CBIS announced shortly after the publication of the consultative document that it would oppose the SDA "if it will increase state ownership in industry",[103] and the day after the left-wing industry secretary Tony W. Benn had proudly described the NEB as a socialist measure that would strengthen the weight of public and employee interests in industrial decision-making,[104] the recently elected Conservative party leader Margaret Thatcher told her party members in Helensburgh that the SDA was "a sort of tartan Wedgwood Benn" and that she would fight "further nationalisations".[105] The Conservative front bench spokesman Teddy Taylor stressed that "[w]e do not argue that the bill in itself will create a marxist nightmare" – a derogative occasionally used about the NEB[106] – but still it was "a further small step along the road to state socialism".[107] According to those on the left supporting the bill, this was indeed what the SDA should be all about,[108] and while government ministers tried to appease their traditional supporters by making friendly noises,[109] they maintained that the SDA should not be able to buy shares in private firms through compulsory acquisition and that any nationalisations would require separate legislation to be passed by parliament.[110]

102 SEPD 1975 p. 1.

103 *Glasgow Herald* 15.2.75.

104 *HCPD* 17.2.75 vol. 886 cols. 935-46.

105 Thatcher 1999: Speech to Helensburgh Conservatives 18.2.75.

106 *HCPD* 18.2.75 vol. 886 col. 1208.

107 *HCPD* 25.6.75 vol. 886 col. 577.

108 E.g. the Labour Party Scottish Council (*Glasgow Herald* 14.3.75), STUC (*Glasgow Herald* 28.2.75), and Scottish left-wing Labour MPs (*HCPD* 25.6.75 vol. 886 col. 513, cf. First Scottish Standing Committee 1975 col. 368).

109 E.g. *Glasgow Herald* 8.3.75, First Scottish Standing Committee 1975 cols. 368ff.

110 First Scottish Standing Committee 1975 col. 463f.

The Conservatives clearly thought that direct investment in industry by the SDA would be more acceptable if they were limited in time, and during the Committee stage of the Bill the opposition tabled an amendment requiring the SDA to dispose of its equity investments "so soon as it is reasonably practicable to do so".[111] This proposal was, however, rejected by the Labour government, arguing that such a requirement would prevent it from getting rewarded from taking financial risks when investing in a project at an early stage of its development.[112] Moreover, the Conservatives did not see a role for proactive public enterprise in revitalising the Scottish economy, and in order to ensure that firms in which the Agency had invested did not subject private firms to 'unfair competition', the opposition tabled amendments both at the committee stage and again at the Third Reading of the bill in the House of Commons which sought to enshrine principles of commercial discipline in the Agency's dealings with individual firms,[113] but Scottish Office minister Bruce Millan argued that such amendments were unnecessary because the new organisation was not intended as a repository for 'lame ducks' firms with no future and that the bill already contained a specific measure of financial discipline which required the investment function to earn a commercial return on its investments.[114] In short, the government expected "the Scottish Development Agency [...] to act in its industrial role competitively and commercially",[115] and by proposing an amendment that restricted Agency investments to entities that in legal terms were companies and thus subject to the same requirements as private firms, the Labour government signalled its intention to make the SDA a player on the same level as private actors rather than an overarching power in the regional economy.

5.3. Scripts for the Future?

In Scotland in the mid-1970s the view that a regional development agency was desirable as a means to promote economic development was widely held, and despite the rather different ideas about what problems should be given priority, the overall impression is still that most interest organisations and political parties shared the belief that Scotland's economic difficulties should be addressed through an arm's-length body

111 First Scottish Standing Committee 1975 col. 453. Similar provisions for compulsory disposal existed in the 1972 Industry Act, cf. *HCPD* 17.2.75 vol. 894 cols. 483ff, First Scottish Standing Committee 1975 cols. 326ff.

112 *HCPD* 25.6.75 vol. 886 col. 581, First Scottish Standing Committee 1975 col. 453.

113 First Scottish Standing Committee 1975 col. 337, *HCPD* 21.10.75 vol. 989 cols. 335ff.

114 *HCPD* 21.10.75 vol. 989 cols. 282-320.

115 *HCPD* 21.10.75 vol. 989 col. 308.

specifically dedicated to this particular task. Demands for a new Scottish development organisation had, however, invariably been formulated as appeals to central government, and while both the potentially governing parties, Labour and the Conservatives, had long pursued preferential regionalist measures,[116] they were the last political actors to embrace the RDA approach, both of them having repeatedly and in no uncertain terms turned down e.g. the persistent STUC demands for a Scottish Development Authority and instead preferred to rely on central government grant schemes and the physical planning powers of the Scottish Office. Only after a decade of gradually intensified regional policy had failed to close the unemployment gap between Scotland and other British regions *and* an escalating political threat from the pragmatic separatist SNP, the idea of an RDA was suddenly adopted by both the major union state parties, and victory in the October 1974 general election meant that it fell to Labour to establish an arm's-length economic development body. On the basis of this neither the internal-evolution nor the external-revolution traditions in the existing literature perspectives would seem to be able to account for the origins of the SDA: the changing by the SNP of the landscape of territorial politics was clearly important in making the major political parties propose a new form of preferential regionalist measure, but in order to understand why Labour and the Conservatives sought to defend the union state by means of a regional development agency, internal learning processes within the administrative system and external pressures from a growing consensus in the Scottish territorial policy network must be taken into account.

The SDA's economic development remit meant that its activities was always likely to be scrutinised from the perspective of its functional effectiveness as a promoter of e.g. employment or industrial activity within the region, but the historical context in which the SDA proposal emerged meant that the new organisation would also be interpreted in the light of British territorial politics, namely as a preferential regionalist measure aiming to contain the SNP (and the Conservatives). Moreover, the comprehensiveness of the tasks combined within the new organisation ensured that the Agency would not only have many, potentially disappointed, 'fathers' but was also likely to be subjected to political controversy along two additional lines: its role in environmental improvement could be interpreted a way of maintaining centralised control of local authority activities, and the introduction of a new policy instrument, public investment in private firms, could be interpreted as an attempt to shift the border in the regional economy between public and private actors further towards the former. Conversely, from the perspec-

[116] Cf. Section 4.1.

tive of the Agency, having to navigate between conflicting views about its future role in the Scottish economy meant that being able to present its own positive and palatable interpretation of its activities through e.g. extensive corporate PR activities would become an essential means of maintaining support from its political sponsors and the wider policy network in Scotland.

As documented above, the initial political debate focused overwhelmingly on the last of these discursive fields and, indeed, the investment function in particular. While left-wing supporters of the SDA saw public ownership as a necessary means to exert influence in private firms, both the Conservatives and CBIS were fiercely critical about what they saw as a form of nationalisation, although the main opposition party seemed to accept that public ownership of equity could be a short-term measure employed to rescue individual crisis-ridden firms, provided that it did not subject other firms to 'unfair competition'. Only the Labour government went much beyond purely normative statements about whether public ownership was inherently desirable or not by arguing that the investment function would have beneficial effects if undertaken proactively and on a commercial basis in emerging firms and industries, for example in order to develop an indigenous supplier base for the rapidly expanding oil-related activities. Even though the Conservatives had not voted against the bill in its entirety after the final parliamentary debate, by associating the proposed agency with anathema like nationalisation and state socialism, the party made it possible for the government and other supporters of the Agency not just to construe the opposition as 'reluctant converts' but also to maintain suspicions about what would happen to the new organisation if the Tories were to regain power.

CHAPTER 6

Resourcing Agency

The capacity of the SDA to influence economic actors depended on the policy instruments at its disposal and the ways that these were put to use, and hence it is necessary to examine the development of both the resources of the Agency and the corporate strategies that guided their employment from the early days in 1975 to the merger into Scottish Enterprise in 1991. Through its sponsorship central government is likely to have exercised some degree of influence on the ways and means of the organisation, and hence the discussion will attempt to establish both the changing position of the Agency and its interaction with the Scottish Office as the sponsoring department. The aim of this and the following chapter is to explore the various forms of resource inter-dependencies between the SDA and its political sponsors (Chapter 6), and to examine how they influenced the general guidelines for employment of resources entailed in the development strategies of the Agency (Chapter 7), thereby illuminating the balance between continuity and change and establishing general features that individual areas of SDA policy would have to take into account, by either observing, challenging or circumventing the corporate-level framework.

Turning first to the question of resources, the present chapter first considers the general relationship between the sponsoring Scottish Office and the SDA on the basis of the statutory authority accorded to the organisation and the formalised interaction on the corporate level between the Agency and its political environment (Section 6.1). This is followed by an analysis of the SDA executive with regard to its organisational and informational resources (Section 6.2), and an examination of the much more politically controversial issues relating to the quantity and quality of the financial resources available to the organisation (Section 6.3). Finally, the development of the Agency with regard to resources is compared with that of other public bodies involved in economic development in order to place the organisation in the broader context of British regional policy (Section 6.4).

6.1. Authority, Sponsorship and Policy Networks

The formal authority of the SDA to promote regional development in Scotland derived from an act passed by the British parliament in 1975 that established the organisation as a statutory body with the purpose of "furthering the development of Scotland's economy and improving its environment".[1] This remit was very wide-ranging indeed – in fact the Agency was given the power to "do anything [...] which is calculated to facilitate the discharge of their functions"[2] – but in practice the 1975 Act also identified a number of activities that the new body would be expected to carry out, namely:[3]

- provide finance for and invest in industrial undertakings,
- provide sites and premises for industrial activity, and
- improve the environment and address the problem of derelict land.

In addition to this, other functions of a less specific nature included "promoting or assisting the establishment, growth, reorganisation, modernisation or development of industry or any undertaking in an industry" and to promote "industrial democracy" in firms controlled by the Agency.[4] In principle the statutory powers of the new organisation could be applied throughout Scotland, but the continued presence of the HIDB meant that in practice the territorial remit of the SDA was limited to the Scottish lowlands except in policy areas and with regard to projects that were outwith the remit or capacity of the HIDB, e.g. environmental improvement or large investments.[5] All in all the original act of parliament clearly gave the new organisation a broad remit while at the same time identifying three core policy areas, and the wide-ranging remit and powers would seem to indicate that the SDA was not just expected to implement central government policies, but also to devise new initiatives, and as the change of government in 1979 only brought about minor and largely symbolic changes in the formal authority of the organisation, from this perspective the SDA would seem to be a politically 'robust' example of the RDA approach to regional policy, a multi-functional body vested with a significant degree of strategic initiative.

In practice, the operational freedom of the new organisation was, however, circumscribed in two ways from the very beginning: by the sponsoring powers of the Scottish Office on the one hand, and the position of the Agency in the British regional policy network on the

---

1 SDA Act 1975 Section 1 Subsection 1.

2 SDA Act 1975 Section 2 Subsection 3.

3 SDA Act 1975 Section 2 Subsection 2.

4 SDA Act 1975 Section 2 Subsection 2.

5 Kirwan 1981, Hogwood 1982 p. 49, cf. Scottish Affairs Committee 1984 pp. 38-44.

other. Beginning with the direct relations of sponsorship between the Scottish Office and the SDA, the institutionalised channels of political influence constitute the formal rules governing the actions of different actors, and the analysis will therefore take its point of departure in statutory sources and regulations. Such rules may be violated and the nature of influence vary, but formal patterns of organisation are nonetheless the starting point for attempts to exert or avoid influence, and thus the following sets the scene for subsequent analyses by establishing the institutional constraints under which corporate and policy developments took place. As sponsor department the Scottish Office exercised its responsibility for ensuring that the Agency used its powers in accordance with its statutory objectives in four complementary ways that remained in place throughout the lifespan of the organisation:[6]

- The Secretary of State for Scotland appointed the chairman and board of the SDA.[7]
- The financial and organisational resources of the Agency were subject to controls: public funding was decided as part of the annual budgetary process of central government, the detailed expenditure plans of the Agency were subject to Scottish Office approval, and key aspects of staffing were subject to central government regulation.[8]
- The policies and development strategies of the SDA could be influenced in various ways: industrial strategies and corporate plans have always been discussed with the Scottish Office prior to adoption,[9] and in 1986 the SDA's general strategic thrust was the subject of a major review by central government.[10]
- The implementation of individual policies has been regulated to varying degrees: in some policy areas major individual projects had to be approved by the Secretary of State for Scotland,[11] guidelines concerning particular policy areas could be agreed between the Scottish Office and the Agency,[12] Agency activities were evaluated as

---

6 References are made to the original 1975 Act as subsequent legislation has not significantly changed the original set-up. For a more detailed survey, see Kirwan 1981.

7 SDA Act 1975 Section 1.

8 SDA Act 1975, especially Section 1 Subsection 6, Section 12-13, Schedule 1 Section 10, and Schedule 2; some of these powers are exercised jointly with UK departments, especially the Treasury, cf. the discussion below.

9 Cf. Chapter 7.

10 IDS 1987a cf. Chapter 7.

11 SDA Act 1975 Section, IDS 1987a pp. 113ff.

12 Kirwan 1981 p. 7.

part of the 1986 government review, and the Secretary of State for Scotland had the power to issue specific instructions to the Agency.[13]

It is, however, also important to note the very indirect nature of parliamentary accountability that the SDA was subjected to: Agency accounts and annual reports were the only information that parliament was to receive on a regular basis,[14] and otherwise the activities of the Agency were subject to, mainly ex-post, scrutiny by parliamentary committees.[15] Although these parliamentary enquiries will have put additional pressure on the SDA to change some of its practices, it is abundantly clear that in practice political sponsorship was overwhelmingly exercised by a government minister and his department. The sponsoring powers of the Scottish Office did not just cover the input of key resources to the Agency – leadership, money and staff – but also more detailed aspects of the SDA's strategies, policies and implementation. These external preferences could present themselves to the SDA either directly from the Secretary of State in the form of e.g. broad strategic guidance, or through the generally much more detailed regulation undertaken by the Industry Department for Scotland (IDS), the Scottish Office's economic development department.[16] This dual-track approach suggests that two types of considerations will have influenced the Agency through the sponsor relations, namely on the one hand party-political and ideological interests, and on the other hand preferences of an organisational or administrative nature, reflecting the Scottish Office position as a central government department with a traditional bureaucratic mode of operation.

All in all the rules governing sponsorship instituted by especially the SDA Act clearly vested the Scottish Office with wide-ranging powers that would allow ministers and civil servants to become involved in most aspects of Agency activity, and as the sponsorship was of the organisation as such – the Agency owed its existence to one political sponsor – the ultimate threat of being reorganised or terminated will undoubtedly have served to reinforce them. *If* these powers *had* been used systematically, then the Agency would have come close to being an executive arm of central government, but explicit general statements by the parties involved about how sponsorship was conducted in prac-

---

[13] SDA Act 1975 Section 4.

[14] SDA Act 1975 Schedule 2 Sections 8 and 9.

[15] Cf. the discussion in Chapters 7, 8, and 10.

[16] The SDA's sponsor department was called the Scottish Economic Planning Department, SEPD, until the Conservative government in 1983 introduced a name more in tune with its own ideological inclinations (Midwinter *et al.* 1991 p. 55). For the sake of brevity IDS will be used for the Agency's sponsor department when referring to the entire period from 1975 onwards, while the old name SEPD will only be used when dealing exclusively with the period before 1983.

tice and what outcomes it produced would appear to be contradictory. The Scottish Office has continuously stressed the importance of maintaining an arm's-length relationship: "[t]he Agency is – and was always intended to be – creative and not just an executive".[17] This is echoed by some Agency top-executives who claim that the Scottish Office had virtually no role in the development of SDA policies,[18] but others emphasise the recurring problems involved in having a sponsor department, not least the more detailed regulation by the IDS.[19] It is, however, also interesting to note that persons responsible for the Agency's operational divisions maintain that the policy-making initiative in most cases has rested with the SDA,[20] and that the 1986 government review recommended that in the future the direct contact between Scottish Office ministers and the Agency boards should be improved and that "Scottish Office monitoring should be conducted on a properly planned basis".[21] Taken together this would seem to suggest that relations between the Agency and its sponsoring department were essentially asymmetrical:

- with regard to strategy and policy development external control has primarily been situated on a fairly general level and conducted in a largely reactive manner,
- implementation of individual policies appears to have been subject to much more detailed control via regulations, joint committees, budgets and staffing.

On this account the Scottish Office appears to have been a screener of Agency activities rather than an instigator of new departures in regional policy, but at the same time keeping a watchful eye on the way in which the Agency employed its policy instruments. Paired with the absence of direct parliamentary accountability, this would seem to suggest that in terms of direct political sponsoring the SDA effectively seems to have operated from an, albeit precarious, arm's-length position.

17 IDS 1987a p. 106, cf. the Scottish Office guidelines for the Agency's industrial investment functions (SDA 1977 pp. 62f, SEPD 1980a), Gavin McCrone 20.6.90, Gregor Mackenzie 15.10.90.

18 E.g. George Mathewson 9.7.90, Edward Cunningham 13.6.90, Sir Robin Duthie 12.7.90.

19 E.g. Iain Robertson 11.7.90, James Williamson 14.6.90, Sir Kenneth Alexander 25.7.90, John Firn 15.6.90.

20 Iain Shirlaw 1.6.90, Mike Sandys 31.5.90, Gerry Murray 30.7.90.

21 IDS 1987a p. 114.

Figure 6.1 Central government and regionally-based economic development policy

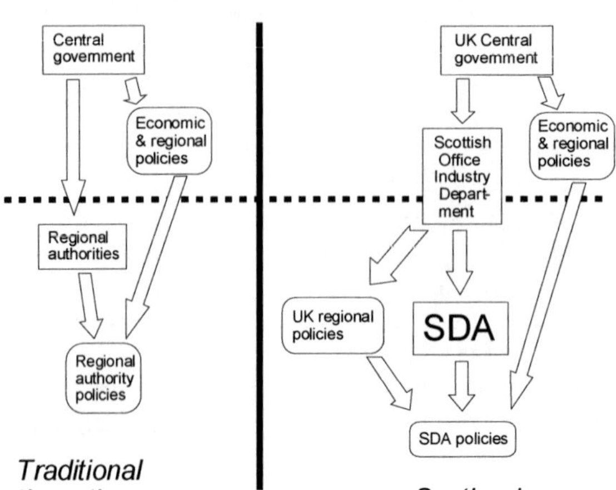

The SDA's room for manoeuvre was, however, furthermore circumscribed through its position in the broader context of British regional policy where, compared with the situation in the context of a traditional three-tier political system, the peculiar institutional set-up in Scotland would appear to have intensified the pressure on regionally-based development actors for coordinating their activities with national policy programmes, as illustrated by Figure 6.1: not only was the SDA sponsored by a territorial department of central government present within the region, but this department was at the same time an implementor of regional policy in its own right, and this created two different types of pressures for policy coordination. On the one hand certain policy areas like inward investment attraction, factory building and industrial investment were subject to *general regulation* on the British level because they involved very direct inter-regional competition and government wanted to maintain 'a level playing field' with equivalent services in England and Wales. In such cases the regulation of Agency activities may have been routed through the Scottish Office but did in fact originate in a more extensive policy network in which departments with an all-British remit like the DTI or the Treasury tended to dominate.[22] On the other hand coordination with regard to *individual projects* was also necessary because British regional policy in the form of central government grant schemes continued to operate in Scotland, and thus Agency activities involving financial policy instruments such as industrial

[22] Cf. Chapters 8, 10 and 12.

investment or inward investment attraction will have needed to take such programmes into account,[23] especially because since 1975 IDS had been administering Regional Selective Assistance grants in Scotland as the Scottish Office department responsible for economic matters.[24] This gave added importance to the IDS-SDA nexus and with regard to inward investment, politically sensitive because of the risk of inter-regional competition for footloose projects, coordination was even institutionalised through the setting up of *Locate in Scotland*, a joint SDA-IDS body responsible for attracting inward investors.[25]

All in all it can be concluded that the institutional setting placed the SDA in a complex position: while the sponsor relation between the Scottish Office and the SDA appears to have worked on an arm's-length basis with the Agency being capable of initiating new policy initiatives, the scope for independent policy-making could still be severely undermined in the long run if the Scottish Office chose to exercise its wide-ranging sponsor powers and exploit its position as the implementer of British policies in Scotland in order to exercise detailed control. The nature of the sponsor relation between the Scottish Office and the SDA can therefore *neither* be determined solely on the corporate level of analysis *nor* once and for all. On the one hand the distribution of formal authority, especially the reactive nature of Scottish Office control, suggests that the degree of operational freedom enjoyed by the Agency could vary between policy areas depending on whether sponsors deemed these to be more or less controversial. On the other hand the degree to which the sponsoring powers of the Scottish Office have been exercised are likely to have varied in time, reflecting shifting concerns of territorial and party politics, bureaucratic positioning strategies and ideological considerations, and thus the way in which these institutional relations were translated into organisational practice must therefore be established through concrete analysis of developments on the corporate level as well as in individual policy areas.

---

23 The original investment guidelines explicitly stated that the function operates in the broader context of British regional policy (SDA 1977 p. 60), and close collaboration is reported to have taken place in many individual cases (Wilson Committee 1977f vol. 6 pp. 190f, SDA 1977a, IDS Head Gavin McCrone in evidence to Committee of Public Accounts (Committee of Public Accounts 1984b p. 26)).

24 Young & Hood 1984 pp. 44ff, cf. McCrone 1986, Hogwood 1982. The original SDA Act had in fact allowed the Agency to function as intermediary for the Scottish Office with regard to Regional Selective Assistance (SDA Act 1975, Section 5), but this option was never used, presumably in order to avoid the risk of competitive bidding between assisted areas for foot-loose investment (McCrone & Randall 1985 p. 235). The option was removed when the SDA Act was amended in 1980.

25 See Chapter 10.

6.2. Organisational and Informational Resources

The SDA and Political Patronage

The analysis of the SDA as an organisation begins 'from the top' by considering the Agency's board and Chief Executives, formally in charge of the organisation, and other external links on the corporate level that could influence SDA activities.

Appointing the members of the SDA's board was part of the statutory role of the sponsor department, but the parliamentary Act layed down fairly detailed criteria as regards the composition of the board: appointees should have relevant experience in "industry, banking, accountancy or finance, environmental matters, local government or the representation of workers".[26] Both the original Labour government appointments and those from 1979 onwards were in line with the letter of the SDA Act in the sense that different organisational and professional backgrounds were represented, and in order to ensure coordination between the two Scottish RDAs, the Chairman of the HIDB was member of the Agency board throughout its existence.[27] The scope for political patronage was, however, still considerable because board members served on account of their personal merits rather than as representatives of organisations or institutions, but although the relative weight of members with a business background increased in the 1980s at the expense of trade union or local government experience,[28] the continued inclusion of the latter should have ensured that also in the 1980s a variety of non-Conservative perspectives on regional development will have been present.[29] The two most significant changes, also symbolically, did, however, occur on the media-exposed positions as Chairman and Deputy Chairman where the original line-up – a Labour politician and the Agency's Chief Executive – was, unsurprisingly, replaced at the end of their terms of office with two industrialists by the new Conservative government, but whether this in itself would be able to bring about more substantial changes in the work of the organisation hinges on what role the Agency board played in the policy-making processes within the organisation. Here both board members[30] and

26 SDA Act 1975 Section 1.

27 Hogwood 1982 p. 49, Scottish Affairs Committee 1984 p. 266, cf. SDA 1976-91.

28 While members with a business background were the largest minority on the original SDA board, they came to constitute a majority from 1981 onwards (SDA 1976-91).

29 Also after the change of government in 1979 local government experience was provided by prominent Labour politicians, reflecting the party's dominance in lowland Scotland.

30 Sir William Gray 19.10.90, Sir Robin Duthie 12.7.90, George Robertson 30.10.90, Sir Kenneth Alexander 25.7.90.

executives[31] agree that the role of the board in the SDA, like a board in a private firm, has primarily been to discuss general issues and strategies while the development of both corporate strategies and individual policy programmes has been undertaken by the Chief Executive and the Agency's functional directorates. This pattern is also confirmed by what is otherwise known about the way in which the board worked, with meetings normally only once a month,[32] a not too uncommon failure of executive directors to put forward alternative options for the board's consideration,[33] and a part-time executive Chairman whose main functions would seem to have been corporate PR and political liaising rather than policy development.[34] The Agency's longest-serving Chairman, Sir Robin Duthie, summed up the position like this:

> The executive put their policies forward and most of the time this was endorsed by the Board, but the Board then supported these policies in the community and the members were publicly very supportive about what the Agency was doing.[35]

All in all the board appears to have performed two roles within the SDA. On the one hand the board functioned as a sounding board for executive proposals which policy-makers within the organisation would have to persuade of the merits of new initiatives, and thus the effectiveness of the board as an instrument of party-political control would ultimately depend on the extent to which SDA executives were able to develop and package policies that fell within the strategic and discursive parameters of the board. On the other hand the board acted as ambassadors for the SDA to the public and in their respective constituencies: here the professional background of board members became valuable sources of informal authority, and thus the more business-oriented composition of the board in the 1980s may well have been an advantage in relation to enhancing the reputation of the Agency with private economic actors.

With board members appointed in a personal capacity, the only thing ensuring a reasonable degree of rapport between board members and

---

31 Lewis Robertson 20.7.90, George Mathewson 9.7.90, Iain Robertson 11.7.90, Edward Cunningham 13.6.90, David Lyle 23.7.90, Frank Kirwan 1.6.90, Douglas Adams 18.7.90, Ian Hart 18.10.91.

32 SDA 1977 p. 33, Sir William Gray 19.10.90, Sir Robin Duthie 12.7.90, Sir Kenneth Alexander 25.7.90.

33 IDS 1987a p. 110, cf. David Lyle 23.7.90, Sir Kenneth Alexander 25.7.90.

34 Sir William Gray 19.10.90 cf. Lewis Robertson 20.7.90; Sir Robin Duthie 12.7.90, cf. George Mathewson 9.7.90. The picture section of the SDA's Annual Reports (1977-90) also provides copious documentation for the Chairmen's efforts in the field of inaugurating Agency sponsored projects.

35 Sir Robin Duthie 12.7.90.

their respective 'constituencies' would seem to be the prominence of the individuals appointed,[36] and some Scottish interest organisations, e.g. the STUC, have clearly resented not having more direct influence on the affairs of the Agency.[37] The SDA's only formalised contact with Scottish interest organisations was through the so-called *Development Consultative Committee* (DCC) which comprised representatives of the SCDI, local authorities, New Towns, employers' organisations, and trade unions. The DCC was set up in 1977 after the Agency assumed the leading role in Scotland's efforts to attract inward investment at the expense of prominent DCC members,[38] replicating the *Scottish Consultative Committee* through which the SCDI had previously attempted to coordinate external promotion efforts,[39] and while in theory the committee was supposed to be an advisory body,[40] oral evidence suggests that the most important function of its quarterly meetings became briefing of the organisations by the SDA rather than the other way around.[41] The creation of the DCC certainly demonstrated the Agency's awareness of the importance of cultivating its relationship with vested interests in the region, but at the same time its limited function also signals an intention to avoid formalised links with interest organisations that could infringe its strategic and operational freedom, and it is interesting to note that the potentially much more powerful advisory committees on individual policy programmes have been established by discretionary Agency appointment of 'suitable notabilities' rather than interest organisation nominees.[42]

With an unrepresentative tripartite board and one-way communication through the DCC, corporate level links between the SDA and interest organisations and public bodies involved in regional development can in other words only be described as very limited and designed to lend informal authority to the Agency through cooption of notables from different social constituencies. By far the most significant inter-organisational relation between the Agency and the political environment in which it operated was still with its sponsors at the Scottish Office.

[36] Moore & Booth 1989 pp. 34ff.

[37] E.g. the Glasgow Herald 28.2.75, cf. Chapter 14.

[38] SDA/DCC 29.6.77, cf. Chapter 10.

[39] Committee on Scottish Affairs 1980 p. 171.

[40] Chief Executive Lewis Robertson's words of welcome at the first DCC meeting (SDA/DCC 29.6.77), cf. Wilson Committee 1977f vol. 6 p. 192.

[41] Cf. Frank Kirwan 1.6.90, Iain Robertson 11.7.90.

[42] Cf. Chapters 8 and 9.

Changing Organisational Structures

Once we move below the level of the SDA board, the degree to which the Agency was able to influence its own organisation generally increased, but an important exception to this rule can be found at the very apex of the organisation where its Chief Executive was appointed with the consent of the Scottish Office,[43] and as he played a pivotal role, not only as an interface between the executive on the one hand and the board and the Scottish Office on the other, but also by being capable of influencing all the operational directorates, it is hardly surprising that the development of the organisation has often been interpreted in the light of changes in this particular position, not least by Agency staff.[44] In practice both Labour and the Conservatives did, however, appoint Chief Executives belonging to the Scottish business community, suggesting that they wanted the Agency to be seen as a non-departmental body and its figurehead as having a business-oriented approach,[45] and the most significant change in their professional profile was that from 1987 onwards Chief Executives had a civil service background,[46] something which could be interpreted as an attempt to increase Scottish Office influence on the SDA, especially in preparation for the Scottish Enterprise merger. If the coming and going of Chief Executives were indeed landmarks in the development of the organisation, this would in other words seem to have been primarily due to their different strategies with regard to either regional development or their positioning of the SDA *vis-à-vis* the Scottish Office.

With regard to its executive organisation the SDA was by and large in charge of developments, but although the Scottish Office generally

---

43 SDA Act 1975 Section 1 Subsection 6. The only exception was the first Chief Executive who was appointed by the Secretary of State for Scotland after consultations with the board.

44 E.g. Frank Kirwan 1.6.90, James Williamson 14.6.90, Gerry Murray 30.7.90, Douglas Adams 18.7.90.

45 The first Chief Executive, Lewis Robertson, was headhunted from a job as Chief Executive in a major Scottish industrial holding company, having served on the board of the employers organisation CBIS, and his replacement in 1981, Dr George Mathewson, was formerly a director with the government-backed venture capital organisation *3i*. It is interesting that all four of the Agency's Chief Executives, including two ex-civil servants, left the Agency for top executive posts in private sector financial organisations, and thus a spell with the Agency seems to have improved rather than undermined their commercial credentials.

46 From 1987 to 1990 the SDA was lead by Iain Robertson, a former civil servant and Director of the joint IDS-SDA inward investment operation, and in its final year of existence the Agency was headed by James Scott, a senior civil servant who had replaced Gavin McCrone as head of the Agency's sponsor department IDS.

refrained from becoming involved on a detailed level,[47] the discretion of the SDA Chief Executive and its executive Directors operated within certain general boundaries defined by its sponsor department. From 1975 to 1991 the organisational structure of the SDA was in a state of almost permanent change: every year except five saw organisational change at Directorate or divisional level, with the 1980s being a particularly turbulent period, and Table 6.1 summarises the most important stages in the development. Apart from the sheer scale of change, three features are particularly important to note, namely 1) the long-standing importance of structures inherited from organisations incorporated into the Agency when it was set up in 1975, 2) the mushrooming of highly specialised divisions, not least within the Planning and Projects Directorate, and 3) the increasing importance of space as an organisational principle. In the following each of these issues will be addressed in turn, especially with regard to their implications for strategic development and the Agency's position *vis-à-vis* its external political environment.

The parliamentary act that established the SDA also incorporated three existing organisations in the new body:[48]

- *Scottish Industrial Estates Corporation* (SIEC), the DTI funded body responsible for building and managing factories in Scotland,
- *Small Industries Council for Rural Areas in Scotland* (SICRAS), the Scottish subsidiary of the Development Commission providing loan and advice to manufacturers in rural areas, and
- the *Derelict Land Clearance Unit* of the Scottish Office's Development Department, administering grants to local authorities for clearance of derelict industrial sites.

---

47 The only major exception being the recommendation of the IDS/Treasury review that the SDA should set up a separate policy unit, cf. the discussion below.

48 SDA Act Section 15, cf. IDS 1987a p. 18, MacDonald & Redpath 1980 p. 104.

Table 6.1 The changing structure of the SDA executive

1976-78	*1982*	*1985*	*1989-91*
Strategic planning			**Policy unit (1987-91)**
Industry * investment * small business * factory policy * management advice **Promotion & information** * inward * PR investment **Environment** * factory building & maintenance * land renewal	**Planning & projects** *corporate plans * sectoral initiat. * health care & biotechnology **Finance & industry services** **Investment** **Small business & electronics** **Locate in Scotland** * inward investment **Estates & environment** * factory policy * factory building & maintenance * land renewal	**Planning & projects** * sectoral initiatives * health care & biotechnology * engineering * Scottish industries * tourism & leisure * service industries * management development * technology transfer * advisory services **Finance & property management** **Investment** **Small business & electronics** **Locate in Scotland** * inward investment **Property development & environment**	**Industry & enterprise development** * electronics * advanced engineering * Scottish resource industries * service industries * management development * technology transfer * advisory services * trade promotion * energy & environment **Investment** **Locate in Scotland** * inward invesment **Property & urban renewal** * construction * urban renewal * land renewal **Enterprise & special initiatives** * *Scottish Enterprise* merger * training
Urban renewal * research & policy	**Area development**	**Area development**	**Regional Directorate (offices from 1988)** * small investments * small business * property * land renewal * area projects

Source: SDA 1976-91. *Note*: Directorates and other primary units given in bold, subdivided according to whether they carry out corporate functions, or provide services along functional or spatial lines.

Incorporating existing organisations made the Agency operational from day one in some policy areas, and given the level of expectations in the political environment, this would certainly have been an asset in its own right. From a long-term perspective the potential drawbacks were, however, also obvious, especially because two of the merged bodies – SIEC and SICRAS – were fairly large and well-established organisations which at the beginning accounted for the greatest part of SDA activity[49] and continued to be major areas of activity within the Agency throughout its existence.[50] The integration of inherited bodies engaged in traditional development activities regulated through the British policy network was probably not helped by the fact that the embryonic SDA at the same time had to develop new types of regional policy like industrial investments,[51] but the most significant problem was the nature of the resources which the incorporated organisations embodied. As civil-service type bodies with predominantly executive tasks, SIEC, SICRAS and the Derelict Land Clearance Unit had organisational structures, personnel and modes of operation that were not ideally suited to become part of a regional development agency of which policy innovation and proactive implementation were expected, and the Agency's difficulties with developing a strategic capacity in the inherited areas of activity were reflected in a rather peculiar organisational set-up in the late 1970s where factory policy was placed in a separate unit in the Industry Directorate rather than in the Environment Directorate responsible for constructing and managing property,[52] and even after factory policy had been united with the executive functions in one directorate in 1979, an inter-directorate top-executive factory policy group was established in order to integrate the property function with other SDA activities.[53] Another, and more discreet, form of organisational heritage may, however, have contributed to alleviating the 'strategic deficit' with regard to factory building and environmental improvement, because less than 6 months after the SDA had formally started to

[49] At the end of the financial year 1976-77 environmental improvement and factory building accounted for 78% of total SDA expenditure (calculated on the basis of SDA 1977), and inherited staff for 87% of the total number employed (calculated on the basis of SDA 1977 p. 3).

[50] Cf. Chapters 11 and 12.

[51] Cf. SDA 1977 p. 8, Lewis Robertson 20.7.90. Although the Agency itself was located in Glasgow, the Edinburgh headquarters of SICRAS more than 40 miles away was maintained as the SDA's Small Business Division (SDA 1976 paragraph 15f).

[52] SDA 1976-79 cf. Page 1977 pp. 71f, 84ff, 111f and Table 6.1. Similarly the development of a long-term strategy for environmental improvement was undertaken in collaboration with the Strategic Planning Unit (SDA 1978 pp. 46f).

[53] Alan Dale 16.10.91.

operate and well before its own organisation and personnel was in place, the Scottish Office had made the Agency responsible for the coordination of a major urban renewal project in Glasgow's deprived East End.[54] This introduced a new breed of staff with a background in planning and urban development, many of which joined from the team preparing the recently disbanded New Town at Stonehouse,[55] and hence a move towards a more integrated and proactive use of physical infrastructure will have been facilitated.

Organisational inheritance and the sheer breadth of the SDA remit always implied that achieving organisational coherence and coordinated policy priorities was bound to be a demanding task, but in addition to this the coherence of the organisation was also potentially undermined by its preferred mode of operation. Unlike the traditional emphasis on hierarchy and rule-oriented policy implementation within the civil service, RDAs were expected to develop an innovative and dynamic approach capable of analysing problems and devising remedial initiatives, and by these standards the SDA was indeed a 'model agency', at least according to its own executives.[56] In the words of James Williamson, Head of the Policy Unit,

> [t]he Agency is predominantly a project-driven organisation. Projects are not passed down from above to the workers, they are developed by project champions and then put up for approval.[57]

The project-driven mode of operation not only implied that the role of long-term planning was fairly limited, but also that 'the movers and shakers' at the sharp end of policy implementation played an important role in the development of the organisation, and other features indicate that this rendering is more than corporate hype.

As can be seen from Table 6.1, the development of the functional directorates was characterised by a constant mushrooming of new specialist units in a process that did not follow any discernable principle, and although a permanent specialist unit was set up with the task of analysing the economic environment in which the Agency operated in the

---

54 See Chapter 12.

55 SDA 1976 paragraph 35, Naylor 1984, John Firn 15.6.90, cf. the discussion in Chapter 12.

56 James Williamson 14.6.90, Edward Cunningham 13.6.90, Neil Hood 1991a, Douglas Adams 18.7.90, Gerry Murray 30.7.90, Iain Shirlaw 1.6.90, John Firn 15.6.90. The SDA's two first Chief Executives both underline their own role in the development of the Agency, but this does not, however, necessarily imply a top-down regime but merely underlines that taking new initiatives was not the prerogative of any one group within the organisation.

57 James Williamson 14.6.90.

early 1980s,[58] the position of corporate planning within the organisation seem to have remained relatively weak. Initially the Agency did have a special unit responsible for Strategic Planning, but under Edward Cunningham the directorate became heavily involved in individual projects too,[59] and while corporate plans were still prepared in what since 1982 had, more aptly, been named the Planning & Projects Directorate, the focus of activity would clearly seem to have been on developing and implementing individual projects rather than establishing long-term goals for the organisation.[60] This is suggested not only by the expanding number of operational divisions within the Directorate, but also by the nature of the planning output in the first half of the 1980s where the Agency's corporate planners, despite setting high standards for the individual divisions by demanding specification of "key objectives, […], criteria for assessing success or failure, […] [and] the means by which the objectives are to be achieved",[61] still produced a document entitled *Corporate Strategy 1985/86 – 1987/88* that was mainly a summary of existing and planned projects – "a budget with words"[62] – within the framework of rather general priorities.[63] Similarly, the 1986 government review of the SDA recognised that it is "staff in operating divisions who are usually best placed to spot changes in the market for their service" and stressed that "[w]e do not wish to imply that the Agency's corporate planning should be entirely 'top-down'",[64] but the report still recommended that a yearly planning procedure be adopted, corporate priorities properly defined, impact evaluation given systematic attention, and that major new policy initiatives should be subject to discussions with IDS.[65] This was to be realised through the setting up of a new Policy Unit, established outwith the existing directorates with reference directly to the Agency's Chief Executive, and it may be tempting to interpret the insistence on formalised long-term corporate planning either as an example of a conflict between government civil servants and arm's-length project champions or as an attempt to pave the way for more extensive adherence to the political preferences of government ministers. When the new-model corporate planning process was implemented in the late 1980s, it still involved extensive consulta-

58 Douglas Adams 18.7.90, cf. Lewis Robertson 20.7.90, Danson 1980, Danson *et al.* 1990b.

59 Cf. Table 6.1.

60 Lochhead 1983 p. 121, Danson *et al.* 1990b, cf. Frank Kirwan 1.6.90.

61 SDA 1984a p. II.

62 Charles Fairley 26.7.90.

63 Cf. the discussion in Chapter 7.

64 IDS 1987a p. 106.

65 IDS 1987a pp. 103-8, 113f.

tions within the Agency,[66] and while the corporate plans produced in accordance with the new procedures contained more detail, they again took the form of general criteria for intervention and policy implementation rather than detailed prescriptions for new policy initiatives.[67] This was in excellent accordance with the intentions within the new Policy Unit because "it is very difficult to sit in the centre and say what must be done",[68] and even though the need to specify aims and evaluate impact will have made life more difficult for project champions, "the Agency is full of ingenious people who believe in getting around things – that's why we employ them".[69]

All in all it can be concluded that a clear-cut dichotomy between a pre-1986 project-driven era and a post-1986 phase dominated by corporate-level planning cannot be sustained. Instead the opportunistic project-driven approach seems to have continued as the Agency's basic mode of operation, tempered since the mid-1980s by an increased role for corporate planning in improving the quality of projects by means of guidance and evaluation rather than prescriptive prioritisation of specific activities where the new Policy Unit turned out to be more of a 'project pruner' rather than a 'prime mover'. This balance between project drive and corporate screening implies that, compared to a departmental hierarchy, the Agency was undoubtedly more difficult to control from above, and thus the SDA's degree of operational freedom *vis-à-vis* the Scottish Office may in other words have been underpinned by its mode of operation. Conversely, the absence of a clear and structured set of priorities may at the same time also have made it more difficult to resist external pressure from e.g. the Scottish Office to engage in specific initiatives of particular political salience.

Informational Resources and the Price of Knowledge

The multi-functional nature of the SDA as a development body meant that it would have to be able to mobilise and process information of many different kinds. By incorporating three existing organisations the Agency had originally made a head start with regard to some of its development activities, but in other policy areas the building of new informational resources was required because some functions, like e.g. industrial investment or sectoral strategies, were novel in the context of

66 SDA 1987a p. 16, SDA Board Papers 17.3.87, Frank Kirwan 1.6.90, James Williamson 14.6.90.

67 SDA 1987a, cf. Chapter 7. New formal scoring systems were introduced for internal assessment of projects in both economic development and environmental improvement (SDA 1987a p. 9, SDA Board Papers 30.7.87, 29.1.87, March 1988).

68 James Williamson 14.6.90.

69 James Williamson 14.6.90.

British regional policy. What was required was a combination of knowledge about particular development issues, ability to interact with private sector firms and organisations, and capacity to develop new policies to address the problems of regional development, and in order to ensure access to adequate informational resources, the SDA took a dual approach, attempting at the same time to enhance in-house expertise and to acquire additional input from external sources. For rather different reasons both of these avenues were subject to Scottish Office regulation.

With regard to human resources within the Agency, a limit applied to the total number of staff, resulting in a rather step-like development from 1975 to 1991,[70] and although its employees were not civil servants, civil service regulations still influenced working conditions: the discrepancy between the salaries offered by the SDA and e.g. private sector consultancies or financial institutions made it difficult to attract specialist staff,[71] and in the early years the wish to prevent the SDA from becoming 'top heavy' was a major area of tension with the London departments in charge of enforcing civil service standards.[72] According to the first Agency Chief Executive

> the Treasury would not have it that there could be a Director of Planning, and the reason in their minds was a very simple one, namely that he would not have enough people under him[73]

and in the Agency's formative period this seems to have been at least part of the explanation for the relatively slow progress towards the more 'innovative' role originally envisaged by its political proponents.[74]

The use of external informational resources by the SDA can be grouped under two headings. With regard to *policy implementation* some information-based services were not implemented directly by the Agency but indirectly via support for other providers. From the mid-1980s advisory services, especially those of a fairly standardised nature aimed at new and small firms, were increasingly carried out via e.g. Enterprise Trusts, local public-private partnerships employing the

---

70 IDS 1987a pp. 109ff. The average number of SDA staff in the three periods 1975-79, 1979-84 and 1984-90 were 546, 724 and 684 respectively (calculated on the basis of SDA 1976-88 and Committee of Public Accounts 1992).

71 Lewis Robertson 20.7.90, Donald Patience 31.5.90, Page 1977 p. 109. The introduction of performance-related pay in 1985 earned George Mathewson the accolade "Shit of the Week" in a civil service union journal (*Scottish Business Insider* Nov. 1985 pp. 4f).

72 Page 1977 p. 109, Lochhead 1983 p. 40, Grant 1982 p. 120.

73 Lewis Robertson 20.7.90.

74 Edward Cunningham, the SDA's Director of Planning, only took up his position in May 1977, i.e. nearly 18 months after the SDA started operating.

expertise and informal authority of private sector managers,[75] but the 1986 government review also encouraged the Agency to withdraw from the provision of other types of advisory services on the grounds that these activities 'crowded out' private sector business consultants.[76] Conversely, the Scottish Office was generally sceptical about the SDA's use of external consultants in connection with *policy development*, and the 1986 review prompted new procedures that established Agency-wide control with the use of consultants and required board approval for consultancy studies costing more than 200.000 pounds.[77] Although the cost of these studies may have been a stumbling block in its own right for the Scottish Office, it was also clear that while civil servants focused on the contents of the studies and found many of them to be questionable 'value for money', Agency executives stressed the importance of the informal authority *vis-à-vis* private sector executives that would be gained by basing policies on the work of private consultants of international standing.[78]

All in all it is clear that the political sponsors of the SDA for a variety of reasons – financial prudence, ideological considerations, civil service values – influenced the position of the Agency with regard to informational resources, and that this regulation appears to have become more extensive from the mid-1980s onwards. While it is difficult to establish at the corporate level to what extent this hampered policy development and implementation in the late 1980s, it clearly instituted additional criteria that 'project champions' within the organisation had to negotiate when devising new initiatives.

Organisation and Space

A significant trend in the development of the organisational structure of the SDA was, as illustrated by Table 6.1, the gradual move towards an increasingly area-oriented approach to regional policy in which resources were allocated to offices responsible for geographical localities rather than specific sectors of industry or particular policy instruments. The introduction of spatial considerations was warranted by the original parliamentary act that enabled the SDA to exercise its functions "in relation to Scotland or any part thereof",[79] but the organisational framework for the spatial aspects of the Agency's activities changed

75 SDA 1985 p. 65, 1987 p. 64, cf. Chapter 11.

76 IDS 1987a Ch. 7, cf. the discussion in Chapter 11.

77 IDS 1987a p. 77, cf. Gavin McCrone 20.6.90, and Chapter 7.

78 Gavin McCrone 20.6.90, George Mathewson 9.7.90, Edward Cunningham 13.6.90, John Firn 15.6.90, Charles Fairley 26.7.90, Sir Robin Duthie 12.7.90, Sir Kenneth Alexander 25.7.90, cf. Moore & Booth 1986a p. 65.

79 SDA Act 1975 Section 2 Subsection 1.

profoundly over the years. The first decade from 1975 to the mid-1980s was characterised by a series of so-called area initiatives,[80] focusing Agency resources in a small number of selected areas for limited periods of time, but these ad-hoc attempts to regenerate some of Scotland's most deprived areas were gradually supplemented by and eventually integrated in a more comprehensive organisational pattern. In 1985 an all-purpose office was established in Aberdeen, delivering the full range of SDA services to the north-east of Scotland, then in 1988 a network of seven regional offices was introduced, covering the entire SDA area and primarily involved in small business support, infrastructure improvement and area initiatives,[81] and finally in 1989 these offices were grouped together in a separate Regional Directorate.

The origins of the increasingly locality-oriented organisation of the SDA can be traced back to a number of different sources. *Firstly*, an important factor, not least in the development of the area initiatives seems to have been a political wish at the Scottish Office – Labour and Conservative alike – to be seen to do something in areas considered as 'black spots' in terms of unemployment or urban dereliction. *Secondly*, the SDA was right from the very beginning acutely aware of the controversial nature of the geographical balance of its activities. The organisation had been set up shortly after an official report recommended the establishing of a similar body for Glasgow and the surrounding Strathclyde region, Agency headquarters were located in Glasgow, and its first chairman was a former Lord Provost of Glasgow, so perhaps it was not all that surprising that rumour had it that the *S* in *SDA* referred to *Strathclyde* rather than *Scotland*.[82] Although in the early years the Agency clearly thought that devoting special attention to west-central Scotland was not entirely inappropriate given the scale of the problems in this part of Scotland,[83] such considerations could easily be overridden by the need to minimise the level of political 'noise' emanating from the generally speaking more prosperous eastern parts of the country.[84] *Finally*, the regionalisation of the delivery of large parts of the SDA's services in the late 1980s would seem to stem from several sources; the official rationale for this major organisational upheaval was to achieve

80 See Chapter 12.

81 This extensive regionalisation had been foreshadowed within the Agency's industrial property function: the idea had been mooted as early as 1982 (SDA 1982 p. 12) and in 1986 seven regional offices were established (SDA 1986 p. 63).

82 E.g. Lewis Robertson 20.7.90, Harry Hood 25.7.90, Alan Dale 16.10.91, David Lyle 23.7.90.

83 SDA 1977a pp. 3f.

84 E.g. Lewis Robertson 20.7.90, David Lyle 23.7.90, Craig Campbell 16.7.90, Ewan Marwick 24.7.90.

greater 'customer closeness' by ensuring an Agency presence at the local level,[85] but the systematic regionalisation could also be interpreted as being a continuation of a long-term trend towards a less ad-hoc and more systematically designed running programme of area initiatives.[86]

The increased emphasis on space within the organisation would seem to have had potentially important consequences for the SDA. Regionalisation apparently succeeded in boosting demand for Agency services,[87] thereby both increasing the impact and the public profile of its policies, but at the same time the successful expansion of regionalised delivery of existing policies threatened to undermine the SDA's long-term capacity to develop new initiatives in some fields of activity because relatively small specialist executive teams were broken up and dispersed to the regions from the Glasgow headquarters.[88] Regionalisation also affected, perhaps not entirely incidentally, the Agency's relations to its external political environment, both as regards the degree of operational freedom enjoyed *vis-à-vis* the Scottish Office and in its dealings with organised interests on the political scene in Scotland. Institutionalising the geographical focus of SDA activity probably could provide some shelter from the ad-hoc pressures of territorial politics and industrial closures by channelling them into a framework that was more manageable from the Agency's point of view,[89] but this had the side-effect of limiting the flexibility and discretion of the SDA by establishing an additional, and not necessarily compatible, standard of success for policy development and implementation.[90] The SDA was no longer 'just' about improving the industrial and environmental fabric of Scotland because the internal distribution of the changes brought about mattered, too, and the more locality-oriented the structure of the Agency became, the more difficult it must have become to redistribute resources to new areas of activity. Moreover, the final irony was that the institutional solution ultimately adopted, establishing a visible regional presence, may in the long run have backfired by increasing the external pressure on the Agency: the political system in Britain has a strong territorial dimension, from MPs to local authorities,[91] and therefore a network of regional offices would seem to be more exposed to local territorial politics than alternative mechanisms of policy delivery organised along e.g. sectoral lines.

---

[85] SDA 1988 pp. 11ff, Iain Robertson 11.7.90.

[86] David Lyle 23.7.90, Hood 1991a, cf. the discussion in Chapter 12.

[87] Iain Robertson 11.7.90, David Lyle 23.7.90, cf. SDA 1989 p. 5, 1990 p. 4.

[88] Alan Dale 16.10.91.

[89] Midwinter *et al.* 1991 pp. 184ff.

[90] See Moore & Booth 1986 pp. 114ff, cf. Chapter 12.

[91] Cf. Section 4.1.

The Agency clearly did attempt to manage local economic deprivation and pressures, but it could not make geography disappear from the political agenda. After all regional policy was still about the distribution in space of economic activity, and the territorial positioning of the SDA could not be reduced to a question of "Scotland versus the rest", no matter how convenient this might have been.

6.3. Funding Agency

For several reasons funding has attracted more political attention than other types of resources involved in the work of the SDA: finance has been the only resource that has been actively transferred to the arm's-length organisation on a regular basis, the question of public expenditure remained high on the political agenda from the mid-1970s onwards, and as most regional policies of central government involved transfer of financial resources to individual firms, the question of expenditure was used in political discourse to distinguish between 'strong' and 'weak' regional policy. This section explores the changing dependencies with regard to financial resources between the SDA and other actors, first with regard to the provision of public funds and then income generated internally within the Agency through private sector contributions. This analysis has required extensive efforts with regard to reconstructing a consistent series of financial data for the Agency[92] and assembling equivalent information of other organisations and policy programmes, and thus the figures presented below have not been presented before in a systematic fashion.

92 In order to construct a detailed breakdown of expenditure covering the entire period of the SDA's existence, the figures published by the Agency in its Annual Reports have been reworked in order to circumvent changing principles of presentation (possible because all annual accounts include the figures for the previous year in the mode of the present conventions) and in order to establish the SDA's commitment of resources to various policy areas capital and current expenditure have been aggregated and subheadings reassigned from composite categories, cf. the detailed comments in the chapters on individual policy areas. In order to enable comparisons in time all figures are given in fixed prices using the deflator available online from the UK Treasury, www.hm-treasury.gov.uk, unless otherwise stated. No figures for the SDA's first 3½ months of existence in 1975/76 are included because of the summary nature of the Agency's first accounts (SDA 1977c), but it is, however, unlikely that their inclusion would have altered the picture greatly: both the overall level of expenditure per months and the share of resources devoted to property development equal those of 1976/77, the SDA's first full year of functioning.

Public Funding

Being set up by an act of parliament, the SDA could be expected to be financed largely by public funds, and in total around two thirds of the Agency's financial resources were provided by central government.[93] The flow of public funds to the Agency was controlled by means of a financial limit on total expenditure as well as annually agreed budgets, so if the incoming Conservative government had wanted to maintain the Agency but limit its capacity to intervene in the Scottish economy, squeezing the total level of funding and/or tightening budgetary controls could have been attractive options. The overall financial limit to SDA expenditure was determined in the act of parliament establishing the Agency as a statutory body,[94] and, as can be seen from Figure 6.2, this upper limit has been progressively increased since 1975 by a series of parliamentary acts which in the early years included the possibility of extending the limit by ministerial order. The frequent changes of the financial limit in the 1979-81 period were surrounded by a good deal of adversarial rhetoric. The incoming Conservatives quickly reduced what was seen as grossly exaggerated limits introduced by the Labour government as "political gimmickry" in the run-up to the 1979 general election,[95] although removing the power to order an extended limit fell a long way short of attempts while in opposition to reduce the ordinary limit from 500 to 350 millions.[96] Nor was the irony of the situation lost on Labour MPs when following an acute industrial crisis only one year later the same Tory ministers proposed to extend the financial limit to a level close to that of the previous Labour administration as an expression of "ministerial confidence in the Agency's future".[97] This new cross-party consensus about the need for an RDA in lowland Scotland made the question of financial resources much less politically sensitive, and hence in 1987, the only situation in which the SDA was threatening to reach its financial limit, the parliamentary debate was conducted in a much more relaxed atmosphere, focusing on general issues of regional development and differences on policy strategies rather than the future of the Agency as such.[98]

---

93 Calculated on the basis of SDA 1977-91.

94 SDA Act 1975 Section 13. The financial limit only covers amounts outstanding "otherwise than by way of interest"; in practice this has amounted to c 80% of total public funding (calculated on the basis of SDA 1977-91).

95 Conservative Scottish Industry Shadow Spokesman Alex Fletcher 23.1.79 (*HCPD* vol. 961 cols. 244ff). For a similar interpretation, see Keating & Boyle 1986 p. 23.

96 Reported in the Glasgow Herald 9.2.79.

97 Scottish Industry Minister Alex Fletcher 11.12.80 (*House of Commons Standing Committee A* 11.12.80 col. 38).

98 *HCPD* vol. 120 cols. 839ff, cf. Chapter 14.

Figure 6.2 SDA financial limits and accumulated counting expenditure

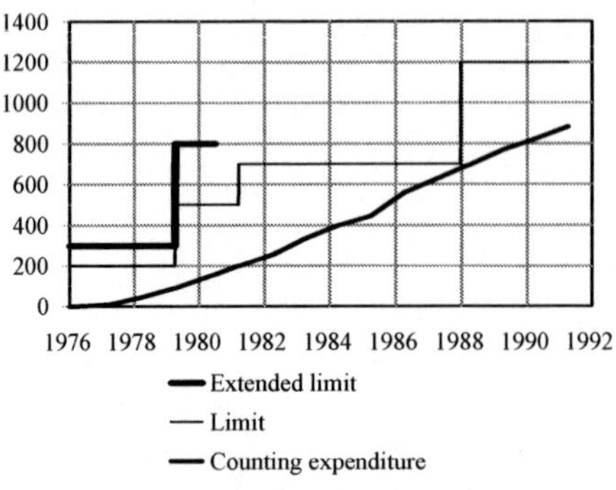

Current prices, £m.
Sources: SDA 1977-91, SDA Acts 1975 & 1987, Industry Acts 1979, 1980 & 1981.

While the political symbolism of changing financial limits cannot have escaped the SDA, the need to agree a detailed budget with the Scottish Office will have been a much more effective way of limiting the Agency's operational freedom, because the budgetary procedures allocated finance on an annual basis for specific areas and sub-areas of activity, the scope for rolling funds over from one financial year to another was very limited, and authorisation from the sponsor department was required if allocated resources should be shifted to another area.[99] These rules established a considerable potential for government influence on the Agency's financial resources from the overall level of expenditure to its distribution between various policy areas, and the evidence available suggests that in the budgetary cycle the Scottish Office did not confine itself to rubber-stamping of SDA proposals. Although the overall financial limit was far from being exhausted, the Agency repeatedly commented in public on its strained resource situation especially in the late 1970s and from mid-1980s onwards,[100] and

99 Page 1977 pp. 89ff, IDS 1987a pp. 111ff, Hood 1991b, Iain Robertson 11.7.90, cf. the specific controls introduced by the guidelines governing individual SDA functions (SDA 1977 p. 64; SEPD 1980a p. 1, 1980b p. 1). It remains unclear as to what extent the forward financial planning included subheadings below the level of broad policy areas such as industrial investments or advisory services (see IDS 1987a 104f, cf. Moore & Booth 1986b p. 110).

100 Chief Executive Lewis Robertson in his oral evidence to the Wilson Committee (1977f, vol. 6, p. 199), and as cited in the Glasgow Herald (24.11.77); the strained resource situation also recurred in the introductions to the Agency's annual reports (1977 pp. 6ff; 1986 p. 7; 1989 pp. 4ff, 1990 p. 8). The only period in which the over-

SDA executives have generally acknowledged limited financial resources as an important reason for the organisation's selective approach to regional policy, both on the corporate level and within individual policy areas.[101] On a more detailed level the Agency's project-driven mode of operation created problems for the Scottish Office because it did not lend itself readily to long-term departmental-style budgeting, and in the 1986 government review of the SDA this dilemma was addressed through a proposal to relax the annuality criteria but at the same time introduce more detailed forward financial planning.[102] This implied less Scottish Office control with the implementation of agreed projects – the exact timing of expenditure would now be a minor problem – but at the price of increased sponsor screening of the initial phase of project and policy development. All in all this indicates that the budgetary powers of the Scottish Office have actually been used to influence the availability and application of financial resources, but while this will have affected the SDA's capacity for developing its own strategic priorities, the outcome will have depended on the strategies pursued by the actors involved, i.e. the balance between the detail of Scottish Office regulation and Agency innovation with regard to economic development policies.

Figure 6.3 Policy inheritance and innovation in SDA expenditure

£m 1985/86.

Sources: Local Employment Act Annual Reports 1971-72, *Industry Act Annual Reports* 1973-76, SDA 1977-91.

all level of funding does not appear to have been a problem was the first half of the 1980s (Sir Robin Duthie 12.7.90), perhaps because government had declared that the Agency's budget would be constant in monetary terms for a four year period (Committee on Scottish Affairs 1980 p. 307).

101 George Mathewson 9.7.90, Edward Cunningham 13.6.90, Iain Robertson 11.7.90, James Williamson 14.6.90, Donald Patience 31.5.90.

102 IDS 1987a pp. 66, 111ff, cf. Standing Commission 1988 pp. 96f.

Having looked at the rules governing the allocation of financial resources to and within the SDA, we can now consider the key outcomes produced by this budgetary process. *Firstly*, the importance of organisational inheritance raised questions about the extent to which SDA funding would be additional to existing central government policies within the region. If the political rationale of the new organisation was to provide preferential treatment for Scotland through inherited and new functions, then it would clearly have to involve expenditure over and above that of the existing bodies which were incorporated in 1975 – otherwise the Labour government would lay itself open to criticism from especially the SNP that the Agency was merely a repackaging of existing policies designed to appease Scottish voters without offending its majority of English backbenchers.[103] Unsurprisingly, Labour ministers appear to have been anxious for the Agency to achieve high levels of expenditure quickly,[104] but as illustrated by Figure 6.3 the levels of expenditure on the inherited capital-intensive functions of environmental improvement and factory building eventually surpassed those of the early 1970s by an increasingly large margin from 1977/78 onwards. The setting up of the SDA would in other words appear to have been associated with a significant increase in the financial resources available in traditional areas of regional policy that now fell within the remit of the Agency, although the degree of preferential treatment involved did of course depend on the level of expenditure in other parts of the UK. *Secondly*, despite attempts to introduce more detailed screening of new policy initiatives, the development of broad patterns of expenditure would not seem to suggest that the SDA became less innovative after the arrival of the Conservative government in 1979, or, indeed, the 1986 government review. As can be seen from Figure 6.3, inherited functions such as environment and property still accounted for more than half of total expenditure in the early 1990s, but at the same time the financial importance of new policy areas had continued to grow in both absolute and relative terms throughout the 1980s, and although real-term expenditure growth was relatively low in the second half of the decade, non-inherited activities increased their share of SDA expenditure from less than 15% in 1979/80 to around 40% in the years leading up to the Scottish Enterprise merger.[105] These aggregate figures do of course not include new initiatives that may have been vetoed by the Scottish Of-

---

103 Page 1977 pp. 54ff, Cunningham 1977 p. 2, cf. *Glasgow Herald* 1.2.75, 28.2.75, 10.12.75, 7.2.76.

104 SDA Chief Executive Lewis Robertson 20.7.90, cf. his evidence to the Wilson Committee (1977f, vol. 6 p. 199) and the defensive note struck in the 1977 annual report (1977 p. 8).

105 Calculated on the basis of SDA 1977-91.

fice, but the overall pattern would certainly seem to suggest that the profile of the Agency's activities continued to develop also after the change of government. *Thirdly*, given the rules of financial sponsoring outlined above, the arrival of a Conservative government committed to setting the market free by rolling back the frontiers of the state could be expected to result in severe pruning of public spending in general and 'interventionist creatures' like the SDA in particular. Such a change is, however, difficult to substantiate, as illustrated by Figure 6.4. On the one hand the real term growth rate of government funding started to decrease significantly already under Labour, and in real terms public funding only peaked in 1982/83 under the Conservative government. On the other hand while the 1980s did see a gradual decrease in public funding,[106] this was largely compensated for by an increase in private sector contributions, and as the budgets of the organisation took expected business income into account, this result must have reflected the intentions of the sponsor department. The explosive growth in the early 1990s in private funding through disposal of industrial property was not translated into additional policy expenditure,[107] and thus the overall result was that the financial resources at the disposal of the Agency remained stable from the late 1970s onwards, as illustrated by Figure 6.3. Some have argued that this level of funding was inadequate[108] – certainly the HIDB was significantly better off in terms of *per capita* expenditure[109] – but although such claims are impossible to assess in detail because they rest upon often implicit assumptions about the nature of complex economic problems and possible remedial strategies, it should, however, be stressed that the financial resources at the disposal of the SDA cannot be described as marginal in relation to private sector manufacturing investment.[110] The financial position of the SDA from 1979 onwards therefore indicates a commitment by the Conservative government well beyond political gesturing to the existence of a regional development agency in Scotland, albeit on the basis of a changing balance between public and private sources of income, and the

---

[106] On average direct government funding from 1986 onwards was around 20% lower than in the first half of the decade (calculated on the basis of SDA 1981-91).

[107] While total SDA income nearly doubled in the last two years of its existence, expenditure increased by only 8% (calculated on the basis of SDA 1977-91).

[108] Standing Commission 1988 pp. 92ff, Alf Young 5.6.90, Helen Liddell 21.6.90, Douglas Harrison 19.7.90.

[109] Todd 1985 p. 46.

[110] SDA funding equalled between around 15 and 30% of gross domestic capital formation in Scottish manufacturing from the late 1970s onwards (calculated on the basis of SDA 1977-91 and CSO *Economic Trends Regional Accounts*). Given the targeted nature of SDA expenditure Danson's comparison with total Scottish GDP (1980 p. 13) would appear to be less relevant.

following pages explores the origins and possible strategic implications of the new funding structure.

Figure 6.4 SDA funding by main source

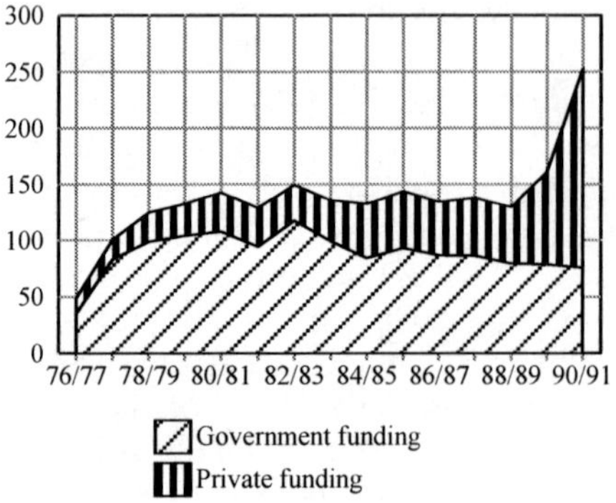

Government funding
Private funding

£m 1985/86.
Source: SDA 1977-91.

Private Sources of Finance

Many SDA activities were capable of generating income: fees could be charged for e.g. advice, occupants of Agency factories paid rent, investments could yield interest, and disposal of buildings and equity would produce capital receipts. As illustrated by Figure 6.4, the absolute level and relative importance of internally generated income increased from 1976 onwards, and the origins of this development can be traced back to a variety of sources. From the outset hand capital-related functions such as factory provision and industrial investment were key functions of the SDA, and the original policy guidelines issued by Labour ministers at the Scottish Office laid down rules that affected their income-generating capacity. On the one hand it was demanded that companies supported by the Agency "are not to be given an unfair competitive advantage" and should be charged "a rate of interest not less than that paid by commercial firms of the highest standard"[111] in order to ensure a reasonably 'level playing field' between supported and non-supported firms within the region as well as between Scotland and the rest of the UK, and thus capital-related functions such as equity and property were clearly not to be provided free of charge to the private sector. On the other hand while the Agency was free to discharge its investment and factory building powers within the general limits set out by Scottish Office guidelines, disposal of existing investments and

[111] Quoted from the original SDA investment guidelines (SEPD 1976b).

factories did require ministerial approval.[112] Although the Agency for rather pragmatic reasons initially appeared to be more keen on using disposals as a means of disciplining invested firms and generate additional income,[113] the different treatment of acquisition and disposal of assets suggests that its original political sponsors favoured the former rather than the latter. Operating under such guidelines the expansion of Agency activities – more factories and a larger investment portfolio – was in itself likely to spark off a process of incremental growth in internally generated income, especially in the form of property rents and interest on industrial investments. The 'earning capacity' of the SDA was, however, also affected by a series of initiatives following the change of government in 1979. *Firstly*, the Agency was actively encouraged to dispose of its capital assets "at the earliest practicable time consistent with its statutory purposes",[114] partly for ideological reasons – Conservative opposition to public ownership – but also as a measure of economy because disposals would allow the recirculation of funds and their use in new development projects.[115] *Secondly*, in order to avoid hidden subsidies to individual firms SDA services had to be delivered 'on a commercial basis':[116] rents and fees should be charged at market level, something which in effect amounted to a (possibly fortified) version of the original level-field policy of the Labour government. And *thirdly*, especially from the mid-1980s private provision of services was to be encouraged in order to avoid that an original 'market failure' was perpetuated by public intervention.[117]

The financial implications of these initiatives were far from unambiguous: disposal of assets would generate one-off income at the expense of long-term earnings, charging rents and fees at 'market level' would presumably generate additional receipts, and a move from direct supply of services towards support of private sector provision would undermine the Agency's possibilities to generate internal income by way of fees. All in all this suggests that the initiatives in the 1980s were governed not so much by a wish to limit public expenditure in the short term – although this was of course a welcome side-effect – but should rather be seen as part of a long-term political project driven by underlying assumptions about the ideal relationship between the private and public sector. In practice the resulting pattern of internal income generation did

112 SEPD 1976b p. 5, cf. Chapter 8.

113 See SDA 1977a, cf. SDA 1977b.

114 SEPD 1980a p. 3, cf. SEPD 1980b p. 1.

115 Secretary of State for Scotland George Younger as quoted in the Glasgow Herald 24.10.79.

116 IDS 1987a p. 30.

117 IDS 1987a p. 31.

of course reflect the extent to which these initiatives were translated into action in individual policy areas, but even so some general features should be noted. *Firstly*, the inherited capacity for income generation was obviously significant: rents from industrial property were the largest single source of private income and accounted for 60% of all receipts up till 1982/83.[118] *Secondly*, there is a sharp contrast between the majority of Agency functions that have generated little or no revenue at all, and the two high-earning areas of industrial property and investment which by providing capital to individual firms accounted for 96% of total receipts from the private sector.[119] *Thirdly*, the upward trend in internally generated income from the mid- and especially late 1980s was largely attained through disposal of property and industrial investments, but whether this reflects an outright privatisation strategy or merely should be seen as an attempt to economise with public resources will of course depend on the extent to which new investments were still undertaken.[120] *Fourthly*, if Agency functions with fee-generating potential like business advice were indeed delivered on an increasingly commercial basis, the effects of this must have been largely negated by the gradual transfer of these activities to the private sector because advisory services recouped little more than 2% of its total expenditure from 1980 onwards.[121] And *finally* the penchant in the late 1980s for publicising the relation between SDA expenditure and associated private investment in the form of 'leverage ratios'[122] was primarily political symbolism – signalling concerns about 'value for money' and adherence to private sector profitability – which had little value as performance indicators due to undisclosed methods of calculation, meaningless headline aggregations and an absence of comparable data for the late 1970s and first half of the 1980s. In financial terms by far the most important development in the 1980s was in other words the increased importance of disposal of existing capital investments, but as the annual budgets agreed with the Scottish Office took account of expected receipts, this must have been due to the general political incentive to accommodate the sponsor

118 Calculated on the basis of SDA 1977-91. Even when the 'anomalous' financial year 1990/91 with large sell-offs of industrial property is included, the property rent share of internally generated income still works out at 34% for the entire existence of the SDA.

119 Calculated on the basis of SDA 1977-91.

120 Disposals accounted for 32% of total receipts from 1976 to 1983 but rose to 66% from 1983 onwards (calculated on the basis of SDA 1977-91). For a detailed discussion, see Chapters 8 and 12.

121 Calculated on the basis of SDA 1977-91. In fact the highest receipts/expenditure ratio in advisory services was recorded before the change of government (8% in 1978/79), cf. Chapter 11.

122 See SDA 1987-91.

department because the Agency's direct financial incentive to boost internal revenues was undermined by the threat of reduced grant-in-aid or increased repayments of government loans.[123]

While the shifting structure of funding was not accompanied by major changes in the total level of financial resources available for SDA activities, it may still have had important strategic implications because the enhanced role of private sector finance effectively installed an additional criterion of success for policy implementation: it was no longer sufficient just to address the perceived problems of the Scottish economy, now policy programmes had to be designed and managed in ways that made the incentives entailed so attractive to the private sector that individual firms were willing either to pay for the services of the Agency or to acquire capital assets hitherto managed by the organisation. In theory this created a somewhat contradictory situation because *if* services could be provided on a commercial basis, then private sector firms – e.g. financial institutions, property developers, or management consultants – should be interested in discharging these functions, and thus the Agency would in effect be reduced to undertaking 'demonstration projects' which could alert private entrepreneurs to hitherto neglected market opportunities. Whether or not this left sufficient scope for a public regional development agency to operate would seem to depend on to what extent and in what ways these public-private assumptions were translated into practice,[124] but the risk of short-termism would certainly seem to be a very real one indeed, simply because a greater reliance on voluntary contributions from the private sector would make it more difficult for the SDA to act in situations where the immediate interests of individual firms did not coincide with the perceived need to address problems of a more long-term nature. In effect, by relying on two very different sources of finance – the Scottish Office and its private patrons – the Agency would have to take into account not only short-term political expediency and the ideological inclination of its public sponsors, but also the short-term financial interests of existing firms within the regional economy.

---

[123] IDS 1987a pp. 111f, cf. Standing Commission on the Scottish Economy 1988 pp. 95ff. One-off government 'claw-backs' of proceeds from realisation of SDA assets have received a good deal of media exposure (e.g. the Glasgow Herald 13.3.86, Standing Commission 1988 p. 66, Alf Young 5.6.90, Ewan Marwick 24.7.90), but in practice they still constituted in average only around 4% of receipts from private sources (calculated on the basis of SDA 1977-91).

[124] This is one of the central issues of contention between the 'external revolution' and 'internal evolution' interpretations of the development of the SDA (compare e.g. Danson *et al.* 1990 p. 175, Moore & Booth 1989 Ch. 7, MacLeod 1998a), and it will be discussed on the corporate level and with regard to individual policy areas in the ensuing chapters.

6.4. The British Context: Resourcing Regional Development

The setting up of a development agency in low-land Scotland in 1975 was clearly a new departure in British regional policy, but more was to come by way of change and innovation over the next decade and a half. In order to understand the extent to which the key features of the SDA identified above were unique or conformed to a broader pattern it is therefore necessary to examine other public bodies, both those responsible for policies on the British level and organisations operating in particular regions, with regard to the resources they could draw on to promote regional development.

A Changing Policy Network

In the mid-1970s the organisations involved in regional development activities were primarily central government departments or arm's-length bodies sponsored by these like the industrial estates corporations and the rural RDAs.[125] In the early 1990s this situation had changed significantly with new actors entering the policy area both from the European level above and the regional level below, and in the following the changing shape of the policy network will be pursued by focusing on the formal authority vested in various bodies.

On the British level responsibility for regional policy remained within central government, with first the Department of Industry and then from 1983 the Department of Trade and Industry taking the lead in policy formulation and implementation.[126] It is, however, also important to note three trends that made central government regional policy more diverse than in the 1960s and early 1970s. *Firstly*, within individual policy schemes authority was gradually delegated to the regional offices of the DTI in England and to the territorial departments for Scotland and Wales, albeit within guidelines applying throughout Britain.[127] *Secondly*, the urban policies sponsored by the Department of the Environment gradually became more and more oriented towards economic development, and thus on the British level two forms of spatially targeted policies for economic development came to be operated by different

[125] Cf. Section 4.3.

[126] Grant 1982 pp. 27-38, Hogwood 1982 pp. 45f, Rhodes 1988 pp. 330-43. The major exception in terms of implementation was the Regional Employment Premium, a wage subsidy operated by the Department of Employment until abolished in 1976 (Wren 1996a p. 108).

[127] McCallum 1979, Hogwood 1982, Begg & McDowall 1986, cf. Chapters 8 and 12. The degree of discretion enjoyed by individual regional offices varied, with those responsible for large assisted areas being able to handle much larger cases than other regions (Hogwood 1982 pp. 39f).

central government departments.[128] And *thirdly*, multi-functional RDAs became a much more prominent feature of British regional policy, as arm's-length bodies with fairly broad remits and a number of different policy instruments at their disposal were established in both lowland Scotland and Wales.[129]

At the same time new actors were also becoming increasingly active in regional economic development in Britain, as indeed elsewhere in Western Europe. On the subnational level local authorities, i.e. the regional councils in Scotland and the counties and metropolitan councils in England and Wales, established economic development units either within their administrative systems or as arm's-length bodies such as Enterprise Boards.[130] While regional-level local governments had a general responsibility for the well-being of their area and therefore potentially a very wide remit with regard to economic development, at the same time their spending on such activities was severely restricted through central government regulation, and the policies pursued were generally of a relatively limited nature.

From 1975 onwards developments on the European level also influenced regional policy in Britain, primarily along two lines. On the one hand central government grant schemes became subject to European regulation, both with regard to designation of assisted areas, the levels of grant support to individual firms, and the form in which subsidies could be given.[131] On the other hand distinct European policy programmes gradually came into being. Having started out as a mechanism of inter-state reimbursement, the 1988 reform of the Structural Funds instituted a programme-based approach that required the submission of multi-annual programmes for areas designated according to separate criteria which were not necessarily identical with the Assisted Areas designated with European approval for British regional support schemes. Central government did, however, generally retain its dominant position by drawing up the strategies to chairing the Monitoring Committees and providing the secretariat for individual programmes,[132] but an important exception to this pattern was found in West Central Scotland where Strathclyde Regional Council had initiated a major programme shortly before the 1988 reform had been approved, and here

[128] Barnekov *et al.* 1989 Ch. 6, Lawless & Haughton 1992.

[129] See Danson *et al.* 1992.

[130] See Goodwin & Duncan 1986, Clarke & Cochrane 1987, Lawless 1988, Totterdill 1989, Pickvance 1990, Cochrane 1990, Cochrane & Clarke 1990, Keating 1991 Ch. 7, McQuaid 1992, Fairley 1999.

[131] See Wishlade 1998, cf. Chapter 8.

[132] See Bachtler 1993, Wishlade 1996, Bache *et al.* 1996.

implementation was placed with an independent programme executive with the Scottish Office involved in a more limited capacity.[133]

Table 6.2 Formal authority in British regional policy in the early 1990s

Remit	*Implementing organisation*		
	England	*Wales*	*Scotland*
State aid regulation	DG IV	DG IV	DG IV
Location control	Ended 1981	Ended 1981	Ended 1981
Financial subsidies	DTI	Welsh Office	Scottish Office
Factory provision	English Estates	RDAs	RDAs
Environmental improvement	DTI	RDAs	RDAs
Regional development	Development Commission Local authorities Structural Funds	RDAs Local authorities Structural Funds	RDAs Local authorities Structural Funds

All in all the period saw major changes in the British regional policy network, and the situation in the early 1990s with regard to formal authority can be summed up as shown in Table 6.2. Compared to the situation 20 years earlier, four important changes can be noted. *Firstly*, the number of actors and tiers of government involved in regional development has increased dramatically, thereby increasing the need for inter-organisational coordination. In 1974 the main actors were two London-based central government departments supplemented by factory-building arm's-length bodies and two non-industrial RDAs, whereas in 1991 this had been supplemented by two territorial departments of central government, three new RDAs, a variety of local authority sponsored activities, and European programmes and regulation. *Secondly*, although limitations were now imposed from the European level, central government continued to be the most important actor in the British regional policy network, not only sponsoring a vast array of organisations and policies, but also regulating the economic development activities of subnational governments and playing a central role in

133 MacLeod 1999, Goodstadt & Clement 1997, Danson *et al.* 1999, Pieda 1992. While the early programmes in the east of Scotland were managed by the Scottish Office, later a Strathclyde-type approach became the norm rather than the exception (Hall Aitken Associates 1995 Ch. 10, McAteer & Mitchell 1996 p. 15).

the implementation of the European Structural Funds. *Thirdly*, within central government itself formal authority to pursue regional policy had become less centralised along two lines: spatially in that the discretion of the regional offices of the DTI and not least the Welsh and Scottish territorial departments had increased, and organisationally in that arm's-length bodies with broad development remits had become a prominent feature in the policy area. And *finally*, the resulting patterns in different parts of Britain clearly differed from one another. The nation-regions Wales and Scotland were now being served by relatively few public bodies with broad powers, primarily the territorial departments and the RDAs they sponsored, covering the entire region and potentially allowing a coordinated and integrated approach to regional development. In contrast to this the situation in the English regions was characterised by less stability and more fragmentation, with changing and overlapping designations of assisted areas and a range of specialised bodies each responsible for particular parts of the regional development remit.[134] In short, while the DTI, a functional department of central government with a general remit covering all of Britain, was still the most important actor, the way formal authority was distributed in the early 1990s implied that the positions of regions in political space, inside or outside England, had clearly become important in its own right.

Arm's-length Economic Development Bodies

The preceding pages have demonstrated that the SDA was part of two broader trends: arm's-length bodies were generally accorded a greater role in regional development, and more specifically the coexistence between major RDAs and a territorial department of central government was replicated in Wales. In the following a closer look will therefore be taken at some key features of these arm's-length development bodies in order to establish the extent to which the SDA has developed in parallel with similar organisations elsewhere in Britain.

In terms of sponsorship most arm's-length economic development bodies have, like the SDA, had only one sponsor, mostly a department of central government, although some of their activities may have been co-sponsored by other actors, and overall their remits were broadly similar to that of the SDA, i.e. to promote economic development and environmental improvement, plus for the rural agencies as a minor complementary aim to support social development and community activities.[135] Publically especially the first Thatcher government was

[134] For a comprehensive overview, see Hogwood 1996.

[135] WDA Act 1975, Welsh Office 1987 Ch. 5, HIDB Act 1965, IDS 1987b p. 14, Tricker & Martin 1984.

more than a little sceptical about the role of arm's-length 'quangos',[136] but in practice the record of the Conservatives in the 1980s demonstrates that pragmatic political considerations took precedence over principled views about public accountability: although the example set by the disappearance of the NEB in industrial policy can hardly have gone unnoticed,[137] the five RDAs were clearly subject to less drastic adjustment,[138] and within urban policy, another form of spatial economic intervention, the setting up of new quangos became a conspicuous mode of operation for central government in the 1980s.[139]

The formal authority of political sponsors to influence arm's-length economic development organisations generally falls under the same headings as those identified with regard to the SDA – appointments, resource allocation, strategies and implementation – although the extent to which some of these powers have been employed in practice appears to vary.[140] On the one hand some organisations have generally been more tightly regulated than others, especially with regard to decision-making powers concerning individual projects,[141] and on the other hand the degree of control exercised over individual organisations have varied over time for e.g. political reasons.[142] The composition of RDA boards have been broadly similar with a range of different interest being present,[143] but in some cases board members have clearly had a more

---

136 See Hood 1988, Harden 1988.

137 See Hague & Wilkinson 1983, Grant 1982 Ch. 5, Kramer 1989, cf. Chapter 8.

138 The 1980 Industry Act epitomised the 'solidity' of the position of the RDAs with regard to changes in formal authority: the NEB lost its main function as a public merchant bank (Wren 1996a pp. 128f), while only minor (parallel) changes were introduced with regard to the two industrial RDAs (cf. Chapter 7), and the position of the three non-industrial RDAs remained unaltered.

139 From 1981 to 1989 a total of ten Urban Development Corporations were established in England (O'Toole 1996 Ch. 1, Lawless 1991, cf. Chapter 12).

140 Cf. Hughes 1998a. On the HIDB see Scottish Affairs Committee 1984; on the WDA, see WDA 1977a, Welsh Office 1987, and Griffiths 1996 pp. 102f.

141 The HIDB's delegated powers with regard to financial support for individual firms were for instance significantly lower than those of the SDA (Scottish Affairs Committee 1984 pp. 266f, cf. Grassie pp. 103ff), and the WDA's discretion in administering industrial property was clearly more limited in the 1970s than that of the SDA (Hogwood 1982 pp. 52f, Committee on Welsh Affairs 1980 p. 82, Eirug 1983).

142 This appears to have been the case with the WDA when in the late 1980s a period of extensive control and political scrutiny followed a series of well-publicised instances in which the arm's-length body appeared to be 'out of control' (Morgan 1994, 1998; Morgan & Henderson 1997, cf. Committee of Public Accounts 1995). Also DBRW, the rural counterpart of the WDA, appears to have become subject of tighter government regulation in the 1980s (Howe 1996).

143 The requirements in the parliamentary acts setting up the WDA and the HIDB are similar to those laid down in the SDA Act.

extensive involvement in the work of the organisation with more full-time members and a more direct role in relation to individual projects.[144] Moreover, all non-English RDAs were in a position similar to the SDA in that they were sponsored by a territorial department of central government that was at the same time implementor of mainstream financial incentives, thus reinforcing the importance of the sponsorship relationship through indirect impacts of other regional policies.[145]

The position of arm's-length economic development bodies vary greatly with regard to organisational heritage. The HIDB had existed for nearly 10 years when it inherited its factory function from SIEC when the latter was being absorbed into the new SDA in 1975,[146] but with regard to both the Development Commission and the Welsh Development Agency (WDA) organisational heritage appears to have played a major role. The former had only acquired a unified RDA-type organisation by 1988 through a very gradual process of organisational mergers and reconfiguration,[147] while the position of the WDA in 1975 was similar to that of its Scottish lowland counterpart in that it incorporated existing organisational structures in the fields of factory building, environmental improvement and services for small businesses in rural areas.[148] Similarly it would appear that a project-driven mode of operation has been a feature of at least all the non-English RDAs in the sense that the organisations do not seem to have evolved exclusively on the basis of top-down corporate planning,[149] although neither the WDA nor

---

[144] The majority of members of the HIDB board was required to be committed on a full-time basis (Scottish Affairs Committee 1984 p. 1, Hughes 1998a), while the degree of discretion which had been delegated from the board to its executive staff was relatively limited (Scottish Affairs Committee 1984 p. 59). It is, however, also interesting to note that several HIDB board members had a civil-service background, although Gavin McCrone of the IDS insisted that "they are not placemen by any means" (evidence to the Scottish Affairs Committee 16.5.84).

[145] The HIDB handled Regional Selective Assistance within its area on behalf of IDS (Scottish Affairs Committee 1984 p. 40), something that suggests that the sponsor department did not worry about competitive outbidding for foot-loose investment from the Scottish Highlands and Islands.

[146] IDS 1987b Ch. 3 and 13, Welsh Office 1987 Ch. 13.

[147] Originally the Development Commission had only been advising the Department of the Environment about the spending of earmarked funds which were then channelled through a number of implementing bodies such as CoSIRA and SICRAS. In 1984 its status was changed to that of a grant-in-aid body with executive powers comparable to other RDAs, but it was only in 1988 that CoSIRA was integrated and a unified organisational structure created (see Rogers 1999 Ch. 8, 10 and 11, cf. Tricker & Martin 1984).

[148] Cooke 1980, Hogwood 1982 p. 53.

[149] See Hughes 1982, 1998a, Scottish Affairs Committee 1984 p. 2, Morgan 1994 & 1998, Morgan & Henderson 1997.

the two non-industrial RDAs seem to match the pace of organisational change that the project-driven approach produced within the SDA in the 1980s. The emphasis on initiative and discretionary decision-making strongly suggests that in terms of staff skills and informational resources the needs of other RDAs will have been parallel to those identified in relation to the SDA, although unfortunately the existing literature does not offer many clues in this respect.[150] The combination of extensive geographical coverage and the use of physical infrastructure for development purposes have helped to ensure that the spatial distribution of RDA services became part of territorial politics within the region and eventually resulted in organisational adjustments:[151] the activities of the Development Commission became increasingly focused on particular target areas from the late 1960s onwards,[152] the HIDB introduced a network of local offices in 1971,[153] and the WDA introduced a regionalised organisational structure in the early 1990s after having operated spatially targeted programmes.[154]

All in all it can be concluded that amongst the central-government sponsored British RDAs, the Development Commission is clearly different from the non-English bodies, both on account of its disjointed spatial coverage, its gradual and late consolidation as a unitary organisation, and the reliance on formalised networks with other sub-national actors in policy implementation. In contrast the Scottish and Welsh RDAs had much more in common with one another as multi-functional development agencies responsible for promoting economic development in consolidated regions. Their respective territorial departments of central government had a wide range of sponsoring powers capable of influencing policy-making and implementation by the arm's-length bodies, but at least with regard to non-financial resources this does not seem to have resulted in great differences between the various organisations or dramatic changes over time. On the one hand the only area in which the SDA could be said to appear to be noticeably different was

---

150 In 1990 the WDA, the second largest British RDA, had around 1/3 less staff than the SDA (calculated on the basis of Committee of Public Accounts 1992 p. 31), while in 1981 the HIDB had a staff complement of 249 (Grassie 1983 p. 132), and around 1990 the Development Commission employed 347 (Rogers 1999 p. 114). In a sample of Western European RDAs surveyed in the early 1990s more than 70% had less than 100 staff (Halkier & Danson 1997 p. 246), and thus British RDAs have been amongst the largest arm's-length development bodies in Western Europe.

151 Cf. Hughes 1998b.

152 Tricker & Martin 1984, Bovaird *et al.* 1989, Rogers 1999 pp. 74ff, Ch. 4.

153 Grassie 1983 pp. 50f. More than a decade later only 20 out of 260 HIDB staff were, however, based outside the Inverness headquarters (Scottish Affairs Committee 1984 p. 20).

154 Rees 1997, Morgan 1994, 1998; Morgan & Henderson 1997.

with regard to the relative importance of a project-driven mode of operation – and this impression may simply reflect the absence of parallel in-depth studies of e.g. the WDA or the HIDB. On the other hand the advent of the new Conservative government in 1979 did not lead to significant adjustments in the formal authority of the non-English RDAs.

Funding Regional Development

Historically central government had been the most important sponsor of regional policy in Britain with funding for individual programmes approved by parliament through its annual budgetary procedures, but although this position was maintained also after 1975, the increasing organisational complexity highlighted above was reflected in the ways in which funding was allocated. With regard to the regional policies sponsored by central government, the inherited patterns can be seen to have gradually changed along several lines. *Firstly*, with regard to financial subsidies the most costly regional policy programmes, the balance shifted decidedly away from automatic schemes entitling firms to public support towards schemes operating on a discretionary basis, thereby facilitating control of public expenditure.[155] *Secondly*, responsibility for discretionary grants had been transferred from Whitehall in London to the Scottish Office and Welsh Office, and thus the allocation of regional policy funding had acquired a more pronounced territorial aspect. *Thirdly*, the setting up of major RDAs in Scotland and Wales meant the emergence of new organisations which were not only placed at arm's-length but also sponsored by territorial departments of central government.[156] *Fourthly*, the increasing importance of urban policy, primarily sponsored by the Department of the Environment rather than the DTI, ensured that funding for spatial economic policy also potentially could become part of wider inter-departmental rivalries.[157] And *finally* the pivotal position of central government with regard to regional policies instigated by sub-national and European actors also applied to the allocation of financial resources. Statutory legislation only allowed elected sub-national governments to use the rather limited proceeds of a particular property tax for economic development activities,[158] and while ERDF monies had originally simply been treated as reimbursement of central government expenditure, the British interpretation of 'additionality' meant that even after the 1988 reform of the Structural Funds Euro-

155 Yuill *et al.* 1989 p. 63, cf. the discussion in Chapter 8.

156 IDS 1987b pp. 132ff, Committee on Welsh Affairs 10.3.80 p. 5, Welsh Office 1987 Ch. 5.

157 Lawless 1991 pp. 18ff, Barnekov *et al.* 1989 Ch. 6, cf. Chapter 12.

158 Miller 1990, Fairley 1999.

pean funding was still effectively used mainly to reimburse British public spending rather than to generate new projects.[159] All in all the period from the late 1970s till the early 1990s was characterised by two basic features, namely that central government continued to be a major sponsor of individual programmes and arm's-length bodies, but at the same time it also influenced the economic development initiatives of other tiers of government through rules governing their access to or employment of financial resources.

Figure 6.5 Grant-in-aid for British RDAs

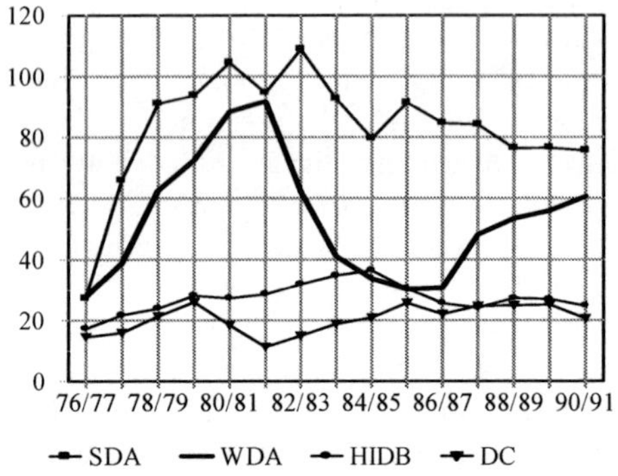

£m 1985/86.
Sources: SDA 1976-91, WDA 1977-91, HIDB 1976-91, Development Commission 1976-91.

The results in terms of financial resources produced by these allocation procedures constituted an important part of the policy environment in which the SDA operated, and this section explores the expenditure trends in the various policy programmes in order to establish the extent to which the Agency reflected broader British patterns and how these developments ultimately affected the position of the Agency within regional policy in Scotland. In order to ensure a reasonable degree of consistency and comparability, the analysis has required recourse to primary sources to a larger extent than what has been the case with other aspects of developments on the British level.[160] In terms of political

[159] For lucid accounts of the vexed question of additionality, see McAleavey 1993 & 1995, cf. Bache *et al.* 1996.

[160] Consistent series of financial information had to be reconstructed from the accounts published by government for all RDAs and for expenditure on central government regional subsidies in Scotland. For central government policies in Scotland, the accounts published in the *Industry Act Annual Reports* have been reworked by both Wren (1996a, 1996b) and Yuill *et al.*, with the former providing greater detail while the latter focuses on the main policy programmes.

commitment to arm's-length bodies an important parameter is the level of public funding provided, and Figure 6.5 shows the changing level of grant-in-aid, the most important form of government funding for British RDAs. This demonstrates that the SDA remained the largest organisation of this type in Britain throughout its existence, accounting for 45% of all central government expenditure by the major RDAs from 1975 to 1991,[161] but also shows that the Agency was clearly not the only development body which experienced a relatively high degree of financial stability in the 1980s:[162] the position of the English Development Commission soon recovered after the lengthy political review undertaken in the early 1980s,[163] and the reduced funding for HIDB from the mid-1980s onwards appears to reflect the effect of additional measures earlier in the decade.[164] Interestingly, the other large 'industrial' development agency, the WDA, deviated from this pattern because its public funding dropped by nearly 65% in real terms from 1981/82 to 1985/86,[165] primarily reflecting reductions in its factory building programme which had been temporarily expanded in the early 1980s.[166] Moreover, all British RDAs had the capacity to generate income from private sources through receipts for services and disposal of investments in equity and property, but in practice the structure of funding developed along rather different lines. For both the large RDAs operating in industrialised regions the relative importance of private funding gradually increased in the 1980s,[167] but at the same time not all sponsors of development bodies adhered rigorously to this general policy, because the RDAs operating in rural parts of Britain maintained broadly the same balance between public and private funding from the mid-1970s through to the early 1990s.[168] These divergences would seem to suggest that the

[161] Calculated in constant prices on the basis of SDA 1976-91, WDA 1977-91, HIDB 1976-91, Development Commission 1976-91.

[162] This tendency can also be seen with regard to other arm's-length bodies with economic development remits, as demonstrated by Harden 1988 pp. 309ff.

[163] Rogers 1999 Ch. 11.

[164] IDS 1987b p. 18.

[165] Calculated on the basis of WDA 1982-85.

[166] Factory building accounted for 80% of WDA expenditure in the early 1980s and was more than halved in real terms from 1981/82 to 1985/86 (calculated on the basis of WDA 1982-85). On the early 1980s 'emergency' programme of factory building in Wales in the wake of industrial closures, see Committee on Welsh Affairs 1980 cf. the discussion in Chapter 12.

[167] While the private share of income for both SDA and WDA averaged around 20% in the late 1970s, this figure had increased to around 45% in the late 1980s (calculated on the basis of SDA 1976-91, WDA 1977-91).

[168] For both rural RDAs the average share of private funding was actually slightly higher in 1976-79 than in 1988-91 (calculated on the basis of HIDB 1977-91 and Development Commission 1977-91), despite political statements about the desirability of in-

budgeting process with regard to RDAs reflected not only general considerations such as a perceived need to limit public expenditure but also concrete circumstances in individual regions.

With regard to main features of the expenditure structures, the extent to which inherited policies continued to dominate varied between RDAs. Both SDA and WDA incorporated existing government programmes of factory building and environmental improvement, but expenditure on non-inherited policy areas was generally higher in Scotland than in Wales, and by the early 1990s it constituted around 40% of total SDA expenditure but only around 20% in its Welsh counterpart.[169] Although both RDAs were sponsored by central government, the degree to which they gave priority to policy innovation would thus appear to differ, and again this would seem to suggest that specific regional conditions, political or otherwise, must have played a role in the policy development of these arm's-length development bodies.

Before 1975 central government enjoyed a near-monopoly on implementation of regional policy in Britain, and in expenditure terms by far the largest programmes had been the financial subsidies granted to individual firms in the designated Assisted Area. From the mid-1970s onwards the shift towards discretionary grants and the termination and downgrading of automatic subsidies resulted in a gradual, but very substantial, decrease in expenditure:[170] in the early 1990s the financial resources committed to British regional grant schemes was less than 25% of the mid-1970s level in real terms, and although Scotland did see individual years with exceptionally high levels of expenditure, the long-term decrease is also in evidence,[171] with the Scottish share on regional investment grants continuing to account for between 20 and 30% of total British expenditure.[172] The continued decline of central government expenditure on regional grants throughout the 1980s contrasted not only with the relatively stable financial circumstances of the RDAs noted above, but also with the growing emphasis on other forms of spatial economic intervention: when the Conservative government took office

creasing the access of individual RDAs to private funding (Scottish Affairs Committee 1985, IDS 1987b Ch. 3 & 4, Rogers 1999 Ch. 12).

169 Calculated on the basis of SDA 1989-91 and WDA 1989-91.

170 For a detailed discussion, see Chapter 8.

171 Expenditure on the REP wage subsidy in Scotland is unknown, but as this programme was terminated in late 1976 the two figures are strictly comparable from 1977-78 onwards. In 1976-77 REP accounted for nearly one-third of total British expenditure on regional grants (calculated on the basis of Wren 1996a p. 81).

172 The most noticeable exceptions are 1982-83 and 1990-91 when Scotland accounted for respectively 51% and 45% of the British total (calculated on the basis of Industry Act Annual Report 1977-91 and Yuill *et al.* 1989 & 1992-93).

the cost of urban policy equalled less than 15% of the value of regional subsidies, and 10 years later the former had overtaken the latter in terms of financial resources by a large margin.[173] The allocation of financial resources by central government would thus seem to suggest changing priorities, both with regard to policies, implementing organisations and the localities targeted.

Expenditure on regional policies sponsored from the sub-national and European levels is less straight-forward to trace because of the multitude of actors involved and complex patterns of co-funding, but still it can be established that for both types of activities the levels of expenditure did increase in the period from 1975 to 1991. On the one hand the level of European expenditure on regional development in Britain grew significantly in the latter half of the 1980s, ultimately reaching a level broadly similar to that of the (by then much reduced) policies sponsored by the British government[174] and with a sizeable share of these monies being spent in Scotland.[175] On the other hand the 1980s was also a period of growth for local authority initiatives in economic development, albeit from a fairly low level,[176] but it is interesting to note that the levels of expenditure reached in Scotland appear to have been lower than those of their English counterparts.[177]

All in all the financial environment in which the SDA operated changed profoundly from the mid-1970s till the early 1990, and the increasingly multi-level nature of regional policy in Britain can be illustrated by means of the adjoining Figures 6.6 and 6.7 which summarise the development with regard to financial resources in Britain and Scotland respectively.[178] On the British level the dominance of central

---

173 Calculated on the basis of Martin 1993 p. 271, Yuill *et al.* 1989 & 1992-93.

174 Calculated on the basis of ERDF 1975-88 and Yuill *et al.* 1989 & 1992-93. While in 1976-77 ERDF expenditure in Britain equalled 15% of the cost of central government regional grant schemes, this figure had risen to 85% in 1987-88.

175 In average 25% of UK ERDF expenditure before the 1988 reform was spent in Scotland (calculated on the basis of ERDF 1975-88), and within the 1989-93 Objective 2 programming period Scotland accounted for around 23% of expenditure in Britain (calculated on the basis of Yuill *et al.* 1992-93 p. 64, Hall Aitken Associates 1996 p. 3).

176 See Goodwin & Duncan 1986, Totterdill 1989, Pickvance 1990, Cochrane 1990, Keating 1991 Ch. 7, McQuaid 1992, Fairley 1999.

177 Fairley 1999 p. 110, McQuaid 1992 p. 30. In the mid-1980s Scottish Regional Councils only spent 38.3% of the maximum allowed by central government, less than half the percentage spend by the Greater London Council and the other English metropolitan councils (Fairley 1999 p. 112).

178 The underlying data represents the most comprehensive set of expenditure figures available, and only three minor corners have been cut: no figures have been included for DBRW in Figure 6.6, Objective 5b funding has not been included in Figure 6.7, and Northern Ireland is included in the ERDF figures in Figure 6.6. In order to allow

government policies gradually eroded while the weight of European programmes and RDA activities grew within a significantly reduced overall level of expenditure.[179] In Scotland the decrease in the aggregate level of expenditure on regional policy was certainly less pronounced,[180] and here the arm's-length RDAs were clearly able to maintain their levels of funding throughout the 1980s while expenditure on traditional programmes of regional subsidies declined, albeit in an uneven way, and thus the period from 1975 to 1991 in other words saw the SDA transformed from being cast in a supporting role to being a major implementor of regional policy in Scotland capable of matching the Scottish Office and the more fragmented European programme authorities, at least in terms of financial resources.

Figure 6.6 Regional policy expenditure in Britain by programme

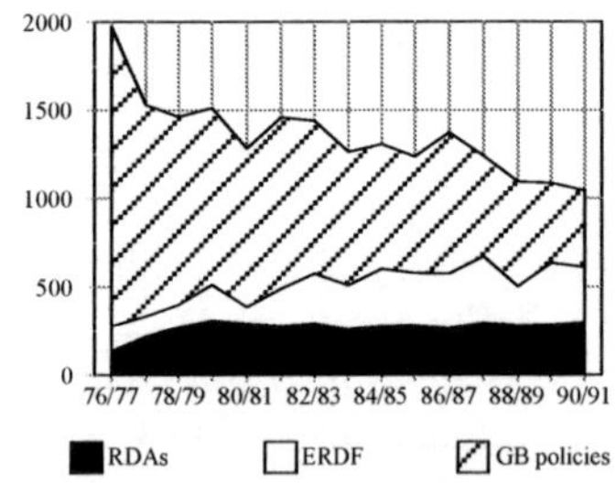

£m 1985/86.
Sources: Yuill *et al.* 1989 & 1992-93, Industry Act Annual Report 1977-91, ERDF 1976-88, Yuill *et al.* 1992-93 p. 64, SDA 1977-91, HIDB 1977-91, WDA 1977-91, Development Commission 1977-91.

Figure 6.7 Regional policy expenditure in Scotland by programme

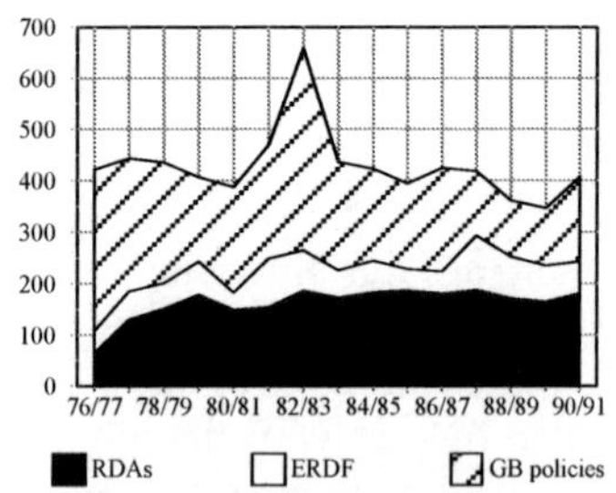

£m 1985/86.
Sources: Industry Act Annual Report 1977-91, ERDF 1975-88, Hall Aitken Associates 1996 p. 3, SDA 1977-91, HIDB 1977-91.

for pre-1988 reimbursement of national regional policy through ERDF, expenditure of the latter under the 'firms' heading has been excluded from the calculation.

179 A comparison of the first and last 3-year periods shows that expenditure was nearly 35% lower around 1990 than around 1978 (calculated on the basis of the sources quoted in Figure 6.6).

180 A calculation similar to that in the previous note shows a difference of only 15% in total expenditure.

6.5. Resource Dependencies and Regional Strategies

Exploring the position of the SDA with regard to policy resources has clearly demonstrated that the SDA must be seen as one of a growing number of arm's-length economic development bodies sponsored by central government, but operating on the sub-national level: beginning with the setting up of RDAs in Scotland and Wales in the mid-1970s, and continuing in a more fragmented form in England in the 1980s with new bodies becoming responsible for urban regeneration and the gradual emergence of the geographically sprawling Development Commission. Like other RDAs, the SDA had been created and continued to be sponsored by central government, and thus the organisation ultimately depended upon British parliamentary majorities rather than its links with the political environment in the region where it operated. Central government continued to be the single most important actor in British regional policy, but the structure of the policy network had changed significantly by the early 1990s: in addition to the new RDAs, the implementation of national programmes had become more decentralised and new actors had emerged in the shape of European programmes and local government economic development initiatives. Although influenced from London, the extension of the policy network with European and local actors as well as the major RDAs all helped to undermine the traditional position of central government as the sole sponsor and implementor of regional policy in Britain – and from a spatial perspective the variation in funding between arm's-length bodies clearly indicates that specific regional factors must have been significant in the process of resource allocation.

As sponsor department the Scottish Office was able to influence SDA activities through a variety of channels ranging from board appointments via strategic reviews to involvement in decisions relating to individual projects, and although these formal rules appear to have changed little over time, the use made of them and hence the ultimate outcome may of course have evolved. In general the Agency did, however, enjoy not only a considerable discretion with regard to implementation, but also continued to hold the strategic initiative in developing new projects and policies, and thus the organisation was arguably placed in an, albeit precarious, arm's-length position *vis-à-vis* its political sponsors. In terms of organisational and informational resources the Agency clearly had capacities that the Scottish Office did not, and thus the pattern of resource dependency was by no means a one-sided one, and the outcome, i.e. the 'length' of the 'sponsoring arm', will therefore ultimately have reflected the balance between the use made by the Scottish Office of its powers of control and review on the one hand and Agency discretion and project-driven initiative on the other, making the mode of operation of the latter a very significant feature indeed. Given its multi-functional nature and the prominence of organisational inheritance within the SDA, the outcome of this interaction is likely to have varied also between different policy areas, but still the study of developments on the corporate level has identified four sets of issues that should be taken into account when examining other aspects of agency activity. *Firstly*, integrating existing civil-service organisations into a new body with a more strategic, proactive and discretionary role will have created intra-organisational tensions, especially because the 'strategic deficit' with regard to inherited activities was concentrated in functions relating to physical infrastructure that involved specialist skills and informational resources that differed from most other areas of Agency activity. *Secondly*, a basically project-driven mode of operation coexisted with a growing emphasis on corporate-level planning, although apparently mainly as a reactive check on fast-lane project drivers. *Thirdly*, like in other British RDAs the increasing importance of space in policy implementation introduced an additional parameter in the work of the SDA: a project could be successful both from a geographical perspective by bringing jobs to an area with high unemployment, and functionally by supporting firms in particular industries throughout Scotland. Needless to say, the two criteria did not always coincide, but the adoption of a dual structure that organised the Agency's activities along both spatial and functional lines would appear to reinforce the dilemma. And *finally* while the overall level of expenditure was maintained in the 1980s, the structure of funding changed with increasing importance being given to attraction of private sector co-financing.

From the perspective of regional development strategy, the consequences of these features of the SDA are ambiguous. In terms of organisation the adoption of a project-driven mode of operation meant that the risk of an uncoordinated and somewhat haphazard effort could not be overlooked, especially when the operational directorates institutionalised potentially conflicting principles of policy delivery along functional and spatial lines, but at the same time pursuing the statutory remit of the Agency by means of an opportunistic search for areas in which its particular blend of resources and experience could also result in the development of initiatives tailored to the specific problems of the region – or make it more difficult to resist short-term political pressures. At the same time the increasingly dual structure of funding subjected the SDA to additional external pressures because the political and ideological inclinations of its public sector sponsors now co-existed with the short-term priorities of existing private sector firms and organisations. Likewise, the increasing relative importance of the SDA within the Scottish regional policy network in terms of financial resources potentially strengthened its position also *vis-à-vis* its political sponsors in the Scottish Office, but from a wider geographical perspective the fairly comfortable survival of the SDA into the 1980s was not unproblematic because the RDA-less industrial regions in England saw their position deteriorate relative to Scotland and Wales, potentially making the organisation more controversial in terms of territorial politics at the British level. In short, the analysis of the inter-dependencies between the Agency and its environment with regard to policy resources has clearly demonstrated the centrality of the nexus between the SDA and its sponsor department while at the same time underlining the importance of additional criteria for successful regional policies: not only will they have to further economic development, but both spatially targeting and the nature of public-private interaction also, and indeed increasing so over time, needed to be taken into account. Navigating the triangle between ideological concerns and territorial politics while ensuring private sector involvement would clearly be a challenge for the development strategies of the SDA, but one that had to be negotiated in order to stay in business in a turbulent political and economic environment.

CHAPTER 7

Corporate Strategies for Regional Development

Having analysed the resources at the disposal of the SDA, we can now consider the strategies through which these resources were employed to influence private firms via regional policies, and this chapter takes a first step by providing an analysis of the statements of strategy for the SDA as a whole which have spelled out the overall aims and methods and served as general guidelines for individual areas of activity. As programmatic statements about the future, such corporate strategies may or may not have been observed after having been produced: the extent to which their priorities influenced individual policies cannot be taken for granted but must be established, and thus an overall assessment of the development of SDA strategies will also have to take into account statements of aims and methods for particular Agency activities as well as the implicit assumptions that can be inferred from implementation in individual policy areas. The situation is furthermore complicated by the fact that corporate strategies are not just for 'internal consumption' but also aimed at an external audience such as political sponsors as a way of generating or maintaining support,[1] and as the sponsoring Scottish Office had been given formal authority to oversee the general strategies of the arm's-length body, statements about the SDA's aims and methods have emanated from both its Glasgow headquarters in the form of e.g. corporate plans and from its sponsor department in Edinburgh in the shape of statutory regulations and guidelines.

Within the existing literature the question of how and why the corporate strategies of the SDA changed is a central area of difference between the external-revolution and internal-evolution interpretations, as demonstrated in Section 2.3 above. While some have argued that the advent of the Thatcher government in 1979 made the Agency abandon its 'interventionist' approach for a much more 'market-oriented' one,[2]

1 Bogason 1992 p. 23.

2 Danson *et al.* 1988 pp. 1f, 1989b pp. 71f, 1989c pp. 4f, 1990a pp. 172ff, 1990b pp. 181f, 1992 pp. 299f; Hood 1991a, Swales 1983, Rich 1983, Midwinter *et al.* 1991 pp. 187ff, Barnekov *et al.* 1989 Ch. 7.

others have argued that the differences between the 1970s and 1980s were in fact rather limited and reflected the learning process of a new type of development body rather than changes forced upon the organisation from outside.[3] These opposing views can also be found among the SDA executives interviewed,[4] where some claim that the SDA's approach in the early years had been that of "an overarching Agency controlling the economy", "protecting ailing companies" and aiming "to own the commanding heights of the economy".[5] This picture is then contrasted with the 1980s where the Agency allegedly adopted a 'commercial' and 'catalytic' approach, working 'with the grain of the market' in trying to make the private sector improve its performance.[6] In contrast to this others claim that the coming of the Thatcher government did not substantially alter the industrial strategies of the Agency:

> we viewed ourselves as being catalytic as opposed to interventionists from the beginning. [...] After 79 I was told the Agency had been interventionist, but then we say: 'Well, what's the difference now?', they say: 'well, you're still interventionist but it's less objectionable because the Conservatives are doing the intervening, not Labour'.[7]

Central to this controversy would seem to be the relationship between public and private actors, something that underlines the potential importance of 'ideological hindsight' in the interpretation of the history of the Agency.

This chapter will attempt to illuminate the corporate strategies of the SDA by going back to the original statements in order to reconstruct the development of the overall aims and methods of the organisation. Three distinct phases can be identified, and for each of these periods documents produced by both the SDA and its political sponsors will be analysed both in terms of development strategies, i.e. the general aims and methods of regional policy, and with regard to the assumptions

---

3 Moore & Booth 1989 pp. 11, 142f, Ch. 7, 1986b; Moore 1994 pp. 53ff, 1995 pp. 234ff, Booth & Pitt 1983, Firn 1982, Keating & Boyle 1986 pp. 24f, Keating & Midwinter 1983 pp. 36ff, Midwinter *et al.* 1991 p. 192, Young & Hood 1984 p. 47, McCrone & Randall 1985, Kellas 1989 p. 225, MacLeod & Jones 1999 pp. 593ff.

4 These conflicting interpretations would seem to reflect the personal history of the interviewees: whereas the first three quotes come from persons who only became involved with the SDA in the 1980s, the fourth quote stems from an executive director which had worked for the Agency from the late 1970s through to the early 1990s.

5 Quotes by George Mathewson 9.7.90, Frank Kirwan 1.6.90, James Williamson 14.6.90. The notion of 'controlling' the 'commanding heights' of the economy is clearly linked to the vocabulary often used in connection with the nationalised industries.

6 Cf. George Mathewson 9.7.90, James Williamson 14.6.90, Frank Kirwan 1.6.90, Iain Robertson 11.7.90.

7 Edward Cunningham 13.6.90.

entailed about inter-regional disparities, remedial procedures, the distribution of roles between public and private actors, and, indeed, the underlying values that were the social rationale for public intervention. This should enable us to draw conclusions about continuity and change in the development of the SDA's corporate priorities with regard to regional development and the degree of bureaucratic autonomy enjoyed by the Agency in these matters – and about the positioning strategies of the arm's-length body *vis-à-vis* its political sponsors. Finally these findings are placed in a broader context through an analysis of the changing strategies entailed in other parts of the British regional policy network in order to illuminate the political environment in which the SDA operated and the extent to which it conformed or deviated from general trends in the thinking about regional development strategies from the mid-1970s to the early 1990s.

7.1. The Early SDA: Statutory Remit and Industrial Strategies

The 1975 SDA Act and the First Operational Guidelines

An obvious point of departure for a discussion of the aims and methods of the SDA is the 1975 Act of Parliament that defined the purposes of the new body as:

- furthering economic development,
- provision, maintenance or safeguarding of employment,
- promotion of industrial efficiency and industrial competitiveness, and
- furthering the improvement of the environment.

These goals were to be obtained by means of a number of functions which allowed the Agency to provide finance, sites and premises to industry, carry on or otherwise promote industrial undertakings, and assist the establishment, growth, reorganisation, modernisation or development of industry, promote industrial democracy in undertakings controlled by the Agency, and bring derelict land into use and improve the environment. In fact, ultimately the SDA was empowered to do "anything, whether in Scotland or elsewhere, which is calculated to facilitate the discharge of their functions".[8]

The description of the purposes and functions of the new body in the original SDA Act defines Scotland's regional problem in terms of slow growth and unemployment caused by uncompetitive firms and inade-

[8] SDA Act 1975 sections 2-4.

quate physical and organisational infrastructure, and although this does not cover each and every way in which regional disparities have been perceived,[9] it was clearly still a very broad remit. The underlying values that made these inter-regional disparities legitimate targets of public policy were stated less clearly, but the dual emphasis on growth/employment on the one hand and industrial efficiency on the other could suggest a combination of traditional concerns about inter-regional equality and a new emphasis on regional competitiveness.[10] In terms of operational development strategies the 1975 Act enabled the SDA to do 'anything', albeit of course within the framework of Scottish Office political sponsorship,[11] but as neither the balance between inherited and new functions nor the expected firm-level impact is specified, the extent to which the new body would emphasise attraction of investment from outside Scotland in line with traditional central government regional policy or primarily support indigenous firms within the region is unclear. The nature of the policy instruments that could be employed by the new organisation was, however, clearly circumscribed by the Act because 'anything' generally only refers to non-mandatory measures which do not legally force private actors to comply with SDA priorities: informational, organisational and financial resources could be offered on more or less attractive terms, but ultimately it would be up to the private actors targeted to decide whether they wanted to become involved.[12] Despite fragments of language associated with interventionism – e.g. 'industrial restructuring' and 'economic democracy' – the power of the SDA to intervene in private firms and industries was limited from the outset, and the Act entailed an important general regulatory mechanism, namely the ability of the sponsor department to institute so-called 'financial duties' for individual policy areas[13] which would function as measures of the extent to which individual firms are subsidised by

---

9 Compared to the issue definitions listed in Table 3.5, the lack of reference to the degree of external ownership would appear to be a deliberate omission, and the only objective not normally considered part of regional policy, promotion of industrial democracy, should probably be seen as a parallel to the powers given to the NEB on the UK level (see Section 2.2 of the Industry Act 1975, cf. Kramer 1989 pp. 9ff).

10 Cf. Moore & Booth 1986e.

11 Cf. Chapter 6.

12 The only exception to this general principle of non-mandatory policy instruments was the Agency's power to acquire land by compulsory means inherited from predecessor organisations (SDA Act 1975 Section 9), and, interestingly, a similar option was not created for the new industrial investment function, perhaps because – in contrast to the NEB – extension of public ownership was not one of the statutory objectives of Scotland's RDA.

13 SDA Act 1975 Section 12.

particular policies – and thus created a platform for turning the discretionary appraisal of individual projects into a political issue.

A more detailed outline of the aims and methods of the SDA can be found in the first Scottish Office guidelines for the investment function,[14] a document which also contains statements of a more general nature. First it is worth noting that the starting point is still a dual perception of the regional problem: the task of the new organisation is defined as "promoting the growth and modernisation of Scottish industry",[15] and thus objectives are present which, at least in the short run, may point in different strategic directions, but, like the 1975 Act, not prioritised. With regard to policy instruments the emphasis on non-mandatory measures found in the Act was underlined when the guidelines insisted that investments should normally be undertaken with the agreement of the board of the private company, and if this agreement was *not* forthcoming the Secretary of State had to be notified so that a political view could be expressed or a specific direction given in advance of the SDA acquiring even a relatively small stake in the company[16] – the Agency was in other words not allowed to use its financial resources against the wishes of existing management unless backed by the political authority of the Scottish Office. Last but not least the investment guidelines specified the role of the financial duty for the investment function as ensuring "an adequate return on the capital employed",[17] and as such a duty will reflect the aggregate performance of all SDA investments, it was translated into implementation criteria according to which individual investment projects would be appraised: the Agency "can, if necessary, wait longer for a return than a private investor would", but "in taking investment decisions it must, however, always have regard to their profitability",[18] and what this meant in practice would of course depend on how 'patient' the Agency was prepared to be and what would be regarded as an 'adequate return'.[19] The considerable discretion vested in the arm's-length body by the guidelines may simply reflect a desire to be seen as both socially considerate and economically responsible, but as 'patience' and 'profitability' could easily be construed to imply incompatible assumptions about the relationship between public and private actors, the juxtaposition was a good starting point for party-political controversy.

14 SEPD 1976b.

15 SEPD 1976b p. 1.

16 SEPD 1976b p. 4, cf. Chapter 8.

17 SEPD 1976b p. 6.

18 SEPD 1976b pp. 6f.

19 Cf. the discussion in Chapter 8.

All in all the original SDA Act and the 1976 industrial investment guidelines clearly gave the new body both an extensive remit, a wide range of powers, and a mixed bag of underlying expectations against which its performance could be judged. The organisation had been given broad and potentially conflicting objectives, reflecting both traditional conceptions of the regional problem as spatially concentrated unemployment and a new focus on modernisation and competitiveness, and this strategic ambiguity was reinforced by the nature of the two major restraints placed upon the new body regarding the relationship between the Agency and individual private actors. It was always perfectly clear what the SDA could *not* do: mandatory policy instruments with powers of compulsion were largely absent, and thus the Agency would have to offer its resources to private actors on conditions that made the latter act in ways more conducive to regional development objectives. At the same time the introduction of financial duties implied that the interaction of the RDA with private actors would be assessed not only on the basis of long-term regional development objectives, but also in terms of the financial performance of individual investments. These two criteria may or may not coincide in any given project, and the dilemma was presented as a need to strike a long-term balance between employment and profitability – a discursive strategy which could be construed either as public subsidies to private firms or as an attempt to turn the Agency into a commercial merchant bank.

'Hidden' Agendas: The SDA's First Industrial Strategies

In the early years the attempts by the arm's-length body to define its corporate strategies within the broad strategic framework established by its political sponsors can be followed through two types of sources, namely the unpublished industrial strategies prepared by the SDA for discussions with the Scottish Office in accordance with the investment guidelines,[20] and the same strategies as they were presented in the SDA's annual reports and elsewhere. Although the industrial strategies were negotiated with the Scottish Office, the initiative still appears to have been with the arm's-length body while the role of the sponsor department was mainly limited to commenting on draft versions prepared by the Agency,[21] and the statements issued by the SDA were generally much more specific than those emanating from its political sponsors. With regard to the way the regional problem was perceived

[20] SDA 1977a, 1978a.

[21] This was the distribution of responsibilities described in the investment guidelines (SEPD 1976b p. 1), recollected by the SDA's first Chief Executive (Lewis Robertson 20.7.90), and reflected in the minutes of a meeting between the SDA and the Scottish Office (SDA 1977b).

the development body introduced a hard-nosed rhetoric which explicitly gave priority to some issues over others: "[t]he Agency is very clearly about jobs, but its attention is fixed on jobs in efficient and progressive companies and industries, and not on the creating of uneconomic employment, jobs for their own sake".[22] The SDA, in other words, stated its task as achieving social goals by economic means:[23] improving the technological efficiency and long-term competitiveness of Scottish firms took precedence over short-term maximisation of regional employment and economic activity. On the basis of this it is hardly surprising that the issue of industrial modernisation was central to the operational strategies of the SDA from the outset, and the increasingly detailed statements show that the Agency concentrated management efforts on developing its 'new' industrial investment function and associated activities.[24] The existing sectoral composition of the Scottish economy was seen as a major weakness, especially the predominance of declining traditional industries such as heavy engineering and textiles,[25] but modernisation could be brought about in several ways, and judging from the Agency's industrial strategies and Annual Reports the emphasis gradually developed. In the beginning the SDA tended to give priority to the revitalisation and restructuring of existing indigenous industries while the proportion of 'technologically based industries' was primarily to be increased by means of venture capital from the investment function,[26] but already in 1978 the first rounds of intense work on corporate planning had, however, resulted in a three-pronged strategy: promoting new firms in high-tech areas by means of venture capital for indigenous entrepreneurs, attraction of inward investment from outwith Scotland, and support for the modernisation of traditional industries.[27]

In the early industrial strategies managing the relationship with private economic actors is a crucial theme in at least three respects. First it can be noted that the Agency prioritised long-term profitability rather than short-term employment effects when undertaking discretionary appraisal of individual projects, but the notion of long-term commercial viability was sufficiently vague to warrant involvement in anything short of outright "'rescue' cases",[28] and thus a considerable degree of discretion was maintained. At the same time it should, however, also be

---

22 SDA 1977 p. 8.

23 Cf. SDA 1978a p. 6; Cunningham 1977 p. 4, 1978 p. 5; *Glasgow Chamber of Commerce Journal* Jan. 1976 p. 19.

24 SDA 1977a, 1978a; cf. Robertson 1978 p. 22.

25 SDA 1977a p. 4, 1978a 9 4; Cunningham 1977 pp. 8f, 1978 pp. 4f.

26 SDA 1977a pp. 3ff, cf. 1977 p. 6.

27 SDA 1978a pp. 26ff, cf. 1978 pp. 6ff.

28 SDA 1976 Section 43, cf. SDA 1977a p. 8.

stressed that a cooperative approach towards key private actors in the regional economy was envisaged with the Agency acting as "a catalytic force" which would "complement rather than substitute or compete against the activities of existing financial institutions" in order to "mobilise complementary resources".[29] The strategic implications for individual projects of co-investing alongside private financial institutions may be uncertain – it could either support the emphasis on commercial viability or lead the Agency into focusing on marginal projects – but politically it signalled an unaggressive positioning strategy *vis-à-vis* private providers of finance, and this was perhaps particularly important because at the same time the SDA clearly intended to change the way development projects were generated. The early strategic statements proclaimed instead of having a largely reactive industrial role – evaluating investment proposals from the private sector – the Agency would soon

> move into an innovatory role, [...] [identifying] elements of industry that can usefully and profitably be created, [...] sometimes as wholly new and independent enterprises, but often in partnership with existing companies.[30]

The nature of this new proactive role did, however, quickly change, at least at the level of corporate strategies, because the following year the innovative role was restated as "the study, often along with others, of market and technical potential, and the relating of that potential to the advantages offered by areas of Scotland".[31] The focus of Agency proactivity in other words shifted from the establishing of new firms – akin to the NEB's remit of extending public ownership to modern profitable industries[32] – to a commitment to identifying new opportunities that could be exploited by or together with private actors,[33] something which soon became synonymous with a sectoral approach where individual industries were examined in order to identify ways in which performance could be improved through various Agency initiatives.[34]

All in all it has been found that the SDA in its first years of existence lived up to its role as an RDA by beginning to develop operational strategies within the broad framework set out by its political sponsors.

---

29 SDA 1978a p. 8, cf. p. 2; see also 1977 p. 4. The implications of the SDA becoming a major industrial holding company was considered seriously only once, but in a document that was not in public domain (SDA 1978a p. 8, cf. the discussion in Chapter 8).

30 SDA 1977 p. 7, cf. Robertson 1978 p. 27.

31 SDA 1978 p. 6.

32 Cf. Chapter 5.

33 SDA 1978a p. 16.

34 Cf. Chapter 9.

Priorities such as industrial modernisation, long-term viability, co-operation with private providers of finance, proactivity and selectivity would obviously depend on how they were translated into practice, but their immediate political significance would appear to be more clear-cut: an attempt to distance the SDA from employment-oriented rescue operations. But although the use of 'interventionist' jargon such as 'restructuring' gradually became less prominent, the relatively vague criteria for 'commercial' firm-level intervention still made it possible to interpret the organisation as just another provider of 'soft money' which would bolster profits or employment without improving the long-term competitiveness and development potential of the firms subsidised.

7.2. Into the 1980s: From Agency to Enterprise?

Conservative Guidelines: Challenging the Past?

The newly elected Conservative government decided to maintain the SDA[35] while modifying the framework within which it was operating, both on a statutory level and by issuing new guidelines from the Scottish Office.

The Industry Act 1980 modified the purpose of the Agency by downgrading "the provision, maintenance or safeguarding of employment" from a purpose in its own right to a specification of the overarching purpose of "furthering economic development", two functions of the SDA were removed – industrial reorganisation and promotion of industrial democracy in invested firms – and a new one added, namely to promote private ownership of industry by disposing of its holdings and restricting the Agency's possibilities of establishing new firms and acquiring shares in existing ones.[36] In one sense this amounted to very limited change because the purposes and functions of the organisation effectively remained unchanged: the SDA had never come around to taking any concerted action with regard to industrial democracy,[37] the commitment to employment was not abandoned all together but only downgraded to a position similar to the one it occupied in the industrial strategies of the Agency, and promotion of private ownership hardly heralded a revolution because extension of public ownership had never in itself been a purpose in the original SDA Act. But still the political significance of the changes was also clear: removing industrial restructuring as a function and encouraging disposal of industrial investments suggested that attention would increasingly focus on the relation be-

---

35 Cf. Chapter 14.

36 Industry Act 1980 Section 1 Subsection 2, and Section 6.

37 SDA 1978a p. 15, cf. Chapter 8.

tween public and private actors and the extent to which Agency activities were compatible with the ideals of the Conservative government.

The amendment of the statutory framework was accompanied by new Scottish Office guidelines for the industrial investments function. This document, later to become a central point of reference for the external-revolution interpretation of the development of the SDA,[38] opens by stating that

> The Agency shall regard as its principal functions, to be discharged in accordance with Government policy: industrial promotion and the attraction of industrial investment to Scotland; [...] the provision of premises for industry; and environmental improvement. The Agency's industrial investment powers are to be used as a complement to these functions.[39]

Following this the sponsor department promised tight budgetary control over the proportion of funds allocated to the industrial investment function, demanded "full regard to the profitability of the enterprise", "maximum private sector participation", and disposal of shares "at the earliest practicable time consistent with its statutory purposes".[40] Like the legislative changes, the 1980 investment guidelines are undoubtedly important as a political signal from the incoming Conservative government. On the one hand the prominence accorded to the attraction of inward investment, hitherto mainly seen as a subsidiary activity by the Agency,[41] was certainly a rearrangement of corporate priorities – perhaps simply reflecting the fast-rising importance of this Agency activity from 1977 onwards[42] – with less emphasis on furthering the development of indigenous industry and more on making Scotland an attractive location for international investors. On the other hand stressing the need to be commercial in the discretionary appraisal of individual projects and downgrading the importance of employment again focused attention on the relationship between the SDA and private actors, but as long as disposals had to be effectuated only when consistent with the Agency's statutory development purposes, considerable room for discretion would still seem to exist, and thus apart from being restricted to taking minority stakes, something which emphasised the non-mandatory character of policy instruments and prevented the Agency from evolving into a state holding company, the investment function could in theory continue without major alterations. Moreover, already the pre-1979 industrial

38 Cf. the discussion in Chapter 8.

39 SEPD 1980a p. 1. Although published in 1980, the text of the guidelines had been finalised in December 1979 (Committee on Scottish Affairs 1980 vol. 1 p. 10).

40 SEPD 1980a pp. 2f.

41 SDA 1978a pp. 26ff, cf. SDA 1978 pp. 6ff.

42 Cf. Chapter 10.

strategies of the SDA had given priority to industrial modernisation over employment objectives, and commercial viability had long been emphasised in Agency statements about individual investment projects, and thus the change of government did not cause a strategic redirection but merely reinforced existing trends within Agency thinking.

All in all the incoming Conservatives pursued two main objectives with regard to the corporate strategies of the SDA: on the one hand additional political authority was lend to the emphasis on commercial viability of individual projects already proclaimed by the Agency, and on the other hand a new hierarchy amongst its policy instruments was instituted which emphasised competition for footloose investment. If this dual manoeuvre was motivated by a political wish to maintain a high-profile role for the SDA without the direct influence on individual firms entailed in the industrial investment function, then, ironically, it could be argued that a new can of worms had been opened, because an extended Scottish promotional effort was bound to intensify competition between the regions for new jobs and investment and hence increase tension within the British regional policy network.

The 3 Cs and Corporate Planning

The response of the SDA to the advent of a Conservative government can be followed in the annual reports and corporate strategies of the early 1980s.[43] According to the corporate strategy drafted in 1984, the central objective of the organisation was "to contribute to the growth of the economy [...] in partnership with the community and co-operation with other institutions" by means of "support for selected new enterprises and existing businesses, geographic areas of potential, and strategically important industrial sectors." Furthermore a climate conducive to economic development should be created by encouraging "a more entrepreneurial culture", "identifying opportunities for others to pursue", and improving the Scottish environment and infrastructure.[44] This general approach led the Agency to select four priority areas, namely *technology* in order to enhance competitiveness, the *service sector* in order to secure new job possibilities, *enterprise development* for indigenous firms, and development of particular *geographical areas*.[45] The organisation's general principles of operation – to improve the function-

43 The 1981-84 strategy is extensively summarised in the 1981 annual report (SDA 1981 pp. 57-63, cf. SDA 1982 p. 69, SDA/DCC February 1985 (2)), while the 1985-88 document has been obtained in a late draft (1984a).

44 All quotes from SDA 1984a, cf. SDA 1981 p. 57.

45 SDA 1984a pp. 2ff, cf. 1984 p. 6, 1985 pp. 4f, 1986 p. 88. The service sector priority was new and replaced the "encouragement of investment" priority (inward, property and industrial) found in the 1981-84 strategy (SDA 1981 pp. 58-62).

ing of a commercially viable private sector by facilitating the adaption of firms to the fast-changing conditions of the market through a catalytic approach that emphasises partnership and strives to maximise the private sector input or 'leverage' – was aptly summed up in the "3 Cs" of George Mathewson, the Chief Executive since 1981, which portrayed the Agency as "Creative, Commercial and Catalytic".[46]

With regard to the perception of the regional problem several things are worth noting. On the one hand industrial modernisation was still seen as the key problem, now couched in terms of adaption to new market conditions, and when for the first time environmental improvement is given a prominent position in a strategic statement, it is seen as a factor contributing to economic development rather than an objective in its own right.[47] On the other hand promotion of employment in Scotland or parts thereof came to play a central role in both the service and area priorities, and this new-found prominence would seem to represent a reordering of the priorities of the SDA when social stability now had become an underlying value that in certain circumstances made public intervention acceptable, especially in connection with area development as a strategy to alleviate local unemployment crises and extreme intra-regional disparities. Moreover, it is also obvious that despite the greater emphasis given to non-financial policy instruments like e.g. advice or changing of managerial attitudes, the activities of the SDA were much more geared towards improving the performance of indigenous firms than had been envisaged in especially the 1980 investment guidelines because most of the priorities evidently targeted existing or new firms within the region. The importance of selectivity, especially with regard to proactive measures, also became more pronounced through the growing importance of sectoral initiatives under the 'technology' priority,[48] and although explicit reference was made to the notion of 'enterprise culture' adopted by the Thatcher government as a metaphor for private-sector led economic development,[49] emphasis was still on proactivity that would enable private economic actors to prosper. In terms of public-private assumptions the Agency had in other words continued to cast itself in the role of the subject, instigating change, yet at the same time, and indeed much more than in the 1970s, the public development body was also presented as "Catalytic", acting in cooperation with private firms and stressing the importance of a positive response from the private sector.

---

46 SDA 1985 p. 5, cf. George Mathewson 9.7.90.

47 In the 1978 industrial strategy environmental renewal is seen as having a "supportive role" (SDA 1978a p. 26, cf. the discussion in Chapter 12).

48 SDA 1984a pp. 2f.

49 Cf. Section 2.2 and the discussion in Section 7.4.

Despite the statutory amendments and new policy guidelines instituted by the incoming Conservative government, the corporate strategies of the Agency of the mid-1980s did in other words display a great deal of continuity with the industrial strategies of the late 1970s because already the pre-Thatcher industrial strategies of the SDA had emphasised the importance of commercial viability, and thus the new political sponsors merely reinforced existing standards for discretionary project appraisal. Perhaps the most striking evidence of the success of this reinforcement can be seen in the return of employment as a major policy goal despite its statutory downgrading in 1980: by the mid-1980s Agency activities must have been 'compartmentalised' to an extent that its political sponsors were willing to accept that employment considerations in e.g. area development projects would not impinge upon the use of profitability criteria in the evaluation of commercially oriented projects. With regard to the second plank of the 1980 investment guidelines, the rearranging of the hierarchy of policy instruments, the impact seems to have been more uneven: the industrial investment function evidently played a more central role in the 1977/78 industrial strategies than in the 1984 corporate strategy, but the expected prominence of inward investment attraction never materialised, presumably because the Agency had managed to convince its sponsors of the merits of other activities aimed at indigenous firms. By the mid-1980s the SDA in other words presented itself as a proactive and selective development body, with the role of private firms and organisations as 'partners' contributing financial and other resources being increasingly recognised. Given the non-mandatory nature of the policy instruments at the SDA's disposal this was of course a truism, but by constantly referring to the need to devise policies entailing a combination of incentives that would bring forward a positive response from targeted firms, the Agency attempted to reposition itself discursively by stressing its dependence on the active contribution of private economic actors.

7.3. Failures of the Market

The 1986 Government Review

One of the most comprehensive statements of the aims and methods of the SDA does not stem from the organisation itself, but can be found in the report of a government review group established in 1986 to carry out a thorough audit of the Agency and its activities as part of a regular programme of reviews of arm's-length bodies. The general purpose of the review was "to assess the functions exercised against the background of the Government's approach to [...] economic policy",[50] and on this background it is a hardly surprising that many saw the review as a potential threat to the organisation.[51] The review group consisted of senior civil servants from the Scottish Office and the Treasury chaired by Gavin McCrone, Head of the sponsoring IDS, and it received extensive support from a small team established within the SDA.[52] The report of the reviewers concluded that "the Agency has substantial achievements to its credit",[53] that its "approach to economic development is consistent with present government philosophy",[54] and that consequently it was not necessary "to change or reduce the scope of the Agency's functions".[55] According to the 1986 review the "ultimate objective" of SDA activity should be defined in terms of "net additional output or employment in the economy",[56] and Agency intervention in pursuit of this would only be warranted in case of "imperfections in the operation of markets", e.g. "where the benefits of an action will accrue to a wider group of people than those participating in the costs and commercial returns," and the assisted activity should be "additional", i.e. "it would not have occurred without Government assistance" and would not "displace another similar activity with no net gain to the aims of the policy".[57] In order to avoid permanent intervention, policy programmes should aim at rectifying the original market failure and to the widest extent possible rely on mechanisms of delivery that favoured private provision instead of public. When services were provided directly by the Agency, commercial rates should be charged wherever feasible, and provision of subsidised services could only be justified as small-scale demonstration projects.[58]

In the 1986 review the regional problem was still perceived in terms of the relatively poor performance of Scotland compared to other regions of the UK with regard to aggregate indicators like output and employment, and thus the traditional tension between 'economic' and 'social' goals was maintained, but by defining the 'ultimate objective' in terms of net growth with regard to these indicators, the review actually contained the most precise statement of Agency objectives yet. More-

[50] IDS 1987a p. III.

[51] George Mathewson 9.7.90, Frank Kirwan 1.6.90, Sir Robin Duthie 12.7.90, cf. John Firn 15.6.90, Alf Young 5.6.90, Helen Liddell 21.6.90.

[52] Frank Kirwan 1.6.90.

[53] IDS 1987a p. III.

[54] IDS 1987a p. 24.

[55] IDS 1987a p. III.

[56] IDS 1987a p. 31.

[57] IDS 1987a p. 28.

[58] IDS 1987a pp. 30f.

over, central concepts in the market-failure approach would clearly seem to require (political) judgements in order to become operational in the context of public policy: deciding which 'market failures' warrant government intervention involves weighing different regional problems and firm-level consequences against each other, and estimates of 'externalities' and 'additionality' depend upon which parts of the socio-economic environment it is decided to include, and thus if its employees became sufficiently adept at justifying projects in the language of market failure, the SDA should still enjoy a considerable degree of leeway. And last but not least, restating the rationale for the Agency in market-failure terms and demonstrating compatibility with Thatcherite principles would provide an 'intellectually coherent' line of defence that could be used by pragmatic Conservatives to fend off hard-line critics.

In terms of operational strategies the 1986 review maintained the requirement that commercial viability was central in the discretionary appraisal of individual projects, but contrary to the vehement insistence found in the 1980 investment guidelines, profitability at the level of individual firms was now to a much larger extent taken for granted, perhaps because this criteria for discretionary assessment of individual projects was no longer seriously disputed in the Scottish context. Instead the central concern of the review group with regard to operational strategies was a new one, namely the choice of policy instruments and especially the risk that subsidised public provision would 'perpetuate market failures' and 'crowd out' private service providers. Restoring or establishing well-functioning markets in various forms of producer services had become the long-term policy objective, and hence the reviewers favoured 'demonstration projects' with a clear 'exit strategy' that should convince private firms about the value of particular services, preferably through subsidised private provision of e.g. advice in order to strengthen the capacity of the private sector, or, if for reasons of e.g. quality or availability public provision was the best alternative, 'commercial rates' reflecting the benefit obtained by the individual firm eventually should be charged. This attempt to minimise the potential market distortion of SDA activities focused on some very real issues relating to the firm-level 'side-effects' of policies in relation to supported firms (subsidy dependency), non-supported firms (non-additionality that dilutes policy impact), and private providers of service (crowding out). At the same time it is also evident that these concerns were underpinned by some basic assumptions about the ideal relationship between public and private actors which condoned limited public intervention but insisted on the superiority of the market as a mechanism of resource allocation. On the one hand it was openly recognised – and indeed acceptable at least in a short-term perspective – that provision of Agency services involved a direct or indirect financial subsidy to private economic actors, thereby giving political legitimacy to the organisation as a proactive public body influencing the choices of private firms by making financial and other resources available on certain conditions. On the other hand the efforts to minimise subsidies and encourage private provision clearly signalled that 'market distortion' should be kept at a minimum, and this would seem to suggest that in practice the scope for Agency policy could be heavily circumscribed because a platform had been created from which political sponsors could question activities in great detail. This was certainly not unproblematic because the methods of implementation preferred by the review group, if adhered to rigorously, would appear to have a number of potential drawbacks[59] like the loss of momentum in policy innovation through outsourcing of services, the unlikeliness of creating a market through subsidised 'demonstration projects' for services with a limited number of potential users,[60] and the difficulties in setting a 'commercial rate' for public services in the absence of comparable private ones.

It can in other words be concluded that while the 1986 government review did not discredit the SDA politically but instead potentially fortified its position by reinterpreting its activities within a market-failure framework acceptable to the Conservative government, the review resembled previous attempts by political sponsors to define the role of the SDA by focusing on the operational strategies and policy implementation, albeit no longer the profitability of individual projects but instead the methods of implementation and their wider impact on private actors within the regional economy, thereby strengthening the position of the Scottish Office for taking a greater interest in detailed policy-making and bring the operational autonomy of the Agency under increasing pressure – and, possibly creating "paralysis through analysis"[61] because under conditions of imperfect information the endeavour to search for solutions with optimal longer-term market-creating potential and minimal short-term displacement could easily be a very time-consuming task.

---

59 Some of these were pointed out by the Agency during the review process, cf. the discussion of individual policy areas below.

60 A point also made by the Standing Commission on the Scottish Economy (1988 p. 32), noting that with regard to e.g. technology and innovation market-type relations are often absent (cf. Section 3.3).

61 Standing Commission 1988 p. 71, quoting the management handbook *In Search of Excellence*, cf. Parsons 1995 pp. 433ff, Hogwood & Gunn 1986 Ch. 10, Ham & Hill 1984.

Adaptive Responsiveness at the SDA

The small Agency team which had been set up to liaise with the review group went on to form the nucleus of the new Policy Unit which became responsible for strategic planning and policy evaluation,[62] and thus staff well-acquainted with government thinking formulated the corporate plan for the period 1988-91[63] which defined the objectives of the SDA as "to increase the aggregate money income of Scotland's residents" and "to improve the physical fabric and environment".[64] Progress against the environmental objective would be measured mainly in terms of output measures – e.g. the areas cleared of dereliction or the number of beneficiaries – while economic development would be judged on the basis of the increase in net labour earnings and employment, and central to the latter was to "remove structural impediments to the development of the market place", "to enhance the competitiveness of the Scottish economy through highlighting opportunities and threats", and attracting "maximum private sector funding to assist in this process".[65] These general principles were then translated into four "strategic thrusts",[66] namely:

- *technology*, aimed at improving economic performance,
- *Scottish enterprise*, needed to be encouraged in both new and existing businesses,
- *internationalisation*, i.e. assisting Scottish firms to operate on international markets, and
- *urban renewal*, upgrading of the physical environment for businesses and residents.

In each of these areas the Agency would be guided by a number of "operating principles" which included aiming for additionality at the level of the Scottish economy unless redistributive goals had been "explicitly endorsed by the Board and executive", policies should have clear objectives and impact would be assessed, private sector input and initiative should be encouraged wherever possible, and delivery of services should take place on a commercial basis unless subsidies were specifically warranted by the scale of externalities, other government policies or temporary 'pump-priming' demonstration projects.[67] Given

---

62 Frank Kirwan 1.6.90.

63 SDA 1987a.

64 SDA 1987a p. 2.

65 All quotes from SDA 1987a.

66 SDA 1987a p. 12.

67 SDA 1987a pp. 8f.

the predominantly reactive screening role adopted by the Policy Unit,[68] it is not surprising that the detailed list of policy priorities addressed a very extensive list of regional problems by means of a wide range of policy instruments and most of all resembled a catalogue of existing SDA activities, much like the previous corporate strategy.[69]

All in all the immediate consequences for the SDA's strategic thinking of the adoption of the market-failure framework would seem to be somewhat paradoxical. With regard to the definition of the regional problem, the adoption of the review's "ultimate objectives" of increased net labour earnings and employment established reasonably clear standards against which policies could be judged but at the same time it also highlighted the inherently contradictory nature of these objectives by drawing attention to the potential tension between maximising output/employment and optimising efficiency/competitiveness, and hence between the underlying values of regional competitiveness and social stability. Still, in terms of problem definition the late 1980s was in many ways the least ambiguous period, and this also translated into a 'cleaning up' of the Agency's sprawling specific strategic priorities: environmental improvement was foregrounded as an objective in its own right instead of being merely a means to promote economic development, the (displacement-prone) service sector priority introduced in the 1985-88 corporate strategy quietly disappeared, and the predominantly redistributive nature of area initiatives was acknowledged. With regard to strategies for policy implementation the picture is, however, a more uneven one. The new 'operating principles' did of course reflect key recommendations made by the review group, but especially issues concerning 'crowding out' of private providers would seem to have been ignored in practice, and its enumeration of planned activities contained a wide variety of policy instruments, including many which relied on direct public provision, clearly maintaining its role as a major implementor rather than the 'provider of last resort' envisaged in the government review. *If* the new language of market imperfection, externalities and additionality *did* have profound consequences for the work of the Agency, this must have been the product of a gradual adjustment of individual programmes through the ongoing process of policy development and evaluation, but less than one year after the document for 1988-91 had been produced, the expression "Scottish enterprise" acquired a whole new meaning (and an additional capital letter) with the proposed merger with the Training Agency, and during the transitionary years that followed the SDA merely reiterated the principles of its current strat-

68 Cf. the discussion in Chapter 6.

69 Compare SDA 1987a with the last pre-review corporate strategy (SDA 1984a).

egy.[70] What came to be the last comprehensive statement of corporate strategy in other words very aptly underlined the tenuous position of corporate planning within the SDA, positioned between the political expectations channelled through the sponsor department and the internal pressures from front-line implementors and 'project champions' within a project-driven arm's-length development agency.

7.4. The British Context: Regional Strategies in Transition

The context in which the SDA operated was characterised by change in terms of the actors involved in regional policy and the resources they committed to spatial economic policy, and in order to clarify whether the development of the strategies of the Scottish RDA was unique or conformed to broader British patterns, this section surveys the changing nature of the regional development strategies pursued within the wider policy network, including the policies conducted from the British level, European programmes, initiatives sponsored by sub-national government and other arm's-length development bodies.

Vacillations and Change in Central Government Policy

Despite the frenetic pace of detailed change in British regional policy in the 1960s and early 1970s, little had been produced in terms of explicit statements of aims and objectives,[71] and thus strategies had to be reconstructed on the basis of the characteristics of the policy programmes themselves – although unemployment was clearly the problem and the remedy to redistribute economic activity by affirming financial subsidies in areas designated by central government – and the Labour governments of the 1970s generally continued the practice of keeping a low profile with regard to strategic declarations about regional policy. The 1976 document *Criteria for Assistance to Industry*[72] did, however, spell out the position with regard to discretionary industrial support which had the "over-riding objective" of making "British industry [...] compete more successfully" through "taxation, planning controls or subsidies" in order to compensate for the fact that the price mechanism and company profits do not reflect the full social cost of e.g. regional unemployment.[73] Supported firms should be "viable" within a "reasonable" period of time (usually three years), and with regard to Regional Selective Assistance projects creating new jobs would normally be

70 See SDA 1989 pp. 8ff, 1990 pp. 8f, 1991 pp. 6f. The only major deviation being the added emphasis given to training in view of the impending merger.

71 See Section 4.3.

72 Department of Industry 1976.

73 Department of Industry 1976 p. 114.

given a higher level of support than investments safeguarding jobs through modernisation.

Here the regional problem was defined in terms of unemployment, and although the ultimately goal was viable and competitive firms, regional unemployment was regarded as something that in itself warranted public intervention, as can be seen from the dependence of financial support on the employment impact of individual projects. When the Labour government implemented extensive cut-backs in public expenditure in December 1977, it still chose to withdraw the labour-related Regional Employment Premium, clearly the most 'conservative' automatic regional subsidy in terms of long-term industrial competitiveness because it favoured labour-intensive low-tech forms of production, possibly in order to comply with European regulation,[74] and thus central government regional policy had in fact become less employment oriented towards the end of the 1970s.

After having been elected in May 1979 the new Conservative administration undertook a review of regional policy, and the new Industry Secretary announced that it intended to continue "a strong – but more selective – regional industrial policy".[75] *Firstly*, the Assisted Areas map was to be adjusted gradually so that by 1982 total coverage had fallen from around 40% to 25% of the British population, concentrating support in the weakest areas through regrading of localities within the existing three-tier system. *Secondly*, the level of automatic subsidy was reduced significantly in the lower tiers of the system, again concentrating efforts in selected areas. *Thirdly*, the policy instruments employed changed: while the various types of financial incentives and the factory building programme in the Assisted Areas were maintained, regulation of expansion in prosperous parts of the country was greatly reduced.[76] *Finally*, access to discretionary forms of support was tightened with "a greater element of self-financing" being introduced with regard to government factories, and applications for Regional Selective Assistance now required private firms to demonstrate both "proof of need" – i.e. that the grant would make a difference in terms of the nature, scale, timing or location of the project[77] – and that projects would "strengthen the regional and national economy" through increased efficiency.[78]

---

[74] McCallum 1979 p. 29, Maclennan 1979, Begg 1983, Damesick 1987 p. 43, Prestwich & Taylor 1990 pp. 223ff, cf. Ashcroft 1978 p. 54.

[75] *HCPD* vol. 970 cols. 1302-21.

[76] *HCPD* vol. 970 cols. 1306f, Begg 1983, Damesick 1987 p. 45, Armstrong & Taylor 1985 p. 316.

[77] *HCPD* vol. 970 col. 1303, cf. Industry Act Annual Report 1980 p. 42.

[78] Industry Act Annual Report 1980 p. 42.

In many ways the first round of changes announced by the Conservative government brought 'less of the same': the regional problem continued to be equated mainly with spatially concentrated unemployment,[79] all the key policy programmes remained in place although the relative importance of financial incentives increased at the expense of instruments based on authority, foreshadowing the discontinuation of IDCs in 1981,[80] financial subsidies were concentrated in smaller areas, and the importance of commercial viability strengthened in the discretionary appraisal process.[81] In terms of political symbolism, entitling the policy statement "Regional *Industrial* Policy"[82] indicated that regional policy should ultimately be seen as part of its general attempt to "encourage national industrial vitality and prosperity",[83] and thus unemployment was seen as the, albeit unfortunate, consequence of an underlying economic malaise against which remedial action was required, i.e. a clearer hierarchy of objectives had been established. All in all by maintaining a slimmed-down version of the existing machinery the new government acknowledged that attempting to redistribute economic activity between the regions was still a legitimate objective, but as the changes were projected to reduce public expenditure by nearly 40%,[84] it was in excellent accordance with the pledge 'to roll back the frontiers of the state' and could therefore be seen as an attempt to subsume regional policy in the general discursive positioning of the Conservative government.

More changes followed from the 1983 White Paper *Regional Industrial Development* which opened by claiming that "imbalances between areas should in principle be corrected by the natural adjustment of labour markets", but went on to note that neither wage flexibility nor labour mobility has hitherto been able to "correct regional imbalances in employment opportunities". While the usefulness of regional subsidies in the competition for internationally mobile investment was acknowledged, it was concluded that "the case for continuing the policy is now principally a social one with the aim of reducing, on a stable long term basis, regional imbalances in employment opportunities" because the traditional economic argument for regional policy, alleviating conges-

---

79 *HCPD* vol. 970 col. 1307, cf. Committee of Public Accounts 1981 pp. 1-4, Armstrong & Taylor 1985 pp. 178ff.

80 Prestwich & Taylor 1990 p. 153, Wren 1996a p. 126, Armstrong & Taylor 1985 p. 316.

81 For a discussion of the difficulties involved in substantiating such claims, see Swales 1989.

82 *HCPD* vol. 970 col. 1302 (italics added).

83 *HCPD* vol. 970 col. 1302.

84 *HCPD* vol. 970 col. 1307.

tion and 'over-heating' in more prosperous areas, was much weaker "when there is no general shortage of labour in any region". The principal task of regional industrial policy would therefore be to stimulate "the development of indigenous potential within the Assisted Areas" with the ultimate goal of creating "self-generating growth".[85] Cost-per-job was seen as important in order to make the most of limited public resources, and the White Paper therefore proposed the introduction of a 'cost-per-job ceiling' on automatic subsidies in order to limit support for capital-intensive projects which involved few jobs and sometimes had little locational choice in practice.[86] Although a limited range of service industries became eligible for support, additional expenditure reductions would flow from making replacement investments ineligible for automatic grants in compliance with European regulations,[87] and by adopting "commercial letting policies and [...] charge market rents" for government factories.[88] In terms of area designation the White Paper was less certain about the benefits of concentrating support in smaller areas,[89] and it also suggested a wider range of criteria so that, in addition to unemployment rates and population changes, factors such as industrial structure, peripherality and long-term unemployment were given more weight.[90] In the ensuing 1984 reform the three-tier Assisted Areas system was replaced by two tiers, Development Areas and Intermediate Areas, and the latter category was extended significantly so that total population coverage rose to 35%,[91] but as Intermediate Area only qualified for discretionary Regional Selective Assistance, the overall outcome was again projected to be substantial reductions in public expenditure.[92]

In the 1983 White Paper the perception of the regional problem is still a dual one, albeit argued in a new and rather different way. While inter-regional differences in unemployment were seen as a symptom of varying degrees of adaptability and competitiveness of firms within the regions, i.e. as a social problem with economic roots, the Conservative

85 All quotes from DTI 1983a pp. 3ff.

86 DTI 1983a p. 7, cf. DTI 1983b. The high levels of support via automatic regional policy grants given to e.g. oil-related developments in especially Scotland had frequently been criticised (Ashcroft 1982, Regional Studies Association 1983 p. 11, Jones 1986, Martin 1986 p. 271, Damesick 1987 p. 55, Begg & McDowall 1987, Townsend 1987, Balchin 1990 pp. 70f, Gudgin 1995 pp. 56f).

87 DTI 1983a p. 4, cf. Martin 1985 p. 380.

88 DTI 1983a p. 13.

89 DTI 1983a p. 9.

90 DTI 1983a p. 17, cf. OECD 1994 pp. 41f.

91 Prestwich & Taylor 1990 pp. 153ff, Begg & McDowall 1986 p. 217.

92 DTI 1983a pp. 7ff.

government decided to interpret regional subsidies as a spatial form of social policy,[93] and this reclassification clearly had ominous overtones when taking place in the context of a liberal economic discourse, especially when combined with the decreasing level of financial resources committed to regional policy. But at the same time increasing the cost-effectiveness with regard to job creation by making automatic regional grants more selective through cost-per-job ceilings and extending geographical coverage of discretionary grants could be seen as a politically adept way of making it possible for the Conservative government to claim to be both financially prudent and socially responsible – while at the same time maximising European reimbursement through ERDF grants.[94] The proposed changes in the operational strategies did, however, also serve to highlight the contradictory objectives of central government policy because two potentially countervailing forms of selectivity were introduced in the main programme of automatic grants: on the one hand exclusion of replacement investment, if adhered to rigorously, would make it relatively more attractive for firms in the Assisted Areas to expand production or innovation, but on the other hand a closer link between financial subsidies and job creation could well undermine regional competitiveness in the longer term and hence the prospects of 'self-sustained growth'.[95] In terms of the underlying assumptions about the ideal relationship between the public and the private sectors, it is tempting to see the 1983 White Paper as signalling the introduction of a market-failure approach, especially when it boldly claims that "in principle" regional differences in unemployment should be ironed out "by the natural adjustments of labour markets", but still the 'technical jargon' is absent and, more importantly, no attempt is made to direct policies closer to the "root causes" of this market failure identified, e.g. by tying grants to wage differentiation or flexible working practices rather than job creation. Instead the 1983 White Paper must be seen as a document in which liberal pro-market rhetoric coexists with pragmatic policy adjustment reflecting the financial and political priorities of central government. By the mid-1980s British firms were in other words facing a slimmed-down and more targeted version of regional policy which for supposedly social reasons redistributed economic activity in order to alleviate inter-regional differences in unemployment

93 The discarding of economic rationales for regional policy in periods of widespread unemployment was repeatedly disputed by regional economists, see e.g. Ashcroft 1982, Armstrong & Taylor 1985 pp. 184f, Damesick 1985, Martin 1993.

94 Anderson 1990, 1992 pp. 131ff; Pickvance 1990, Rees & Morgan 1991, Prestwich & Taylor 1990 p. 153, Morgan 1985 pp. 567ff, Damesick 1987 p. 59, Begg & McDowall 1986, Martin 1985, Jones 1986, Parsons 1988 p. 199, Atkinson & Moon 1994 p. 112, Hogwood 1995 p. 271.

95 For a parallel argument, see Morgan 1985 pp. 573ff.

and which did not constitute a system of incentives that in a consistent way would point private economic actors in any particular direction – except, of course, in a geographical sense.

The next phase of change in British regional policy was announced in the 1988 White Paper *DTI – the Department for Enterprise*.[96] This document was based on the belief that while "sensible economic decisions are best taken by those competing in the market place", public policy could still act as a catalyst "by stimulating individual initiative and enterprise and by promoting an understanding of market opportunities combined with the ability to exploit them".[97] In some regions the decline of traditional industries had not produced an adequate response in terms of new firm formation and innovation, resulting in "higher unemployment and poorer growth", and hence a central objective of regional policy was "to encourage the development of indigenous potential" in order to achieve self-generating growth in the long term.[98] In order to reflect more effectively the Conservative government's general approach to enterprise, regional programmes should therefore focus increasingly on business development, i.e. improving the managerial skills and business strategies "which are so essential but which would generally not have been achieved through reliance on open markets".[99] In practice this meant that the Regional Development Grant was terminated on the grounds that automatic subsidies did not necessarily bring about changes in the projects supported and instead the main financial instrument of regional policy would be the discretionary Regional Selective Assistance where additionality had been required since 1979. The document specifically points out that "international mobile investment projects will because of their nature" frequently fulfill the general appraisal criteria set out for this particular programme,[100] but at the same time two new indigenously-oriented schemes were introduced: the Enterprise Initiative which provided grants towards consultancy fees to SMEs throughout Britain but at higher rates in the Assisted Areas, and the Regional Enterprise Grants which subsidised investment and innovation in small firms.[101]

In the late 1980s central government in other words continued to identify problem regions in terms of a relatively weak position with

96 DTI 1988. For a brief statement of the case for change and the main proposals as they affected Scotland, see IDS 1988.

97 DTI 1988 p. iii.

98 DTI 1988 p. 29.

99 DTI 1988 p. 29.

100 DTI 1988 p. 30. For a detailed meditation on the white paper, see Fairclough 1991.

101 DTI 1988 p. 30, cf. Wren 1996a pp. 186ff.

regard to unemployment and growth, but the emphasis on indigenous potential and self-generating growth suggests that the underlying value transforming spatial difference into legitimate objects of public policy was now concern about the competitiveness of the Assisted Areas rather than a quest for inter-regional social equality. Moreover, the remedial strategies also changed significantly: the Assisted Areas map remained unchanged and hence largely defined on the basis of unemployment, but policies now clearly focused on improving the competitiveness of firms within these localities, either through attraction of inward investment from abroad or by making indigenous SMEs more competitive through financial support for innovative projects and for upgrading of management capabilities through external consultants. Not only did financial support now operate in an entirely discretionary fashion, but it had also become much more selective: while the automatic 'hardware-oriented' investment grants were discontinued, the new programmes targeted only SMEs or small firms and focused on innovation and knowledge-oriented projects, and thus the emphasis had moved decidedly away from redistribution of economic activity within Britain. The 1988 White Paper in other words brought central government much closer to a market-failure approach: policies now attempted to target what was perceived as the key market failures impeding growth in the regions, namely the access of small and medium to funds for investment and specialised information. Nearly ten years after the Conservative government first officially embraced the term "regional *industrial* policy", it had eventually been translated into a significantly different strategy for regional development, and private actors now faced a new set of incentives which supported distinct types of economic activities rather than any form of manufacturing.

All in all central government strategies with regard to regional policy from 1975 to 1991 displayed an interesting combination of continuity and change. The regional problem was throughout perceived as having both social and economic aspects, but the underlying values transforming inter-regional inequalities to legitimate objects of public policy changed from being mainly social equality within the national territory – in the mid-1980s actually very much so – to being concerns about regional and national economic competitiveness from 1988 onwards. Despite party-political changes and the introduction of new approaches to economic policy in general, the main policy instrument remained financial subsidies operating in a reactive manner, but other modes of implementation changed profoundly:

- automatic grants were gradually succeeded by discretionary ones, and

- selectivity changed from manufacturing investment to particular types of firms and came to include services and knowledge-based projects.

Although these changes were gradual and would appear to have resulted from a process of experimentation influenced by many different political considerations, the result was a significant change in regional development from redistribution of manufacturing activity to a two-pronged strategy aiming to improve the competitiveness of indigenous SMEs and attract inward investment from abroad.

The development of central government regional policy in Britain from the mid-1970s to the late 1980s has often been described in terms of weakening and decline,[102] but on the basis of the above analysis such an interpretation is difficult to sustain: certainly less financial resources were committed, but as they were spent in a different, more targeted and selective, way, a comparison between what constitutes two different development strategies would require a much deeper understanding of how firms have reacted to the new set of regional incentives. Most of the changes were clearly brought about by a Conservative government which extolled the virtues of the market and could not be expected to support the traditional redistributive form of regional policy except for reasons of e.g. territorial politics,[103] but the new form of regional policy does, however, underline the complex nature of Thatcherism because it does not fit neatly into liberal assumptions about the ideal relationship between public and private actors: while the traditional approach with unselective subsidies had left private actors to pursue their own strategies within the Assisted Areas, the selective and discretionary policies of the late 1980s had greatly increased the role of public actors in influencing the investment decisions of private firms.

New Actors, New Strategies?

Although the European Structural Funds were involved in development policies in Britain from 1975 onwards, little needs to be said about the strategies involved in the early years because, as noted in Chapter 6, until 1988 the role of the ERDF was to reimburse the expenditure of other actors, primarily local government expenditure on infrastructure but also central government financial subsidies,[104] and thus European funding will simply have supported the strategies development prevail-

---

102 Cf. Section 2.1.

103 A similar theme can be found in the concluding chapter of Parson's discursively oriented history of British regional policy (1988 Ch. 8).

104 The breakdown between these two purposes was 4:1 in favour of local government expenditure (DTI 1983a p. 6, Committee of Public Accounts 1984b p. 42, Pickvance 1990 p. 28).

ing at these levels of government. In the late 1980s, especially after the 1988 reform of the Structural Funds, a new programmatic approach meant that ERDF resources became tied to the drawing up of specific development plans for the areas designated for European support and could be expected to acquire a strategic content in its own right, but the first round of programming was still closely controlled by central government, and thus the first Objective 2 programmes could in many ways be seen as a direct extension of the pre-1988 approach, although support for various forms of advisory services also began to emerge.[105] Strathclyde in Scotland, the largest and least central-government dominated European programme, presents a similar picture in the late 1980s: its official aim was to build "a sound base for long-term self-sustaining growth" via a range of policies focusing on indigenous SMEs, attraction of inward investment, and improvement of the infrastructure and image of the region,[106] but in practice the programme was, however, dominated by infrastructure projects, with less than 10% of total expenditure allocated to firm-oriented activities and basic infrastructure such as transport, water and sewerage accounting for 64%.[107] In terms of regional development strategies the role of the European programmes would in the late 1980s and early 1990s have remained largely an additional source of funding for especially local government activities, and while the emphasis on basic infrastructure may have made regions more accessible and attractive, also for investors from abroad, it was largely left to private actors to decide whether and how to make use of these additional resources, much like the traditional regional policies of central government.

The 1980s also saw the rise of local government as an actor in economic development, and especially in some English conurbations such activities were not just about jobs but took on a wider significance, namely to demonstrate the existence of practical alternatives to the market-oriented policies of the Thatcher government. Under the banner *Restructuring for labour* some Labour-run local authorities and the enterprise boards sponsored by them provided financial and other forms of support for local business which often involved conditions about ongoing workforce and community involvement.[108] Such initiatives

---

105 Pieda 1992; Hall Aitken Associates 1995 Ch. 4, 1996 pp. 22ff; EPRC 1997, cf. Hudson 1989a Ch. 4, Burton & Smith 1996 pp. 84ff, Bache *et al.* 1996 p. 307.

106 Hall Aitken Associates 1996 p. 9, cf. Goodstadt & Clement 1997 pp. 167ff, MacLeod 1999 pp. 243ff.

107 Strathclyde IDO 1994.

108 See Clarke & Cochrane 1987, Lawless 1988, Totterdill 1989, Moore & Booth 1989 Ch. 7, Pickvance 1990, Cochrane & Clarke 1990, Cochrane 1990, Armstrong & Twomey 1993, Tickell *et al.* 1995.

were, however, neither long-lived nor widespread – in Scotland policies continued to be informed by a more pragmatic form of Labour politics – and in practice most local governments tended to pursue more traditional forms of economic development activity, focusing mainly on development of infrastructure and property supplemented by advisory and other services for small firms in the locality.[109] Local economic policies, like the European programmes to which they became increasingly linked, attempted to maintain a framework, mostly in physical terms, in which local businesses could develop and incoming investment could be accommodated, and hence local government effectively left the initiative in regional development with other actors, i.e. private firms and central government agencies.

The strategies of the three major RDAs present interesting contrasts to that of the SDA. The WDA had been set up in parallel with the Scottish Agency, issued with similar statutory powers, entrusted with a similar organisational inheritance, and subjected to equivalent sponsorship arrangements,[110] but in practice priorities have been rather different.[111] The point of departure was clearly the region's poor performance with regard to employment and inter-regional equality the underlying driving force,[112] but while the new industrial investment function had initially been given some attention, from the late 1970s the main strategic concern of the organisation became property-based attraction of inward investment, and this was further reinforced by a special government initiative in 1980 in response to a series of local unemployment crises in the wake of the contraction of steel production.[113] It was only in the late 1980s that indigenous development again became a major priority, especially in relation to the sizeable presence of inward investment within the region in the form of e.g. subcontractor networks.[114] Unsurprisingly, a major government audit undertaken in parallel with the 1986 review of the SDA reiterated the market failure approach favoured by the Conservative government, but in practice the Welsh

[109] Fairley 1999, McQuaid 1992, Lloyd & Rowan-Robinson 1988, Moore & Booth 1986c, Todd 1985 Ch. 4, cf. the previous footnote.

[110] Cf. Chapter 6.

[111] In terms of strategies the DBRW would seem to have reflected its combined remit as rural development agency and new town corporation, with property-driven inward investment and various services oriented towards small firms (Committee on Welsh Affairs 1980 pp. 220-44, cf. Howe 1996 and Chapter 6).

[112] WDA 1981a p. 3.

[113] WDA 1977a, 1981; Wilson Committee 1977f vol. 8 pp. 5f, Committee on Welsh Affairs 1980 p. xxi, Cooke 1980, Eirug 1983, Danson *et al.* 1992 p. 300, Morgan 1994 p. 13, Morgan & Henderson 1997 p. 102.

[114] Rees & Morgan 1991 pp. 169ff, Morgan 1994 pp. 13-28, Morgan 1996, Morgan & Henderson 1997 p. 85.

review mainly concerned itself with organisational issues and performance monitoring,[115] and it was only towards the end of the 1980s that "a more 'entrepreneurial' style"[116] of operation became evident and the sponsoring Welsh Office seriously began to explore the possibilities of an extensive privatisation of especially the industrial property owned by the organisation.[117] As development strategies had relied on attraction of inward investment via industrial property, the issue of public-private relations in other words emerged in a different form, focusing mainly on public ownership rather than the broader issues of market failures and subsidised provision.

The strategy development of the HIDB took its point of departure in the relatively weak economic and social position of the region with depopulation, unemployment and peripherality being the main problems. The original strategies appear to have been at least partly founded on the notion of 'growth poles', attempting to locate major industrial undertakings around two of the largest settlements in the region, Inverness and Fort William, but the conspicuous failure of major flagship developments lead to a new strategy, formally announced in 1982, which emphasised the importance of spreading development projects across the region and the crucial importance of indigenous enterprise, a trend reinforced by a government review undertaken in 1987.[118] The latter insisted, much like the parallel review of the SDA, that the operation of the arm's-length body should take place on the basis of the market-failure thinking then favoured by the Conservative government,[119] but although this later became part of HIDB corporate strategy,[120] the implications of the rather vague specific recommendations of the review in terms of policy changes would seem to be more than usually uncertain.[121]

---

115 Welsh Office 1987.

116 Danson *et al.* (1992 p. 300) quoting Phil Cooke.

117 Griffiths 1996 pp. 101ff, cf. Chapter 12. This indicates that there were limits to the freedom to be 'interventionist' that Peter Walker (1991 pp. 202f) claims to have been granted by Margaret Thatcher when he was appointed Secretary of State for Wales.

118 Grassie 1983 pp. 44ff, Scottish Affairs Committee 1984 pp. 2f, Danson & Lloyd 1991, Danson *et al.* 1993 p. 166. The HIDB continued to have social development as part of its statutory remit, something which in practice involved small-scale support for community facilities and activities in remote parts of the Highlands and Islands and, perhaps more importantly, taking peripherality into account when assessing individual development projects in rural areas.

119 IDS 1987b Ch. 3, cf. Danson *et al.* 1993 p. 168.

120 Lloyd 1997 p. 117.

121 Also with regard to the HIDB it is noticeable that some authors occasionally assume a very direct link between intentional statements of corporate strategy on the one

Despite the ongoing process of organisational consolidation outlined in Chapter 6, the objectives of the English Development Commission from the mid-1970s to the early 1990s remained stable: alleviating the relative deprivation of rural areas in terms of depopulation, economic and social structure, and unemployment.[122] The organisation had been given responsibility for around one third of England in terms of area and 5% in terms of population, and within this spatial framework the Commission operated a system of spatial targeting which concentrated support in areas of special need, identified on the basis on primarily unemployment figures.[123] The development strategies of the Commission focused strongly on the promotion of indigenous growth through the provision of industrial property, advisory services and financial support for small firms and rural community activities, and it was only in the early 1990s that central government pressure for greater involvement of the private sector in development activities and the privatisation of parts of the Development Commission's industrial property became evident.[124] Perhaps a number of factors made it possible for the organisation to maintain its strategies and policies despite party-political changes: being socially motivated was in line with the conception of regional policy favoured by the Conservative government until 1988, being relatively small meant that the potential for spending cuts and income from privatisation was fairly limited, being rural and English meant that the organisation operated in fertile areas in party-political terms from the perspective of central government, and going through organisational change implied that the organisation was attempting to improve already. The Development Commission was perhaps not *The Most Revolutionary Measure*,[125] but arguably in strategic terms the most stable of the major British RDAs.

The strategic development of the major British RDAs from 1975 to 1991 can thus be summed up as follows. On the one hand their strategies appear to be sufficiently different to warrant the claim that this approach to regional policy, even when sponsored by central government departments, allows policies to be adapted to specific regional circumstances – i.e. rural/urban and more or less peripheral – and thus

---

hand and policy innovation on the other (e.g. Danson *et al.* 1993 pp. 168f, Lloyd 1997 pp. 117f).

122 Tricker & Martin 1984 p. 507, Danson *et al.* 1992 p. 298, Development Commission 1984 p. 2.

123 Martin *et al.* 1990 p. 270, Tricker & Martin 1984 p. 512.

124 Tricker & Martin 1984, Bovaird *et al.* 1989 p. 13, Rogers 1999 Ch. 10-12, cf. Development Commission 1984. Already in 1980 the Development Commission described itself as a catalyst for enterprise (Chisholm 1984b p. 514).

125 The title of Rogers 1999.

involved strategic diversification and policy innovation. On the other hand it is also clear that these organisations through their sponsorship by central government were subjected to similar pressures from their political sponsors from the 1980s onwards, revolving around the relationship between public and private actors, which could be construed as the introduction of a more 'market-oriented' approach to regional development. The existing literature does, however, also suggest that this pressure has been exerted in rather different ways, ranging from decreasing financial resources via specific demands for privatisation of publicly-owned factories to general prescriptions about the virtues of a 'market-failure approach', and thus in order to assess the importance of this trend the reactions of arm's-length bodies to sponsor preferences must be taken into account, not just on the level of general strategic statements but in terms of possible changes in their interaction with private actors.

7.5. Corporate Developments – A Reappraisal

The period from 1975 to 1991 saw changes of strategic significance in British regional policy, both within the SDA and in the wider policy network, but the analyses in this and the previous chapter strongly suggest that these developments cannot be accounted for in terms of simple dichotomies like "before/after 1979" or "more/less market". This section sums up the development of the Agency at the corporate level in terms of resources and strategies in order to establish the balance between continuity and change, and situates these findings in the broader British context and in relation to the existing literature.

Statements about the aims and methods of the SDA have been issued by both the organisation itself and its political sponsors, and the adjoining Table 7.1 summarises their key features as they, some more than others, have changed over the years.

Table 7.1 SDA corporate strategy statements 1975-91

Period & source		*Regional problem & underlying values*	*Main policy Instrument & firm-level impact*	*Modes of implementation*
1975 to 1979	**Sponsors**	Slow growth Unemployment Uncompetitive firms Poor infrastructure Inter-regional equality Competitiveness	Non-mandatory Expand/modernise/ preserve indigenous	Patient investment
	SDA	Uncompetitive firms Sectoral structure Unemployment Competitiveness	Non-mandatory No rescues Expand/modernise indigenous	Long-term viability Catalytic Proactive Sectorally selective
1979 to 1986	**Sponsors**	Uncompetitive firms Unemployment Poor infrastructure Competitiveness	Non-mandatory Incoming location No rescues Expand/modernise indigenous	Viability Early disinvestment Catalytic
	SDA	Uncompetitive firms Unemployment Poor infrastructure Competitiveness Social stability	Non-mandatory Expand/modernise indigenous Incoming location	Viability Catalytic Proactive Sectorally selective Spatially selective
1987 to 1991	**Sponsors**	Slow growth Unemployment Competitiveness Social stability	Non-mandatory Expand/modernise indigenous Incoming location	Viability Additionality Less direct provision Temporary subsidies Commercial charges Early disinvestment Catalytic Spatially selective
	SDA	Slow growth Unemployment Uncompetitive firms Poor infrastructure Competitiveness Social stability	Non-mandatory Expand/modernise indigenous Incoming location	Viability Additionality Temporary subsidies Commercial charges Catalytic Proactive Sectorally selective Spatially selective

The problem which the Agency aimed to address was from the outset defined in very broad and potentially conflicting terms – employment and efficiency – and this situation was maintained also after the introduction of a more precise 'technical' definition of objectives following the 1986 government review. A similar coexistence of elements often associated with different paradigms in regional policy can be seen in the underlying values transforming inter-regional differences into legitimate objects of public policy: here regional competitiveness was present throughout the period but mostly existed in parallel with other, more 'social', values, i.e. inter-regional equality or social stability. The only period where 'social' values appear to have been absent is around 1979, but this is in fact a trick of the light caused by the nature of the documents analysed – industrial strategies and investment guidelines – which have little to say about the Agency's involvement in alleviating local unemployment crises through area development. The organisation had in other words been vested with objectives of a potentially conflicting nature, and although this did of course increase the type of activities the arm's-length body could engage in, it could also serve to sustain expectations in the political environment which could be difficult to reconcile.

With regard to policy instruments continuity clearly prevails from a very early point too, as no major changes were undertaken in this part of the statutory framework, but the desired firm-level impact of SDA activities proved to be more disputed, especially until the early 1980s. The Agency emphatically distanced itself from rescues of ailing firms in the late 1970s, something which the Scottish Office investment guidelines only did after the change of government, and the priority accorded to inward investment attraction by the incoming Conservative government never managed to replace the focus on indigenous firms in the corporate strategies of the arm's-length body.

By far the most extensive changes took place with regard to statements about the ways in which policy instruments should or should not be employed, i.e. the modes of implementation. From the 1970s onwards the number of aspects that had to be taken into account by the SDA in its discretionary appraisal continued to grow, focusing first on the (commercial) nature of individual projects and the desirability of private sector involvement, then becoming increasingly selective along sectoral and spatial lines, and finally in the late 1980s emphasising the expected impact (additionality) and the organisation of policy delivery. As SDA policies operated in an exclusively discretionary manner, it is hardly surprising that the sponsor department attempted to provide guidance in order to 'hedge' the activities of the arm's-length body and increase accountability – or indeed that the Agency itself would want to make clear to other actors what standards it applied when appraising projects and programmes. But the increasingly detailed guidance for discretionary activities also meant that it would be possible for the sponsor department to examine SDA policies much more closely than had hitherto been the case, and thus the changes expanded the platform from which the Scottish Office could attempt to influence the arm's-length body. Because the Agency exercised its discretion primarily *vis-à-vis* firms operating in Scotland, the question of policy implementation was intrinsically linked to the relations between public and private actors, an issue which was heavily politicised in Britain in this period, but here the analysis of corporate strategy statements has demonstrated that the core assumptions remained fairly constant: the SDA was cast as a subject in the process of regional development, while private firms were seen not only as objects of public policy but also as potential co-subjects. Nonetheless the rhetoric employed by the SDA did indeed evolve – there was less talk about intervention and much more talk about entrepreneurialism in the 1980s than in the 1970s – and thus the strategies of the organisation could be construed as being associated with particular political trends and discursive positions.

The analysis of the historical development of corporate priorities has shown that both the SDA and its political sponsors have initiated new developments: the development agency did, for example, step up the viability requirement and began sectoral selectivity while the Scottish Office introduced the notion of additionality and commercial charging for services. At the same time it is also clear that both sides occasionally espoused ideas that were not subsequently given the same priority by the other organisation: proactivity was primarily stressed by the development body, while disinvestment and less public provision were only prominent in sponsor department statements. All in all this distribution of roles would suggest that with regard to corporate strategies the relationship between the Agency and its sponsor department was of an arm's-length nature, i.e. that the development body enjoyed a significant degree of strategic initiative within a relative broad framework established by the Scottish Office.

This form of sponsorship enabled the SDA to become distinctly different from traditional central government departments in terms of organisational and informational resources, with a project-driven organisation emphasising the importance of initiative and creativity amongst its front-line specialist staff which should make it possible to devise and implement policies addressing the problems perceived to be specific to the individual region. The direct influence of the sponsor department was more pronounced with regard to financial resources, especially the sources of funding for SDA activities: while the overall level of expenditure was maintained also under the Conservative government, the importance of private sector funding gradually increased through the 1980s, and thus the Agency came under increasing pressure to reconcile long-term development objectives with the potentially more short-term considerations of private firms in order to maintain the support of its political sponsors. It is, however, at the same time interesting to note that the changing structure of finance does not appear to have had any clear-cut impact on the values underlying SDA corporate strategies because the importance of regional competitiveness and the commercial viability of individual projects had been emphasised by the Agency already in the late 1970s, and the selective and relatively proactive strategy adopted from the very beginning continued also after the role of private funding began to increase.

All in all the corporate development of the SDA would appear to have been characterised by a combination of continuity and change. On a large number of analytical dimensions the situation in the early 1990s was not qualitatively different from that in the late 1970s: continuity prevailed with regard to the authority delegated through political sponsorship, the organisational and informational resources at the disposal of the Agency, their translation into non-mandatory policy instruments, and the underlying assumptions about regional problems and remedial strategies which resulted in the emphasis on pursuing regional competitiveness through support for commercially viable projects. At the same time a number of significant changes are also evident, namely the introduction in the late 1970s of intra-regional equality as an additional measure of success, and the intensified attempts from the mid-1980s to hedge the discretionary powers of the Agency, directly through more detailed sponsor department regulation and indirectly through the increased reliance on private sources of finance.

While this analysis on the corporate level does establish the general position with regard to resources, rules, assumptions and strategies on the basis of which the SDA operated, the extent to which these were reflected in concrete activities on the ground can only be established through the ensuing analysis of individual policy areas. It is, however, worth pointing out that from a corporate-level vantage point the development of the organisation does not seem to conform to either of the main interpretations of the history of the Agency as outlined in Table 5.1. On the one hand some key assumptions of the *external-revolution* paradigm have clearly been undermined:

- the strategies of the Agency did not change suddenly or radically from the 1970s to the 1980s, and thus if the party-political change at the Scottish Office in 1979 did have an impact, it will have been of a more long-term or gradual nature akin to the 'gradual revolution' paradigm identified in Section 2.2,[126] and

- as the resources at the SDA's disposal in terms of authority and finance were broadly the same, the overall capacity of the organisation to influence private actors through its policies remained intact, and it would be difficult to argue that a fundamental shift from an 'interventionist' to a 'market-oriented' strategy had taken place.

At the same time, while a high degree of continuity is in evidence, it should also be stressed that contrary to what is assumed by the *internal-evolution* paradigm some of the key strategic developments – the introduction of spatial selectivity in the late 1970s and early 1980s, the increased hedging of the commercial discretion of the Agency in the late 1980s, and the growing importance of private source of finance – did not result from a gradual evolution of strategic thinking within the organisation but had been handed down from its political sponsors. While these changes were perhaps unsurprising in party-political terms – a Labour government introducing intra-regional equality as an additional objective and Conservative sponsors emphasising the commercial nature of Agency activities – the picture is complicated by the high degree of continuity: area initiatives remained a priority well into the 1980s, and it was only after more than five years of Conservative rule at the Scottish Office that the added emphasis on commercial operation trumpeted after the change of government was translated into more detailed attempts to regulate the commercial discretion of the SDA.

The corporate-level analysis of the history of the Agency does in other words suggest that its development cannot be satisfactorily accounted for either in terms of party-political ideology or in isolation from its political environment. The SDA was an arm's-length public body, and its relationship to the Scottish Office and the wider policy network was therefore not fixed once and for all but potentially contested on an ongoing basis, both by the Agency and its political sponsors. The outcome – the 'length' of the arm and the 'grip' of the hand – will therefore have varied, both over time and between policy areas, and although the analysis of corporate developments has certainly illuminated important aspects of the history of the SDA, a fuller picture will clearly require studies of individual policy areas and the broader political environment in which the organisation operated.

---

126 It is noticeable that the 'external-revolution' account by Danson *et al.* of the history of the SDA relies heavily on sponsor department documents and tends to eschew SDA corporate strategies or closer examination of individual policy areas (see e.g. Danson *et al.* 1990a pp. 174ff, 1993 169ff).

In order to provide an overview of the position of the Agency in the British policy network, Table 7.2 sums up the changing positions with regard to resources, assumptions and development strategies of the SDA, the DTI as the lead department in British regional policy, and the wider policy network consisting of other RDAs as well as European and local-government initiatives.

Table 7.2 Key dimensions of regional policy in Britain 1975-91

Dimensions	*SDA*	*DTI*	*Wider policy network*
RESOURCES			
Authority	Scottish Office arm's-length sponsorship	Direct parliamentary sponsorship	New actors/ sponsors: * central government arm's-length RDAs * multi-level ERDF * local authorities
Organisation	Project-driven	Bureaucratic	Variable
Information	Specialist	Generalist	Variable
Finance	Stable quantity, increasingly private	Significant decrease, overwhelmingly public	* RDA quantity stable, increasingly private * local authority limited * ERDF additionality limited till late 1980s
ASSUMPTIONS			
Regional problem	Uncompetitive firms, unemployment, later also poor infrastructure	Unemployment, first sectoral structure then uncompetitive firms	Unemployment, poor infrastructure, rural depopulation
Underlying values	Competitiveness, intra-regional equality	Inter-regional equality, then competitiveness	Inter-regional equality, competitiveness
OPERATIONAL STRATEGIES			
Policy instruments	Non-mandatory	Financial	Non-mandatory
Modes of implementation	Selectivity increasing Discretion regulated Proactive/ reactive	Selectivity increasing Automatic phased out Reactive	Selectivity variable Discretion variable Proactive/reactive

Turning first to the regional policies pursued on the British level, mainstream DTI policies also displayed a combination of continuity and change from 1975 to 1991, albeit in a rather different manner. Continuity prevailed with regard to several analytical dimensions: national-level policies remained sponsored directly by parliament with the London-based DTI as the lead department, financial policy instruments operating in a reactive manner continued to predominate, and programmes were administered by civil service generalists within a traditional bureaucratic hierarchy including a network of deconcentrated regional offices in England and the territorial departments in Scotland and Wales. But at the same time significant changes also took place:

- equality, traditionally the underlying value transforming regional disparities into legitimate objects of public policy, was replaced by concerns for regional competitiveness from the late 1980s,
- the dominance of policies operating in an automatic manner was gradually eroded, and by the late-1980s central government regional policies had become exclusively discretionary,
- policies became increasingly selective, especially with regard to spatial targeting as well as the types of firms and projects eligible for support, and
- expenditure was significantly reduced and constituted by the early 1990s only around 40% of the level in the late 1970s.

Although remnants of the traditional concern with inter-regional equality persisted – especially in the designation of assisted areas where unemployment was retained as a key variable throughout – the period from 1975 to 1991 clearly saw central government move from one regional policy paradigm to another. Instead of the traditional strategy of redistributing economic growth between the regions in a quest for equality within the national territory, the new approach attempted to strengthen the competitiveness of the weaker regions *vis-à-vis* the rest of the world by targeting support on innovative investments and projects. The corporate-level analysis of these changes cannot readily be interpreted along the lines of a rise-and-fall scenario favoured in parts of the literature because the changes taking place at the British level in the 1980s – whether by design or not and for better or worse in terms of objectives like regional competitiveness and employment – did not lead to the 'end' of central government regional policy but simply instituted a new form of top-down intervention. Moreover, it can also be concluded that the development of SDA strategies did not simply mirror changes in programmes operated from the British level. The similarity between central government and Agency strategies for regional development obviously increased greatly from the mid-1970s to the early 1990s, but this convergence reflects changes taking place on the British rather than the Scottish level and, of course, their very dissimilar starting points: Agency policies had been discretionary, selective and oriented towards viable projects from the beginning, but these characteristics only came to dominate the regional programmes of the DTI from the late 1980s when the last automatic grant scheme had been discontinued. But even then the two sets of policies remained different in strategic terms: while British-level policies continued to operate in a reactive manner and were subject to well-defined and largely static regulation of their spatial and sectoral selectivity, the SDA maintained a significantly larger degree of flexibility by having the capacity to be proactive and develop new targeted initiatives.

The period from 1975 to 1991 also saw important changes in the wider regional policy in Britain with the advent of new actors such as central government sponsored RDAs, the sub-national economic development activities of local authorities, and the European ERDF programmes. In terms of resources it has been noted that the other major RDAs appear to have enjoyed arm's-length positions from their sponsors in central government similar to that of the SDA, in most cases reinforced by the need to coordinate especially their financial policy instruments with the regional grants administered by the sponsor department. A corporate-level analysis of the British RDAs does, however, produce a sufficiently heterogenous picture to warrant the conclusion that the development of their resources and strategies can have reflected the perceived specific problems of individual regions at particular points in time. Diversity also characterised their response to parallel pressures from sponsor departments for a more 'market-oriented' approach in the 1980s, and although most RDAs would seem to operate according to the indigenously-oriented strategies associated with bottom-up policies, regional competitiveness and indigenous development, also here important exceptions have been identified, e.g. the Welsh emphasis on attraction of inward investment. Determining to what extent this diversity was a product of the arm's-length position of the RDAs or reflected the priorities of their various political sponsors within central government is difficult on the basis of the existing literature, but whatever the distribution of roles in the policy-making process has been, the fact remains that the new organisational set-up had clearly resulted in a broader, much more diverse, and potentially better targeted array of development programmes than the uniform central government schemes of the 1960s and early 1970s.

The concurrent rise of local government and European initiatives clearly added to the complexity in British regional policy, but although these new activities were politically sponsored by other tiers of government, they were still subject to extensive central government regulation.

In terms of resources the national level established the general rules governing the activities of other actors, set the overall financial limits for individual activities, and positioned itself centrally in the administration of ERDF monies both before and after the 1988 reforms of the Structural Funds. Within this framework the financial importance of local authority economic policies grew, albeit from a very low starting point, but despite the radical anti-Thatcherite rhetoric of some English metropolitan authorities in the early 1980s, the operational strategies generally retained the traditional emphasis on infrastructure development while adding new elements oriented towards indigenous firms such as advisory services and small-scale venture capital. In this sense local authority policies had key elements in common with national urban policy and indeed RDA activities, and thus their greatest strategic importance would seem to have been to contribute to the geographical widening of the areas covered by economic development initiatives. Contrary to this it must remain doubtful whether the pre-1988 ERDF made much difference at all in terms of implemented development projects because the handling of the additionality issue by the British government strongly suggests that European monies primarily co-funded ongoing public capital expenditure in the Assisted Areas, and even after the introduction of a programme-based approach in 1988, the continued emphasis on basic infrastructure would seem to indicate that, at best, ERDF-funded activities contributed to strengthening the general suitability of the designated areas for economic activity while leaving the question of the nature of these activities to other public policies and, not least, private actors.

All in all the key characteristics of British regional policy did in other words change significantly from the SDA started operating in 1975 until the organisation was merged into Scottish Enterprise in 1991. Authority had become divided between more actors in a multi-level governance setting: a setting in which central government continued to play a pivotal role as a regulator of policy activity but where other actors were capable of pursuing their own objectives to an extent that would have been difficult to imagine in the preceding national phase of regional policy. These changes greatly increased the complexity in terms of the resources involved: the number of organisations within the policy network and their level of interaction escalated, and the informational capacity needed to devise and implement selective and discretionary programmes greatly exceeded that of e.g. automatic schemes of financial support. It is, however, interesting to note that although the exchange of financial resources via co-funding of programmes and projects increased, central government not only regulated this exchange but also continued to be the largest single funder of development through DTI policies and the RDAs in Scotland, Wales and rural England, and thus while the role of the national level as a direct implementor of regional policies had diminished greatly by the early 1990s, its roles as regulator and sponsor ensured that central government was still a key actor within the policy network. In terms of the policies these resources were eventually translated into, notable changes took place in three areas: regional competitiveness replaced inter-regional equality as the main *raison d'être* of public intervention, new policy instruments were introduced many of which were information-based, and the modes of implementation became increasingly discretionary, proactive, and selective, not only with regard to the types of sectors, firms, and projects supported but also in terms of spatial coverage. The latter development is in a sense emblematic for the transition from the national to the multi-level phase of regional policy: this form of public intervention had traditionally targeted a hierarchy of Assisted Areas defined by central government as being in need of support, but by the early 1990s this relatively fixed situation had been eroded along both spatial and temporal dimensions when a mosaic of overlapping designations had been instituted by different tiers of government, partly through self-designation by individual regions and subject to periodical reviews from the European level.

While these changes have undoubtedly at least to some extent been driven by considerations other than the immediate efficiency of regional policy – e.g. British territorial politics, general budgetary restraints, or European high politics – it is also clear that RDAs, and perhaps the SDA in particular, have played an important role in this process of policy innovation: the emphasis on selectivity and discretion came to dominate the strategies of arm's-length bodies before DTI strategies changed, and thus their development should not be interpreted as a mirror image developments on the national level. Although RDAs have been established by central government, the degree of experimentations allowed by their sponsor departments enabled at least some arm's-length bodies to launch new departures in terms of the ways and means of regional development. Add to this the growing relative importance of RDAs in regional policy, both in Britain at large and in Scotland in particular, and the case for continuing to pursue the history of the SDA seems to be a convincing one, not just from a Scottish but also from a British perspective. Having established the general framework through an analysis of key aspects of the Agency's development at the corporate level, such an undertaking does, however, for several reasons require that the focus on analysis shifts to the individual policy areas in which the SDA engaged. On the one hand the development of Agency activities may have been uneven, e.g. depending on the degree of regulation by its political sponsors or simply because the project-driven organisation could resemble what the Head of the Policy Unit James Williamson tongue-in-cheek called "a collection of feudal baronies".[127] On the other hand statements of corporate strategy may have had a significant element of discursive wrapping – "explaining the same projects in another language without changing the fundamental project"[128] – in an attempt to position the SDA in the discursive terrain in a way thought to be to the liking of its sponsors and thereby 'shelter' its activities from political scrutiny and controversy. Of course the analytical challenges of organisational diversity and discursive creativity also exist within individual policy areas, but by subjecting its most important fields of activity to systematic scrutiny, an even more subtle interpretation of the development of the Agency should begin to emerge.

---

[127] James Williamson 14.6.90.

[128] James Williamson 14.6.90.

Part III

Developing Policies

CHAPTER 8

Industrial Investment

The next five chapters examine the development of key areas of SDA policy under the headings industrial investment (Chapter 8), sectoral initiatives (Chapter 9), inward investment (Chapter 10), advisory services (Chapter 11), and infrastructure improvement (Chapter 12). Based on the internal structure of the organisation, some of the chapters will discuss a number of related activities rather than one specific policy programme, but each chapter will establish the relationship between continuity and change with regard to policy design and implementation, and analyse the interaction between the development body and its political and economic environment. Here it should perhaps be underlined that no systematic evaluation of the economic impact of policy programmes will be attempted, although the *perceived* consequences of policies are, of course, considered in the policy chapters because the image of success or otherwise may play an important role in Agency strategies and government sponsorship, but that the role of the wider political environment, especially political parties and organised interests, with regard to individual policy areas will primarily be discussed in the ensuing Chapter 14.

Investment Contested

Public investment in private firms was not in itself new in the context of British regional policy because a number of organisations had used various types of loans in support of development,[1] but the prospects of large-scale equity investment in a heavily industrialised region was undoubtedly innovative,[2] especially because the SDA would have the capacity to initiate the setting up of new industrial operations and very significant financial resources were to be dedicated to this particular function. The statutory investment powers of the Agency were trans-

---

1 Wren 1996a pp. 78ff, Rogers 199 pp. 76ff, Grassie 1983 p. 64, cf. Chapter 6 and the discussion in Section 8.3.

2 The HIDB had also been given powers of equity investment (Danson *et al.* 1992 p. 165), but these were exercised in a completely different economic context, namely the remote and predominantly rural Highlands & Islands.

lated into three distinct activities: *industrial investment*, a new function established in 1975, *concessionary loans* for small firms, a function partly inherited from the predecessor body SICRAS, and *development funding* for experimental projects and high-risk ventures in e.g. technological infrastructure, gaining importance in the 1980s. All of them involved investment of public money in private sector projects, but as commitments were undertaken for specific purposes, on varying financial terms, in separate organisational settings, and under different degrees of political scrutiny, it is necessary to cover all three functions in order to produce a comprehensive picture.

Specific organisational and political circumstances did, however, ensure that one of these financial instruments, industrial investments, completely overshadowed the other two. On the one hand industrial investment was the only major new activity instituted by the SDA act and could thus readily be construed as the *raison d'être* of the Agency, while the concessionary loans had been inherited from a predecessor organisation and the use of development funding evolved on a low-profile incremental basis. On the other hand the introduction of large-scale investment in the form of equity as a policy instrument introduced a qualitative new relationship between public and private actors in regional policy because in contrast to the one-off nature of grants, holding equity would give a government-funded body a direct say in the affairs of private firms on an ongoing basis, and thus the investment role of the SDA could easily be interpreted in the light of general views of the virtues or otherwise of public ownership. This chapter will therefore be organised around an examination of the industrial investment function while other types of investments by the SDA will mainly serve as a useful contrast and complement to their high-profile relative.

Given the centrality of public-private relations in the polarised party politics in the late 1970s, it would probably have been surprising if the change of government in 1979 had not been followed by a new set of Scottish Office guidelines which, famously, relegated industrial investments to a last-resort option and insisted on a more commercial approach, and both outside observers and SDA executives agree that profound changes *did* take place in the 1980s:[3] while earlier the organisation leapt "into situations where angels would fear to tread and trying to rescue things which were unrescueable, industries which were dying",[4] it later adopted a more commercial venture capitalist approach, focusing especially on small, new high-tech firms. Some approve of

3 See e.g. Rich 1983, McCrone and Randall 1985, STUC 1987a, Danson *et al.* 1988 & 1993, STUC 1988, Draper *et al.* 1988 pp. 280f, Standing Commission 1988.

4 Donald Patience 31.5.90, cf. Neil Hood 22.6.90, Craig Campbell 16.7.90, Alf Young 5.6.90, Gregor Mackenzie 15.10.90.

these changes and others deplore them, but it is widely agreed that from the late 1970s to the 1980s a transformation took place of five aspects of the industrial investment function, namely a change of policy *objectives* from support of regional employment to promotion of sectoral change, a move away from using the *authority* of equity ownership to restructure firms or industrial sectors, a shift of *firm-level* focus away from existing firms towards the promotion of new ventures, the introduction of an increasingly commercial *appraisal* of individual projects, and a reduction of the overall level of *activity*.

The widespread belief in 1979 as a watershed with regard to the industrial investment function should, however, for several reasons be treated with caution: the extent to which the radical rhetoric of the Thatcher administration was translated into policy changes has been called into question, both in Britain in general and in Scotland in particular,[5] the preceding chapters revealed important continuities in SDA resources and strategies at the corporate level, and the consensus about the investment function only covers the nature but not the origins of the alleged transformation, with some observers seeing external factors as being paramount[6] but others claiming that the changes in the 1980s had been foreshadowed by an internal learning-by-doing process on the basis of the first years of practical experience.[7] All in all a dichotomous interpretation built around party politics is clearly not the only possibility, but in practice an in-depth study of the SDA's investment function will require more than a re-examination of the incoherent and selective information published by the Agency which has hitherto been available to external observers,[8] and therefore a new database, SDAINV,[9] has been assembled which covers key aspects of all major Agency invested firms from 1975 through to 31 March 1990 and incorporates unpublished SDA data.[10]

5 See e.g. Booth & Pitt 1983, Moore & Booth 1989, Holliday 1992, cf. Section 2.2.

6 George Mathewson 9.7.90, Donald Patience 31.5.90, Frank Kirwan 1.6.90, Standing Commission 1988, Danson *et al.* 1988.

7 McCrone & Randall 1985, Rich 1983, Parsons 1988 p. 183, Hood 1991a, Edward Cunningham 13.6.90, Gerry Murray 30.7.90, Ewan Marwick 24.7.90.

8 In the early years of the Agency's existence a brief description of the nature of invested firms and the purpose of individual investments was included in the annual reports, but towards the end of the 1980s just c 25% of the firms in the portfolio were listed and only basic financial information provided.

9 I gratefully acknowledge the assistance of Donald Patience, Director, and Anne McGovern of the Agency's Head Office Investment Division in providing information without which the SDAINV database could not have been constructed.

10 The SDAINV database was constructed by combining information from the computerised database within the Agency, covering the period from 1982 onwards, with other data supplied by the Agency's Investment Directorate and information pub-

The chapter is divided into three main sections. First the process of policy design is examined, focusing on the origins, resources and the operational strategies of the industrial investment function (Section 8.1), then the process of policy implementation will be analysed, especially with regard to levels of activity, modes of implementation, and outcomes both on the level of individual firms and in terms of institutional change (Section 8.2), and finally these developments will be situated in a broader context through an examination of investment-oriented policy programmes in British regional policy at large (Section 8.3). This will make it possible to assess the development of the SDA's investment function not only against the specific hypotheses regarding the investment function in relation to the external-revolution and internal-evolution interpretations of the development of the Agency, but also from a British perspective as a case study of the complex relationship between political rhetoric and policy change during the Thatcher years, and in a European context as an example of an arm's-length RDA engaged in innovative forms of regional policy of a politically controversial nature.

8.1. Designing Industrial Investment

The analysis of the investment activities of the SDA commences by exploring the origins of the industrial investment function, and then proceeds to an examination of the resources committed and the strategies through which they were employed.

The Origins of Industrial Investment

The use of the plural 'origin*s*' in the heading is particularly appropriate here because the inclusion of this function in the policy profile of the SDA can be traced back to not one, but several sources. Each of these embodied a specific political or economic rationale that may or may not bear resemblance to what eventually became SDA strategy, and thus the historical origins of the function are likely to have influenced expectations within the political environment.

---

lished in the annual reports. Data covers key aspects of all major investments from the setting up of the Agency till 31 March 1990 (consistent data about the last 12 months of operation have not been possible to obtain), but aggregate data on expenditure and receipts relating to the function does not suggest that this year was markedly different from e.g. 1988/89 or 1989/90. The database covers all investments undertaken by the Head Office Investment Directorate (HOID), and for each investment information is provided on the geographical location and industrial sector in which the firm operates, the date and purpose of the original investment, the date and method of disinvestment if applicable, and – where available – the amount invested.

Among the institutions and organisations promoting the idea of a regional development agency for Scotland in the early 1970s, it is possible to identify three types of arguments supporting the introduction of industrial investment as a function of new RDA. *Firstly*, especially in the thinking of the Scottish labour movement bringing about changes in the relationship between public and private actors was prominent. Here the persistence of inter-regional disparities led to the conclusion that future regional policies would have to entail "greatly extended public ownership", especially in sectors of the economy with "growth prospects in employment terms", because of the need for "relocation of existing establishments which private enterprise is unlikely to accept" and for "taxpayers [to] get a claim on future profits" when firms are given "large public handouts".[11] Support for instituting a policy instrument that entailed the possibility of the ongoing exercise of public authority over the strategic decisions of individual firms in the name of equity was also used in connection with the nationalisation of other industries considered to be of strategic importance, and hence this particular feature of the bill arose fierce Conservative opposition.[12] *Secondly*, the case for industrial investment as an instrument of regional policy was argued in terms of ensuring an adequate supply of capital for especially small and medium-sized industrial firms as highlighted by several central government enquiries,[13] and thus industrial investments as regional policy were seen as remedying a perceived weakness of the Scottish economy, namely an inadequate supply of risk capital for indigenous firms. *Thirdly*, the proposal for an industrial investment function had also been associated with attempts to change the balance between the British and Scottish levels of governance by enhancing the authority and financial resources available to Scottish actors within the field of regional policy. This had for some time been considered important, and in 1974 the detailed blueprint produced by the Research Institute of the SCDI proposed what later turned out to be the eventual outcome, namely industrial investment by a Scotland-wide RDA and Scottish Office control with mainstream regional grants,[14] something which had the dual advantages of moving resources and decision-

11 Labour Party Scottish Council 1973 p. 6, cf. STUC 1973 p. 277. Similar arguments were later used to criticise the proposal eventually put forward by the Labour government (*Glasgow Herald* 28.2.75, 29.8.74)

12 Cf. the discussion in Chapters 5 and 14.

13 Labour Party Scottish Council 1973 p. 10, STUC 1973 p. 277.

14 SCRI 1974 Ch. 5.

making capacity north of the border without challenging British-level coordination with regard to mainstream grant schemes.[15]

The broad consensus around vesting the new RDA with powers to invest in private firms may well explain the eventual inclusion of this feature in the Labour government's official proposal in 1974, but in the long run the diversity of arguments could become a mixed blessing because rather different expectations had to be met, and opposition could be provoked in many quarters at the same time. Conservatives and the Scottish business community could fear the threat of creeping nationalisation, the Scottish financial sector could interpret the function as unwanted public sector competition, Scottish territorial interests could focus on the extent to which the investment function brought additional financial resources to the region, and, conversely, English territorial interests could be incensed by the advantage given to Scotland over equally crisis-ridden regions south of the border. The future performance of the industrial investment function could be judged on the basis of criteria which may not be compatible – e.g. the degree of influence exercised in individual firms, its ability to act as a risk-taking investor, or simply the sheer volume of investment – and thus positioning the new industrial investment function was likely to become a delicate and controversial affair.

Sponsoring Industrial Investment

The way in which the SDA's investment activities were regulated by the Scottish Office suggests that many of the issues aired in the public debate in the mid-1970s also came to influence the approach to political sponsorship in this particular area. The guidelines governing the industrial investment function contained statements in favour of maintaining the operational freedom of the SDA in day-to-day matters,[16] but first and foremost specified the limits within which the arm's-length body could operate.

In terms of delegation of authority from the Scottish Office to the SDA with regard to individual investments the original 1976 guidelines subjected major investments, majority shareholding and hostile takeovers to Scottish Office control and thus effectively limited the field in which the SDA could exercise discretion to provision of capital on a smaller scale as a minority investor – hardly encouraging for those who

15 From an administrative perspective the fact that the HIDB already had invested by means of equity, albeit in a limited number of cases on a fairly small scale (HIDB 1974 p. 93), had demonstrated that it was feasible to vest a semi-autonomous public organisation with investment powers and maintain administrative and political supervision.

16 SEPD 1976b p. 11, 1980a p. 5.

had seen the function as a significant source of additional funding or, indeed, control with individual firms. The revised guidelines issued by the Conservative government in 1980 instituted a slightly more selective approach: hostile acquisitions were prohibited, and by insisting that the SDA contribution could be no more than 50% of total project cost except in very small investment cases, collaboration with private sources of finance had in effect been made compulsory. Add to this the lowering of the maximum financial commitment to individual firms and the ambiguous encouragement to dispose of shares "at the earliest practicable time consistent with its statutory purposes",[17] and the future of the investment function would now clearly seem to resemble that of a public merchant bank for Scottish SMEs rather than a major industrial holding company. The notion of the SDA as a risk-taking investor was, however, underlined in both the 1976 and 1980 guidelines when it was stipulated that loans should be provided on terms and at rates equivalent to those of "firms of the highest standing",[18] and loans could thus be provided to high-risk projects on relatively favourable terms.

Financial resources for industrial investments were made available through the annual budgetary cycle, mainly by central government,[19] but as no information is available on neither the budgets proposed by the SDA nor those eventually agreed to by the Scottish Office, it is only possible to establish the actual and not the planned expenditure pattern.[20] It is, however, worth noting that early SDA projections expected around 40% of total expenditure to be allocated to this particular function,[21] and also that for the 1980s it has been claimed that funding for industrial investments was not a major issue.[22]

The Scottish Office maintained that its role was to monitor individual investment cases "by exception",[23] and systematic monitoring of the industrial investment function was instead situated on the aggregate level.[24] The SDA Act enabled the sponsor department to establish financial duties for individual activities, and the original version required "at least 15% return for the companies in which the Agency

---

17 SEPD 1980a p. 3.

18 SEPD 1976b p. 6, 1980a p. 4.

19 European funding for projects in areas designated following coal and steel closures constituted a minor supplement (Lochhead 1983 p. 56).

20 The actual pattern of expenditure will be discussed in Section 8.2.

21 SDA 1977 p. 6, 1977d pp. 3ff.

22 Donald Patience 31.5.90.

23 As reported in National Audit Office 1985 p. 5.

24 The first draft of the original guidelines even envisaged a regular role for parliament in assessing performance, albeit not on a detailed level (SEPD 1976a p. 4), but this was removed in the document eventually adopted (SEPD 1976b).

invested" but at the same time introduced a measure of 'financial patience' by making it possible to include profits reinvested internally in the calculation.[25] An obvious drawback of this approach was its insensitivity to the macro-economic environment, and a new and more 'macro-sensitive' formula was introduced from 1981/82 which expected the function to "achieve over a rolling five year period a cash return at least equal to the cost of Government borrowing over the same period".[26] The new duty was based on the assumption that the average life of Agency investments would be five years, thereby creating an incentive to dispose profitably of the average investment after such a period, and at the same time the new duty artificially depressed the Agency's results because the revised formula included provisions for potential losses but excluded appreciations in the value of equity holdings, thereby punishing failure without rewarding success. The new duty was widely criticised for favouring an investment strategy aimed at securing short-term financial gains to meet Scottish Office targets at the expense of long-term development objectives,[27] and the inadequacy of the revised formula was tacitly recognised by the sponsor department from the very beginning because in practice the performance of the function was not measured against the cost of government borrowing, but against much lower targets agreed between the Scottish Office and the Agency.[28] It is, however, even more important to note that the financial duty was consistently calculated in relation to the capital directly employed in investments and thus effectively ignored the administrative cost involved in appraising new projects and supervising existing investments – expenditure which could potentially constitute a fairly substantial subsidy in its own right.

With investment guidelines effectively limiting the discretion of the SDA to smaller minority investments and a monitoring regime follow-

---

25 SEPD 1976b p. 7, cf. SDA 1978 pp. 7f. Both the SDA and the Scottish Office recognised "that it would be inappropriate to attach a strict target of return to be met by the Agency" (SDA 1977b p. 2), something which perhaps indicates that the duty was mainly in place because of the Treasury.

26 Reproduced in Committee of Public Accounts 1982a (pp. 18f).

27 Chief Executive George Mathewson in his evidence to the Committee of Public Accounts 2.12.81 (1982a p. 4); SDA 1984 p. 10, 1986 pp. 42ff; Committee of Public Accounts 1982a p. VIII, 1985 pp. X & 18; National Audit Office 1985 p. 10, IDS 1987a p. 50, Donald Patience 31.5.90, Edward Cunningham 13.6.90, Love 1984 p. 83, Draper *et al.* 1988 p. 283.

28 The 1986 government review recommended a revision of the duty and the introduction of a set of additional performance indicators (IDS 1987a p. 51), but while the first two versions of the financial duty merely had taken a long time to emerge (cf. SDA 1978 p. A6, Committee of Public Accounts 1982a pp. VI & 18), a third, and more suitable, formula never materialised.

ing the aggregate financial performance of the function, the Agency's arm's-length freedom of operation would appear to have been rather limited, but the eventual outcome would, however, depend on the position adopted by the Scottish Office in practice when considering individual proposals for e.g. large investments, and the pragmatic approach taken in the question of financial duties could perhaps indicate that despite the rhetoric of the 1980s for some reason the political sponsors were after all less inclined to deploy the institutional levers at their disposal.

Organisational and Informational Resources

When considering organisational and informational resources, distinguishing between the three types of investments undertaken by the SDA is essential.

Responsibility for *concessionary loans* had been inherited from SICRAS and was first transferred to the Agency's Small Business Division (SBD), and then in 1988 to the new regional SDA offices in order to take support for industry closer to the small firms targeted.[29] *Development funding*, on the other hand, was not placed in one particular organisational unit but gradually became available to operational divisions throughout the SDA[30] in line with the project-driven mode of operation in e.g. sectoral initiatives. The *industrial investment* function was virtually started from scratch in 1975, and a division of labour was instituted where investment cases in excess of £50,000 continued to be handled centrally by the Head Office Investment Directorate (HOID),[31] while minor investments became part of a broader, separate and increasingly decentralised effort to support small firms by the SBD. This state of affairs had the advantage of allowing the development of specialist knowledge through the integration of industrial investments with other forms of support for individual firms, either via the sectoral teams in the case of larger projects or within the SBD for smaller ones, but it may of course also have enabled diverging approaches to policy implementation to evolve within the two parts of the Agency.[32]

Undertaking industrial investments required discretionary assessment of a wide range of aspects relating to individual project proposals, from financial packages and production technologies to market prospects and management capabilities, and therefore recruiting appropriate staff was

29 SDA 1976 Section 15, 1988 p. 12.

30 See IDS 1987a p. 73.

31 SDA 1976 sect. 20, 1979 p. 42; IDS 1987a p. 44, Donald Patience 31.5.90.

32 The 1986 government review criticised the investments undertaken by the SBD for being uncommercial (IDS 1987a pp. 50f).

essential. The first two chief executives, Lewis Robertson and George Mathewson, both had investment-oriented backgrounds, but at lower levels in the organisation the innovative nature of the activity was in itself a problem in the early years because staff with relevant qualifications were difficult to obtain, and the Agency's room for manoeuvre was furthermore circumscribed by Treasury limits on the number of staff and the civil-service tradition inherited from SICRAS with regard to small-business support.[33] The growth of private sector venture capital in the 1980s increased the number of persons with relevant experience – indeed the new chief executive and the new investment director were recruited from this part of the financial community – but at the same time this also subjected the Agency to fierce competition on salaries etc. for scarce staff with very specialised skills.[34]

The only major organisational change brought upon the investment function from outside was the provision in the 1980 guidelines which required the SDA to set up an "Investment Subsidiary" headed by a board of directors with relevant private sector experience. According to the guidelines the subsidiary would be "responsible for the assessment of investment proposals and for the management of all investments",[35] and thus policy implementation would effectively appear to have come under direct private sector control.[36] When *Scottish Development Finance Ltd.* eventually was established almost two years later its role had, however, effectively been downgraded to that of a sounding board for the SDA executive,[37] and the 1986 government review even implied that this limited role was perhaps not taken seriously enough by the Agency.[38] The investment subsidiary would thus appear to have reinforced the SDA's network of contacts in the Scottish financial sector and provided additional information about particular projects and funding possibilities, but perhaps its most significant role was to embody the commitment of the Agency to cooperation with private sector financial institutions in investment matters.

All in all the organisational and informational resources allocated to investment activities would seem to display a combination of continuity and change: the basic pattern of organisational specialisation in relation

33 SDA 1976 sect. 17, 1977 p. 8, 1978 p. 9; Lewis Robertson 20.7.90.

34 George Mathewson 9.7.90, Donald Patience 31.5.90. According to Eaton (1987 p. 43) the number of persons employed by HOID had remained stable during the 1980s.

35 SEPD 1980a p. 2.

36 Cf. Keating 1988b p. 49.

37 SDA 1982 p. 40, 1984 p. 41, 1985 p. 46, 1986 pp. 42ff, 1987 p. 56, Donald Patience 31.5.90, cf. Finnie 1982.

38 IDS 1987a p. 49.

to different financial instruments and project sizes remained in place and the most efficient form of Scottish Office regulation appears to have been of a general nature, i.e. staff numbers and pay scales, but at the same time the new industrial investment function does, however, also seem to have embarked on a steep learning curve from 1975 onwards with regard to discretionary project appraisal, and hence it still remains to be seen whether the advent in the early 1980s of a new management team brought about radical change or merely reinforced ongoing development trends.

Changing Aims and Methods

The first official consultative document from the Scottish Office about the new RDA hinted that concerns about private sector underinvestment were seen as a reason for establishing a public investment bank on the regional level,[39] but the original SDA Act did not specify particular goals for the investment function, although the absence of a NEB-style obligation to extend public ownership was conspicuous.[40] The first guidelines issued by the Scottish Office announced that the concessionary loans inherited from SICRAS was to continue "on its present lines" with areas of coverage and rates of interest subject to specific approval by the Scottish Office and the Treasury in order to ensure continued coordination throughout the UK policy network.[41] In contrast to this industrial investment was subjected to a separate and much more elaborate regime of regulation where sponsor department consent was required for particular types of investments, something which instituted a weak form of selectivity where larger investments were not ruled out but more complicated to implement. The overall aim of the function was merely defined by the guidelines as "promoting the growth and modernisation of Scottish industry by means of new investment",[42] suggesting that underinvestment was a general problem in the regional economy and inadequate investment in new technology a specific weakness. Like the SDA Act, the investment guidelines in other words contained an inherent tension between output and employment on the one hand and technology and competitiveness on the other, and this ambiguity was replicated in the financial duty which on the one hand introduced profitability as a criterion in the appraisal and monitoring of individual investment projects, but at the same time underlined that patience could be required.[43] However, by making industrial investments

39 SEPD 1975 p. 1.

40 Cf. the discussion in Chapter 7.

41 SEPD 1976b p. 6.

42 SEPD 1976b p. 1.

43 SEPD 1976b pp. 6f.

the first function for which Scottish Office guidelines issued and the only one for which guidelines were published in the annual reports of the Agency,[44] the Labour government also showed that great importance was attached to this function and implicitly recognised its controversial nature.

For more elaborate strategic statements from the late 1970s, it is necessary to turn to the documents produced by the SDA itself, and from the very beginning the relative importance of the new function was underlined when the organisation repeatedly pointed out that over the first three years its share of total expenditure was expected to be between 1/3 and 2/5.[45] At the same time the Agency also developed a much more elaborate rationale for the industrial investment function, dissociating itself from some objectives and embracing others vociferously. The negative delimitation efforts of the SDA involved two parallel lines of argument: the message of *not* being "in the business of rescuing lame ducks" was reiterated time and time again[46] – publically renouncing protection of employment or existing firms as an overriding preoccupation – and the investment function was *not* seen as a vehicle for extending public control over the private sector.[47] Instead the positive *raison d'être* for the industrial investment function was a perceived weakness of the Scottish economy, namely the existence of "an equity gap, with uneven edges which become sharper with small size, unconvincing management, apparent risk, or short history",[48] i.e. specific problems related to particular sectors and types of firms. The organisation itself estimated that only 6-700 hundred existing Scottish firms were potential targets for SDA investments,[49] but at the same time it was also accepted that due to its responsibility for both regional employment

44 Cf. Page 1977 p. 87, Eaton 1987 p. 8. The guidelines appeared in SDA 1977 (pp. 60ff).

45 SDA 1977 p. 6, 1977d pp. 3 & 7, 1978a p. 20; Cunningham 1977 p. 3, Robertson 1978 pp. 22f. Some of the early press coverage suggested by discreetly aggregating funding for industrial investments and factory provision (e.g. *Glasgow Herald* 15.4.75).

46 E.g. SDA 1976 Section 43, 1977 p. 8, 1977a p. 8, 1978 p. 6, Wilson 1977f vol. 6 p. 194.

47 Wilson Committee 1977f vol. 6 pp. 205f. According to a former SDA investment executive, Chief Executive Lewis Robertson proclaimed the intention of having "a share-holding in every company in Scotland, and a director on every board of every company in Scotland" (Charles Fairley 26.7.90), and if the level of expenditure on investment originally envisaged had been attained, internal discussions of the implication of becoming a major industrial holding company (SDA 1978a p. 20) would have been more than a one-off event.

48 SDA 1977d p. 4, cf. Wilson Committee 1977f vol. 6 p. 209, Cunningham 1977 p. 10, 1978 p. 12.

49 SDA 1977a pp. 7f.

and long-term profitability Agency investments would by no means be restricted to SMEs involved in high-risk activities.[50] Furthermore, its ability to provide patient funds for existing private firms, capability to create new firms, and willingness to intervene in the internal affairs of invested companies in order to rectify perceived management weaknesses was also believed to set the organisation apart from private institutional investors.[51] At the same time the limits to both intervention, proactivity and financial patience were, however, also repeatedly stressed by the SDA: involvement on a day-to-day basis in invested companies should be avoided,[52] restructuring of entire industrial sectors was discussed without becoming a priority,[53] the creation by the Agency of new firms was not its "preferred *modus operandi*" due to the "much greater effort [...] required",[54] risks were to be "reasonable and considered",[55] and at least initially the SDA was fairly reluctant to embrace the idea of acting as a venture capitalist in high-risk high-tech areas.[56] Although the Agency's time horizons were generally longer than private sector investors, "long-run profitability is an essential of sound business and a necessary and salutary discipline on management",[57] and the organisation was seen as a catalyst for development among Scottish-based companies, working along-side private sector investors in order to make the most of limited resources and "strengthen the Agency's credibility as a non-political source of finance".[58] As the equity gap was "in part perceived rather than real; [...] a gap of information rather than of true availability",[59] the Agency could ultimately hope to change the role of private financial institutions from 'risk-averse' adversaries to assistants in the promotion of regional development.

On the basis of this it is clear that the detailed clarification of the aims of the new industrial investment function was undertaken by the

---

50 SDA 1977b.

51 SDA 1977d p. 17, cf. 1976 Section 43, 1977 pp. 11ff, 1977a pp. 8ff; Cunningham 1977 pp. 3f, Robertson 1978 p. 27.

52 SDA 1977d p. 17, cf. 1977 p. 13, 1977a pp. 8ff.

53 SDA 1977a pp. 6f, 1977e.

54 Chief Executive Lewis Robertson in response to trade union representative in meeting of the Scottish Economic Council, a consultative body to the Secretary of State for Scotland (as reported in SDA 1977e). The possibility was occasionally referred to throughout the period of Labour government, e.g. SDA 1977 p. 11, *Glasgow Herald* 16.6.76, 14.4.79.

55 SDA 1977 p. 12, cf. 1977a 4f and senior SDA staff quoted in the *Glasgow Herald* 25.8.76.

56 SDA 1977a pp. 4f, 1977b p. 4, 1978 p. 51, 1978a pp. 12f.

57 SDA 1977 p. 13 cf. 1976 Section 49, 1977a p. 8, 1978a pp. 2f.

58 SDA 1977a p. 11.

59 SDA 1977d p. 13.

SDA rather than the Scottish Office. In terms of underlying values this involved giving priority to long-term efficiency and competitiveness at the expense of employment considerations and, equally important, adopting a selective approach to industrial investments that targeted specific sectors and types of firms on the basis of a perceived problem within the regional economy, the 'equity gap'. The main contribution of the Scottish Office was the attempt to put implementation on a semi-commercial footing by influencing the discretionary appraisal of individual projects – thereby formally introducing an additional standard for policy implementation that did not necessarily correspond to long-term regional development objectives – but still the way in which the SDA intended to implement its investment policy was sufficiently ambiguous to fuel suspicions about its role *vis-à-vis* private-sector actors: some of the limits to patience and intervention were not exactly trumpeted left, right and centre,[60] and while intervention in the management of invested firms could easily be construed as the first step in a process of creeping nationalisation, patience could be seen as the slippery road that would eventually turn the Agency into a shelter for crippled waterfowl.

After the change of government in May 1979 the first draft version of a new set of investment guidelines appeared only two months after the general election,[61] later fortified through changes in the SDA Act in the Spring of 1980.[62] Apart from relegating industrial investment to an auxiliary function, the commercial nature of the activity was underlined by new and presumably more stringent appraisal criteria for individual projects according to which the Agency

> [...] should always have full regard to the profitability of the (invested) enterprise, [...]. seek to encourage maximum private sector participation, [...] and not invest in any enterprise for which sufficient and appropriate private sector money [...] is available.[63]

The revised guidelines clearly signalled what the new occupants at the Scottish Office had decided this activity was *not* about, namely extending public ownership or rescuing lame ducks, but at the same time the new guidelines actually encouraged smaller investments and cooperation with private financial institutions, and it is therefore particu-

---

60 Reference to the catalytic role of the Agency has only been found in the industrial strategies and the submission to the Wilson Committee, i.e. in contexts that were essentially internal to the political system or only subjected to a very limited degree of publicity.

61 Committee of Public Accounts 1980 pp. 8f.

62 Industry Act 1980 Section 1 Subsection 2.

63 SEPD 1980a p. 2.

larly interesting to note that the sponsor department not only defended[64] but actually extended the investment capability of the SDA in the early 1980s by instituting development funding as a new financial instrument under the heading of 'venture capital', aimed especially at projects especially in areas of new technology and operating *outside* the commercial discipline of the financial duty.[65]

The response of the SDA to this was, at least on the level of strategic statements, not particularly submissive. On the back of a new three-year corporate strategy the first annual report published under the new Chief Executive George Mathewson announced a threefold real-term increase in expenditure on investment for the period 1981/82-1983/84,[66] and although the term 'equity gap' was rarely used, the overall objective of the function would appear to be by and large the same, namely "to encourage things to happen which would not happen without the Agency".[67] The sectors and types of firms singled out for special attention were also mainly in line with the priorities of the late 1970s with small and medium-sized companies, high-tech industries and management restructuring all recurring, albeit the latter was rather less prominent than earlier.[68] The only major additions compared to the late 1970s would appear to be a commitment, in line with the investment guidelines, to "encouragement of private sector financial participation", and – perhaps more surprising – pledges to support "the revival and change of companies facing either industrial or financial problems",[69] "to retain a high proportion of decision-making within the Scottish economy",[70] and the increased prominence of high-risk technology projects escaping the discipline of the financial duty through the use of development finding.[71] The Agency's approach to industrial investments in the early 1980s would in other words appear to be dual: continuing and expanding its selective approach and even suggesting the possibility of what looks like rescue operations, and at the same time enthusiastically embracing the hard-nosed commercial rhetoric of the Scottish Office and the ideas of private co-investment.

---

64 Gavin McCrone in his evidence to the Public Accounts Committee 2.12.81 (1982a pp. 6f).

65 Kirwan 1981 pp. 30f, cf. IDS 1987a p. 73.

66 SDA 1981 p. 59, cf. *Glasgow Herald* 24.11.81.

67 SDA/DCC 1983 (4) p. 3.

68 SDA/DCC 1983 (4) p. 3.

69 SDA/DCC 1983 (4) p. 3.

70 George Mathewson in his evidence to the Public Accounts Committee 2.12.81 (1982a p. 4).

71 SDA 1984 p. 37, SDA/DCC 18.9.84.

Compared to the first investment guidelines of the Labour government, the changes introduced by the Conservatives in the early 1980s were definitely noticeable, but compared to the strategic thinking of the Agency before 1979, the similarities are striking and changes more limited: the SDA had already publically committed itself to commercial viability, and thus the genuinely innovative element in the new Scottish Office approach was, apart from the symbolic downgrading of the function, in effect the specific requirement of maximum private sector involvement which was designed to reinforce commercial appraisal criteria through indirect pressure from other providers of finance. Whether the two-pronged response of the Agency to the challenges of the early 1980s should be interpreted as a move in the direction of a more traditional merchant bank approach does, however, depend on the eventual balance between commercial project appraisal and a selective approach giving priority to investment projects that would otherwise have had difficulties in obtaining external funding. Whereas the former obviously was politically vital to communicate loud and clear in the new political environment, the more low-key existence of the latter should not be overlooked, and thus the organisation would still seem to have the potential to differ from private sector institutional investors.

In the second half of the 1980s the central strategic statement of the Scottish Office can be found in the report of the 1986 government review which recommended the continuation of the function because "the Agency's investment activity in recent years has made a worthwhile and cost-effective contribution to the Scottish economy"[72] and formulated a positive rationale by pointing to market failures where inadequate access to capital affected especially high-tech projects and small expanding businesses. These problems occurred either because risk aversion and inadequate information made private investors reluctant, or due to externalities "where the company involved would not be able to appropriate a significant part of the benefits that would accrue to the Scottish economy as a whole".[73] In the first scenario the role of the SDA would be to convince the private sector of the scope for profitable investments by means of demonstration projects, whereas in the second scenario a more permanent role was envisaged with the Agency bearing the cost of externalities associated with strategically important investments by filling a gap in the market.[74] In terms of implementation, the review echoed the commercial appraisal criteria of the investment 1980 guidelines in the language of market failure and recommended a more structured approach to disposal of the portfolio, although the report also

72 IDS 1987a p. 50.

73 IDS 1987a p. 46.

74 IDS 1987a p. 46.

warned that premature disposal of investments could involve "forgoing potential financial or development gains".[75] The reviewers themselves in other words appeared to be caught between three ambitions that were not necessarily compatible – remedying particular regional market failures while at the same time limiting public ownership and securing maximal financial benefits to the public purse – and this strategic ambiguity was underlined by their suggestion that the two other, uncommercial, types of SDA investments, concessionary loans for rural enterprises and development funding for technology projects, could be extended in close conjunction with the Scottish Office.[76]

As was generally the case in the late 1980s,[77] the SDA responded to the review by adopting a market-failure framework and now defined the objectives of the industrial investment function as "filling a perceived gap in the availability of finance, especially for small companies and developments carrying a high degree of risk",[78] and although the requirement for commercial implementation left only a narrow scope for the Agency as a "direct provider of last resort",[79] viability of projects was still to be achieved "over a larger timescale than the private financial sector would sometimes be prepared to accept".[80] Disposal of the more mature parts of the SDA's investment portfolio did, however, never become a strategic priority, and the Agency also maintained that its investment powers played an important "catalytic role" because the ability to put its own money on the line gave the organisation "a much wider influence and credibility in the eyes of both companies and financial institutions"[81] when negotiating financial packages for new investment projects.[82]

The adoption of market-failure terminology has sometimes been interpreted as the logical extension of the 1980 investment guidelines which created a catch-22 situation where every Agency investment had to meet two conflicting criteria:

---

75 IDS 1987a p. 51.

76 IDS 1987a pp. 51, 81f.

77 Cf. Section 7.3.

78 SDA 1988 p. 54, cf. Agency views reported in National Audit Office 1988 (p. 6).

79 Both quotes from SDA 1990 p. 32. Interestingly, although financial subsidies for private venture capitalists involved in particular types of projects could have been a distinct possibility, this option was not considered by the 1986 government review.

80 SDA 1987 p. 54.

81 The SDA as paraphrased in IDS 1987a (p 47), cf. Donald Patience 31.5.90 and Sir Robin Duthie 12.7.90.

82 National Audit Office 1988 p. 6. Along similar lines development funding in the guise of "seed-corn and pre-venture finance" also became a priority in the 1988-91 corporate strategy (SDA 1987a p. 8).

> Agency interventions must be increasingly commercial, but must simultaneously avoid areas where there is the prospect of the private sector filling gaps in market provision (either actually or potentially).[83]

Such an interpretation would, however, appear to be based on an interpretation of the general approach to economic policy of the review group, but in the case of the Agency's investment function specific forms of market failure had been identified and the general requirement to avoid anything that could deter potential private sector activity was actually played down.[84] An alternative reading of the industrial investment strategies of the late 1980s would therefore suggest that the criteria for discretionary project appraisal formulated by the government review and largely adopted by the SDA did in fact *not* differ significantly from existing Scottish Office guidelines or Agency rhetoric in the first part of the decade which both stressed the long-term viability of invested firms, support for projects that would otherwise not have happened, and maximum private sector leverage. Moreover, the objectives for the industrial investment function set out by the review were the most elaborate interpretation ever endorsed by the Scottish Office because it defined what was earlier known as the 'equity gap' in more precise terms such as information barriers, risk aversion and externalities, and thus the review strengthened the case for Agency action by allowing not only for temporary demonstration projects but also for quasi-permanent government intervention and subsidies.[85] But what was perhaps even more important was that the sponsor department was no longer solely concerned with the financial performance of Agency invested firms but clearly accepted that the industrial investment function should be assessed in relation to specific problems within the regional economy – and even produced an unreserved statement of support for this hitherto controversial function from the Scottish Office.[86]

The above analysis of the strategies for the investment function has demonstrated the crucial strategic importance attached to the industrial investment function relative to other instruments of financial support

---

83 Standing Commission 1988 p. 88, cf. Isobel Lindsay 6.6.90. For similar lines of reasoning, see Davies 1978, Radice 1978, National Audit Office 1988 pp. 5f.

84 IDS 1987a p. 46.

85 Perhaps the most surprising novelty of the late 1980s was that the strategic initiative would appear to belong to the Scottish Office rather than the development agency, but this may simply be an optical illusion because both the UK Treasury and the review team of the SDA had also made significant contributions to the 1986 review (Frank Kirwan 1.6.90, cf. Section 7.3).

86 Unlike other SDA functions, industrial investment was not subjected to a triannual programme of external evaluation in the wake of the 1986 review (SDA Board Papers 29.7.87).

and clarified the development of three key areas aspects. *Firstly*, the nature of the regional problem targeted remained surprisingly stable, with the equity gap of the 1970s re-emerging in market-failure terms in the 1980s. *Secondly*, the expected firm-level impact seems to have been somewhat more contentious, with the Scottish Office and the SDA appearing to have different (and shifting) priorities with regard to rescues until the mid-1980s, with only the latter stressing the importance of targeting particular sectors of industry. *Finally*, with regard to modes of implementation continuity clearly prevailed with the insistence on appraisal of investment projects on a (long-term) commercial basis, with regard to the persistent focus on SMEs, and in the Agency's long-standing commitment to priority sectors. Given the importance attached to the investment function in the early years of the SDA's existence it is hardly surprising that the industrial investment strategy developed much in parallel with the overall corporate strategies of the Agency, but despite this, a number of important differences between investment strategies and corporate strategies are, however, also evident. *Firstly*, the emphasis on proactivity was never particularly prominent in industrial investments, and largely absent in the 1980s, perhaps because this role had been taken over by development funding. *Secondly*, unlike criteria related to size and sector, the spatial location within Scotland of individual investments seems to have been of limited concern. And *finally*, industrial investments were not subjected to the general requirement of involving the private sector directly in the delivery of policy programmes in the late 1980s.

All in all the immediate effect of the change of government in 1979 would not appear to have been a dramatic shift of direction with regard to the development strategy of the industrial investment function, but instead seems to have reinforced existing trends, both within the SDA and in terms of sponsor department regulation. Add to this that the 1986 government review strengthened rather than undermined the position of what used to be the most controversial Agency function, and the external-revolution paradigm would not seem to be a particularly useful interpretation of the developing strategies in this particular area of activity. Given the British discursive terrain of the 1980s, emphasising a commercial approach could of course readily be interpreted by the wider Scottish environment as the kowtowing of a central government sponsored RDA, but still understanding the full complexity of the development of the Agency's investment strategies requires both a longer time perspective and a systematic consideration of non-discursive phenomena: Agency investment appraisal may have changed radically in practice and so may the willingness of the sponsor department to accept projects above guideline limits, and thus the possibility of 'back-door

Thatcherisation' of the function cannot be discounted before an inquiry into policy implementation has been undertaken.

8.2. Implementing Industrial Investment

In order to illuminate the way in which the investment strategies have been translated into concrete action within the regional economy, this section explores key aspects of the rules, resource exchanges and outcomes of policy implementation, namely the overall level of activity, patterns of resource exchange related to policy instruments and modes of implementation, and measures of aggregate and institutional impacts. The text below will first examine industrial investment in detail before discussing the implementation of the two other types of investments under a separate heading, but first a general note of caution must be struck with regard to the interpretation of information about investment activities. With conditional finance as the main policy instrument, the level of activity will have depended not only upon the specific conditions attached to individual investments, but also on how private firms responded to these initiatives, and thus the response to this particular SDA policy would also reflect broader trends in the economic environment such as macro-economic conditions and the availability of alternative sources of external funding. The link between strategic decisions by public actors, and the eventual outcome in terms of completed investments and the performance of invested firms is in other words complicated by its dependence on the strategies of private actors and their wider economic environment, and thus the interpretation of activity indicators will require some caution, especially with regard to possible causal links to the political priorities of the Agency and its political sponsors. Given the fairly crude nature of some of the existing interpretations of the history of the industrial investment function and the limited amount of analysis hitherto undertaken, a systematic and careful exploration of output measures should, however, still be able to make a significant contribution to the understanding of what was undoubtedly the most controversial function of the SDA.

Financial Resources and Investment Activity

The introduction to this chapter noted that the level of activity in the field of industrial investment was generally assumed to have decreased in the 1980s following the new guidelines issued by the incoming Conservatives, and as investment involves the commitment of public funds to individual private sector projects, expenditure and the number of investments undertaken are obvious measures of activity on the aggregate level. As can be seen from Figure 8.1, industrial investment was by far the most important of the three functions in financial terms,[87] and the total amount committed to investment has evidently fluctuated with the late 1970s and mid-1980s as peaks and the early 1980s as a conspicuous trough, but the average yearly spending in real terms was actually higher after 1979 than before,[88] and thus the idea of a permanent and radical downgrading of the role of investment following the change of government is clearly not supported by aggregate expenditure data. Instead the overall importance of investment in SDA expenditure remained fairly constant, with a share of total SDA expenditure remaining close to the average of 11% in most years,[89] and a closer look at the industrial investment expenditure reveals that the overall development has generally been linked to the amounts actually committed to major investment projects handled by HOID,[90] while expenditure on administration – a hidden subsidy deliberately kept of out the financial duty – was significantly higher in the 1980s than the 1970s, probably reflecting both the growing size of the portfolio administered and the front-end costs involved in evaluating potential projects.[91] Investments had an obvious capacity for generating income through interest or capital receipts, and as the SDA's portfolio matured and the Scottish Office started to encourage disposal of investments this potential certainly came to be realised, as illustrated by Figure 8.2: proceeds from disposal of equity accounting for 75% of internally generated investment income,[92] and this compliance undoubtedly made the industrial investment function less controversial while at the same time reinforcing the demand for commercially viable investments.

---

87 No figures are available for concessionary loans in rural areas in the early years, but even on the (unrealistic) assumption that all small business investment belonged under this heading, such expenditure would still have accounted for less than 15% of the total (calculated on the basis of SDA 1977-91). Likewise development funding is known to have existed as a financial instrument from the early 1980s (Kirwan 1981 pp. 30f, SDA/DCC 18.9.84 p. 9, IDS 1987a p. 73), although it was not stated separately in the accounts before 1986/87.

88 The average yearly real-term spending (1985/86 prices) was less than 14 £m before and more than 15 £m after the change of government (calculated on the basis of SDA 1977-91).

89 Calculated on the basis of SDA 1977-91.

90 The peak in expenditure in the late 1980s was mainly the result of increased commercial provisions for risky investments (calculated on the basis of SDA 1977-91).

91 Administrative costs accounted for 6% of investment expenditure in 1976-80 but had risen to 23% in 1989-91 (calculated on the basis of SDA 1976-91). These figures probably underestimate the true cost of appraisal and supervision because they exclude administrative costs incurred in SBD and the Agency's sectoral divisions.

92 Calculated on the basis of SDA 1977-91.

Figure 8.1 SDA expenditure on investment by function

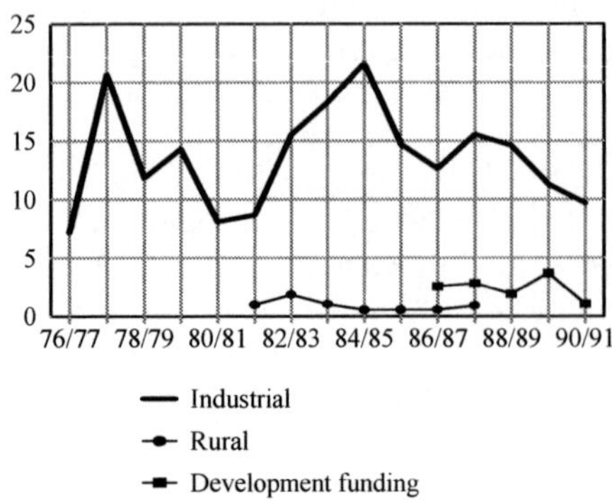

£m 1985/86.
Source: SDA 1977-91.

The continued importance of industrial investment in the 1980s is even more pronounced if the number of major investments undertaken by the SDA is used to measure the level of activity, as demonstrated by Figure 8.3.[93] The financial contraction following the change of government in 1979 was clearly not the result of a significantly smaller number of investment cases but would instead seem to reflect a pronounced decrease in the average size of the commitments undertaken, and after 1982 the average number of transactions each year is more than double of that of the alleged glory days of the late 1970s,[94] although the number of investments undertaken each year fluctuated greatly.

[93] Like SDAINV in general, Figure 8.3 does not include second-round investments, i.e. provision of additional funds to SDA invested firms, but as becoming involved with new firms is likely to be more demanding in terms of Agency management resources than extension of existing commitments, the figure would still seem to be a reasonable indicator of the changing levels of investment activity.

[94] Calculated on the basis of SDAINV.

Figure 8.2 Investment-related income as percentage of expenditure on investment

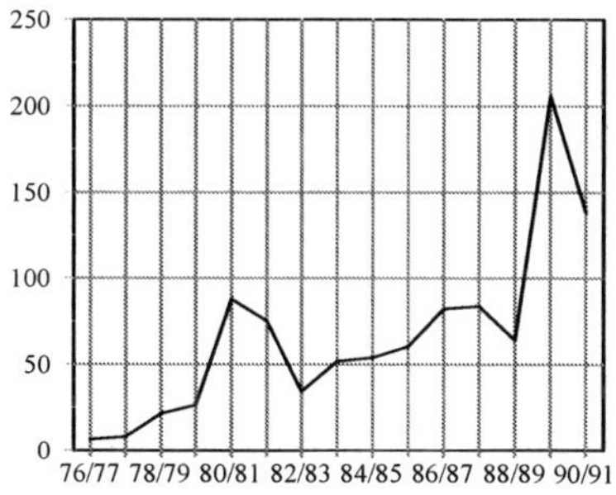

Source: calculated on the basis of SDA 1977-91.

Figure 8.3 Indexes of SDA investment activity

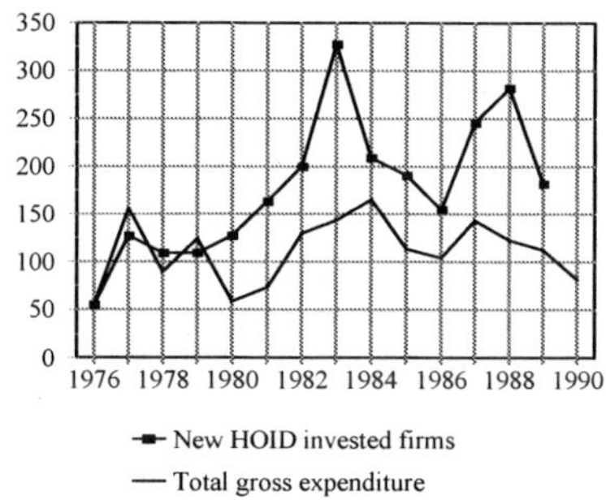

100 = average 1976-79,
real term expenditure figures relate to financial years, investment projects to calendar years.
Sources: calculated on the basis of SDAINV and SDA 1977-91.

All in all it can be concluded that neither in terms of financial resources nor investment projects can 1979 be regarded as the crucial watershed which clearly separates two distinct phases of operation from one another. Instead of a dichotomous interpretation, the aggregate-level activity indicators would seem to suggest that it would be more meaningful to distinguish between four periods: the *start-up* phase (1975-78) where activity in this new policy area expanded rapidly, the years of *stagnation* (1979-81) where the number of investments were stable while spending was halved and disposals increased, the years of *expansion* (1982-84) where the function regained its prominence, both in absolute and relative terms, and a period of *consolidation* (1985-91) where the level of activity settled on a somewhat lower level. The contrast between the two first periods could easily be construed as a reflection of party-political differences, and later the steady increase in internally generated income allowed the Scottish Office to maintain this impression in the 1986 government review by presenting a rather imaginative calculation that minimised expenditure on industrial investment in the 1980s,[95] something which the Agency's critics readily accepted perhaps because it corresponded neatly to what one would suspect a Thatcherite government to do.[96] Such a partial interpretation along external-revolution lines does, however, disregard the expansionary third and consolidary fourth periods, and as such an expansion will have required the consent of the sponsor department, the adherence of the Scottish Office to proclaimed aims of the 1980 guidelines of making industrial investment a last-resort measure would seem to have been rather limited in practice. It should, however, also be noted that beneath these fluctuations a long-term trend has been identified on the basis of aggregate level activity indicators, namely that the Agency invested alongside private actors in an increasing number of firms and that these investments became increasingly successful in financial terms. This could be construed as a move towards that of a more traditional institutional investor with private-sector standards of success, and the ability of the SDA to make a difference – i.e. influence regional development by acting differently from other investors – would therefore seem to hinge on a more detailed assessment of the quality of its investments, e.g. the added value to individual proposals in terms of information and organisation when appraising the initial project proposals and supervising the performance of invested firms.

Policy Instruments and Modes of Implementation

Using industrial investment as a policy instrument in pursuit of regional development objectives involved making financial resources available on certain conditions in order to make private actors engage in projects that promote economic development within the region, and the nature of these conditions can be illuminated by analysing the modes of implementation because they constitute the detailed rules according to which individual investment proposals may be translated into financial commitments. The balance between finance as an incentive and the conditional options created through the modes of implementation may be more or less attractive to private firms, and thus these modes may not only influence the volume of the function, but also promote particular forms of economic activity, and because of the ongoing nature of the investment relation the Agency would be able to employ its organisational and informational resources to influence individual firms not only

95 By substituting gross expenditure figures with net figures, the share of investment is reduced from more than 10% to about 1% (gross figure calculated on the basis of SDA 1977-91, net figure from IDS 1987a p. 44), producing an indicator of the overall strain on the public purse rather than a measure of the potential impact of Agency activity on firms in the regional economy.

96 E.g. Standing Commission 1988, Douglas Harrison 19.7.90.

when appraising a new project but also when supervising – and disposing of – existing investments. In the following each of the three key modes of implementation within regional policy are examined in turn: the general practices of *project generation*, issues of *selectivity* is explored with regard to the size, type and sectoral distribution of projects, and finally *project appraisal*, both with regard to pre-investment scrutiny and post-investment monitoring.

Project Generation

In terms of project generation the industrial investment function operated primarily in a reactive manner with proposals being brought to the Agency by private firms and financial institutions.[97] Although the late 1970s rhetoric about the need for new initiatives vaned in the 1980s, the SDA appears to have acted in a pro-active way in a number of investment cases based on the work of e.g. its New Ventures Unit or the sectoral divisions also after the change of government.[98] Still, most projects were initiated by private economic actors, and the Agency's high public profile and co-operation with private financial institutions ensured that it was unnecessary to publicise this particular function despite its reactive mode of operation.[99] In terms of development strategies this state of affairs implies that while the SDA retained a *capacity* to act proactively, most projects must have been designed by private actors and the main roles of the Agency will have been to assess to quality and potential of individual projects and, if necessary and possible, to improve what was being proposed by other economic actors.

Investment Selectivity: Size, Type and Sector

In formal terms the degree of selectivity involved in the industrial investment function was limited and remained so from the late 1970s through to the early 1990s. In principle the SDA was permitted to invest in nearly any form of economic activity – the only exceptions being media interests in the early years[100] and firms opposing Agency investment from 1980s onwards[101] – but in practice the playing field was not even with regard to investment projects because the investment guidelines effectively gave preference to projects under a certain size, and widely published SDA strategies identified certain priority sectors and preferred types of investment projects. Individual firms would have a

97 Lewis Robertson 20.7.90, Donald Patience 31.5.90.

98 E.g. SDA 1981 pp. 26f, 1985 p. 44, 1988 pp. 54ff, 1989 pp. 34f, 1990 pp. 32f; Donald Patience 31.5.90.

99 Jennifer Forbes 15.6.90, Harry Hood 25.7.90.

100 SEPD 1976b pp. 10f.

101 SEPD 1980a p. 3.

fairly clear idea about whether they were situated within these areas in advance of approaching the Agency – unlike the demand for commercially viable projects which would seem to be less well-defined and hence will be discussed under the heading of project appraisal – and thus in the following selectivity will be examined in relation to three strategic priorities, namely the size of investment projects and the invested firms, the type of investment projects, and the sectoral composition of the invested firms.[102] All three areas had been the subject of SDA strategic thinking, and it could be expected that the average size of financial commitments have decreased, that the importance of investment in new firms has increased, and that the sectoral focus of the function has shifted from traditional towards modern industries.

Figure 8.4 Indexes of average HOID investment size

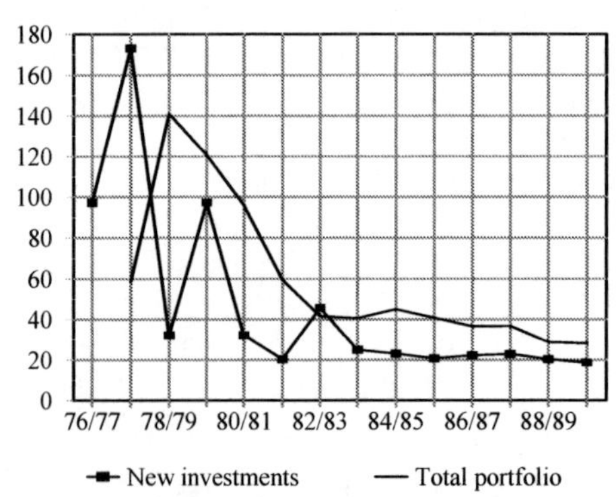

100 = average 1976-79.
Sources: Calculated on the basis of SDA 1976-90 and SDAINV.

The new guidelines introduced by the Scottish Office in 1980 suggest that in comparison to the late 1970s the average size of the investments undertaken will have decreased, and by implication therefore also the size of the invested firms because smaller investment projects are likely to be of greater interest to smaller firms. Regarding the latter the Agency had, however, been cautious about involvement with large firms

[102] It is worth noting that in this policy area the spatial distribution of activity seems not to have been an issue, except perhaps in the very early days where the SDA hinted vaguely about a possible bias towards Strathclyde (e.g. 1977 pp. 12f), but in practice the geographical distribution of firms within the portfolio have been roughly equal to that of activity in the Scottish economy at large, with Strathclyde accounting for around half in both (calculated basis of SDAINV and *Economic Trends Regional Accounts* November 1990).

also in the early years,[103] and this pragmatic view would appear to have prevailed throughout: the SDA predominantly invested in relatively small firms from the very beginning, with only around 1/3 of HOID invested firms having more than 100 employees,[104] and in the mid-1980s half of the firms invested in by the Small Business Division had five or fewer employees.[105] With regard to investment project size the difference between the late 1970s and the 1980s is, however, noticeable, as illustrated by Figure 8.4, with the average size of SDA investments, both new and existing, being significantly lower in the 1980s than in the 1970s.[106] While living up to the requirement of the new investment guidelines also had the additional advantage of making the organisation less exposed to the failure of individual firms within the portfolio,[107] it is, however, interesting to note that the contrast in average investment size in the portfolio between the late 1970s and the late 1980s – the former is 6-7 times the latter in real terms – is to a very large extent the result of the presence in the early years of a small number of very large investments in a relative small portfolio.[108] If these large investments are discounted, the average size of HOID investments prior to the change of government is only twice that found in the second half of the 1980s,[109] and as a handful of major investments also featured in the late 1980s,[110] the main contrast in terms of size would in other words seem to be the extent to which large investments came to be cushioned by a much larger number of smaller investments.[111] In terms of size selectivity we have in other words found that the emphasis in terms of projects changed towards smaller investments, much as expected, but that this would not appear to have taken place at the expense of larger invest-

---

103 SDA 1978a pp. 7f.

104 SDA 1977d p. 21, IDS 1987a pp. 44f.

105 Non-HOID investments accounted for around 29% of the total amount invested (IDS 1987a p. 44).

106 The size of individual investments is the only area in which it has not been possible to establish a comprehensive series of data: the amount committed is available for 88% of investments undertaken until 1984 but only for 28% of investments undertaken from 1985 onwards.

107 Donald Patience 31.5.90.

108 When the average investment size within the portfolio culminated in 1978-79, the six largest investments (all above the 1 £m threshold later instituted by the 1980 guidelines) accounted for 71% of HOID invested funds on the basis of less that 20% of the invested firms (calculated on the basis of SDA 1979 pp. 83f).

109 Calculated on the basis of SDA 1979 pp. 83f, SDA 1976-80 and SDAINV.

110 From 1988 onwards, SDA accounts list all investment above the 1 £m threshold and consistently show 8-10 large investments in the portfolio (SDA 1988-91).

111 In 1990 HOID had 8 industrial investments about the 1 £m threshold, constituting only 6% of invested firms and 21% of the total value of the SDA portfolio (calculated on the basis of SDA 1990 and SDAINV).

ments, although the relative position of the latter was reduced as the number of small firms in the portfolio increased.

Figure 8.5 SDA investments by project type

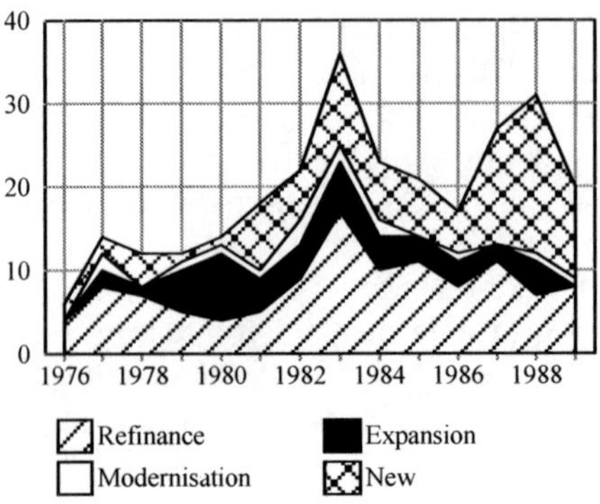

Number of investments undertaken each year.
Source: SDAINV.

The second aspect of selectivity to be explored, the preferences with regard to project types and firm-level impact, was also subject to strategic considerations throughout the SDA's existence. From the very beginning rescues as a way of preserving individual firms had been ruled out by the Agency, and while this seemed to be at least partly reversed after the change of government, added emphasis was clearly given to supporting new firms in order to promote entrepreneurialism within the regional economy, and thus overall the expectation would be to find a shift towards investment projects in new firms. In order to illuminate the development of project-type selectivity within the function, firms in the SDAINV database have been grouped under four headings according to the firm-level purpose of the original investment project – technological modernisation, expansion or refinancing of existing firms, or support for new firms[112] – which each entail different possibilities for the policy-making organisation with regard to influencing the strategies of private actors, i.e. the uncertainty surrounding firms with little or no track-record or the organisational focus when trying to find new owners and/or investors for a firm experiencing a cash-flow crisis in the wake of previous modernisation or expansion.

[112] The four categories have been derived from Table 3.11 by excluding options which were not relevant in connection with this particular function and collapsing two others (ownership, preservation) which proved difficult to distinguish in the empirical data. Individual projects have been assigned on the basis of a review of internal SDA classifications.

As can be seen from Figure 8.5, the project-type characteristics of HOID investment projects present an interesting pattern where the relative importance of different firm-level impacts change over time. *Firstly*, more than one third of all investments have been classified as new, and it is obvious that the importance of this category increased immensely during the period. The average number of investments in new firms per year in the late 1980s was five times that in the late 1970s,[113] but still investments in existing firms consistently outnumbered new ventures until 1987 after when a rough balance between the two types of investments can be observed. *Secondly*, it is clear that expansion and especially modernisation of existing productive facilities have played only minor roles in the overall picture, accounting for little more than 20% total investments between them.[114] *Thirdly*, the most common type of investment, accounting for more than 40% of the overall total,[115] was aimed at improving the financial or organisational foundation of manufacturing rather than the productive capacity. The refinance/ownership type of investment comprises both rescues of ailing firms and management buy-outs (MBOs) where (parts of) the existing management team takes control backed by external finance, and during the period the number of rescues fell dramatically while the number of MBOs increased even more.[116] This shift may of course simply reflect the general change in the discourse surrounding economic and regional policy – what used to be regarded as rescue operations in 1970s may in the 1980s have been classified under other headings such as e.g. MBOs – but it could also be seen as being in line with a new approach to investment appraisal on part of the Agency where having a new management team in place was seen as a prerequisite for financial commitment to ailing firms.[117] At the same time MBOs could be seen as an unconfrontational way of bolstering Scottish control of the regional

[113] Calculated on the basis of SDAINV.

[114] Calculated on the basis of SDAINV. The only period in which expansion was relatively prominent was in the early 1980s, perhaps because a combination of severe economic crisis and political uncertainty made expansion the 'safest' type of project both for firms and the Agency.

[115] Calculated on the basis of SDAINV.

[116] The average share of rescues in the refinance/ownership category fell from 25% to less than 5% from the late 1970s to the late 1980s, while that of MBOs increased from below 10% to around 50% (calculated on the basis of SDAINV), but also after the appointment of George Mathewson as Chief Executive in 1981 the SDA was involved in major rescues, and e.g. in 1981 the reconstruction of the Weir firm was officially hailed as an "ideal illustration" of how the industrial investment function could be employed (SDA 1981 p. 28, cf. Young & Reeves pp. 156f).

[117] Cf. the discussion below. MBOs also had the advantage of being considerably less risky than other forms of investment (SDA/DCC September 1986 p. 1).

economy,[118] a controversial issue where some have maintained that "the SDA has acted to protect promising Scottish companies from 'foreign predators'"[119] while others have seen the Agency as ineffective in this respect.[120] The available evidence does, however, suggest that while the SDA has certainly in some cases spoken publically against major external take-over bids,[121] this has rarely been followed by financial commitments,[122] perhaps because the Agency believed that "the evidence suggests a net positive effect to the individual Scottish companies taken over".[123]

On the basis of the analysis of the types of investment projects undertaken by the SDA and their firm-level implications two conclusions can be drawn. First and foremost it has become clear that the industrial investment function has predominantly targeted the indigenous sector of the economy and that the change of government in 1979 was *not* followed by a radical shift of focus from existing to new firms. The growth in new-type investments, especially in the late 1980s, supplemented rather than replaced financial commitments to existing firms, and thus perhaps the most significant development may well be a much more low-key change, namely the rapidly increasing role of MBOs in the Agency's investment portfolio.

[118] An argument used by the SDA itself (SDA/DCC September 1986 p. 1).

[119] Keating & Midwinter 1983 p. 178.

[120] STUC 1987a p. 16, Standing Commission 1989 pp. 40f.

[121] See SDA 1981b, cf. Cameron 1985, Leruez 1982 Ch. 5, Craig Campbell 16.7.90.

[122] The SDA did not invest in industries in which government was involved through nationalisations, and this kept, for better or for worse, the organisation from becoming involved in an aspect of industrial decline that dominated public debate in the 1980s, cf. the discussion in Chapter 14.

[123] SDA/DCC September 1988.

Figure 8.6 SDA investments by sector

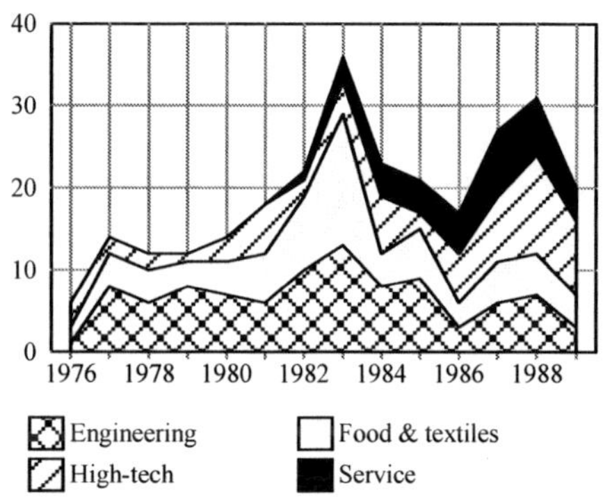

Number of investments undertaken each year.
Source: SDAINV.

Turning now to the role of sectoral priorities, the external-revolution interpretation of its development would seem to imply a shift of emphasis from trying to restructure ailing 'sunset' industries to attempting to promote what was perceived as the 'sunrise' industries of tomorrow. In the SDAINV database investments have been classified to one of four headings on the basis of historical patterns of employment and technology which position particular forms of economic activity in the discursive terrain as belonging to the past or the future.[124] This separates 'traditional' industries such as engineering, food and textiles with a long but declining presence within the region from 'modern' high-tech and service industries with recent high rates of growth.

On the basis of the information held in the SDAINV database, a straightforward dichotomy between the 1970s and the 1980s would hardly seem to be the best way to account for the changing sectoral composition of the investments of the SDA, as illustrated by Figure 8.6. *Firstly*, the average yearly number of new invested firms in traditional sectors was actually nearly 50% higher in the 1980s than in the late 1970s, and until 1985 such projects constituted around 75% of all investments undertaken,[125] with the conscious investment drive of the new management team in the first half of the 1980s actually resulting in record levels of firms in traditional sectors being supported by the

124 Individual firms have been classified on the basis of information published in annual reports (SDA 1977-82) and by simplifying the internal classification system used in the HOID database from 1982 onwards.

125 Calculated on the basis of SDAINV.

function. *Secondly*, the two traditional areas develop in very different ways: while investment in engineering has been fairly stable, albeit gradually decreasing after the mid-1980s, food and textiles were largely responsible for the first of the two peaks in investment activity, especially through a large number projects in textiles and clothing. *Thirdly*, investment in service and high-tech industries increased in the 1980s. No service sector commitments had been undertaken before 1982, and in high-tech industries such as electronics and bio-technology the average level of activity in the 1980s was three times that of the late 1970s.[126] All in all the relative weight of traditional and modern elements within the portfolio clearly starts to change in the mid-1980s with the latter gaining in importance, but it would be difficult to argue that this had taken place at the expense of the former. A more reasonable interpretation of the sectoral profile would be to see the growing involvement of HOID in high-tech and services as the adoption of additional priorities, especially in the second half of the 1980s.

The third and final mode of implementation to be considered in relation to the industrial investment function is the way in which the appraisal of individual projects was undertaken, and here the situation is very clear-cut, at least at the general level: Scottish firms did not have a statutory right to Agency investment at certain rates, and instead individual proposals were assessed by SDA staff against non-overt criteria and resulted in financial support tailored to the individual situation rather than calculated on the basis of a fixed formula.[127] The function did in other words operate in a discretionary manner, and in contrast to other financial instruments within regional policy this discretion was not exercised only at one point in time: in addition to the initial screening on the basis of which it would be decided whether or not to support a particular project, discretion was also an integrated part of the ongoing relationship between the Agency and the invested firm because in both phases the public body could attempt to influence the strategic decisions of the firm drawing on technical or organisational knowledge within the Agency. The analysis begins by looking at the procedures concerning pre-investment appraisal, then turns to the way in which the SDA conducted its relations with invested firms, and finally examines a number of indicators which could illuminate the extent to which the central discretionary appraisal criteria, commercial viability, has been adhered to in practice.

---

126 Calculated on the basis of SDAINV.

127 The non-automatic nature of the function is evident from the investment guidelines (SEPD 1976b, 1980a) and the persistent sponsor interest in appraisal criteria.

With regard to pre-investment appraisal of investment proposals it is clear that the procedures applied by the SDA have gradually been formalised over the years. Given that the function started from scratch in a situation where little or no experience with this type of policy instrument was at hand, it is hardly surprising that the extensive 'check-list' originally produced by Chief Executive Lewis Robertson[128] had to be modified along the way. Following the change of government, the Agency's investment procedures were severely criticised by the parliamentary Committee of Public Accounts,[129] and this resulted in changes in the first half of the 1980s which were later described as considerable improvements.[130] Only the late 1970s check-list has, however, been published, and already here scrutiny of a wide range of issues is prescribed,[131] including organisation, finances, technology, production and markets – areas that continued to be of concern in the early 1980s.[132] The precise balance between these factors has probably varied between individual cases, but the questions of ensuring long-term profitability and the presence of the right quality of management were certainly very prominent, both in the late 1970s and in the 1980s.[133] From the beginning the organisation made great play of its high rejection rate in the early years in order to drive home its commitment to "the discipline of bedrock tests of viability",[134] and in the early 1980s the Agency's new investment director informed the members of the DCC that "many of the cases which are brought to the Agency will require a substantial amount of structuring".[135] A consultancy study undertaken in connection with the 1986 government review found that 40% of the firms surveyed claimed that the SDA had demanded changes in management or strategies in connection with the investment negotiations,[136] and thus the Agency appears to have lived up to its bullish rhetoric, at least in the 1980s. Unfortunately, no comparable survey from e.g. 1978/79 has been

[128] SDA/DCC 22.11.77, John Firn 15.6.90.

[129] Committee of Public Accounts 1980 p. IX.

[130] National Audit Office 1985 p. 14, Committee of Public Accounts 1985 p. VII.

[131] SDA 1977d pp. 28ff. An interesting testimony of the organisation's awareness of the possible political pressure from elected politicians was the requirement to check whether local MPs had expressed interest in the fate of a particular firm (SDA 1977d p. 32).

[132] SDA 1981 p. 28.

[133] See e.g. SDA 1976 sect. 49, 1977a p. 9, 1977d p. 17, SDA/DCC 83(4) p. 2, National Audit Office 1985 p. 11, SDA 1989 p. 40, cf. Lewis Robertson 20.7.90.

[134] SDA 1977 p. 6, cf. SDA 1976 Section 17, oral evidence to the Wilson Committee (1977f vol. 6 p. 197), SDA/DCC 22.11.77.

[135] SDA/DCC(83)4 p. 3, cf. SDA 1985 p. 44.

[136] IDS 1987a p. 49, cf. Donald Patience 31.5.90, Sir Robin Duthie 12.7.90, Gavin McCrone 20.6.90.

located, and a direct comparison between the practice before and after the change of government can therefore not be undertaken.

Discretionary appraisal was also involved in post-investment monitoring, and like has been shown to be the case with pre-investment appraisal, a gradual formalisation of procedures took place in this area too.[137] From the beginning the official policy of the SDA was to "avoid interference in the day-to-day management of companies, which are expected to operate at arm's length from the Agency",[138] but senior SDA executives have maintained that in fact most of the early investments were financial commitments followed by attempts to improve the existing organisation,[139] and this is supported by the fact that in the early years the Agency committed considerable resources to invested firms in a way what can perhaps best be described as 'part-time secondment' where SDA staff were assigned to particular investments and sometimes spent one day each week with 'their' company.[140] Apart from the obvious risk of starting to identify with the short-term interests of the invested firm rather than the long-term goals of the public development body,[141] this practice of 'filling management gaps' was a considerable strain on the Agency's limited resources, and the emphasis changed fairly soon based on the experience with the first investments undertaken because in 1978-79 a specialised unit, Management Services, was established to provide advice and consultancy primarily to SDA invested firms,[142] something which marked the beginning of a new strategy aiming to upgrade the skills of the existing management team.

All in all the quality of management would clearly seem to have been a major preoccupation of the SDA's discretionary appraisal of investment projects, both in the initial screening of new proposals and the subsequent supervision of invested firms, but at the same time it is also evident that the means employed to reach this end gradually changed: the importance of pre-investment screening of the existing management within the firm increased and the criteria applied appear to

137 National Audit Office 1985 pp. 6ff.

138 Robertson 1978 p. 27, cf. Lewis Robertson 20.7.90.

139 Edward Cunningham 13.6.90, Sir Robin Duthie 12.7.90, Gavin McCrone 20.6.90, cf. SDA 1977b p. 2.

140 SDA 1977a p. 9, Gerry Murray 30.7.90, Charles Fairley 26.7.90, Sir Robin Duthie 12.7.90.

141 Charles Fairley 26.7.90.

142 SDA 1979 p. 24, Gerry Murray 30.7.90. The precise date remains unclear because references to Management Services as a part of the Industry Directorate can actually be found in the first organisation chart of the SDA (1977 p. 35), but the formulation in the annual reports for 1978 and 1979 suggest that the importance of this activity is (foreseen to be) growing significantly.

have become more stringent (i.e. only investing when the right management team was in place), and post-investment efforts to remedy perceived management deficiencies moved from a hands-on approach with part-time secondment in order to 'fill gaps' to a more 'arm's-length' approach which through consultancy and advisory services attempted to strengthen the management team emerging from the pre-investment screening of the project. Especially the timing of the establishing of the Management Services unit would seem to suggest that the Agency approach had already started to change under the Labour government, initially driven by internal organisational considerations like economising with scarce staff resources, but the advent of a government with a sceptical attitude to permanent public sector involvement with private firms undoubtedly fortified the trend. Ironically, this of course implied that the Agency became more heavily involved in long-term strategic issues through the initial screening of the management team behind the investment proposal and the strategically oriented consultancy support of the Management Services unit, but although in the first years the resource transfers to individual firms were clearly more visible, the new and more discreet approach may well have resulted in greater changes on the level of individual firms in the long run.

Figure 8.7 Financial duty on industrial investments

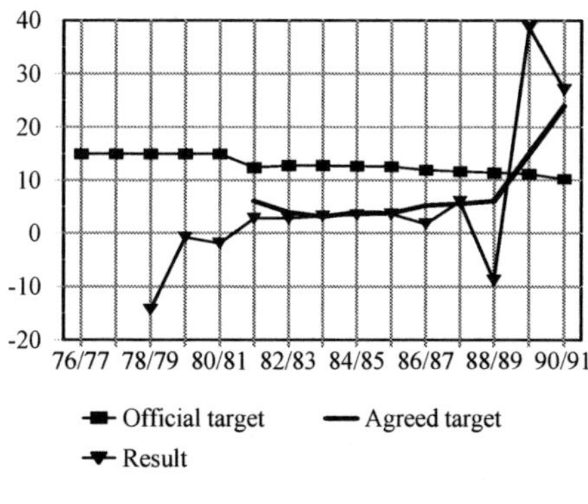

Per cent.
Source: SDA 1977-91.

According to the strategic statements of the SDA and the Scottish Office the key standard against which the ongoing performance of individual investments would be judged was commercial viability, and a number of factors combine to create the expectation, also inherent in the external-revolution interpretation, that the function became more commercially oriented in the 1980s: intense media scrutiny in the late 1970s had publicised some major investment failures,[143] the new investment guidelines appeared to demand radical change in the discretionary assessment of industrial investments, and the Conservative government abandoned some of the major pre-1979 Agency investment projects in a very public manner.[144] In the following a number of possibilities for illuminating the extent to which more commercial appraisal criteria have been adopted are explored, first through the investment duty and then via the SDAINV database.

As can be seen from Figure 8.7, until the early 1990s the performance of the industrial investment function when seen through the financial duty fell more than a little short of the official targets,[145] and the fact that such underperformance was tolerated also in the 1980s can be seen as a "tribute to the political value of the SDA to the government".[146] The 'agreed targets' introduced in the early 1980s also underlines the symbolic nature of the financial duty because the accuracy with which these were determined in the 1980s would seem to presuppose a close cooperation between the Agency and its sponsor department, making the agreed targets resemble joint projections rather than external performance control. It would of course be tempting to contrast the negative results of the early years with the steady, if slow, progress made in the 1980s in order to demonstrate the increasingly commercial approach of the SDA, but this would be more than a little misleading because the new financial duty introduced in 1981/82 changed the way in which the results were calculated,[147] and the trend observed in the 1980s may simply reflect the improving macro-economic environment and the gradual maturing of the portfolio,[148] and thus the financial duty is of little use even as an indirect measure of commercial viability criteria.

---

143 See *Glasgow Herald* 10.2.77, 23.3.78, 20.4.78, 14.2.79, 22.9.79, 26.9.80.

144 Frank Kirwan 1.6.90. The most prominent of these was Stonefield Vehicles, promoted by the SDA as an example of its proactive approach to regional development (e.g. SDA 1978 pp. 23, 27), cf. Lewis Robertson to the Committee of Public Accounts (1980), and STUC 1981 p. 61.

145 The negative return for 1988/89 was offset by a massive 38.67% return the following year due to timing of the disposal of a particular shareholding (SDA 1989 p. 58), and the 1988/89 result did therefore not represent a new downward trend in the calculated results.

146 Moore & Booth 1984 p. 15.

147 Committee of Public Accounts 1982a pp. 9, 18f.

148 The young age of the portfolio was frequently referred to by the SDA and its executives when questioned about its poor performance in relation to the financial duty (e.g. Committee of Public Accounts 1980 p. X, SDA 1985 p. 45, 1988 pp. 54ff, Hood 1991a).

Figure 8.8 Financial measures of success and failure in industrial investments

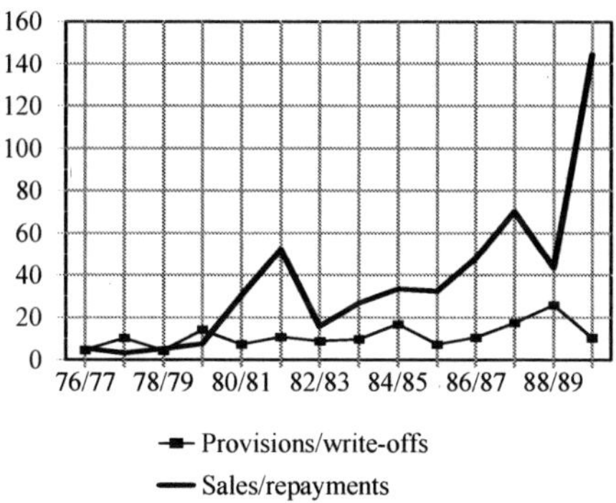

Provisions/write-offs and sales as percentage of value of portfolio.
Source: SDA 1977-91 and SDAINV.

An alternative way of illuminating whether the industrial investment function had indeed become more commercial in its orientation over the years would be to focus on those firms in which the involvement of the SDA was eventually terminated: either because the receiver had to be called in when a firm failed to perform adequately, or because stakes in a more successful firm could be sold. As illustrated by Figure 8.8, the financial liabilities incurred by the SDA through less successful investments remained fairly stable around c 10% of the value of the portfolio until the late 1980s,[149] and thus the advent of a new management team in the early 1980s did clearly *not* result in a large share of the portfolio being handed over to the receiver. It is equally clear from Figure 8.8 that receipts from sales and repayments increased significantly after the change of government, culminating in 1981/82 and again in the late 1980s, but this does not necessarily imply a more commercial approach on part of the development agency because when the both SDA and the private sector knew that there was a strong political demand for sell-offs, this could easily depress prices and hence make it attractive for private investors to acquire even firms with a less convincing track record,[150] and hence the available evidence can hardly in itself support a

149 The relative stable level of failures in financial terms recorded was the product of a few major losses before the early 1980s, while a large number of smaller firms ending up in receivership from then on (calculated on the basis of SDAINV).

150 The first surge in Agency income from successful investments in 1982/83 was created by a small number of large investments among which firms controlled by the Agency were over-represented – 4 out of 6 firms sold in the period 1.4.79 to 30.3.82

clear dichotomy between periods in which the SDA has adopted a more or less commercial approach to investment appraisal.

All in all the evidence with regard to the three modes of implementation has not lent unqualified support to the idea that the approach of the SDA was transformed dramatically following the change of government in 1979:

- project generation remained predominantly reactive although a proactive capacity was retained,
- in terms of selectivity an increasing number of small investments in new firms and modern industries occurred, but at the same time large projects, investments in existing firms, and investments in traditional industries continued, and
- in terms of appraisal the way in which discretion was exercised changed, with the focus moving away from post-investment organisational support towards pre-investment screening in combination with post-investment informational support, but whether this improved the performance of the invested firms could not to be established on the basis of the indicators examined.

These conclusions – that the profile of the investments changed by branching into additional areas and that it remains to be demonstrated that the function came to operate on the basis of a more commercial form of discretionary assessment – do not sit easily with the external-revolution interpretation. Rather than a complete shift of focus and approach, the capacity of the Agency for influencing the strategic decisions of individual firms would appear to have remained intact despite the new political rhetoric and discursive 'wrapping', and, ironically, the changes hailed by the Conservative Scottish Office may even have increased the long-term strategic imprint of the SDA on its invested firms. In the 1980s the investment function came to target a segment of firms, new SMEs, that because of their size and age could be seen as particularly receptive, and at the same time the tying together financial support with organisational and informational resources appears to have become increasingly concentrated in an area of crucial importance for the development of these firms, namely the strategic capabilities of their management teams. Although such changes made the resource exchanges involved in the function less visible – relatively

were either subsidiaries or associated companies while only one-third of HOID investments fell in these categories – but although this could point towards a politically motivated clear-out, it could also reflect a certain economic logic (cf. Donald Patience 31.5.90), namely that the projects in which the organisation had ended up as the main source of external finance were the most risky investments undertaken in the late 1970s.

fewer large investments and no staff secondment – and politically less contentious, at the same time they will have increased the role of public priorities within the regional economy.

Industrial Investment and Regional Economic Change

The last dimension of policy implementation to be explored is the outcome of the industrial investment function in terms of its impact on the regional economy, because although a comprehensive evaluation of the economic impact of SDA policies in terms of e.g. jobs or competitiveness falls outside the scope of this study, some aspects of implementation outcome become part of a success/failure discourse and are therefore clearly relevant from the perspective of the politics of regional policy. In the following two lines of inquiry with regard to policy outcomes are therefore pursued: first resource exchanges and organisational impact on the level of individual firms, and then the wider impact of the function in terms of institutional change within the regional economy.

Resource Exchange and Firm-level Outcomes

The stated objectives of the industrial investment function with regard to individual firms can be summarised under five headings:[151] provide investment finance, increase employment, improve competitiveness, strengthen small and new indigenous firms, and strengthen firms in 'modern' sectors of the regional economy.

Looking first at the industrial investment function as a means of making investment finance available for firms operating in Scotland, SDA expenditure on investment in the period 1976-91 amounted in average to no more than 2% of gross domestic fixed capital formation in manufacturing,[152] but even with an average leverage of e.g. 4[153] this would still imply that the organisation was directly involved in around 10% of manufacturing investment in Scotland, something which could hardly be dismissed as marginal if efforts were concentrated in particular segments of the Scottish economy. This, of course, presupposes that these resource exchanges – investment capital for particular projects sometimes with additional organisational conditions attached – actually influenced the strategies of private economic actors, and this is some-

---

151 Cf. Section 8.1. Despite being the most controversial area of SDA activity, the unpublished study by consultants Coopers & Lybrand summarised in the 1986 government review report is apparently the only evaluation undertaken (IDS 1987a pp. 44-52).

152 Calculated on the basis of SDA 1977-91 and the Regional Accounts of *Economic Trends* for the same period.

153 The lowest figure given in the SDA's annual report was 4.9 for HOID investments, and 2.6 for SBD investments (SDA 1985-91).

thing that is difficult to establish in the absence of in-depth studies of individual investment cases,[154] but in terms of visibility for public and private actors in Scotland this may, however, well have been less important than more tangible fact that the SDA had been involved in quite a number of investment projects within the region.

The extent to which the industrial investments of the SDA increased employment within the Scottish economy is more difficult to ascertain, not least because of the uncertainty surrounding the question of what would have happened in the absence of intervention. The number of jobs 'safeguarded or created' through the function has been one of the 'key figures' reported in the Agency's annual reports since the late 1970s, and credit was taken for more than 4000 jobs in most years through to the early 1990s.[155] Unsurprisingly, these gross figures are higher than the findings of the consultancy study carried out in connection with the 1986 review which estimated net job creation at the level of the Scottish economy for the period 1981-85 "to about 10,400 additional jobs".[156] While this net figure is certainly relevant from the perspective of assessing the overall effectiveness and efficiency of the function, gross figures – assuming that they bear any relation to actual developments within the invested firms – are still relevant as a measure of the number of employees affected directly by the function in a positive manner, either in the form of creation of new jobs or safeguarding of existing ones,[157] and if this is accepted, then the importance of the function in terms of employment can hardly be dismissed as insignificant, being equivalent to around 1.5% of the workforce or 8.5% of the unemployed in Scotland in the first half of the 1980s.[158]

---

[154] The consultancy study undertaken in connection with the government showed that in average 50% of the HOID invested firms surveyed claimed that their projects would have been abandoned in the absence of Agency funding (reported in SDA/DCC(87)5 p. 3), but in the absence of access to the original study the reliability of these findings is difficult to gauge, especially because the catalytic role of HOID in persuading private financial institutions to act as co-investors can easily be underestimated.

[155] The average number of jobs claimed was around 4600 for the whole period and nearly 6000 from 1982/83 onwards (calculated on the basis of SDA 1977-91).

[156] IDS 1987a p. 48. In the same period the total number of gross jobs claimed by the Agency added up to nearly 26,000 (calculated on the basis of SDA 1982-86).

[157] The balance between 'creating' and 'safeguarding' jobs in the gross figures published by the SDA is not known, but among HOID invested firms the balance may have been fairly even throughout the years. Based on the breakdown according to project type, and assuming that 'new' or 'expansion' projects create jobs while 'modernisation' and 'refinance/ownership' safeguard jobs, the overall balance is nearly even in SDAINV, also in most years throughout the period.

[158] Calculated on the basis of SDA annual report figures for job creation 1981-85 and December 1987 figures for employment and unemployment (*Quarterly Economic Commentary* 14, 4, pp. 39-44).

Turning now to competitiveness, this concept can be defined and measured in numerous ways,[159] but as the SDA never endorsed one particular perspective and as 1986 government review concentrated mainly on financial aspects of the function, the best way forward will be to use indicators which could illuminate the degree to which new technology and organisational modernisation are likely to have been central to SDA investment projects. By combining the classifications of HOID investments according to sector and project type it is possible to produce a synthetic measure of the role of competitiveness-oriented projects,[160] and as can be seen from Figure 8.9 competitiveness would appear to have been a significant consideration within the investment function throughout, although it was perhaps pursued most consistently in the late 1980s, perhaps by an internal learning-by-doing process of the new management team within the Agency or prompted by the 1986 government review's comments about its risk averse approach.

Figure 8.9 Competitiveness-oriented HOID investments

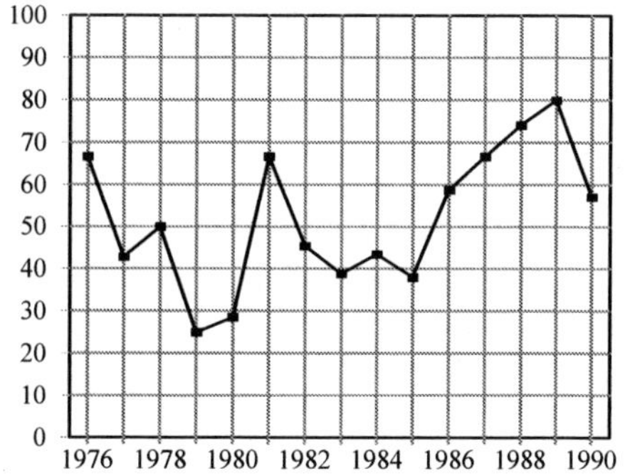

Per cent of investments undertaken each year.
Source: SDAINV.

With regard to the objective of strengthening small and/or new indigenous firms, it has already been established that new and/or relative small indigenous firms were the main recipients of investment, much in accordance with the official corporate priorities of the Agency, and the profile of the portfolio, HOID as well as SBD, also differed from that of firms within the Scottish economy at large, perhaps not so much in

159 See e.g. Maskell & Malmberg 1999, Budd & Hirmis 2004.

160 The index is calculated as the share of HOID invested firms classified either as belonging to high-tech sectors or involving the project types 'new' or 'modernisation', as such investments are assumed to be associated with promotion of productivity and new technology.

terms of size and age,[161] but certainly with regard to in terms of ownership where the investment function did not target externally owned firms which accounted for 19% of employment in Scottish manufacturing.[162]

Moreover, in terms of pursuing the objective of strengthening 'modern' sectors of the regional economy, HOID clearly gave priority to manufacturing at the expense of services: while the share of services in private sector employment was around 50% in the mid-1980s, it had only risen to around 25% of HOID investments in the same period.[163] Within manufacturing the sectoral profile of Agency investments also diverged from the existing employment structure: already in its first decade of operation investments in high-tech firms were clearly over-represented, and in the second half of the 1980s even more so.[164] This outcome will of course have been influenced by the sectoral composition of the projects submitted for consideration, but may also reflect that from an early point the proactive efforts in investment acquired a very distinct high-tech dimension through the work of the New Ventures Unit,[165] and thus the SDA's industrial investments will have contributed to changing the sectoral structure of the economy by giving priority to 'modern manufacturing' at the expense of services and 'traditional manufacturing'.

All in all the firm-level outcomes of the SDA's industrial investment function would seem to have been characterised by continuity in terms of the quality of the objectives pursued, but some degree of change with regard to the intensity with which some of these were pursued. The mainstay of the operation was clearly to improve the provision of investment finance for indigenous manufacturing establishments, and over the years the function became increasingly focused on small-to-medium sized firms in high-tech sectors. Employment was an obvious consequence of many investments, but the constant, and perhaps increasing, emphasis on organisational and technological improvement of the invested firms in combination with the limited and decreasing number

---

161 Although the methods of calculation differ, the share of HOID investment in new firms, rising from 22% in the late 1970s to 44% in the late 1980s, would seem to be roughly at the same level as the share of new manufacturing firms within the Scottish economy at large according to VAT registrations compiled by the DTI's Small Business Services (www.sbs.gov.uk).

162 Scottish Office 1988 p. 37.

163 Calculated on the basis of Scottish Office 1988 and SDAINV.

164 In 1976-79 the share of high-tech projects was 1.7 times greater than the share of high-tech employment, and in 1986-89 this figure had increased to 3.9 (calculated on the basis of SDAINV and Scottish Office 1988 p. 18).

165 SDA 1981 pp. 26ff.

of outright rescues suggest that short-term employment considerations can hardly have been pursued systematically at the expense of other aims. Instead the persistent attempt to use the industrial investment function to address what was seen as structural problems within the economy, the weakness of indigenous segments and high-tech sectors within the regional economy, indicate that promotion of efficiency and competitiveness must have been a central concern. Although it has been shown that traditional industry was not eschewed by the function and the priorities pursued may have been sensible from the perspective of long-term economic competitiveness, this specific way of implementing industrial investment could be construed as promotion of (often un-unionised) white-collar and female employment at the expense of those skilled male workers who historically had dominated traditional industries and formed the backbone of the trade union movement – and therefore firm-level outcomes along these lines would help to position the Agency in the Scottish discursive terrain as an agent of Thatcherite modernisation.

Industrial Investment and Institutional Change

The introduction of the industrial investment function as an instrument of regional policy in Scotland has been interpreted as a major institutional innovation in two rather different ways: as an attempt to extend public ownership, and as provider of long-term external finance for particular segments of firms within the regional economy. Both these roles would have involved instituting new rules and incentives which could affect the strategies of private economic actors, and in the following the extent to which these two institutional outcomes have manifested themselves are examined in turn.

The notion that industrial investment was a function that was designed to redraw the borderline between the public and the private sector dates back to the mid-1970s and was also incorporated into the external-revolution interpretation, but acting as a 'back-door nationaliser' was in many ways at odds with both the formal rules governing the function – the organisation had no powers of compulsory purchase – and with the way in which its more limited powers were used in practice where no hostile acquisitions were undertaken and buy-back clauses were introduced from an early point to give private owners the right to reassert their control.[166] But *if* extension of public ownership *had* been a goal in its own right, then it could be expected that its investment strategy

[166] Wilson Committee 1977f vol. 6 p. 206, cf. Lochhead 1983 p. 181. Such arrangements would, however, have been subject to the approval of the Secretary of State for Scotland under the original investment guidelines (SEPD 1976b p. 5, cf. SDA 1977b 11), and appear to have been potentially controversial in the late 1970s (SDA 1977b p. 3).

attempted to maximise control in particular firms, and therefore it needs to be established to what extent the portfolio contained firms that were either controlled by or heavily depending on SDA capital. Firms within the HOID portfolio can be classified according to the formal influence in the form of voting rights the Agency was able to exercise,[167] and in practice it was only in the first two years of operation that the Agency acquired a small number of subsidiaries while the growth in non-subsidiaries was much faster, and as of 31.3.79 only 22% of the invested firms were outright controlled by the development body.[168] Despite repeated surges in the number of associated companies that took place both in the mid- and again in the late 1980s, the late 1970s was the only period in which the investment function could have been used to e.g. restructure sectors of industry, but a sectoral breakdown of the fifteen subsidiaries and associated companies in which the SDA invested in the period 1976-78 shows that these firms were distributed across a wide range of activities, and thus control even of subsectors would seem to be difficult to achieve.[169] Still, the Agency made no secret of the fact that it had been the most important source of external finance for around half of its invested firms in the first years,[170] and the organisation apparently wielded enough power over a significant part of its invested firms in the late 1970s to sustain the suspicion that it was in the same business as the NEB, namely to redraw the boundary between public and private. This was of course clearly *not* the case in the 1980s, and not even a growing number of associated companies in the late 1980s[171] were able to make the issue of public ownership central to the 1986 government review where 'disposal strategy' had become a pragmatic question about not "forgoing financial or developmental gains" through early disposal rather than associated with the principle of 'rolling back the state'.[172] All in all it can in other words be concluded that the issue of public owner-

---

167 Subsidiaries are firms in which the SDA holds 50% or more of voting rights, associated firms are non-subsidiaries where 20% or more of voting rights are held. This classification was used in SDA annual reports from the beginning, but the associated category disappeared after 1981/82, perhaps in order to avoid drawing attention to the fact that the Agency still was a major investor in some firms.

168 Calculated on the basis of SDAINV.

169 Calculated on the basis of SDAINV. For a similar argument, see Page 1977 p. 52. Some sources suggest that carpet-making could be the one area in which the Agency via its investments had a good deal of leverage via its investments (SDA 1977a pp. 5ff, Gerry Murray 30.7.90).

170 In the early years nearly half of the projects did not involve co-funding from private financial institutions (SDA 1977d p. 9).

171 The average share of associated companies in the period 1982-90 was 30% (calculated on the basis of SDA 1983-91 and SDAINV).

172 IDS 1987a p. 51.

ship had been exaggerated in the late 1970s and then withered away, in the portfolio as well as an issue in relation to the sponsoring department. The influence of the industrial investment function was in other words *not* achieved through outright ownership but rather through a combination of financial, organisational and informational resources which were employed in order to influence the strategies of private firms which came to depend on the SDA as a source of external funding.

The alternative perspective on the institutional role of the SDA's investment function within the Scottish economy sees it as a way of addressing the difficulties in raising long-term capital experienced by relative small and/or new firms, but the extent to which the SDA actually filled the 'equity gap' has been called into question, not least from an external-revolution perspective, because post-1979 sponsor demands for commercial project appraisal and private co-investment could have made the Agency's investment function operate in the same conservative and risk-averse manner as the private financial institutions it was supposed to complement.

It would have been good to know what Scottish firms thought of the way the industrial investment function was being implemented and especially how this related to the alleged equity gap, but in the absence of such information it is interesting to note that private sector financial institutions surveyed by Coopers & Lybrand in connection with the 1986 review maintained that an 'equity gap' did exist, especially with regard to smaller and more risky projects, although "some institutions [...] saw the Agency as being too commercial and risk-averse in its approach".[173] Most of the private actors who would have been in direct competition with the Agency if the latter had essentially acted as a public sector merchant bank did in other words suggest that the SDA had indeed 'made a difference' also in the 1980s, and this view can be substantiated by a closer look at the way in which the function operated because *despite* the commercial rhetoric a considerable element of subsidy continued to be an integrated and indeed essential part of the function. *Firstly*, it is clear that from the perspective of firms seeking external finance the Agency could offer relatively better financial terms compared to private providers of loans and equity, partly by being a more patient investor capable of waiting longer for a return,[174] and partly because it was capable of lending to new firms or risky ventures on

173 IDS 1987a pp. 47f.

174 SDA 1977d p. 17, 1984 p. 36, Donald Patience 31.5.90, Edward Cunningham 13.6.90, cf. Draper *et al.* 1988 pp. 282ff. Nearly 2/3 of the funds were committed in the form of equity (IDS 1987a p. 44), a form of investment with no set date of repayment.

terms akin to those of "firms of the highest commercial standing",[175] thereby effectively subsidising the invested firm by upgrading its credit rating.[176] *Secondly*, the SDA bore the cost of often extensive front-end 'due-diligence' investigations for relatively small and risky investment projects, both through HOID itself and the sectoral divisions,[177] even if occasionally only a symbolic financial stake was eventually taken in order to maintain credibility with co-investing private financial institutions (or staff morale within HOID).[178] The Agency also appears to have maintained closer links with invested firms in terms of monitoring and advice than was usual among private venture capitalists,[179] and perhaps the best measure of this subsidy is the administrative expenditure of the investment function, covering both portfolio administration and (parts of) the front-end investigations, which was *not* included in the investment duty[180] but equalled around one third of the amount invested in the late 1980s.[181]

All in all this leads to the conclusion that also in the 1980s the SDA would seem to have continued to address the issue of the equity gap through its industrial investment function by evaluating projects and putting their management teams into a viable shape by subsidising extensive front-end investigations, thereby acting as a focal point for an informal regional venture capital network which included the three Scottish-based clearing banks as well as the new venture capitalists which emerged in Britain in the 1980s.[182] Moreover, while addressing a specific equity gap was the economic outcome of the industrial investment function in institutional terms, the SDA also managed to maintain its political acceptability through an ingenious form of double entry book-keeping and dual-track discourse that separated the profitability of

[175] The wording of both the 1977 and 1980 Industrial Investment Guidelines (SDA 1977 p. 62, SEPD 1980 p. 4).

[176] The Wilson Committee had estimated that small firms paid on average 2% more for bank loans, partly due to the fixed cost element in assessing the applications (Hall & Lewis 1988 pp. 1699f), and SBD investments persistently showed very low returns and hence involved a considerable element of subsidy (IDS 1987a pp. 47ff).

[177] Donald Patience 31.5.90, cf. SDA 1986 p. 42, IDS 1987a p. 47, Draper *et al.* 1988 pp. 282ff.

[178] Donald Patience 31.5.90, National Audit Office 1988 p. 6, cf. IDS 1987a p. 47.

[179] Cf. Harrison & Mason 1992 pp. 3ff.

[180] Committee of Public Accounts 1982a pp. 18f, cf. National Audit Office 1985 p. 10.

[181] Calculated on the basis of SDA 1983-91.

[182] On the Scottish financial sector, see Draper *et al.* 1988, McKillop & Hutchinson 1990 Ch. 6, Bain & Reid 1984. On British venture capitalism, see Advisory Council on Science and Technology 1990 Ch. 5, Wright *et al.* 1994, Minns 1992, Harrison & Mason 1992, Martin 1992a.

invested firms from the administrative cost of ensuring that they were profitable.

Concessionary Loans and Development Funding

Industrial investment was undoubtedly the most conspicuous policy function relying primarily on the provision of financial resources, but analysing the way in which the other investment functions of the SDA operated may serve to underline the broad array of options and the scope for flexible handling of individual projects.

Provision of loans at concessionary rates of interest in designated rural areas was inherited by the Agency from SICRAS in 1975,[183] and the programme continued to be coordinated with similar ones in England and Wales through the British policy network.[184] In terms of modes of implementation concessionary loans operated in a reactive manner with rates of interest determined by the sponsor department,[185] and thus while the appraisal of individual projects was discretionary, the rate of assistance for approved projects was not. Employment was officially recognised to be more important as an objective than in the industrial investment function,[186] and given the economic structure of the rural areas targeted and the administration of the programme by SBD, small indigenous firms will have been the most likely beneficiaries. Unlike industrial investments, concessionary loans were spatially selective in that the programme only operated in designated rural areas,[187] but as the amounts involved were limited,[188] this geographical bias may in practice have been offset by the less overt subsidisation of small firms in other parts of Scotland through the industrial investments of the SBD.[189] All in all the way in which the SDA employed concessionary loans to promote rural development bore strong resemblance to the traditional regional policy grant schemes of central government with its reactive assessment of applications from private actors in designated areas, but with the important qualification that the administration of concessionary loans

183 SEPD 1976b, SDA 1976 par. 15f, cf. Rogers 1999 pp. 76ff.

184 SDA 1977 p. 62; IDS 1987a pp. 46f, 51; cf. Kirwan 1981 p. 27.

185 IDS 1987a pp. 47ff, Lochhead 1983 pp. 174f.

186 IDS 1987a pp. 46f.

187 From 1982 to 1986 the SBD administered a small subsidised loans scheme on behalf of the European Coal and Steel Community in designated areas affected by coal and steel closures.

188 Concessionary loans accounted for less than 5% of investment expenditure in most years in the 1980s for which data is available (calculated on the basis of SDA 1982-88).

189 The SDA also ran a small grant scheme for craft-based firms (Lochhead 1983 pp. 168ff).

was situated on the regional level in an organisation that gave small rural firms in Scotland access to the even wider range of policy measures of the SBD.

In contrast to this fairly traditional form of regional policy, development funding appears to have been at the other end of the spectrum as an innovative instrument for high-risk ventures used by the SDA for a diverse series of high-tech and infrastructural projects involving various forms of co-operation with private actors.[190] Development funding was governed by a separate set of guidelines stressing its long-term economic objectives *without* instituting financial performance targets, and as this type of investment projects could involve extensive externalities or greater risk than ordinary industrial investments, it was handled through a process in which discretion was exercised jointly between the Agency and its sponsor department.[191] Most of the projects were initiated by the Agency on the basis of work undertaken in its sectoral or other divisions,[192] and the terms and conditions appears to have been tailor-made with the SDA as a very patient investor. The scale on which development funding was used in the second half of the 1980s came to be significant,[193] and although the appraisal and monitoring procedures in some major projects were criticised by the National Audit Office,[194] the Agency's approach seems to have been in accordance with the official rationale of the function, namely to support projects which "although not capable of earning an adequate financial return, offer wider economic benefits which warrant public funding".[195] Development funding has in other words complemented mainstream functions such as industrial investment by catering for some of the more high-risk situations without being restricted by commercial criteria for project appraisal.

While industrial investment remained the major activity predominantly based on conditional access to financial resources – exploiting the 'double standards' set by the sponsor department which required invested firms to adhere to commercial profitability criteria but accepted

190 National Audit Office 1988 pp. 8ff, cf. IDS 1987a p. 73.

191 Kirwan 1981 pp. 30f, SDA/DCC 18.9.84, IDS 1987a p. 73 & 82, National Audit Office 1988 p. 8.

192 Examples of this include ventures exploiting university R&D commercially, the Scottish Exhibition and Conference Centre in Glasgow and the National Hyperbaric Centre in Aberdeen (National Audit Office 1988 p. 9, SDA Board Papers January 1988).

193 In the years 1986-91 development funding accounted for 15% of investment expenditure in average (calculated on the basis of SDA 1986-91).

194 1988 pp. 10ff.

195 IDS 1987a p. 73.

substantial subsidisation of project appraisal and monitoring – the two other functions would appear to have provided useful complements in each their way, also from a political perspective. The inherited concessionary rural loans programmes was, on the one hand, a visible sign of commitment to parts of Scotland outwith the industrialised central belt,[196] while development funding, on the other, allowed proactive initiatives with regard to more complex or technologically advanced forms of infrastructure outside the discipline of the financial duties, thereby creating a high profile for the Agency but with the sponsor department more directly involved in the appraisal and implementation of individual projects. With financial policy instruments like these, the SDA would certainly seem to represent a departure from the paradigm that had characterised the national phase of regional policy in Britain till the mid-1970s.

8.3. The British Context: Investing in Regional Development

Finance in support of investment had long been at the core of British regional policy and continued to be a central element in the multi-level phases from the mid-1970s onwards, both in terms of expenditure and with regard to the political significance accorded, and it is therefore hardly surprising that it was developments in this field that prompted general interpretations of British regional policy in the 1980s as being 'run down', 'dismantled' or 'abandoned'. This section first examines the development of financial policy instruments employed by central government, and then considers investment-oriented schemes operated from the subnational and European levels respectively.

[196] Cf. e.g. the robust line taken by chief executive George Mathewson when questioned by the Scottish Affairs Committee (1984).

Figure 8.10 Central government regional policy grant expenditure in Britain

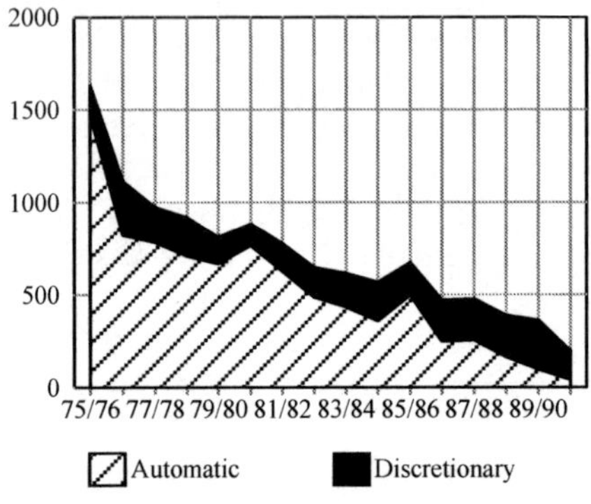

£m 1985/86.

Source: Yuill *et al.* 1989 & 1993, Wren 1996a p. 81.

Central Government Grants in Transition

On the British level investment through equity never became a major activity in its own right,[197] and grant support continued to be the primary means of financial support for regional development. Moreover, given the continued centrality of investment grants to British regional policy, basic features of their development have already been outlined:[198] the perception of the regional problem as unemployment caused by economic weaknesses remained, but the underlying values changed from social concerns about spatial inequality in life opportunities to economic concerns about regional competitiveness, and programmes continued to operate in a reactive manner but became less exclusively oriented towards manufacturing, and a gradual shift occurred with regard to project appraisal away from automatic towards discretionary grants. As illustrated by Figure 8.10, the overall level of real-term expenditure

---

197 Loans constituted nearly half of Regional Selective Assistance expenditure until 1980, but was then substituted by grants. Support could also be given in the form of equity, but even before 1980 this was done only very rarely (*Industry Act Annual Reports* 1973-91, statistical appendix on Section 7 support). The NEB did establish regional boards in the north of England which dealt with investments below 0.5 million pounds (Hogwood 1982 p. 42, Kramer 1989 p. 18), supposedly to placate the English regions in the wake of the setting up of RDAs in Scotland and Wales, but the NEB was apparently not entirely happy about becoming entangled in what it saw as diversionary territorial politics and the activities of the regional boards appear to have been limited (Grant 1982 pp. 115f, Hudson 1989a pp. 116ff).

198 Based on Sections 6.4 and 7.4.

declined no less than 78% from 1975/76 to 1990/91,[199] and most of this was accounted for by the move away from automatic grants where eligibility criteria in terms of designated areas and claimable items changed before, ultimately, this form of support was discontinued. Contrary to this discretionary grants, especially Regional Selective Assistance, was broadly maintained in real terms throughout the period,[200] and thus the relative importance of discretionary financial instruments increased dramatically.

Looking at policy output in greater detail, the difference between automatic and discretionary grants is a crucial factor. While issues of industrial sector and project type do not need to be investigated for automatic grants because their distribution will have reflected the composition of investment activity in the Assisted Areas, the output with regard to discretionary grants will to some extent have reflected the criteria on the basis of which applications were assessed. In the case of Regional Selective Assistance,[201] the main criteria were the viability of project and firms, proof of need, strengthening of competitiveness on both the regional and national level, and creation or safeguarding of jobs. Establishing whether an individual project complied with these criteria would require staff with competence and information concerning a wide range of issues across manufacturing and other sectors of the regional economy – from finance to markets and technology[202] – but some projects were, however, seen by the DTI fulfilling these criteria more readily than others, namely new green-field investment by foreign companies,[203] presumably because they in a very visible way would create new employment, introduce modern technology and organisation.[204] Publication of these appraisal criteria and the publicity surrounding individual inward investment projects may even have created the impression that discretionary grants were primarily aimed at foreign

[199] Calculated on the basis of Yuill *et al.* 1989 & 1993, Wren 1996a p. 81.

[200] Average real term expenditure in the years 1975-79 and 1986-91 was 205 and 206 million (1985-86 pounds) respectively, while the level in the intervening years had been around 30% lower (calculated on the basis of Yuill *et al.* 1989 & 1993). The two minor discretionary grant programmes, the original Office and Service Industry Grant and later the Regional Enterprise Grant, only accounted for in average 6% of expenditure on Regional Selective Assistance (calculated on the basis of Yuill *et al.* 1989 and 1993).

[201] The main criteria are listed in the appendix to the *Industry Act Annual Reports*.

[202] Whether generalist civil service organisations like central government departments were well-equipped to meet this challenge has often been questioned, e.g. Swales 1989, Allen *et al.* 1988 pp. 19ff, Ashcroft & Love 1988, cf. Beckman & Carling 1989 pp. 12ff.

[203] DTI 1988 p. 30.

[204] Cf. Raines & Wishlade 1997 pp. 157ff.

multinationals, and especially in the 1980s a disproportionally high share of discretionary regional assistance was awarded to foreign firms, in average nearly three times that of the existing stock of foreign investment in the UK.[205] Although support for very small innovative indigenous firms in the Assisted Areas was given additional emphasis through the introduction of the Regional Enterprise Grant in 1988,[206] this does not alter the general picture, namely that in the 1980s central government regional grant schemes became biased towards large-scale employment creating high-tech green-field ventures on the basis of the official interpretation of the guidelines for discretionary project appraisal, thereby linking regional development with central government attempts to position Britain in the international competition for footloose investment projects.

All in all it is clear that in terms of investment-related measures central government regional policy had changed profoundly between 1975 and 1991: support for indigenous firms within the Assisted Areas decreased while in practice the main emphasis of the officially two-pronged strategy seemed to be on incoming foreign firms rather than SMEs based within the region. Discontinuation of the automatic grant schemes undoubtedly looms large in the interpretations of the 1980s as a period in which regional policy was 'dismantled' or 'phased out' of regional policy, presumably because automatic investment grant programmes had hitherto been the main source of new jobs in the Assisted areas according to government sponsored evaluations,[207] but this should not be allowed to obscure the fact that what amounted to a new strategy was being implemented in a fairly consistent manner, and while change may have been driven by short-term financial and political considerations, it still constituted a form of regional policy, albeit of course a rather different one.

---

205 While the average share of FDI-related Regional Selective Assistance values from 1979 to 1991 was 46%, FDI only accounted for 16% of GDP on average (calculated on the basis of *Industry Act Annual Reports* 1979-91 (no data given for 1978 and earlier) and UNCTAD 1997, 2000).

206 It has sometimes been argued that discretionary grants favour large companies at the expense of SMEs because the former can better afford to invest management time in lengthy and potentially abortive grant negotiations (see e.g. Allen *et al.* 1986 & 1988), but detailed case-studies would seem to suggest that this effect is at least limited (Walker & Krist 1980, McGreevy & Thomson 1983), and the existence of a streamlined procedure for minor applications shows that central government was aware of this potential barrier (Committee of Public Accounts 1988 p. X).

207 Moore *et al.* 1986 estimated that by 1981 the Regional Development Grant accounted for around 2/3 of the net jobs created in the Assisted Areas through regional policy, while the impact of Regional Selective Assistance had been much more limited, and nowhere near the gross number of jobs claimed in the *Industry Act Annual Reports*.

New Actors, New Investment?

Investment in the form of loans and equity formed part of the armoury of all the major RDAs sponsored by central government, and thus investment functions much like those of the SDA could be found in other British regions, and indeed across Europe.[208] Given the general similarities in the way in which these arm's-length bodies were sponsored,[209] it is hardly surprising that the handling of financial instruments was also subject to fairly uniform guidelines, not only in terms of administrative authority where statutory powers were near identical and similar financial duties in operation,[210] but also with regard to political scrutiny in the sense that on most occasions the investment functions of more than one RDA was scrutinised at the same time.[211] In practice both the strategies and implementation of the investment functions did, however, differ in ways that would seem to reflect specific economic and political circumstances and thus a closer look at individual RDAs will be necessary.

As the RDA for rural England, the Development Commission continued to provide investment finance to small firms in designated areas, and although taking equity was an option, concessionary loans at rates determined by central government continued to be its most important financial policy instrument. Finance on attractive terms was often seen as a means of getting in contact with individual firms which could then benefit from other forms of support such as advisory services,[212] and hence the importance of an integrated development strategy was emphasised.

In contrast to this the RDA responsible for the remotest and least industrialised parts of Britain, the HIDB, had a much wider range of financial policy instruments at its disposal, in effect combining elements from both their rural and 'industrial' counterparts: like the Development Commission grants could be given for 'social purposes' in designated remote and 'fragile' areas,[213] like the SDA and the WDA investment on quasi-commercial terms in the form of concessionary loans and equity

---

[208] A survey undertaken in the early 1990s found that equity investment and/or grant-giving formed part of RDA activity in 60% of the organisations surveyed (calculated on the basis of the data on which Halkier & Danson 1997 was based).

[209] Cf. the discussion in Section 6.4.

[210] Committee of Public Accounts 1982a pp. Vff.

[211] Examples of this include Committee of Public Accounts (1982a, 1982b, 1985) and National Audit Office (1985).

[212] Development Commission 1984, Tricker & Martin 1984, Rogers 1999 Ch. 10-12.

[213] IDS 1987b Ch. 9. Expenditure on social development, individual projects as well as aggregate, was subject to Scottish Office regulation, and this function amounted to less than 4% of HIDB expenditure in the late 1980s.

played a significant role,[214] but unlike any other arm's-length body the Board effectively carried out functions within a mainstream central government grant scheme within its area, namely Regional Selective Assistance under what was referred to as 'the Arrangements',[215] and thus firms within the Boards area of operation benefited from what was in a very real sense a one-door approach to financial support for regional development which even accepted higher levels of public support for individual projects than in other Assisted Areas.[216] Policy instruments based primarily on financial resources accounted for in average nearly 60% of HIDB expenditure from 1975 to 1991 with around half of this being grants, and when the sparsely populated nature of the Highlands & Islands is taken into account, in relation to GDP the level of RDA-based financial support will have been more than 20 times higher than in lowland Scotland.[217] The economic structure of the region meant that the majority of investments was undertaken in small indigenous firms, and contrary to most other RDAs financial support was not primarily given to manufacturing: in fact the latter only amounted to 35% of expenditure in the period 1976-85 while primary production and tourism accounted for 30% and 35% respectively.[218] Although the possibility of introducing a financial duty in line with that applied to the SDA's industrial investment function was discussed on several occasions,[219] the absence of progress would seem to suggest that the question of viability never became an acute issue, perhaps because it was generally expected to be more difficult to invest profitably in a remote rural area.

The investment functions of the WDA did in many ways come closer to those of the SDA in that they had been set up with the same statutory powers, operated in regions of a mainly industrial nature, and were sponsored by a territorial department of central government. Also the Welsh body had inherited the concessionary loans scheme from the Development Commission, but its main investment activity aimed, like it Scottish counterpart, to address what was perceived to be an equity

[214] Committee of Public Accounts 1982b pp. Vf, IDS 197b Ch. 4. An attempt to set up a venture capital organisation for the Highlands & Islands failed because of lack of suitable investment projects (Grassie 1983 pp. 50ff, Scottish Affairs Committee 1984 p. 48).

[215] Scottish Affairs Committee 1984 p. 40, IDS 1987b p. 43.

[216] Committee of Public Accounts 1982b p. 10.

[217] Calculated on the basis of SDA 1977-91 and HIDB 1977-91 by using the 1987 GDP figures in the Regional Accounts in *Economic Trends*. From 1976 to 1991 average HIDB expenditure on investment and grants was more than 20% above the spending on investment by the SDA.

[218] IDS 1987b Table 4.2, cf. Committee of Public Accounts 1982b p. 10, Scottish Affairs Committee 1984 p. 6.

[219] Committee of Public Accounts 1982b p. VII, IDS 1987b p. 57.

gap within the regional economy.[220] In terms of activity the level of ambition of the WDA with respect to the function was significantly lower than that of the SDA in the early years, citing a dearth of promising projects in Wales and predicting fairly limited levels of expenditure for a function that was essentially seen as a complement to the main business of building and managing industrial property.[221] In practice the investment expenditure of the two RDAs did, however, develop along virtually parallel lines:[222] in both cases the average share of total expenditure amounted to 11% from 1976 to 1991. The WDA was also subjected to sponsor department regulation which should ensure that considerations of commercial viability formed a central part of discretionary appraisal of individual projects, with the financial duty being the core symbol and recurring parliamentary investigations in the 1980s by the Public Accounts Committee bringing added political urgency to the matter,[223] and also in the Welsh context doubts were expressed, at least in the early years, about the relevance from regional development perspective of a quasi-commercial merchant-bank approach.[224] In the absence of detailed information about individual investment projects akin to those in the SDAINV database it is more difficult to gauge whether implementation entailed e.g. sectoral priorities aimed at addressing specific weaknesses of the regional economy,[225] or whether criteria of financial profitability came to play a larger role in discretionary project appraisal after the change of government in 1979 and the issuing of new guidelines,[226] but the 1986 review by the Welsh Office did recommend – in addition to improved monitoring procedures – that more attention should be given to ensuring private sector co-funding,[227] apparently to little avail as the same demand was repeated a decade later.[228] However, what can be established is the fact that like in the case

220 Wilson Committee 1977f vol. 8 pp. 4ff, Welsh Office 1987 Ch. 11, Morgan 1994 p. 25.

221 See Wilson Committee 1977f vol. 8 pp. 5f, cf. Committee on Welsh Affairs 1980 pp. 91ff. The expected annual expenditure of the WDA on investment in the early years was 6 million pounds, a puny figure compared to the brash projections of its Scottish counterpart which were 4 times larger, cf. Section 8.1.

222 Calculated on the basis of SDA 1977-91 and WDA 1977-91.

223 WDA 1977a, Eirug 1983, Committee of Public Accounts 1982a & 1985, National Audit Office 1985.

224 Committee on Welsh Affairs 1980 pp. 87ff, Cooke 1980.

225 WDA operated a Small Business Unit with finance available on broadly similar terms to that of the SDA's SBD (Lochhead 1983 p. 17), and hence it must be assumed that also in the Welsh case priority was accorded to relative small investments and firms.

226 Cf. Committee on Welsh Affairs 1980 p. 91.

227 Welsh Office 1987 Ch. 11.

228 Welsh Office & WDA 1995.

of the investment function gradually became capable of generating a quite significant level of receipts,[229] perhaps suggesting that the function became increasingly subject to commercial pressures in much the same manner as its Scottish counterpart.

Turning now to development activities sponsored by other actors than central government, the role of the European level in providing direct financial support for productive investment was very limited in the period under consideration. Before the reform of the Structural Funds in 1988, the funding officially claimed to be in support of individual firms was in effect only reimbursement of existing central government schemes,[230] and after 1988 infrastructure came to dominate the new programmes.[231] Contrary to this, subnational initiatives in economic development did to some extent include provision of finance to individual firms as a policy instrument, and the controversy surrounding some of the major urban enterprise boards did indeed help to give this type of activity a relatively high political profile.[232] There is, however, still a dearth of systematic research in the area, but the picture emerging appears to be a dual one, with the majority of local authorities providing especially loan finance for small firms,[233] while in a limited number of English conurbations arm's-length enterprise boards undertook equity investment on a smaller scale but on terms not dissimilar to that offered by the SDA.[234] While the central-government sponsored RDAs no longer enjoyed a monopoly on providing long-term finance for productive investment, the efforts of sub-national actors remained more limited in scope, both in terms of the funding involved and indeed in terms of achieving the comprehensive spatial coverage provided outwith England.

8.4. Investment and Regional Development Strategies

Having analysed the SDA's strategies and implementation and the wider context in which the investment function operated, it is now possible to bring the different strands of inquiry together in order to illuminate their interplay.

---

229 From 1984/85 to 1990/91 investment-related receipts equalled in average 65% of expenditure in the case of the WDA while the equivalent figure for the SDA was 99% (calculated on the basis of SDA 1985-91 and WDA 1985-91).

230 Cf. Section 7.4.

231 Hall Aitken Associates 1996, cf. Minns 1992.

232 Cf. the discussion in Section 7.4.

233 In Scotland financial support to individual companies constituted 8% of local authority expenditure on economic development in 1990/91 (McQuaid 1992 Table 4), the equivalent of around half the gross expenditure of the SDA on its investment functions (SDA 1991).

234 See Hall & Lewis 1988, Cochrane & Clarke 1990, cf. Lawless 1988.

First it is important to note that from a purely 'internal' Agency perspective a high degree of correspondence has existed between overall corporate strategies and the specific objectives set out for the industrial investment function, something which is hardly surprising given the political significance accorded to the function. At the same time it is, however, also evident that in practice some aspects of policy implementation have differed from the official strategic pronouncements in the late 1970s: expenditure never came close to the levels originally foreseen by the Agency, and while the shift to a more proactive approach to industrial investment eventually did materialise in the form of projects from the New Ventures unit, this was on a much less grand scale than apparently envisaged in the early years, possibly caused by the quantity and quality of projects being brought forward which did not live up to the impatient quest of the original political sponsors for tangible results. In contrast to this, after the change of government in 1979 the implementation of the industrial investment function followed the strategic statements of the SDA quite closely, and at least in the early 1980s the discrepancy was now *between* the sponsor department and the arm's-length body, both in retrospect with regard to the unfounded implications of the 1980 guidelines about the back-door nationalising past of the Agency, and, notably, with regard to levels of expected activity which were maintained after the new guidelines had declared industrial investment a 'measure of last resort'. In practice the SDA's commitment to goals of regional economic competitiveness appears to have been both earlier and more consistent than the Scottish Office: management improvement in invested firms was introduced already in the late 1970s while employment still appeared to be a major aim according to the sponsor department, and sectoral modernisation was pursued nearly a decade before this aspect was officially recognised by the Scottish Office. In terms of modes of implementation, continuity would seem to prevail with regard to project generation, predominantly reactive but retaining a capacity for proactive ventures, but some of the 'soft' forms of selectivity changed over time: whereas the function persistently focused on relatively small firms and project sizes, additional emphasis was given to high-tech activities and new firms in the 1980s. Discretionary appraisal of individual investment projects was an area which the 1980 investment guidelines had used as a demarcation between the 'employment-oriented' 1970s and the new 'commercial' era of the 1980s, but in reality the Scottish Office had persistently demanded that the investment function operate in a commercial manner, and the main contribution of the new guidelines was the requirement for private sector co-investment which is likely to have reinforced the demand for investment in viable projects much more effectively than various versions of the official financial duty. Whether project appraisal did in fact become more commercial is, however, unclear because the available performance indicators are open to manipulation and display little by way of consistent trends. What has been established, however, is that the methods employed in the pursuit of organisational modernisation and firm-level competitiveness began to change from post-investment secondment to intense pre-investment screening even before 1979: although the latter approach may have entailed an even greater public sector influence on strategic decision-making in the private sector, it could still be construed as a more hands-off approach because of the less visible nature of Agency input.

All in all the way in which the industrial investment function had been implemented does not comply particularly well with the hypothesis that the change of government in 1979 was the major turning point because several of the pillars of the alleged Thatcherite revamping had in fact at least been in the making before 1979, and the reminder either proved to be short-lived, additional, or difficult to substantiate. This is not to say that the investment function did not go through different phases or that the advent of a Conservative government was without consequence, but it does underline the need for an alternative to the prevailing dichotomised interpretation, although such a reconstruction does, admittedly, result in a much more complex picture of the changing features of the function, as will be evident in the following.

When the SDA was set up in 1975, the concept of direct investments in industry as an instrument of regional policy was not only controversial, but also innovative, and therefore its first years of operation involved a good deal of experimentation in a situation with great political expectations and limited organisational resources available. This produced a portfolio which in the early years consisted of relatively few firms, including some fairly large financial commitments, some subsidiaries, some very difficult rescue cases, and an over-representation of high-tech industries. The Agency was keenly aware of management shortfalls in its invested firms, but the attempt to achieve organisational modernisation through part-time secondment of staff was not enough and the inevitable outcome was a number of well-publicised failures, and thus already before the change of government the SDA was in the process of adjusting its approach: no new subsidiaries were acquired, the average investment size declined, and the emphasis in dealings with invested firms started to move from technical support towards advisory services of a more strategic nature. Having established pre-1979 precursors of some features that were later glorified by the Conservative government could suggest that these were motivated by an internal learning-by-doing process, but these developments could also have been prompted by e.g. the regulatory regime set up by the Labour government such as the financial duty and staffing restrictions.

Having said this, it would, however, still be premature to discard the change of government in general as being of minor significance. 1979 definitely was followed by a short-term decrease in activity and unleashed a good deal of political gesturing, but although most of the latter, like the emphasis on commercially viable projects, was 'old hat' by SDA standards, the importance of the institutionalisation of the demand for leverage in the form of private sector co-investment should not be underestimated. In effect this circumscribed the Agency's scope of action in individual investment cases and propelled the development body towards becoming the organiser of a venture capitalist network on the regional level in which its role was to subsidise, and occasionally initiate, individual projects by undertaking extensive front-end investigations into ventures that by the standards of private venture capitalists were relatively small and risky. Letting the private sector participants in the network underwrite commercial standards was of course a political *coup* in the ideological climate of the 1980s because it bestowed much-needed legitimacy on the activity, but at the same time it also effectively limited the impact of the industrial investment function to lowering the size and extending the sectoral coverage of the existing private provision of venture capital within Scotland. Compared to the situation in other peripheral regions of the UK this was in itself no mean achievement, but it is also obvious that even after the emergence of development funding as a separate financial instrument, the Agency was in effect prevented from engaging in major one-off rescues of individual firms, large-scale attempts of sectoral restructuring or being the sole provider of external finance for major innovative projects – the kind of activities that had played an, albeit minor part, in the very early days of the function. Instead activity grew in other areas: investments in new firms and modern sectors of the economy emerged as more than just a supplement to support for existing firms in traditional industries, and MBOs became prominent as an instrument of organisational modernisation and perhaps also functioned as a form of piecemeal new-style rescues. Although the policy programme continued to operate in a largely reactive manner also in the 1980s, it did in other words still entail the possibility of a significant, but more discreet, public sector input into the strategies of a considerable number of indigenous firms.

Table 8.1 SDA investments and DTI financial policy instruments 1975-91

	SDA	*DTI*
Priority	Strategic decrease, implementation maintained	Decreasing
Regional problem	Equity gap Uncompetitive firms	Unemployment, increasingly uncompetitive firms
Firm-level impact	Expand/modernise indigenous, increasingly new indigenous	Gradual shift from indigenous expansion/preservation towards incoming location and modernisation of indigenous
Project generation	Reactive (proactive capacity)	Reactive
Appraisal	Discretionary (viability)	Gradual shift from automatic to discretionary appraisal (viability, additionality)
Selectivity	Size: mainly small Sector: increasingly high-tech Project type: investment	Spatial: varying Size: Increasingly small from late 1980s Sector: limited and broadening Project types: investment, from late 1980s also knowledge

The development of the industrial investment function of the SDA can be compared with that of the financial instruments employed on the British level by central government in order to promote regional development, and as illustrated by Table 8.1 these policies developed in a rather different way. While the Agency's investment function was broadly maintained both in quantitative and qualitative terms, the period from the mid-1970s to the early 1990s saw major changes in the use of regional financial incentives: expenditure was reduced drastically, project appraisal shifted from automatic to discretionary, and the importance of support for foreign inward investment in the Assisted Areas increased – all in all something which in strategic terms contrasted sharply with the pre-1975 emphasis on redistribution of economic activity *within* the UK through automatic subsidies and relocation of branch plants. Although DTI policies *did* come closer to SDA-style implementation through the shift towards discretionary project appraisal, the emphasis on incoming investors in the former was very different from the persistent focus on indigenous firms by the latter. Moreover, while the discretionary criteria according to which individual projects were appraised would seem to be similar – according to programmatic statements commercial profitability was central to both – in practice they may well have diverged because of the different resource dependencies involved. For its part the SDA was dealing with relatively small firms and its decision-making procedures were supported by the expertise in other parts of the Agency as well as private financial institutions, whereas the DTI was often dealing with large multinational firms and its ability to draw on external informational resources in the appraisal process may well have been overshadowed by the national political priority accorded to attracting inward investment to Britain in competition with other European countries. When the political dimension is added – individual Agency investments were undertaken under the veil of commercial secrecy while large inward investment cases often became very public affairs – it is clear that in many cases the arm's-length agency must have been better placed to exercise commercial discretion than the civil servants of central government.

From a Scottish perspective it is also interesting to note that the relative importance of SDA investments gradually increased *vis-à-vis* central government regional grants in real financial terms from 1975 onwards, simply because the former maintained its level of expenditure while the latter was reduced significantly, but even towards the end of the decade the investment function still only accounted for around 12% of total regional policy expenditure on financial instruments in Scotland.[235] Given the indigenous orientation and discretionary nature of the industrial investment function, it would, however, be more reasonable to compare Agency investment expenditure with the share of central government regional subsidies in Scotland paid to indigenous firms on the basis of a discretionary appraisal process because this would give an idea of the extent to which the firms targeted by this particular SDA programmes would have had their access to this type of external funding increased. As can be seen from Figure 8.11,[236] the presence of the investment function increased the availability of discretionary financial support for indigenous projects of importance to regional development with around 50%,[237] something which amounted to a considerable degree of preferential treatment of Scotland over the Assisted Areas in

235 Calculated on the basis of SDA 1977-91, HIDB 1977-91, *Industry Act Annual Reports* 1977-91, and estimates based on McQuaid 1992 and Fairley 1999.

236 No breakdown of British expenditure on Regional Selective Assistance according to foreign/indigenous ownership was given in the *Industry Act Annual Reports* before 1979. Figures for expenditure on Regional Selective Assistance in Scotland are not broken down according to external/indigenous ownership, but an estimate can be arrived at by applying the British shares to the Scottish total.

237 For the entire period 1978-91 SDA investment expenditure equalled 46% of that on Regional Selective Assistance for indigenous Scottish firms, but if the exceptional year 1990-91 is discounted, the figure rises to 56% (calculated on the basis of the sources given in Figure 8.11 cf. the footnote above).

England – especially because financial resources were tied in with informational and organisational forms of support that may have improved the quality of individual projects and hence in the longer term the competitiveness of the regional economy.

Figure 8.11 Discretionary regional investment support in Scotland

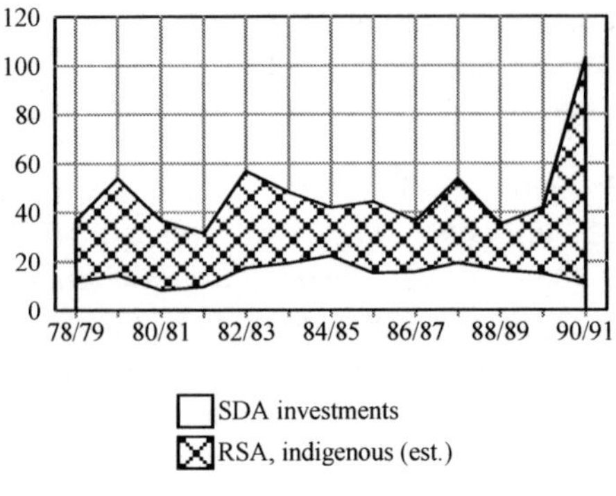

£m 1985/86.

Source: SDA 1979-91 and *Industry Act Annual Reports* 1979-91.

Presenting itself to the world in commercial, business-oriented terms much in line with the Conservative government's general discourse on economic (and from the late 1980s also regional) policy have undoubtedly helped to sustain the external-revolution perspective as a framework for interpreting the development of the SDA, but these discursive features, driven by the Agency's sponsors (and readily accepted by the political opposition), should not disguise the remarkable degree of continuity in implementation which was achieved by the SDA in a turbulent political environment: strategies and implementation did not change radically from the late 1970s to the 1980s, the capacity of the Agency to influence private actors was not eroded but took new forms, and the gradual and limited nature of change contrasted sharply with the comprehensive overhaul central government regional policy was subjected to in the same period. Whether the limited and gradual changes which have been identified were the product of internal learning-by-doing or external pressures from the sponsor department would, however, be good to know. On the one hand the timing of change certainly points towards the presence of internal causes, but at the same time it is also clear that initiatives by the sponsor department, notably the requirement for private sector co-investment and the introduction of a new management with venture capital experience, are likely to have reinforced internal trends and hence actually provided continuity rather than radical change.

Ideologically motivated Conservative politicians with limited concerns about Scottish territorial politics may have thought that 'profitable public investment' was a contradiction in terms that would soon render the industrial investment function unoperational,[238] but in practice the new guidelines had clearly left sufficient room for discursive manoeuvre and implementory ingenuity for the Agency to continue to employ its most controversial policy instrument as long as this was not perceived to be operating in ways that were at odds with the key assumptions and interests of central government in Edinburgh and, indeed, in London. Remaking the rules and playing the game would still seem to be two different things.

238 As suggested by George Mathewson (9.7.90).

CHAPTER 9

Sectoral Initiatives

A prominent feature of the SDA was what came to be known as the sectoral initiatives: policies aimed at promoting economic development by focusing on particular sectors of industry.[1] A sectoral initiative typically involved a study of the specific position of a particular industry in Scotland which then formed the basis for tailor-made policy measures attempting to remedy identified weaknesses and pursue perceived opportunities, drawing on the full range of instruments at the Agency's disposal and hence constituting an integrated approach to regional policy. A tailor-made integrated development strategy will always involve some degree of selectivity, simply because different policy measures would apply to individual firms according to how they were positioned *vis-à-vis* the underlying structuring principle, and sectoral initiatives would result in different forms of attention being given to various industries, perhaps resources being concentrated in particular sectors at the expense of other forms in order to e.g. 'pick winners' or 'cushion' declining industries, and a major issue is therefore likely to be the selectivity involved in focusing policies on particular activities in the regional economy.

The sectoral initiatives of the SDA would certainly seem to live up to the expectation of the prominence of the question of selectivity because both in the public debate, amongst key policy actors, and in parts of the academic literature it is claimed that the Agency gave priority to new high-tech industries while relatively little attention was given to reinvigorating traditional industries.[2] Especially those who see this as an unwarranted 'high-tech bias' have sometimes alleged that this sectoral priority did not reflect extensive analysis of the structural weaknesses of the Scottish economy but instead appeared to be guided by the image-making of its PR consultants for whom "electronics […] is much more

[1] For a concise introduction, see IDS 1987a pp. 72f.

[2] E.g. Young & Reeves 1984 p. 144, Danson *et al.* 1989a p. 562, *Glasgow Herald* 5.2.86, Helen Liddell 21.6.90, Ewan Marwick 24.7.90, Craig Campbell 16.7.90, cf. Gavin McCrone 20.6.90, Alf Young 5.6.90, Bruce Millan 23.7.90, Sir William Gray 19.10.90, Douglas Adams 18.7.90, Douglas Harrison 19.7.90, cf. Leruez 1982 p. 141, Standing Commission 1988 Ch. 2, *Glasgow Herald* 5.2.86.

sexy".[3] At the British level the Labour government of the late 1970s had already set its sight on new industries and made this a priority area of the NEB,[4] and during the Conservative 1980s most new industrial policy initiatives by central government focused on high-tech areas,[5] and thus the pressure to 'go high-tech' may have been a permanent feature of the Agency's political environment throughout its existence. But in Scotland the incoming Conservative government had specifically made promotion of inward investment a prime function of the SDA while downgrading industrial investment to a last-resort auxiliary measure, and this could easily be construed as a shift in sectoral priorities: industrial investment had near-exclusively targeted indigenous firms including those in traditional industries, whereas inward investment was likely to bring in primarily firms in new high-tech industries akin to the foreign-owned plants already located within the region. On the basis of this it is hardly surprising that the external-revolution perspective on the development of the SDA acquired a sectoral dimension, arguing that the Conservative government brought about a 'high-tech bias' which lead the Agency to neglect traditional areas of economic activity which were undergoing a process of contraction and restructuring. But again this contrasts with the internal-evolution perspective, as exemplified by the Agency's Director of Planning Edward Cunningham who maintained that the organisation had taken a dual approach: "we recognised that the traditional industries involved a large number of employees and, parallel with emphasising technology sectors, also gave priority to certain traditional sectors."[6]

The analysis will be based on a combination of published materials, especially the sectoral strategies and annual reports of the SDA, a range of unpublished sources including DCC and other papers on corporate and industrial strategies, a series of interviews with senior executives involved in the sectoral work of the organisation,[7] supplemented by the existing, small but very useful, academic literature.[8] In order to keep

---

3 Douglas Harrison 19.7.90, cf. Hargrave 1985 p. 67, Standing Commission 1988 Ch. 2, Danson *et al.* 1989c pp. 9f, 1993.

4 Grant 1982 p. 109, Kramer 1989 Ch. 1, cf. Section 8.3.

5 Cf. Section 9.3.

6 Edward Cunningham 13.6.90. The robust statement of the SDA's first chairman – "we did not see a future in metal bashing" (Sir William Gray 19.10.90) – is atypical, because other key executives reiterated the two-track approach (e.g George Mathewson 9.7.90, Neil Hood 22.6.90, Charles Fairley 26.7.90, cf. Gregor Mackenzie 15.10.90, Gavin McCrone 20.6.90).

7 Six interviews were conducted with sector personnel, covering the overall strategic dimension in the entire period of study and nearly two thirds of all major sectoral initiatives undertaken.

8 Especially Hood & Young (eds.) 1984, Moore & Booth 1989.

track of the development of a policy area which contains more than forty individual measures covering eleven different sectors and around half the manufacturing workforce in Scotland, a database, SDASECT, has been constructed covering key characteristics of the individual measures entailed in each of the eleven sectors targeted by the SDA from 1975 to 1991.[9] The chapter falls in three parts: first the design of sectoral strategies is explored (Section 9.1), then implementation of the key initiatives is followed in some detail (Section 9.2), and finally the broader British context is examined, both with regard to developments within regional policy as such and the adjoining area of sectoral policies pursued by central government throughout Britain (Section 9.3).

9.1. Design

Origins of a Sectoral Approach

After more than two decades of bottom-up regional policy, including the 1990s which saw the spread of interest in role of clusters as drivers of regional economic development across the western world,[10] it is worth underlining the fact that before 1975 sectoral considerations played only a very marginal role in British regional policy. The grant schemes operated by central government were sectorally selective 'only' in the sense that they excluded most service activities, but they did not discriminate between manufacturing sectors and dealt with firms on an individual basis, and hence these reactive and unselective grant programmes could hardly be expected to address specific weaknesses of individual regions.[11] Contrary to this the existing rural RDAs had pursued policies which could be described as sectoral in as far as specific types of economic activity such as crafts, crofting and fishery were

9 The eleven sectoral initiatives analysed are those identified by the Agency itself, and within each all policy measure have been classified according to duration and prominence – the distinction between primary and secondary measures reflect the relative weight accorded to measures in SDA publications and in some cases involves quite a degree of analytical discretion – and its characteristics with regard to policy instruments, modes of implementation and firm-level impact.

10 The work of Phil Cooke and Kevin Morgan (e.g. Cooke & Morgan 1993, 1998; Morgan 1997) are central contributions. For useful overviews, see Mariussen 2001, Asheim 2001, Raines 2001; for salient scepticism, see Lagendijk & Cornford 2000, Lagendijk 2003.

11 Cf. Section 4.3. Already in 1969 Gavin McCrone – who later came to play a central role in the setting up of the SDA as Chief Economic Advisor at the Scottish Office – that it would be "valuable for each region to have a view on the type of development for which it thought it was best suited ... (although it) would involve much research to determine the type of industrial growth that any given area is best suited for" (McCrone 1969 p. 217).

targeted,[12] and as from the mid-1960s sectorally-oriented industrial policies had been pursued in Britain – from the mergers of the Industrial Reorganisation Corporation (IRC) via sectoral Industry Act grants to the mid-1970s tripartite advisory Sectoral Working Parties[13] – it is, perhaps, less surprising that the notion of introducing a sectoral approach on a smaller spatial scale via a proactive regional development body arose.

Some of the major blueprints for a new development body for low-land Scotland specifically included a sectoral approach: the team behind the West Central Scotland Plan envisaged 'industrial reorganisation' along sectoral lines as a "subsidiary function" of its proposed development agency,[14] and the Labour proposal for a National Enterprises Board was partly inspired by the IRC and explicitly linked the new body with public management of ailing industries.[15] A transposition of the sectoral approach from the UK to the regional level had in other words been heralded by major players on the Scottish political scene, but it is also worth noting that thinking within the party setting up the SDA had placed sectoral initiatives in a specific political context which could readily be construed as part of a strategy which involved public-sector reorganisation of ownership patterns in private industries.

Sponsoring Sectoral Initiatives

The original SDA Act of 1975 did not mention sectoral initiatives explicitly, but the wording of the industrial 'catch-all' function – "otherwise promoting or assisting the establishment, growth, reorganisation or development of industry or any undertaking in an industry"[16] – could suggest that activities along sectoral lines were in fact being envisaged: in the political discourse of the 1970s "reorganisation […] of industry" was a reference to the IRC which had operated along sectoral lines, and the adding of 'in an industry' to 'any undertaking' would seem superfluous if policies were not otherwise to be aimed at particular industrial sectors.

Once the Agency had been established sectoral initiatives were subjected to a relatively low degree of formalised regulation by the sponsor department compared to other policy areas: no official guidelines were issued by the Scottish Office, but in practice the interest of the sponsor department appears to have gradually grown. From the very beginning the sponsor department was involved in a regular cycle of discussions

12 Grassie 1983, Rogers Ch. 8, cf. John Firn 15.6.90.

13 Young & Lowe 1974 Ch. 2, Grant 1982 pp. 51, 63ff.

14 West Central Scotland Plan Team 1974a p. 71.

15 Labour Party Scottish Council 1973 pp. 8ff.

16 SDA Act 1975 Section 2.2.c.

about first the industrial and then the corporate strategies of the SDA,[17] and sectoral initiatives were an integrated part of these planning documents. Then in 1984 this general scrutiny was supplemented by the setting up of a high-level SDA-IDS industrial policy study group in 1984 in order to ensure that "government policies were properly reflected in the industrial sectoral work [...] and similarly that government expertise was actually drawn upon".[18] Finally the sectoral initiatives were covered by the 1986 government review and emerged as an area in which the reviewers found room for improvement. The procedure for selecting sectoral priorities was criticised for being "too informal",[19] and the report insisted that "there is scope for a closer partnership between the Agency and IDS in industrial sector work" also with regard to "the commissioning and supervision of studies", although this "need not [...] imply any inhibition on the creativity of the Agency".[20]

All in all this would seem to suggest that at least until the late 1980s the SDA have clearly enjoyed an arm's-length degree of bureaucratic autonomy with regard to having the authority to undertake the overall planning and implementation, although it remains to be seen whether the steps towards more detailed regulation were capable of reversing the original distribution of roles which cast the Agency as the initiator of sectoral initiatives.

Organisational and Informational Resources

As none of the SDA's predecessor bodies had been engaged in sectoral initiatives, an organisational capacity for designing and implementing such policies had to be developed from scratch, and significant results could therefore hardly be expected overnight.

As illustrated by Table 6.1, the organisational framework for sectoral work within the SDA gradually changed, and it is possible to identify three main trends:

- a rapid increase in the number of units involved in the early 1980s followed by some rationalisation through mergers in the second half of the decade,

---

17 Cf. Chapters 6 and 7.

18 Iain Robertson 11.7.90. Tangible results in the form of Scottish Office input in the study phase of sectoral initiatives continued to be largely limited to participation in steering groups following consultancy studies of individual industries (SDA Papers 27.10.86). The only example of a joint SDA/Scottish Office study was even undertaken by external consultants (SDA/DCC 87(11)).

19 IDS 1987a p. 77.

20 Quotes from IDS 1987a pp. 80f.

- an increasing degree of specialisation in the 1980s compared to the early years where most initiatives were managed within multi-purpose units,[21] and
- a gradual concentration of the sectoral work within the Agency's Planning & Projects Directorate,[22] probably in order to achieve synergy and limit the risk of fragmentation.

The size of most sectoral divisions was fairly small,[23] and together with their cross-regional purpose this accounts for the fact that the sector work was maintained as part of the central functions after the major decentralisation in 1988.

This organisational specialisation was connected with public relations because specialised units were perceived as signalling professional credibility and a way to make the efforts of the SDA more visible, both to the industries targeted and the general public.[24] But the organisational patterns also reflected the changing motive forces behind the sectoral work: while sectoral work in the late 1970s was often connected with the needs of the industrial investment function for a sound basis for investment decisions,[25] it became a corporate priority in its own right in the 1980s.[26] Finally, the mushrooming of specialist divisions is also an excellent example of the project-driven mode of operation of the Agency: in the absence of a master plan opportunities were pursued as they were identified by executives,[27] and it is hardly surprising that the civil servants undertaking the 1986 government review questioned the cost effectiveness of the SDA's internal division of labour.[28]

The informational resources available for sectoral work were extended by means of external input, especially in the policy design phase. In a number of cases advisory committees were established, providing additional sectoral business expertise and enhancing the credibility of

21 New Ventures was a division within the Industry Directorate responsible for investments in new high-tech firms, while the Industrial Programme Development division specialised in sector work (John Firn 15.6.90, Charles Fairley 26.7.90).

22 Headed by Edward Cunningham, the name of the directorate changed several times, cf. Table 6.1.

23 In 1984/85 the 7 specialist divisions employed 43 persons, of which 12 in electronics (IDS 1987a p. 72), and thus the rest had in average 5 staff.

24 SDA Board Papers 87(5).

25 SDA 1978a p. 16, 1979 p. 26; John Firn 15.6.90, cf. *Glasgow Herald* 16.6.76, Danson 1980 p. 14.

26 Cf. Chapter 7.

27 George Mathewson 9.7.90, Iain Shirlaw 1.6.90, Charles Fairley 26.7.90, Frank Kirwan 1.6.90.

28 IDS 1987a pp. 81f.

policy initiatives with the private actors targeted.[29] Moreover, the majority of industry studies were undertaken by private consultancy firms rather than Agency staff or the Scottish Office, not just because of lack of adequate internal resources,[30] but also because major international consulting firms were seen as more independent and impartial sources of information[31] and thereby further boosted the standing of ensuing policy initiatives within the business community. While the SDA clearly appreciated external contributions in terms of information and informal authority, these aspects of the sectoral approach were by no means uncontroversial in the Scottish political environment: the handpicked and informal nature of the advisory committees was not to the liking of especially the labour movement,[32] the prominence of external consultants has been construed as an abdication of the Agency from responsibility for the overall strategic direction of its policies,[33] and the need to buy expensive external consultants instead of using in-house resources (or Scottish Office economists) continued to cause friction with the sponsor department,[34] eventually resulting in the 1986 review recommendation that expenditure on studies above a certain limit should be approved by the SDA board.[35]

All in all the development of the organisational and informational resources devoted to the SDA's sectoral initiatives has clearly been primarily internally driven and would seem to allow for the high degree of flexibility required when development activities are being tailor-made to tackle specific problems in the regional economy. Especially when combined with extensive use of external consultants, the conspicuously project-driven mode of operation did, however, also tend to make the overall strategic direction of the Agency's sectoral work look more tenuous than perhaps was the case.

---

29 Edward Cunningham 13.6.90, Iain Shirlaw 1.6.90, Douglas Adams 18.7.90.

30 Charles Fairley 26.7.90, cf. Bellini 2002 pp. 52ff.

31 Edward Cunningham 13.6.90, Charles Fairley 26.7.90, cf. IDS 1987a p. 79.

32 Standing Commission 1989 p. 45.

33 Danson *et al.* 1989c, 1990b; Standing Commission 1988 Ch. 2, 1989 p. 43, cf. Firn 1982 p. 15.

34 Gavin McCrone 20.6.90. As early as in 1982 the SDA's Annual Report contained an elaborate defence of its use of external consultants (1982 pp. 10f).

35 IDS 1987a p. 77. The pragmatic response of the Agency appears to have been to break down major consultancy studies into smaller, less costly tranches (Charles Fairley 26.7.90, cf. Hood 1991a p. 15), although this may of course simply reflect that the organisation now had obtained a suitable overview of the situation in the main sectors of the Scottish economy and therefore could begin to look at selected subsectors in more detail (Edward Cunningham 13.6.90).

Changing Aims and Methods

The SDA Act was generally vague in relation to sectoral initiatives – the most specific reference was 'industrial reorganisation' – but early on it was clear that such an approach would become central because the success of SDA activities such as industrial investment and more proactive projects was seen to depend on developing a "more specific focus through detailed industry studies" that would "provide the essential groundwork for the development of specific strategies for individual industries"[36] and allow the organisation to work proactively with industries as "their corporate planning unit".[37] This indicates the open-ended nature of the sectoral work in which modernisation – of individual firms, particular sectors, and the sectoral make-up of the Scottish economy – was the overriding goal,[38] and while the Agency also had undertaken cross-sectoral studies of e.g. how to "encourage entrepreneurial business skills",[39] studies along sectoral lines were at the very core of strategic planning.[40] In the late 1970s the SDA clearly attached much greater importance to sectoral initiatives and had a much broader understanding of the notion than its sponsor department which apparently focused on problems that could be addressed by means of the industrial investment function.

The change of government in 1979 altered the legal framework for the SDA's sectoral work because the 1980 Industry Act symbolically removed "industrial reorganisation" as a statutory function,[41] but as the Agency's sectoral approach had been much broader, the practical consequences were rather limited. In the 'interim' period between the change of government in 1979 and the advent of a new management team at the Agency in 1981 the public profile of the ongoing sector work was low,[42] but a sectoral approach in the broader sense of the word continued to be

36 SDA/DCC 15.3.78 p. 1. See also SDA 1977 pp. 5ff, 1978a p. 16; Cunningham 1977 pp. 8ff, Robertson 1978 p. 27.

37 John Firn 15.6.90, cf. Robertson 1978 p. 27, Lewis Robertson 20.7.90.

38 SDA/DCC 15.3.78 p. 1, cf. the discussion in Chapter 7.

39 SDA 1977 p. 32, cf. 1978a p. 30.

40 SDA 1976 par. 47, 1977 p. 32, 1978 p. 50, 1978a p. 30; cf. Lewis Robertson 20.7.90.

41 Industry act 1980 Section 1.2.c, cf. Section 8.1.

42 The 1979 annual report merely records the production of the first corporate plan of the SDA and the completion of "several important sectoral studies" (SDA 1979 p. 16), while in the 1980 edition the only mentioning of the electronics study, later to achieve paradigmatic status, is in a caption in the section on inward investment (SDA 1980 p. 24), possibly because the initial electronics strategy proposed just before the change of government in 1979 also entailed a separate high-tech holding company (*Glasgow Herald* 12.4.79).

at the core of the strategic planning of the organisation,[43] and the first half of the 1980s actually became the heyday of sectoral initiatives in Scottish regional policy. The underlying rationale for the approach remained unchanged,[44] standard procedures involved in the tackling of individual sectors were publicised,[45] under the heading "support for technology" the 1985-88 corporate strategy designated four broad industrial sectors for support,[46] and the annual reports of the period contained detailed reports on the progress of Agency work in relation to particular sectors.[47]

The 1986 government review endorsed sectoral studies as a means of "understanding [...] which sectors are most likely to succeed in Scottish conditions" and "add to the information available about marketing and investment opportunities for a particular industry",[48] but in many ways the report was critical of both the prominence of the sectoral approach within the Agency and of particular aspects of the way in which these initiatives were being pursued by the SDA. In addition to specific comments on individual initiatives and the issue of costly external consultants discussed above, the review group also insisted that the effectiveness of the sectoral approach as a management tool for resource allocation within the Agency remained unproven and recommended the development of, alternative or supplementary, non-sectoral ways of structuring policy delivery.[49] Although the SDA board officially "affirmed its confidence in the value of the Agency's involvement in sector studies and initiatives"[50] and the informational advantages of the sectoral approach continued to be publically extolled,[51] the official statements of the development body in the late 1980s were in effect mostly in line with those of its political sponsors: the importance of cross-sectoral initiatives was stressed,[52] and its new corporate strategy no longer included sectoral priorities despite the fact that the Planning and Projects Directorate was still organised along sectoral lines.[53] But perhaps the most important development was that the overall objective of the

43 SDA 1981 p. 61, 1982 pp. 10f; cf. Firn 1982 pp. 17f.

44 Iain Shirlaw 1.6.90, Edward Cunningham 13.6.90.

45 SDA 1982 p. 34, 1983 p. 29.

46 SDA/DCC 83(11), SDA 1984a pp. 2f.

47 SDA 1982 pp. 34-38, 1983 pp. 29-34, 1984 pp. 29-35, 1985 pp. 36-39, 1986 pp. 34-38.

48 IDS 1987a pp. 74f.

49 IDS 1987a pp. 80f.

50 IDS 1987a p. 77, SDA Board Papers 30.7.87.

51 SDA 1987 pp. 6, 38ff, cf. SDA Board Papers 1987 (5).

52 SDA 1987 p. 39.

53 SDA 1987a.

sectoral initiatives seemed to have been broadened, perhaps partly inspired by the importance attached to incomplete information in the market-failure approach: instead of technology the new focus on access to market opportunities brought a much wider range of issues influencing the competitiveness of private firms to the fore.[54]

The main findings of the above analysis can be summarised as follows. As formulated by the SDA – and largely accepted by the Scottish Office – the underlying rationale behind the sectoral approach remained fairly constant, emphasising the capacity to take informed decisions in order to play a proactive role and establish the Agency as a serious partner for the private sector. In contrast to this, the specific objectives have been more changeable and disputed: sector-level reorganisation through public ownership had effectively been sidelined already in the late 1970s when the Agency adopted a much broader dual strategy aiming to modernise individual firms and sectors as well as change the overall sectoral balance, the more narrow focus on technology in the first half of the 1980s perhaps appeases high-tech oriented Conservatives, but, conversely, in the late 1980s the adoption of the market-failure framework would seem to have widened the scope of problems that could be tackled by sectoral initiatives, perhaps even more so than in the late 1970s, because focus would now seem to be on the end result – market access – rather than modernisation as a particular means to achieving this. Finally, with regard to prominence the importance of the sectoral approach can by and large be traced back to the SDA itself while the role of the Scottish Office was one of direct or indirect restraint, something which in the late 1980s created a situation where the *de facto* sectoral organisation of significant parts of the Agency's policies was no longer reflected in the official corporate strategies of the organisation.

All in all this points towards three conclusions with regard to the strategic thinking concerning sectoral initiatives. *Firstly*, the approach must be seen as a way of structuring policy delivery, a 'meta-policy' based on two principles: selectivity in terms of focusing on specific industries, and a mode of operation in which detailed information about the selected sectors was a central resource because it enabled the RDA to develop tailor-made, integrated and proactive policies and at the same time enhance its informal authority *vis-à-vis* private economic actors targeted by the initiative. *Secondly*, the discrepancy between developments on the strategic and organisational levels in the late 1980s – downgrading and continuity respectively – underlines the need to take a closer look at the way in which sectoral policies were implemented. And

[54] SDA 1987 p. 6, cf. Iain Shirlaw 1.6.90.

finally there appears to be a possible tension between the commitment to selective tailor-made policies and the changing general objectives – reorganisation, modernisation, technology, market access – which constituted the discursive context of individual initiatives, and the extent to which the latter influenced the former must be established empirically because, for instance, focusing on modernisation and technology could easily be construed as evidence of a 'high-tech bias'.

9.2. Implementation

The analysis of the implementation the SDA sectoral initiatives starts with an analysis of resource commitments and levels of activity that illuminates both the prominence of this policy area in the Agency's overall profile and the controversial question of sectoral priorities, and then undertakes a detailed and cross-sectoral examination of policy instruments, modes of implementation, firm-level impact, and regional economic change. Given that tailor-making policies to specific problems lies at the heart of this policy area, a high level of diversity between sectors can be expected, and pursuing these differences will be particularly important because the preceding analysis of policy design has uncovered a combination of continuity and change with regard to objectives and priorities in the context of what appears to be ongoing friction between the arm's-length body and the Scottish Office.

Resources and Sectoral Priorities

Two types of indicators will be employed in order to gauge the changing levels of activity with regard to the SDA's sectoral initiatives, namely input in terms of financial and organisational resources and output in the form of policy measures. In addition to charting the aggregate levels of activity, this should also be able to illuminate the overall balance to the initiatives with regard to sectoral selectivity.

Figure 9.1 Expenditure on sectoral initiatives by type

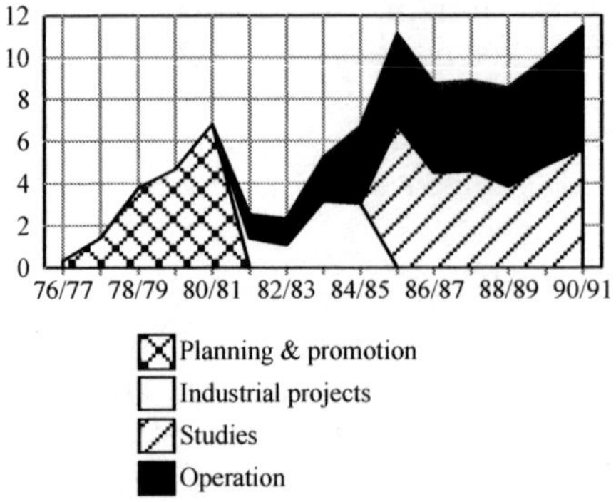

£m 1985/86.
Source: SDA 1977-91.

As illustrated by Figure 9.1,[55] the overall level of financial resources committed increased until the mid-1980s,[56] and thereafter dropped somewhat in the late 1980s only to reach a new high in 1990/91.[57] While operational cost increased gradually, expenditure on projects/studies has clearly been a more variable component, and although a drop of 33% in the very year the IDS review was being prepared is hardly a coincident, it is also evident that very considerable sums continued to be spent on sectoral studies: the average level of expenditure in the post-review years of 1986/87 to 1990/91 was 50% higher in real terms than in the 'golden sectoral age' from 1982/83 to 1985/86,[58] and procuring information still accounted for around half of total sectoral expenditure.[59] In relation to other SDA activities the importance of the sectoral work also remained fairly stable, accounting for around 6-7% of Agency gross

[55] Measuring sectoral expenditure is complicated by changing accounting conventions: before 1981/82 promotion of inward investment was included, and the change from 'industrial projects' to 'studies' also appears to imply some degree of relocation between the two main categories.

[56] The inclusion of inward investment promotion in the early years explains the impression of a sudden reduction of expenditure in 1981/82.

[57] Some sectoral studies have been co-funded by private industry associations (Lochhead 1983 p. 123), but in general this policy area has been overwhelmingly funded by public monies: only in the late 1980s receipts in the order of 2-3% of gross expenditure is recorded (calculated on the basis of SDA 1977-91), and a sudden increase in the last two years of the Agency's existence appears to reflect profitable sales of property in connection with development projects (SDA 1990 pp. 53f, 1991 pp. 46f).

[58] Calculated on the basis of SDA 1982-91.

[59] A much higher figure than the 23% which can be calculated on the basis of the report of the government reviewers for the 1978-86 period (IDS 1987a p. 84).

expenditure from 1984/85 onwards,[60] and thus the expenditure figures would seem to suggest that despite pressure from its political sponsors the prominence of the sectoral approach was maintained, both in terms of continuation of existing policies and with regard to undertaking studies which could inform new initiatives.

Figure 9.2 The development of the sector work of the SDA

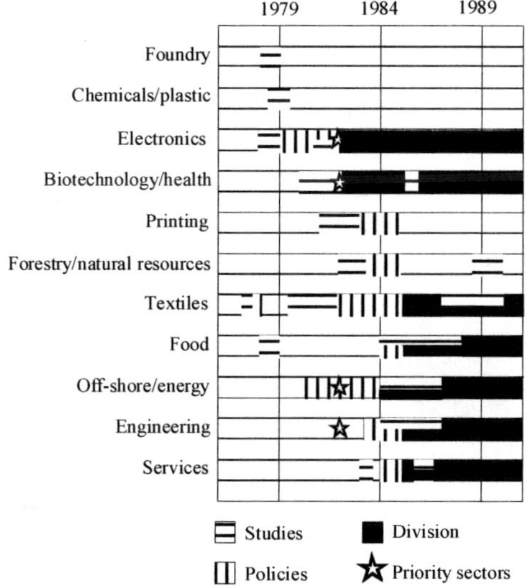

Source: SDA 1976-91, 1977a, 1978a, 1984a, 1987a; SDA/DCC 1977-89, IDS 1987.

Devising detailed measures of output in relation to a heterogenous policy area is difficult, but as sectoral initiatives according to the axiomatic approach should contain information procurement through *sectoral studies*, the building of organisational resources in the form of *sectoral divisions* within the Agency, and *policies* aimed at particular sectors, and recording the occurrence of these activities on a sectoral basis will provide a first overview of developments. The results are summarised in Figure 9.2,[61] and a striking feature is the continuity in terms of levels of activity since 1978: at any one time except the very early 1980s between five and eight different industries were being attended to. Still, an equally obvious observation would be that the

60 Calculated on the basis of SDA 1977-91.

61 The policy signature has only been used when no corresponding sectoral division was in existence. The priority sectors were announced in 1982 but had been singled out as 'growth sectors' in the outline of the 1981-84 corporate strategy already in 1981 (SDA 1981 p. 61).

Agency's approach has differed between industries: some were only scrutinised briefly in order to appraise the need for more thorough studies or in support of the industrial investment function while in other cases the ensuing policy initiatives were relatively short-lived,[62] but still the majority of sectors conform more or less to the official axiom that a specific sectoral initiative should be based on an in-depth study of the current situation and future prospects of a particular industry and be supported by the setting up of a sectoral division.[63] The heterogenous approach could be seen as further evidence of the project-driven nature of the sectoral initiatives, but a closer look at Figure 9.2 suggests that these differences are not randomly distributed over time but can be grouped into a number of 'generations' on the basis of the nature and timing of the activities. Two generations of *major initiatives* are immediately obvious: the first generation of large-scale initiatives, launched in the late 1970s and becoming a permanent feature of the Agency's policy profile, and the second generation of large-scale initiatives, introduced in the mid-1980s. In addition to these paradigmatic exercises, three generations of *minor initiatives* can be identified, namely the industry surveys undertaken in the late 1970s, sectoral work in the early 1980s which produced one-off policy measures, and a number of minor initiatives undertaken in the late 1980s, including sub-sectoral studies.[64] The existence of such 'generations' would seem to indicate that the preferred approach to sectoral work has changed over the years, either as part of an experimental learning process through which the axiomatic approach gradually emerged, and/or perhaps a 'generational bundling' of sectors took place because the resources available to the Agency at different points in time made some industries seem 'doable'.

62 These differences were also reflected in the distribution of expenditure on studies, administration, and other items (IDS 1987a p. 84 table 10.3).

63 Some of the exceptions were also noted by the government reviewers (IDS 1987a p. 72 note 1).

64 For a number of reasons this last 'generation' is less visible in Figure 9.2: much of the sub-sectoral work related to sectors already covered by earlier initiatives and were of a relatively small scale, and because the last two SDA annual reports were much less comprehensive.

Table 9.1 SDA sectoral initiatives summarised

Sector (priority)	*Regional problem*	*Policy measures*	
		Primary	*Secondary*
Textiles (minor 1977, major 1979-91)	Uncompetitive firms	Investment Generic marketing Management adv.	Specialised advice Technical centres
Foundry (minor 1978)	Uncompetitive firms	(Study, no policies)	
Chemicals (minor 1978-79)	Traditional sectoral structure	(Study, no policies)	(Identified biotech as potential focus)
Food (minor 1978, major 1984-91)	Uncompetitive firms	(Preliminary study) Generic marketing	Specialised advice Sector organisations Seed-corn capital Graduate placement Management advice
Electronics (major 1978-91)	Traditional sectoral structure Few indigo. firms Inward investment low-grade	Inward investment Venture capital	Micro-application Sector organisation Training Just-In-Time Club Market intelligence Research commerc.
Biotech& health (major 1980-91)	Traditional sectoral structure Few indig. firms Underutilised public research	Venture capital	Management advice Inward investment Sector organisation Research commercialisation
Off-shore/energy (major 1980-91)	Uncompetitive firms	Technology centres Market intelligence	Specialised advice Trade promotion
Printing (minor 1981-84)	Uncompetitive firms	Specialised advice	
Natural resources (minor 1982-84, 1989)	Uncompetitive firms Underutilised natural resources	Investment Inward investment	
Engineering (major 1983-91)	Uncompetitive firms	Management advice	Technology centres Specialised advice Market intelligence Training

In order to bridge the gap in a heterogenous policy area between the development of the three key components of individual initiatives and the ensuing more compressed cross-sectoral discussion, Table 9.1[65] provides a summary of the eleven sectoral initiatives with regard to the perception of the regional problem and the remedial measures involved.[66] A high degree of variation is immediately evident, and thus while the general thinking behind the sectoral approach has been fairly constant, this has nonetheless been compatible with what looks like a pervasive strategic pluralism. Assumptions about the nature of the regional problems entailed in the sectoral initiatives ranged from concerns about the overall balance between sectors in the regional economy ('modern' sectors too small), via the structure of individual industries (some traditional industries too fragmented, limited interaction with public research institutions), to the characteristics of individual firms (uncompetitive, lacking in strategic capacity). And, unsurprisingly, this diverse diagnosis was translated into equally varied policy initiatives in terms of the actors targeted and the policy instruments employed, and in the following sections these measures will be analysed in greater detail.

The number of policy measures targeting private actors within the individual sectoral initiatives can serve as an additional measure of activity, and Figure 9.3 summarises the development, distinguishing between primary and secondary measures. Again a distinctly phased development is in evidence: until the early 1980s few measures were in place, then rapid growth occurred from 1983 to 1986, followed by a period of consolidation in the late 1980s and early 1990s. Also at this level of analysis the 1980s rather than the 1970s does in other words come across as the 'sectoral' decade, and while the 1986 government review may have slowed down the momentum in terms of new measures, it clearly did not lead to a significant reduction in the level of activity.

---

65 Constructed on the basis of sources underlying SDASECT, i.e. SDA 1985-91, 1979a, 1981a, 1982a, 1982b, 1985a, 1986a, 1988a, 1988b, 1988c, 1988d, 1989a; SDA/DCC 78(4), 78(17), 83(11) 84(8), 88(3), 120 Magazine 2/1984, 5/1985; Edward Cunningham 13.6.90, Iain Shirlaw 1.6.90, Charles Fairley 26.7.90, Mike Sandys 31.5.90, John Firn 15.6.90, Frank Kirwan 1.6.90, Haug *et al.* 1983, Firn & Roberts 1984, Crichton 1984, Mackay 1984, Henderson 1984, Hargrave 1985, Gibbons 1989, McKillop & Hutchinson 1990.

66 Like the sectors themselves, the individual measures are those presented by the SDA as distinct policy instruments, while the distinction between primary and secondary measures reflect the relative weight accorded to measures in SDA publications, something which in most cases is obvious but sometimes involves a degree of analytical discretion.

Figure 9.3 Sectoral policies by priority and broad sector

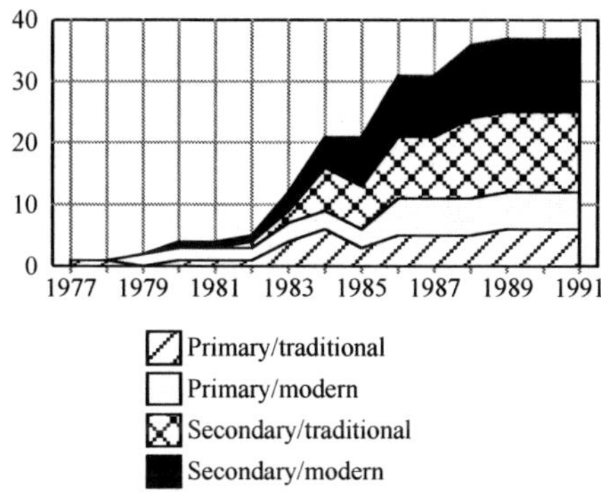

Number of policy measures in operation.
Source: Calculated on SDASECT.

Turning now to the question of sectoral selectivity, Figure 9.2 showed that more than half of the sectors in which sectoral policies were implemented by the SDA had a history in Scotland that stretched back towards the first industrial revolution, and this picture is replicated in Figure 9.3 where in the heyday of sectoral initiatives traditional industries persistently accounted for around half of the total number of policy measures in force from 1983 onwards.[67] On the basis of this it can hardly be argued that traditional industries have been systematically bypassed, but still a more detailed analysis may well lend support to the high-tech bias hypothesis because, as illustrated by Figure 9.2, the resources committed to various industries have certainly differed. Traditional industries were strongly represented in the short-lived initiatives in the late 1970s and early 1980s, while conversely new industries like electronics, biotechnology & healthcare were tended to through sectoral divisions and long-standing policy measures. Similarly the Agency also attended rather differently to the four priority sectors announced in 1982:[68] while the two high-tech sectors, electronics and

---

[67] Calculated on the basis of SDASECT.

[68] SDA 1982 p. 34. The same sectors had already in 1981 been singled out as "growth sectors" (SDA 1981 p. 61). The report of the 1986 review showed that high-tech and service industries accounted for nearly 2/3 of total sectoral expenditure from 1978 to 1986 (calculated on the basis of IDS 1987a p. 85) and that the level of expenditure per employee was 23 times higher in high-tech sectoral initiatives than in those targeting traditional industries (calculated on the basis of IDS 1987a tables 10.1 & 10.3), but when the review was undertaken the second and predominantly traditional

biotechnology/health care, continued to get a 'full treatment' with studies, strategies and specialist divisions, more limited efforts were undertaken in more traditional areas like food and textiles until the arrival of the second generation of major sectoral initiatives.

All in all it is in other words clear that until the mid-1980s the sectoral work of the SDA did have a leaning towards high-tech areas but hereafter became more evenly balanced. Moreover, it is evident that this evolving pattern of sectoral selectivity cannot be seen as an effect of the change of government in 1979, simply because the major first-generation high-tech initiative, electronics, had been initiated already in 1977[69] – all the major initiatives in traditional sectors were developed in the late 1980s so in fact the 'high-tech bias' became far less pronounced under the Conservative government – and thus factors internal to the arm's-length body clearly also need to be considered when trying to explain the development of this policy area. Starting in high-tech areas and only including more traditional industries from the mid-1980s onwards could of course reflect underlying priorities – changing the sectoral structure was the primary objective – but the sequence of sectoral priorities also seems to have been guided by considerations of a rather more pragmatic nature. The early industrial strategies announced that priority would be given to industries where its policy instruments were likely to have a sizeable impact, and the specific arguments used to justify the sequence of sectoral initiatives clearly demonstrates the presence of pragmatic considerations: high-tech sectors were seen as "easier to get your hands round" because of their limited size,[70] while traditional Scottish areas of economic strength were seen as more difficult to deal with because the possibilities of making an impact were limited by the large size and fragmented structure of the industries and what was perceived as old-fashioned and inflexible thinking of managers and trade unions.[71] It could even be argued that the early work done in high-tech sectors was effectively a precondition for the subsequent move into traditional areas because it established the professional credibility of the SDA *vis-à-vis* a business community,[72] especially in industries where the organisation had been involved through its investment function in the late 1970s and thereby possibly antagonised other

---

generation of large-scale sectoral initiatives was in an early stage and much of the expenditure on programmes and projects had not yet been incurred.

69 SDA/DCC 78(4) p. 3, Sir Robin Duthie 12.7.90.

70 Iain Shirlaw 1.6.90.

71 Edward Cunningham 13.6.90, Sir Kenneth Alexander 25.7.90, Iain Robertson 11.7.90, Neil Hood 22.6.90, Iain Shirlaw 1.6.90, Mike Sandys 31.5.90, James Williamson 14.6.90.

72 Edward Cunningham 13.6.90.

firms competing in the same markets.[73] While the long-term strategic perspective generally appears to have been a dual, albeit not even-handed, one, pragmatic reasons did in other words point the Agency towards concentrating its initial efforts in modern sectors of industry in order to achieve results, something which its political sponsors would no doubt appreciate but which also gave the Agency a high-tech image, an image that was further strengthened by the policy profile of the individual initiatives which will be discussed in the following sections.

Policy Instruments and Modes of Implementation

Unlike industrial investment where the primary policy instrument was self-evident – conditional finance – the heterogenous nature of the sectoral initiatives makes it necessary first to establish the character of the policy instruments and the modes of implementation involved before discussing their development and strategic implications.

The policy instruments employed in the sectoral initiatives of the SDA can be analysed according to their specific combinations of incentives and options, and this produces a number of interesting findings, also with regard to the question of the sectoral balance of the activities. Looking first at the incentives employed in order to make private actors adopt new patterns of behaviour, Figure 9.4 charts the use of four policy resources in the sectoral initiatives.[74] Given the limited degree of authority delegated to the Agency by the Scottish Office, the complete absence of authority is hardly surprising, and instead the most common incentives used have clearly been information and organisation. In quantitative terms this dominance has increased throughout the 1980s,[75] reflecting the importance in the sectoral initiatives of specialist advisory services, inward investment promotion, technology centres and sectoral business organisations, with informational instruments coming to the fore already in the early 1980s and the rapid increase in the use of organisational measures occurring in connection with the second generation of large-scale initiatives. This does, however, not mean that financially-based measures merely played a marginal supplementary role, because they were in fact concentrated in primary measures targeting modern industries in the guise of venture capital for electronics and

73 Neil Hood 22.6.90, Frank Kirwan 1.6.90, Crichton 1984 p. 229.

74 Measures relying on a combination of incentives have been classified according to the one assumed to be most prominent one from the point of view of the private actors targeted.

75 From 1986 onwards measures using information or organisation as the primary incentive accounted for more than 90% of the total (calculated on the basis of SDASECT).

biotechnology firms and grant support for users of private consultants.[76] In contrast to this traditional industries had been targeted overwhelmingly by informational and organisational measures, except in the early years where restructuring was sought in parts of the Scottish textiles through the industrial investment function.[77]

Figure 9.4 Sectoral policies by incentive

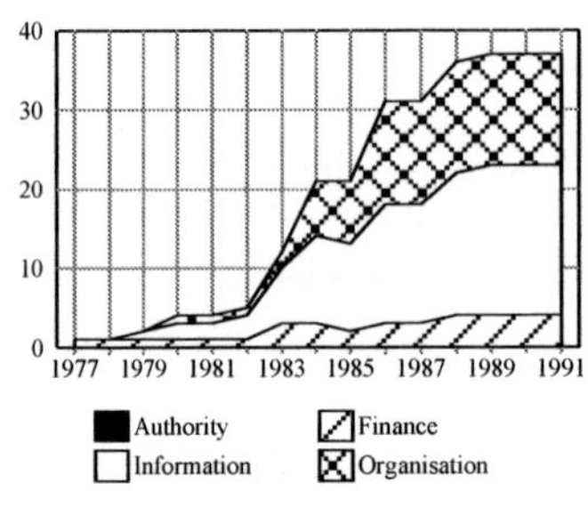

Number of policy measures in operation.
Source: SDASECT

Figure 9.5 Sectoral policies by options

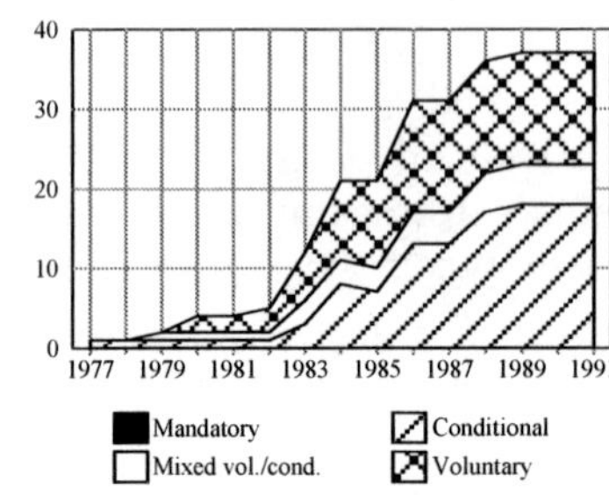

Number of policy measures in operation.
Source: SDASECT

Turning to the options created for private actors via sectoral policy instruments, Figure 9.5 demonstrates that a fairly even balance has existed between measures in which access to resources required the firm to comply with particular conditions, typically to undertake a specific project, and measures of a voluntary nature where the firm itself could decide what use to make of the resources made available by the Agency.[78] Except for the late 1970s where only a very limited number of sectoral initiatives had come to be implemented, such a balance was maintained until the organisation was merged into Scottish Enterprise in 1991, but it is worth noting that the conditional access to resources was clearly more common amongst the primary sectoral measures than the

76 Direct financial incentives constituted 25% of primary policy measures targeting modern industries (calculated on the basis of SDASECT).

77 In the late 1980s all primary policy measures targeting traditional industries were based primarily on organisational or informational incentives (source: SDASECT).

78 Some policy measures have involved both voluntary and conditional elements, usually in combination with different incentives used as part of a set sequence, e.g. voluntary information followed by conditional finance in relation to inward investment promotion.

secondary ones,[79] but evenly spread between modern and traditional areas of economic activity.[80] Finally when combining incentives and options into policy instruments, some combinations are, unsurprisingly, more common than others: financial resources are only made available on a conditional basis, conditions have been imposed in nearly two-thirds of the measures which use organisation as their core incentive, and relatively few constraints apply to the access to informational resources.[81]

The modes of implementation governing the way in which these policy instruments have been used as part of the sectoral initiatives also generally underline the high degree of continuity which characterises many aspects of this area of SDA activity. By definition a sectoral initiative will be selective along sectoral lines, but in addition to this many of the policy measures employed entailed additional forms of selectivity. Throughout the single most important form of selectivity was a focus on particular sub-sectors of the broader industry under which a particular measure operates, but the presence of firm-type and other criteria gradually become more pronounced as this area of Agency activity expand in the 1980s.[82]

In terms of how individual projects were generated, the picture emerging is in several senses of the word a much more mixed one. Measures operating in a purely reactive manner were clearly marginal throughout,[83] and the majority of policy measures would appear to have operated in both proactive and reactive ways depending on circumstances,[84] but it is striking that the number of proactive measures grew rapidly from 1983 to 1986 and then stabilised.[85] Moreover, to the extent

79 While in the early 1990s only 40% of secondary sectoral measures have been classified as conditional, the figure for primary measures is 66% (calculated on the basis of SDASECT).

80 In the early 1990s 47% of the measures targeting traditional industries were conditional, while the same figure was 50% for modern industries (calculated on the basis of SDASECT).

81 While only 36% of all measures based on organisational resources have been classified as voluntary, the figure for information-based measures is 52% (calculated on the basis of SDASECT).

82 Sub-sectoral criteria accounted for around 60% of additional selectivity in sectoral policy measures from 1984 onwards (calculated on the basis of SDASECT).

83 The two policy measures which would seem to have operated in a reactive manner were a basic information service about application of micro-electronics and grants for firms using business advisory services, both measures which aimed to stimulate particular sectors by increasing demand rather than influencing producers.

84 Measures operating in both reactive and proactive manners constituted at least half of the total throughout the period (calculated on the basis of SDASECT).

85 In the period from 1982 to 1986 the share of proactive measures increased from nil to 45% (calculated on the basis of SDASECT).

that sectoral measures involved appraising individual projects – this was of course not the case with policy instruments which made resources available on an unconditional basis – policies operated in an almost completely discretionary way,[86] and thus when combined with the increasing number of conditional and proactive measures it seems safe to conclude that the extent to which sectoral policy measures have been capable of impressing public priorities on private actors will have increased rather than decreased in the 1980s compared to the 1970s.

All in all the policy instruments and the associated modes of implementation have displayed a remarkable degree of continuity: informational and organisational incentives have dominated, a roughly equal balance has existed between conditional and voluntary measures, additional selectivity has primarily focused on sub-sectors within the targeted industries, and project appraisal has been predominantly discretionary. It would in other words be difficult to point towards a clear-cut *general* change of direction in terms of sectoral policy instruments at any one point – neither the party-political change in 1979 nor the government review in 1986 seem to have made much impact – but when the focus of attention turns to *individual sectors* of industry, the analysis has highlighted differences in the way in which modern and traditional industries have been approached. While the former have relatively often been targeted by policy measures involving conditional access to finance and other resources, voluntary and information-based measures would seem to have played a greater role in relation to the latter, and this contrast could easily be construed as evidence that more efforts were being made to make new industries grow than to rekindle crisis-ridden traditional areas of economic activity because the measures employed in relation to modern industries came closer to traditional regional policy measures.

Sectoral Initiatives and Regional Economic Change

The final aspects of policy implementation to be explored are the outcomes of the sectoral initiatives of the SDA in terms of resource exchanges and organisational impact on the level of individual firms, and the wider impact in terms of institutional change within the regional economy. The Agency had given priority to sectoral initiatives since the late 1970s, and a dual approach had been proclaimed in the sense that the general aims included both support for modernisation of firms within traditional industries in order to make them more competitive *and* support for new industries in order to modernise the overall sectoral make-up of the Scottish economy. On the basis of this it could be ex-

86 The only non-voluntary measures in SDASECT which have not been classified as discretionary are grant-aid for use of private consultants and training measures.

pected that the resource exchanges involved in policy implementation would develop in a gradual manner, and that policies would target both new and existing firms, and the extent to which this has been reflected in policy implementation will be pursued in the following. In the absence of detailed and consistent information about the implementation of individual policy measures,[87] the argument will be based on a reinterpretation of the findings in the previous sections and additional analysis of the SDASECT database.

With regard to resource exchanges in sectoral initiatives, it has been shown that resource *input* in the form of expenditure and organisation on this new policy area increased until the mid-1980s and then levelled out,[88] and as around half of this expenditure was incurred by information gathering through e.g. consultancy studies,[89] the sectoral initiatives have been characterised by relatively low levels of resource exchange[90] which were then targeted very accurately in order to address specific issues regarding particular groups of firms in an intensive manner.

For each of the policy measures the type of firms targeted has been estimated, and Figure 9.6 provides an overview of the development of the sectoral initiatives from the perspective of their firm-level impact. It would appear that the nature of the firms targeted by sectoral measures have passed through several phases, starting with a focus on ownership in the 'restructuring' days of the late 1970s, then primarily being concerned with new firms and modernisation of existing ones in the early 1980s, while from the mid-1980s onwards expansion of existing activities was the major area of policy growth, eventually becoming the most common firm-level target of sectoral policy measures, although with new indigenous firms and modernisation of existing firms as major supplementary targets.[91] Unsurprisingly, the emphasis on new firms was particularly pronounced in modern industries while expansionary and modernising measures were more likely to be aimed at traditional

87 Neither on expenditure, staffing nor activity indicators has information been available on the level of individual policy measures.

88 Cf. Figures 9.1 & 9.2.

89 In the period for which information is available in the SDA's annual reports, studies accounted for 48% of total sectoral expenditure (calculated on the basis of SDA 1986-91). For the 1978-86 period covered by the government review, figures show that information gathering accounted for 23% of expenditure on sectoral initiatives (IDS 1987a p. 84).

90 Average expenditure per policy measure remained relatively stable in real terms from the early 1980s onwards, between 0.1 and 0.15 million pounds (1985/86 prices, calculated on the basis of SDA 1977-91 and SDASECT).

91 In the early 1990s expansion accounted for 57% of the total number of measures, while new indigenous and modernisation for 16% and 22% respectively.

industries,[92] and perhaps the most important finding is that the attraction of new incoming firms played only a minor role, even in relation to modern industries.[93] This clearly demonstrates that the sectoral initiatives of the SDA did not ignore existing firms within the region, and thus in terms of firm-level impact the sectoral initiatives would clearly seem to constitute at least a dual approach, targeting indigenous as well as incoming firms, although of course individual inward investment cases may have a much higher public profile than other forms of resource exchanges in support of e.g. modernisation or expansion of existing firms.

Figure 9.6 Sectoral policies by firm-level impact

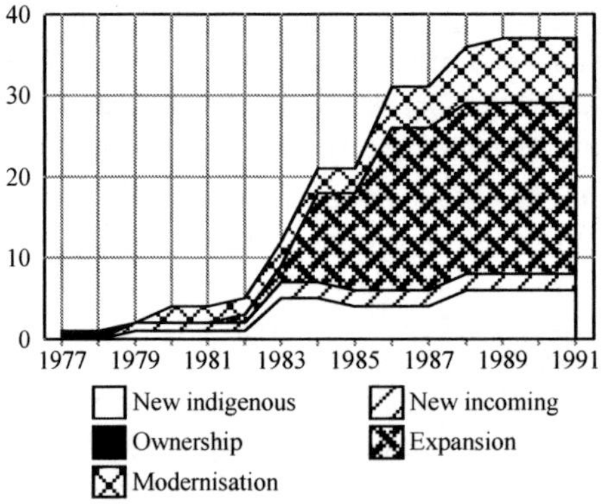

Number of policy measures in operation.
Source: SDASECT.

When looking at the impact of the SDA's sectoral initiatives in terms of bringing about institutional change in the Scottish economy, it should first be emphasised that the introduction of sector-specific policies on the regional level *in itself* can be seen as a major institutional innovation. The essence of this approach to regional development was to use public funds to gather information and develop strategies for particular areas of economic activity in which private actors could participate if they thought it to be advantageous, and thus through the sectoral initiatives the Agency instituted a capacity to act as a provider of corporate

92 In modern industries 39% of policy measures targeted new (indigenous and incoming) firms, while expansionary and modernising measures accounted for 63% and 32% respectively within traditional industries.

93 Attraction of new incoming firms from outwith Scotland was the prime firm-level target of only 5% of the total measures in the early 1990s, and 11% of those targeting modern industries.

strategic direction, not only for individual firms, but for groups of firms and entire industries. Whether or not the actual strategies devised were more or less appropriate and successful is unimportant from this perspective: the point is that the sectoral initiatives instituted something that neither public nor private actors had hitherto engaged in and thereby modified the rules of the game. Contrary to previous forms of regional policy which primarily made additional financial resources available for firms to use within their existing strategies, the Agency's sectoral initiatives provided new strategic suggestions which private firms could then decide whether to follow or not. And contrary to private sector business consultancy firms the Agency was capable of working proactively with entire industries and distribute information and strategic suggestions to firms which would otherwise have been reluctant or unable to foot the bill of expensive international consultants.

Turning to the specific objectives set for the sectoral initiatives, one of the two overriding goals was sectoral change, i.e. promotion of the relative importance of new industries in the Scottish economy. Although employment figures depend on how sectors are defined, there is little doubt that the relative weight of the 'modern' industries has increased from the 1960s onwards, but the near-doubling of the share of electronics from 9% to 14% from 1966 to 1990 took place in a period where overall manufacturing employment was nearly halved, and thus in absolute figures even this high-profile growth industry merely maintained roughly the same number of jobs throughout the period.[94] In contrast to this overall employment in services has increased by around one third in the same period, with banking, finance and business services experiencing even faster growth rates and increasing their share of service employment from 7.7% to 13.8%.[95] It would of course be tempting to interpret these figures as a sign of the success of the SDA's priorities, but in the absence of detailed evaluations it is difficult to know which role its broad array of policy measures played in the ongoing process of economic restructuring,[96] simply because these changes may also reflect general economic trends on the British, European or global level, but the combination of actual employment trends with the publicity possibilities associated with inward investment promotion as a policy measure has no doubt contributed significantly to the image of especially the electronics initiative as being more successful than measures targeting traditional sectors within the Scottish economy.

---

94 Scottish Office 1991 p. 49.

95 Calculated on the basis of Scottish Office 1991 p. 59.

96 The 1986 government review only deals with individual sectoral initiatives in a rather summary fashion, despite having strong views about their relative merits.

While the absence of systematic evaluations also makes it difficult to assess the impact of the SDA's sectoral initiatives on the aggregate performance of individual industries, it is still possible to establish the extent to which its policies can have contributed to enhancing the position of Scottish-based actors by categorising sectoral policy measures according to the modes of social coordination which they aim to modify. Promotion of Scotland as a location for inward investment attempted to enhance the region's position *vis-à-vis* other possible locations in the increasingly global competition for footloose investment projects, and this would seem to involve a market-type relationship between the Agency and individual multinational investors because the latter are capable of 'shopping around' for the best deal in terms of e.g. public subsidies or closeness to suppliers and customers. The various measures aimed at promoting Scottish products and services as a generic brand with particular qualities – e.g. clean and healthy food, or prudent financial services – operated by creating a network between private firms characterised by long-term cooperation and mutual dependency within the industry. And finally many measures aimed at enhancing the strategic options available to individual firms through access to additional resources, something which would strengthen the firm as a hierarchical organisation *vis-à-vis* its economic environment and the position of the existing management team within it.

Figure 9.7 Social modes of coordination and the primary policy measures of the major sectoral initiatives of the SDA

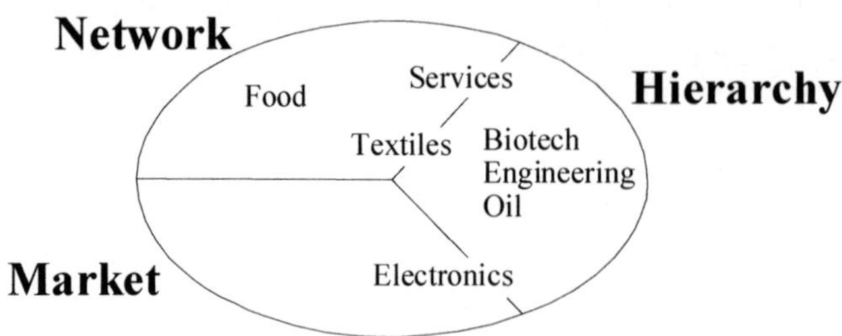

On the basis of this it is clear that both markets, networks and hierarchies have been targeted, and that, as illustrated by Figure 9.7, the primary policy measures employed in the large-scale sectoral initiatives differ with regard to the social modes of coordination which they aim to modify. While efforts in relation to some sectors have concentrated on institutional innovation in relation to one mode of coordination, others have combined hierarchy-oriented measures with network- or market-oriented ones, but as no particular pattern would seem to emerge with regard to modern versus traditional sectors,[97] the different approaches would rather seem to reflect the position of *individual* industries,[98] e.g. the preponderance of multinational firms in electronics, the fragmented structure of indigenous firms in food and textiles, the presence of a strong research base in biotechnology, and weak strategic capabilities in many local engineering firms. No matter which form of institutional innovation the SDA introduced, this would always involve devising a combination of policy instruments which would entice private economic actors to respond by adapting their own strategies accordingly, and in practice this is known not always to have happened:

- at least one attempt to establish a sectoral marketing organisation failed,[99]
- often inward investment organisations in other regions or countries persuaded individual firms to locate in e.g. Wales or the Irish Republic rather than in Scotland, and
- attempts to make foreign investors locate high-grade functions in Scotland or increase their use of local subcontractors appear to have fought largely in vain against the organisational patterns of multinational firms.[100]

Some of these less successful measures would seem to reflect the limited resources – financial or otherwise – available to the Agency, but still they could be seen as a part of an internal learning process where the development body gradually attempted to enhance its capacity to influence private actors by refining its operational strategies.

Apart from confirming the impression of the tailor-made nature of the sectoral initiatives, the variation in institutional innovation is, however, also important with regard to understanding the role of the SDA in general – and its public profile in particular – in relation to individual policy measures because different forms of public intervention would seem to entail rather different requirements:

- *market-oriented* measures are likely to require an ongoing and public effort in order to attract the attention of private economic actors that

97 None of the large-scale initiatives in relation to traditional sectors have been primarily market-oriented, but the forestry initiative inward investment played a major role.

98 Moore & Booth (1989 pp. 82ff) build their argument about the varying degrees to which the SDA was able to influence different industries on a similar analysis of resource dependencies between the Agency and private economic actors on the basis of e.g. the firm structure and market position of Scottish firms in a particular sector.

99 It proved difficult to get support from the Scottish raspberry producers, at that point enjoying a dominant position in the European market, for a sectoral organisation (Moore & Booth 1989 pp. 81f).

100 Cf. Chapter 10.

are not necessarily committed in the long-term to any particular location and capable of moving activities elsewhere,

- *network-based* initiatives may in themselves have a high public profile but require less ongoing effort by a development body because private actors are encouraged to take collective ownership of e.g. a sectoral organisation, and
- the *hierarchy-oriented* measures induce internal changes within individual firms which are less likely to enjoy a high public profile.

The very fact that the SDA's sectoral initiatives attempted to modify different modes of social coordination in order to promote economic development within the region may in other words provide part of the explanation for the perceived success of individual sectoral initiatives because some of them were more likely to entail or attract publicity – something which of course may or may not have reflected their contribution to changing the rules of the game in accordance with what was perceived to be in the best interest of Scotland.

9.3. The British Context: Sectoral Approaches to Regional Development

Before 1975 the only sectoral dimension of British regional policy was its near-exclusive focus on promotion of manufacturing rather than service sector activities,[101] and thus from a historical perspective the much more specific sectoral initiatives of the SDA would appear to have been innovative. What looks like a new Scottish departure may, however, simply have reflected broader trends in e.g. central government policy, and therefore the following pages will first pursue sectoral aspects of regional and industrial policies conducted from the British level, and then provide an outline of the sector-based activities of other regionally-based development bodies.

Central Government Regional and Industrial Policies

In the period from 1975 to 1991 the regional grant programmes of central government went through quite a number of alterations with regard to area designation and the volume of financial resources involved, and, indeed, the forms of economic activity eligible, but in terms of sectoral selectivity grant schemes continued to focus predominantly on manufacturing and the main extension of support for tradeable services was an indirect one, namely the introduction in 1988 of the Enterprise Initiative which supported the use of consultants by small firms which could benefit private consultancy firms based in the desig-

---

[101] Cf. Section 4.3.

nated localities. Within manufacturing the increasingly discretionary policies will, however, not automatically have translated into a sectorally neutral distribution of grant support, but in practice the preference given to firms in modern industries was fairly limited,[102] and thus the outcome of central government regional policy would not appear to have been heavily skewed against firms in traditional industries.

In the adjoining – or according to the 1988 DTI white paper overarching[103] – area of industrial policy a much more pronounced shift of sectoral preferences would seem to be in evidence in the same period.[104] In the mid-1970s British industrial policy had acquired a distinct sectoral dimension through the setting up of a series of temporary Industry Act measures which made investment subsidies on a 'semi-automatic' basis available to firms for modernisation purposes in particular sectors of industry,[105] and the vast majority of these programmes targeted segments of traditional industries and had been terminated according to schedule before the change of government in 1979.[106] Especially after the liberal Keith Joseph had been replaced by Patrick Jenkin in 1981 as Secretary of State at DTI, the Conservative government gave renewed emphasis to industrial policy in the guise of 'constructive interventions' in the form of support for new technologies,[107] but while the sectoral programmes had subsidised investment in rather narrowly selected industries, the new technology-oriented schemes focused more on increasing the market for particular products through financial support for investments and, increasingly, information procurement.[108] 'Con-

102 Cf. Section 8.3.

103 From being the topic of a separate white paper in 1983 (DTI 1983a), regional policy was just a dimension of the 1988 enterprise-oriented white paper (DTI 1988).

104 For overviews, see Grant 1982, Wilks 1985, 1987; Coates 1996, Wren 1996a.

105 Industry Act Annual Report 1972-81, Wren 1996a pp. 98ff, Imberg & Northcott 1981 pp. 10f.

106 Wren 1996a table 6.1. Only 10.7% of total expenditure on the sectoral programmes were allocated to modern industries such as electronic components and instruments/automation (calculated on the basis of Wren 1996a table 6.1). It has often been noted that support on technology and innovation would exacerbate regional differences because firms in e.g. the prosperous South East England were better placed in terms of financial and human resources to pursue extensive modernisation strategies (IDS 1987a pp. 131f, Amin & Pywell 1989 p. 470, cf. Rothwell & Zegveld 1985 Ch. 7), and with only one exception, the Scottish share of Industry Act sectoral support was far lower than the relative weight of the region within the British economy (Young & Reeves 1984 table 5.14).

107 Wilks 1985 pp. 128-35, Hudson 1989a p. 118, Theakston 1996, Wren 1996a pp. 141ff, Cortell 1997. Grant-equivalent expenditure on technology-related measures overtook that on sectoral schemes in 1981/82, culminated in 1984/85 and decreased rapidly from 1986/87 onwards (Wren 1996b Tables 1 & 3).

108 Wren 1996a pp. 121ff, 145ff, 164ff.

structive interventions' thus only promoted the development of high-tech industries within Britain to the extent that individual firms chose British providers of goods and services, and these measures can therefore at best only be seen as sectorally targeted measures in a *very* loose sense of the word.

All in all it is in other words clear that while in the 1980s central government policies, industrial more than regional, gradually came to focus more on modern technologies, this did not involve any attempts to develop a coherent strategic framework for particular sectors of industry but instead reflected a general commitment to competitiveness by making technological change more attractive to individual firms through policy instruments based on financial and informational resources.

New Actors, New Sectoral Strategies?

In the 1990s a sectoral approach to regional policy had been adopted by some Western Europe RDAs, although it was clearly not generally the dominant way of organising development activity,[109] and this picture is in many ways not unlike the situation in Britain from 1975 to 1991. The only other arm's-length development body sponsored by central government which operated in a predominantly industrial economic environment was the WDA, and here the evidence clearly suggests that the use of sectors as the organising principle for policy development only began on a limited and experimental basis in the late 1980s,[110] possibly because of the weakness of the indigenous manufacturing base and the prevailing strategic focus on inward investment.[111] As the RDA for rural England, the Development Commission had a limited manufacturing base within its area, and hence the only field in which something akin to a sector-based approach could be said to have been in operation was in relation to the promotion of crafts in rural communities.[112] In Scotland the work of the HIDB had originally been organised along broad sectoral lines with units focusing on agriculture, fishery, tourism and industry, but from 1985 onwards this was replaced by a project-based mode of organisation. The latter did, however, still contain a number of sector-based projects, especially related to enhancing the marketing capacities of small firms through support for networking

109 See e.g. Bellini & Pasquini 1998, cf. Halkier & Danson 1997.

110 See WDA 1981a, Welsh Affairs Committee 1989, cf. Rees & Morgan 1991, Henderson 1998, Izushi 1999. It is also noticeable that the accounts of the WDA did not, unlike those of the SDA, have a separate heading for expenditure on sectoral programmes (WDA 1977-91).

111 Cf. Sections 7.4 and 10.3.

112 Tricker & Martin 1984, Rogers 1999 Ch. 10-12.

arrangements,[113] and thus the HIDB was the only example among the RDAs sponsored by central government – apart from the SDA – of an arm's-length body which continued to organise a sizeable part of its activities along sectoral lines.

Contrary to this little evidence has been found of a sector-based approach in activities co-sponsored from the European level, originally because the ERDF was mainly used as a mechanism for reimbursing central government for expenditure on its own programmes, and from the late 1980s onwards because the first generations of regionally-based programmes were still predominantly oriented towards provision of fairly basic physical infrastructure.[114] In the late 1980s some large local authorities in England attempted to develop a sectoral approach through e.g. enterprise boards,[115] but their limited geographical scope would appear to have made such an approach difficult and in practice made support for individual firms the overriding concern.

All in all it is therefore clear that of the new actors introduced into British regional policy from the 1970s onwards, only the major central-government sponsored RDAs operated under conditions which made a sectoral approach to regional development feasible, and even among this small band of organisations it was only the two Scottish ones – and indeed the SDA in particular – which structured their activities sectorally to any significant extent.

9.4. Sectoral Strategies for Regional Development

The sectoral initiatives of the SDA constituted a meta-policy in the sense that they reflected a particular way of organising the development and implementation of regional policy: information gathering in selected industries formed the basis for tailor-made strategies which combined a range of different policy instruments in order to address sector-specific problems. But at the same time the sectoral initiatives were also seen as an integrated part of the Agency's overarching project of modernisation of the sectoral structure of the Scottish economy, and hence it would not be unreasonable to expect that the SDA had primarily nurtured modern high-tech areas while traditional manufacturing industries such as engineering and textiles had been given less attention. Add to this the general attitude of the Thatcher governments towards traditional (unionised) forms of manufacturing, and it could be expected that this high-tech bias

113 IDS 1987b Ch. 8.

114 Cf. Section 7.4. Expenditure on e.g. environmental upgrading of particular localities for tourism purposes could be seen as an exception (cf. Hall Aitken Associates 1995 pp. 72ff).

115 Lawless 1988, Totterdill 1989, Urry 1990, Cochrane & Clarke 1990.

would become even more pronounced in the 1980s. In short, the original modernising impetus of the Agency was allegedly amplified by party-political preferences in the 1980s, and thus by combining well-rehearsed discursive positions from the SDA and its political sponsors a perfectly believable picture emerges in which the sectoral initiatives were bound to neglect the plight of the traditional backbone of the Scottish economy. The aim of this concluding section will be to review the balance between continuity and change, and indeed between external and internal influences, in the development of the Agency's sectoral initiatives.

One of the most conspicuous features of this area of SDA activity has been the high degree of continuity which is in evidence: a sectoral approach was introduced in the industrial strategies of the Agency in the late 1970s, and it remained an important principle of policy organisation all the way through to the early 1990s. For external consumption the importance of the approach was downplayed on certain occasions, i.e. in the wake of the change of government in 1979 and again following the critical 1986 government review, but still the incremental development of sector-based organisational structures within the SDA continued while the financial resources allocated to sector work kept growing until the late 1980s, and the continuous priority accorded to this meta-policy is noteworthy for several reasons. *Firstly*, it makes the Agency stand out in a British context because it was the most prominent example of an RDA which not only adopted a sectoral approach, so widespread within industrial policy at large in the 1970s, but also stuck to it when it had clearly fallen out of favour with central government. Moreover, neither central nor local government was capable of pursuing similar policies in the English regions – indeed the 1986 government reviewers record a laconic "NIL" in their listing of corresponding policy measures in regions south of the border[116] – and as the WDA had decided to organise its activities along other lines, Scotland was in effect the only part of Britain in which a sectoral approach played a major role in regional development policy. And *secondly*, these activities continued despite the major reservations aired by its sponsor department in the 1986 review, and thus sectoral initiatives must be seen as a policy area in which the development body was capable of defending its arm's-length position in general and, indeed, in particular its project-driven mode of operation and specifically the importance of confidence-building expenditure on external consultants.

A similar picture of continuity also emerged when the sectoral initiatives were analysed on a more detailed level. On the one hand strategies were always very diverse, and even the 'generations' of initiatives

[116] IDS 1987a p. 136.

identified on the basis of organisational features did not bring with them identical policy prescriptions: electronics and biotechnology targeted different types of firms, the advisory services devised for printing were different from the one-off projects initiated in relation to forestry, and attempting to bring the extremely heterogenous group of large-scale initiatives in the mid-1980s onto one formula would clearly be more than a little difficult. On the other hand no clear-cut breaks have been identified with regard to either of the sectors targeted or the policy instruments employed; in practice both the initial selection of industries and the ensuing policy measures would appear to have been driven by concerns for the specific threats and opportunities faced by firms in particular sectors.

Continuity is also the best way to describe the sectoral priorities entailed in this area of SDA activity because while both modern and traditional industries were attended to throughout the period from 1975 to 1991, the analysis has also shown that the sectoral initiatives of the SDA *did* in fact entail a pattern of resource allocation which to some extent prioritised modern industries rather than traditional ones. This was of course fully in accordance with the Agency's overriding aim of modernising not just individual sectors but also the sectoral structure of the Scottish economy as a whole, but it was also a form of selectivity that was likely to be politically controversial because in employment terms priority was given to generating future jobs in nascent sectors rather than to safeguarding existing jobs in declining sectors. In practice this sectoral bias did clearly *not* imply that traditional industries were being neglected altogether, merely that the allocation of organisational and financial resources was skewed in favour of modern industries, but a variety of mutually reinforcing factors ensured that the sectoral bias could be perceived or construed as being much more pronounced. From the outset the sequence of major sectoral initiatives meant that modern industries were pushed to the forefront in the late 1970s and early 1980s, and electronics in particular came to embody this new departure in regional policy, not just because it was the first major sectoral initiative which achieved some measure of success, but also because in institutional terms the electronics strategy was a market-based one which required constant high-profile publicity in order to market Scotland as a premier location for inward investment. Contrary to this the strategies pursued in most other sectors, modern as well as traditional, relied primarily on attempts to establish inter-firm networks or improve the internal hierarchy within individual firms in order to strengthen competitiveness. Moreover, in addition to being inherently more visible, the electronics strategy also relied on policy instruments – financial subsidies and place promotion – that historically had been at the very core of British regional policy and hence had a reputation for some degree of effectiveness. Contrary to this the more innovative informational and organisational instruments employed *vis-à-vis* traditional industries – e.g. networks for generic marketing of textile products or development of the strategic capabilities of engineering firms – not only had a lower public profile within Scotland but also still remained to prove their effectiveness as tools of regional policy. Finally the existence of a pronounced sectoral bias did of course sit well with the dominant party-political discourse of especially the 1980s which attempted to paint Labour as the party of the past in general and of past industries in particular while in contrast the Conservatives portrayed themselves as the party of new economic departures, and thus inherent features of the development of SDA policies could readily be interpreted in a way that fitted the dominant political perspective from 1979 onwards.

All in all a detailed analysis of the sectoral initiatives does, however, not lend support to an external-revolution interpretation of the history of the SDA: continuity dominated policy design and implementation within this field of activity, and attempts to exert external influence were belated, rather limited in reach and as much inspired by clashing organisational cultures as by party-political discourse – and, of course, ultimately not particularly successful in making the Agency change its ways. On the contrary the sectoral initiatives would seem to be an area of activity which can best be accounted from an internal-evolution perspective: through the gradual extension of this meta-policy from the first major initiative in electronics to the much more comprehensive coverage achieved in the late 1980s, the basic idea of extensive sectoral analysis leading to the development of tailorised strategies was still maintained. But given the context of the sectoral initiatives – the changing discursive terrain, party-political positioning, and developments in other areas of Agency policy such as the symbolic downgrading of industrial investment at the expense of inward investment promotion – it is perhaps less surprising that the illustrious high-tech beginnings of the sectoral approach eventually for many observers came to overshadow the extensive but more low-profile efforts in relation to more traditional segments of the Scottish economy.

CHAPTER 10

Promotion of Inward Investment

Traditional central government regional policies had primarily used financial incentives in their attempt to redistribute investment and jobs, and in the prolonged period of economic growth after the Second World War the UK, itself a major overseas investor, received a major share of US subsidiaries, many of which were set up in areas designated for regional support.[1] American investment came 'voluntarily', driven by access to British and wider European markets, making use of existing pools of available industrial labour and whatever government subsidies were available, as no major concerted effort was made at the national level to promote either Britain or the designated Assisted Areas to potential foreign investors. In the 1970s this passive promotional stance was, however, given up under the combined pressures of persistent economic crisis and the increasing international competition for what was now a decreasing number of footloose investment projects, and in order to maintain Britain's status as the 'preferred location' for especially American foreign direct investment (FDI), national promotional efforts increased while at the same time regional actors also began to see overseas promotion as an important and legitimate activity, offering to assist potential investors with the many practicalities involved in setting up production. The 1970s saw the birth of what has been dubbed the 'inward investment service class':[2] an expanding network of professionals engaged in trying to ensure their part of the world – be it Britain, Scotland, or Cumbernauld New Town – the largest possible share of investment from abroad.

Also the SDA became engaged in promotion of inward investment, and this particular policy area became an integral part of the external-revolution interpretation of the Agency's development. The 1980 guidelines issued by the Conservative government officially downgraded industrial investment to an auxiliary status and instead elevated promo-

[1] See e.g. Hamilton 1987, Young *et al.* 1988 pp. 46ff, Bailey *et al.* 1994, Brown & Raines 1999, Raines 1999.

[2] Phelps & Tewdwr-Jones 2001.

tion to its primary industrial function,[3] and this document was therefore seen as heralding a shift away from focusing on indigenous firms towards a happy-go-lucky 'strategy' which was based on international 'glamour hunting'[4] and introduced "a perverse bias towards multinational branch plant investment in the electronics industry" that attempted to "out-Thatcher Thatcher".[5] In contrast to this external-revolution perspective, others have argued that the involvement of the SDA in inward investment has been characterised by continuity and evolutionary change, and, more specifically, that the role of promotion in the 1980s has been grossly exaggerated and that changes taking place have primarily been driven by an internal learning-by-doing process that gradually refined the Agency's approach.[6] In short, the extent and direction of change in this policy area is disputed among existing interpretations of the development of the SDA.

In addition to these general issues there are, however, controversies specific to this policy area which are likely to have influenced both its development and the ways in which it was interpreted. On the one hand the physical nature of inward investment – a new productive facility locates in one particular geographical locality – points towards the potential for conflicts between different regions and tiers of government. On the other hand the desirability of inward investment has also frequently been questioned, either because external and especially foreign ownership was seen as unattractive *per se*,[7] because 'branch plants' were believed to be more vulnerable in times of economic crises, or because the plants are seen as impoverishing the quality of the Scottish economy through the preponderance of semi-skilled 'screw-driver' assembly operations and the absence of higher-order functions such as R&D or marketing.[8] Either way, it has often been stressed that the prominence of inward-investment related questions has been enhanced by their 'media-friendliness' and 'marketability':[9] success in this policy area tends to be

3 SEPD 1980a p. 1.

4 An expression used by Ewan Marwick (24.7.90).

5 Danson *et al.* 1989a p. 562. More temperately worded versions of the same argument can be found in Danson 1980 p. 13, Dunford 1989 p. 116, Danson *et al.* 1990b, Rich 1983 p. 282, Douglas Harrison 19.7.90, Helen Liddell 21.6.90, Alf Young 5.6.90, cf. Ewan Marwick 24.7.90.

6 Todd 1985, Hood 1991a, Neil Hood 22.6.90, Douglas Adams 18.7.90.

7 See e.g. The Scottish Labour Party's defence of the indigenous profile of the proposed RDA (*Glasgow Herald* 29.8.74), the evidence of STUC to Committee on Scottish Affairs (1980 vol. 2 pp. 261ff), Standing Commission 1989 pp. 46ff, cf. Firn 1975, McDermott 1979.

8 See e.g. Hood & Young 1976, Young *et al.* 1994a, Amin *et al.* 1994.

9 Todd 1985 p. 35, Neil Hood 22.6.90, George Mathewson 9.7.90, James Williamson 14.6.90, Isobel Lindsay 6.6.90.

lumpy – involving large number of jobs and tangible change like erection of new factory space – and this makes it attractive as a platform for media exposure in order to raise the profile of the promotional body and its political sponsors.

The analysis of the SDA's promotion of inward investment will be based on published materials, including the annual reports of the Agency and its associated promotional body *Locate in Scotland* (LIS), complemented by interviews with senior executives within the organisation involved with promotion of inward investment. In illuminating Scottish developments, and indeed its British context, two extensive parliamentary enquiries investigating key aspects of inward investment have been particularly useful because they have generated a series of detailed policy statements and contain extensive cross-examination of key members of the 'inward investment service class' both in the beginning and towards the end of the 1980s.[10] The existing academic literature focuses primarily on economic aspects of inward investment, especially trends in the spatial distribution of FDI and the long-term merits of the presence of 'branch plants' in peripheral regions, and hence it will be most useful in the latter sections of this chapter. Again the text falls in three parts: first the design of inward investment promotion in Scotland is explored, including the resources and strategies involved (Section 10.1), then the implementation of promotional strategies is examined (Section 10.2), and finally the broader British policy context is surveyed, both in terms of inward investment promotion on the national level and the activities of RDAs and other sub-national actors (Section 10.3).

10.1. Design

Origins of Inward Investment Promotion

While the UK had a long history of investing abroad, the country itself turned into a major recipient of FDI after the Second World War when the US became the world's largest outward investor by some distance.[11] Britain assumed the role of the 'preferred location' of American FDI, accounting for nearly 60% of US investment in Europe in 1961 and falling gradually to just below 40% in the following decade.[12] In the

10 Committee on Scottish Affairs 1980, Welsh Affairs Committee 1989.

11 Britain is estimated to have been the source of 45.5% of the world's FDI stock in 1914, while the USA is estimated to have been the source of 52% in 1960 (Hamilton 1987 p. 169).

12 Research by Hood & Young quoted in Committee on Scottish Affairs 1980 (vol. 2 p. 33).

light of this long-standing importance of cross-border direct investment for the British economy, it is hardly surprising that the policy stance taken by British government in the postwar period, Conservative as well as Labour, was a liberal one, supporting both inward and outward movement of capital,[13] and by the mid-1970s foreign-owned firms accounted for nearly one million manufacturing jobs in the UK, equal to more than 10% of employment in this sector and predominantly of US origin.[14] Proactive selling of Britain as a location for inward investment had not been undertaken:[15] investors themselves would have to discover the attractions of e.g. a skilled industrial work force and government subsidies in the Assisted Areas.

From the 1940s onwards FDI came to play a significant role also in the Scottish economy. Growing fast especially in the 1960s and 1970s, the Scottish share of inward investment into the UK was more than 35% in the mid-1970s, and by 1975 non-UK firms accounted for more than 15% of manufacturing employment in Scotland, including high-tech multinationals such as IBM, Hewlett Packard, Motorola, and National Semiconductor.[16] If Scotland received a relatively large share of foreign investments into Britain, this may at least in part have been due to the fact that the region adopted a more proactive promotional stance than the national level, because since the 1950s the tripartite SCDI had promoted Scotland as an investment location, especially to American business executives with Scottish family links.[17] While SCDI had originally enjoyed the advantage of being one of the first organisations in Western Europe to engage in this type of activities, the 1970s saw the rise of especially the Irish Development Authority as a major government-sponsored competitor,[18] and as an increasing number of American investment projects began to end up on the other side of the Irish Sea, improving the promotion of Scotland came to be seen as important by many actors. Moreover, even the Irish organisational set-up attracted interest because its 'one-door approach', combining promotional and grant-giving powers within one organisation, was perceived to be able to operate both faster and more efficiently than the Scottish set-up where a voluntary organisation was responsible for informing potential investors

---

13 For a perceptive historical analysis of British policy towards FDI, see Bailey *et al.* 1994.

14 Hamilton 1987 p. 182.

15 Cf. Section 10.4.

16 Young 1984 pp. 96f, Randall 1985 pp. 252f, Taylor 1986 pp. 20f, Walker 1987 pp. 57f, Dunford 1989 pp. 81ff.

17 SCDI evidence to Committee on Scottish Affairs (1980 vol. 2 p. 168).

18 See e.g. SDA/DCC (78)29 pp. 1ff, Rich 1983 p. 282, Young 1984 p. 105, Love & Stevens 1988, Iain Robertson 11.7.90, Neil Hood 22.6.90.

about the attractions of the region alongside sub-regional actors extolling the virtues of particular localities, grants had to be negotiated with the Scottish Office, and the availability of industrial property was the responsibility of the Scottish Industrial Estates Corporation. In short, the prominence of foreign investment within a regional economy where indigenous manufacturing was contracting meant that a slow-down in the number of incoming firms in the mid-1970s[19] could readily be translated into political demands for a more efficient handling of inward investment in Scotland.

The various blueprints for a new RDA produced in the 1970s did, however, differ with regard to their attitude towards inward investment. On the one hand the proposal for a state holding company put forward by the Scottish Council of the Labour Party cast multinational companies in the role of the villain trying to "evade" attempts to direct investment into the designated Assisted and actively promoting FDI was not contemplated.[20] In contrast to this most other actors saw inward investment as largely beneficial to the regional economy. Even the West Central Scotland Plan, primarily focusing on measures aiming to make indigenous firms more competitive, also recognised the importance of making promotion of inward investment more effective through coordination between stakeholders,[21] and the SCDI, the organisation which in the early 1970s was the main promoter of Scotland overseas, specifically argued for the need to create an Irish-style one-door approach with publicity and grant-giving powers placed within one organisation in order to succeed in an increasingly competitive environment.[22] When the Labour Party returned to the Scottish Office in 1974, its position was closer to the SCDI than to its own party organisation: the manifesto for the February general election had included a commitment to using inward investment as a means to create additional employment,[23] and the consultation paper on an RDA for lowland Scotland proposed one of its functions to be "in concert with existing development bodies, promoting Scotland as a location for new investment".[24] From the outset the role of inward investment promotion within the new organisation was in other words potentially controversial as a strategic priority, and those who saw inward investment as beneficial to the regional economy generally

19 While the growth of employment in non-UK firms in Scotland had grown by around 50% in the five-year periods 1958-63 and 1963-68, the equivalent figure for 1973-78 was only 3% (Randall 1985 pp. 252f).

20 Labour Party Scottish Council 1973 p. 6.

21 West Central Scotland Plan Team 1974a p. 55.

22 Scottish Council Research Institute 1974 pp. 37ff.

23 *Glasgow Herald* 15.2.74.

24 SEPD 1975 p. 2, cf. *Glasgow Herald* 1.2.75.

recognised the organisational difficulties involved in bringing about a more competitive Scottish effort.

Authority to Organise Promotion

The 1975 parliamentary act did not explicitly mention promotion of inward investment as a function of the new development body, but as the Agency was allowed to operate "in Scotland or elsewhere" in pursuit of its functions and objectives,[25] and the possibility of being given responsibility for administration of regional selective assistance was included in the original act,[26] the contours of an Irish-style one-door approach would seem to emerge.

Despite having been authorised by parliament, the Agency proceeded rather cautiously in the early years. Promotional efforts had hitherto been undertaken by organisations that could not readily be incorporated, namely local authorities which had gained a statutory right to undertake international promotion only a few years earlier, and the tripartite SCDI that promoted Scotland as a location for inward investment supported by a government grant for this particular purpose. In November 1976 the Scottish Office announced the transfer of its promotional funding from the latter to the SDA,[27] and at the same time it was agreed that for a transitionary period of two years the Agency would provide financial support for the continuation of the SCDI's promotional efforts while its own capabilities were gradually being developed.[28] Moreover, the SDA also inherited a system of inter-organisational coordination amongst Scottish actors with an active interest in FDI, relaunched as the Development Consultancy Committee (DCC) at the political level and with a parallel meeting of development officers at the administrative level,[29] and, perhaps in order to build informal authority through tangible action, the Agency opened an office in New York in March 1979 and announced plans to establish a physical presence elsewhere in the US and Europe.[30] Although this move had been cleared with its political sponsors at the Scottish Office,[31] an independent Scottish presence overseas was unlikely to be popular on the UK level where the DTI and the

---

25 SDA Act 1975 Section 2 Subsection 3.

26 SDA Act 1975 Section 5. This possibility was never used in practice, cf. Chapter 8.

27 *HCPD* 19.11.76 cols. 765ff, cf. *Glasgow Herald* 18.11.76, 20.11.76.

28 SDA 1977 p. 31, cf. SDA and SCDI evidence to the Committee on Scottish Affairs (1980 vol. 2 pp. 3f, 35, 163ff).

29 SDA and SCDI evidence to the Committee on Scottish Affairs (1980 vol. 2 pp. 38, 163, 171), cf. Section 6.2.

30 Committee on Scottish Affairs 1980 vol. 2 pp. 37ff, SDA 1979 p. 45.

31 Scottish Industry Minister Alex Fletcher in evidence to the Committee on Scottish Affairs (1980 vol. 2 pp. 322f), Gregor Mackenzie 15.10.90.

Foreign Office had recently increased their promotional efforts through the London-based *Invest in Britain Bureau* (IBB) and the consular services abroad.[32] At the same time the Agency's publically advocated an Irish-style one-door approach,[33] combining promotional and grant-negotiating powers within the same organisation, something which was likely to have raised eyebrows in the SEPD and the Scottish Office which handled the financial negotiations with inward investors concerning discretionary regional grant support. Despite formally having the power to promote Scotland as a location for international investment, its attempts to improve the Scottish market position would by the end of the 1970s seem to have brought the SDA into conflict with just about everyone else with a stake in promotional activities: from individual localities wanting to highlight their particular attractions, via its sponsors in the Scottish Office, to the DTI as the lead department in the UK regional policy network, not to mention the Foreign Office and its world-wide network of consular offices.

From the perspective of the Agency the fact that the new Conservative government designated promotion as one of the Agency's "principal functions"[34] could be a double-edged sword: with more importance attached to inward investment, the full authority of government could be brought to bear on simplifying a complicated institutional set-up, but at the same time technocratic considerations about how best to tackle foreign investors were also more likely to be influenced by territorial politics such as assumptions about the ideal relationship between the UK and its constituent parts. Fortunately, this process of organisational realignment within the governance of inward investment promotion came to be an unusually transparent one, because in October 1979 the recently reconstituted Select Committee on Scottish Affairs decided to subject the attraction of inward investment to Scotland to parliamentary scrutiny,[35] thereby maintaining a fluid situation regarding the SDA's authority to undertake overseas promotional activities until the issue was finally settled by the Secretary of State for Scotland in early 1981.[36]

---

32 Sir William Gray 19.10.90, Hogwood 1982. On IBB, see Section 10.3.

33 See e.g. SDA 1979 pp. 44f.

34 SEPD 1980a p. 1.

35 The reasons given for launching an enquiry were concerns over the limited results in terms of incoming firms achieved in the late 1970s under the existing institutional set-up (Committee on Scottish Affairs 1980 vol. 1 p. 1, Neil Hood 22.6.90, cf. Gavin McCrone 20.6.90), although it has also been seen as an attempt to limit the SDA's scope for action (e.g. Alf Young 5.6.90).

36 Formally announced in the beginning of March (Committee on Scottish Affairs 1981), the decision had leaked to the press already in January (Hood & Young 1982).

In the first half of 1980 the Committee on Scottish Affairs conducted an extensive inquiry which produced a detailed catalogue of, and a remarkable degree of consensus about,[37] the weaknesses of the promotional set-up which had developed in Scotland in the late 1970s.[38] *Firstly*, it was generally agreed that Scottish efforts would benefit from improved coordination at the regional level, because fragmentation through the uncoordinated efforts of individual localities was not only wasteful duplication but could potentially scare off investors through relentless doorstepping by e.g. local authorities.[39] *Secondly*, the working relations between the regional and national level were seen as unsatisfactory, albeit for rather different reasons. Especially those situated on the UK level argued that separate Scottish promotion through the SDA would undermine the efforts of IBB to exercise 'fairness' and 'impartiality' towards regions across Britain and could ultimately damage the UK's position *vis-à-vis* e.g. the Republic of Ireland,[40] in effect applying the fragmentation argument at a higher spatial level. Others maintained that "Scotland is different" and hence merited separate promotion,[41] and the SDA itself argued that, unlike e.g. a new town or even the UK, Scotland was a "marketable unit abroad [...] with an international identity".[42] *Thirdly*, concerns were also voiced about the quality of UK-level efforts in the field of inward investment: the length of the line of communication from a British consulate in the US to the SDA – having to pass through the IBB/DTI and the Scottish Office on the way – was seen as hampering an effective Scottish response to expressions of interest by American firms,[43] and the marketing skills of the front-line civil servants in the consular service were repeatedly called into ques-

37 The report of the committee included not only the conclusions adopted by its Conservative majority, but also the draft conclusion proposed by its Labour chairman, and thus the parallel nature of the argument can readily be established, cf. below.

38 For overviews of the committee inquiry, see Hood & Young 1982, Hogwood 1981, 1982, 1987a.

39 The Committee on Scottish Affairs came to this conclusion (1980 vol. 1 p. 33) after hearing parallel evidence from SCDI, COSLA, the Scottish CBI, and the SDA (1980 vol. 2 pp. 38, 46, 165, 369ff) – while the local authority sponsored development body appearing before the committee strengthened the impression of territorial parochialism (1980 vol. 2 pp. 275ff).

40 Committee on Scottish Affairs 1980 vol. 2 pp. 67-88.

41 SCDI in evidence to the Committee on Scottish Affairs (1980 vol. 2. pp. 163ff, 183). Civil servants from the Scottish Office also argued in favour of the Agency maintaining its offices abroad as a 'supplement' to its use of the UK consular services (1980 vol. 2 pp. 16ff).

42 Chief Executive Lewis Robertson in evidence to the Committee on Scottish Affairs (1980 vol. 2 p. 203).

43 Committee on Scottish Affairs 1980 vol. 2 pp. 55, 69, cf. Hogwood 1987 pp. 5ff.

tion,[44] even by central government representatives, although the effect of 'gold-braided' consuls and government authority with some investors was recognised even by the SDA.[45] And *finally*, the practical arrangements governing the handling of investment cases within Scotland also left a good deal to be desired: SDA promotional staff could not discuss issues concerning the availability of discretionary regional grants with prospective investors because this was to be handled exclusively by the Scottish Office, confusion allegedly resulted from foreign executives being introduced to a host of different actors responsible for minor parts of the process, and again the perceived deficiencies of generalist civil servants with regard to marketing and business-related skills were emphasised.[46]

Despite political differences and positioning, the members of the committee also agreed about the nature of the measures needed to improve the international promotion of Scotland as an investment location:

- extensive planning and research in order to target relevant firms, and front-line staff on the ground abroad with marketing and industry-relevant skills,
- internal Scottish coordination should be improved, both between the SDA and local actors with regard to external promotion, and between the SDA and the Scottish Office when handling of individual investment cases, and
- the distribution of roles between the UK and Scottish level should be resolved in order to create a constructive working relationship between the SDA and higher-tier organisations.[47]

While the enquiry was under way progress was being made with regard to some of these issues: procedures for handling individual investors considering a Scottish location were improved through the institution of regular 'case meetings' between the Scottish Office and the Agency, and at the UK level coordination between the central government departments and agencies on the one hand and regional bodies like the SDA on the other was formalised through the setting up of the Committee on Overseas Promotion which would oversee both promotional efforts abroad and ensure that British regions were not competing against each

44 Chief Executive Lewis Robertson in evidence to the Committee on Scottish Affairs (1980 vol. 2. p. 203). DTI Under Secretary Binning even declared that "the particular skills required are not ones which come easily to civil servants" (1980 vol. 2 p. 68).

45 Chief Executive Lewis Robertson in evidence to the Committee on Scottish Affairs (1980 vol. 2 p. 45), cf. the parallel views of the Scottish New Towns (1980 vol. 2 pp. 213ff).

46 Committee on Scottish Affairs 1980 vol. 2 pp. 190ff.

47 Committee on Scottish Affairs 1980 vol. 1 pp. 33f, 39-45.

other by means of discretionary regional grants.[48] Still, a major outstanding issue was how the SDA should operate abroad, and here the Committee on Scottish Affairs split along party lines. The Conservative majority, and indeed Scottish industry minister Alex Fletcher, preferred secondment of Agency staff through the IBB to consular posts abroad in order not to infringe the territorial integrity of the UK, "not least the need for a coordinated UK regional policy",[49] while the Labour minority promoted an improved version of status quo with a continued independent presence of the SDA overseas through separate offices and the setting up of a "permanent joint body of SDA and SEPD staff for handling would-be investors in Scotland".[50] After seven months the Conservative Secretary of State George Younger overruled his parliamentary colleagues and decided in favour of a solution much like the one advocated by the Labour members of the committee: *Locate in Scotland* (LIS) was established as a joint venture between the Scottish Office and the SDA, guided by the Secretary of State in conjunction with top executives from the Agency and the Scottish Office, its first director being a civil servant on secondment from SEPD, and, presumably in order to soothe unionist sensibilities, Agency overseas offices to be reviewed after an initial trial period of two years.[51] The reasons for deciding to maintain a separate Scottish presence on the international market for FDI attraction were undoubtedly many: the government itself claimed its decision reflected the need to counter increased international competition for footloose investment – echoing the public and vociferous campaign of the SDA[52] – and that, an ingenious and almost federalist argument, the new institutional set-up would allow Scotland to make a fuller contribution to the task of maximising the flow of inward investment into the UK.[53] But in weighing up technical efficiency against perceived threats to national unity and eventually supporting a solution advocated by the Labour opposition, George Younger also positioned himself as a pragmatic defender of Scottish interests, something which could be useful in a situation where the electorate in this part of the UK had voted overwhelmingly for parties other than the Conservatives.

48 Committee on Scottish Affairs 1980 vol. 2 pp. 190ff, 207f, 338, cf. the discussion below.

49 Committee on Scottish Affairs 1980 vol. 1 p. 33.

50 Committee on Scottish Affairs 1980 vol. 1 p. 45. The minority proposal took up many of the ideas put forward by professor Neil Hood of Strathclyde University in his evidence to the committee (1980 vol. 2 pp. 135ff, 150-62).

51 Committee on Scottish Affairs 1981.

52 E.g. SDA Chairman Robin Duthie's evidence to the Committee on Scottish Affairs (1980 vol. 2 pp. 290ff), Glasgow Herald 29.8.1980, cf. Hogwood 1981, 1987a.

53 Committee on Scottish Affairs 1981, cf. Neil Hood 22.6.90.

Unsurprisingly, the review of the SDA's offices abroad turned out favourably, and thus since February 1982 no serious questions appear to have been raised about the Agency's authority to operate abroad in pursuit of foreign investment.[54] In relation to sub-regional actors, local authority promotion abroad effectively became legally subject to control by LIS,[55] and although the New Towns retained an international remit, being sponsored by the Scottish Office undoubtedly helped to ensure cooperative behaviour. Future government reviews concentrated exclusively on technical or strategic issues, and thus the architecture of Scottish inward investment promotion remained intact until the early 1990s. Acceptance of the UK-level coordinating framework of the Committee of Overseas Promotion and closer cooperation with its sponsor department within LIS had eventually bolstered the SDA's international promotion authority, albeit at the price of closer liaising with central government departments and agencies, not least its political sponsors at the Scottish Office.

Organisational and Informational Resources

Attention can now turn to exploring the ways in which organisational and informational resources were mobilised in the quest to bring more FDI to Scotland, distinguishing between the early years of the late 1970s and the LIS period from 1981 onwards.

In the late 1970s the SDA was gradually assuming the role as the main promoter of Scotland to foreign firms and formally in charge of coordinating the efforts of subregional actors through the inherited voluntary set-up relaunched as the DCC,[56] as illustrated by Figure 10.1. Although increased publicity was seen as important in the beginning, a more targeted approach gradually began to emerge, partly inspired by the electronics sectoral initiative and involving extensive research in order to identify dynamic industries with firms that might be thinking about investing abroad and could conceivably consider a Scottish location,[57] and the decision to establish a permanent presence abroad could

54 SDA 1982 pp. 18, 70. In mid-1984 an inter-departmental review, presumably within the framework of the UK-level Committee of Overseas Promotion, again recommended that a separate SDA presence be maintained, but it is doubtful whether the future of the offices was genuinely at stake at this point, and the newspaper reporting comes across as either rather sensationalist or orchestrated by the Secretary of State to underline his Scottish credentials (see *Glasgow Herald* 4.7.84, 5.7.84, 11.7.84, 27.7.84, 1.8.84, cf. Naughtie 1985 p. 27).

55 Keating & Boyle 1986 p. 23, cf. Neil Hood 22.6.90.

56 Committee on Scottish Affairs 1980 vol. 2 pp. 163ff, cf. Craig Campbell 16.7.90 and evidence by the Scottish New Towns to the Committee on Scottish Affairs (1980 vol. 2 p. 223).

57 John Firn 15.6.90.

be interpreted as a way of raising the Agency's general profile as well as being able to target specific potential investors.

Figure 10.1 The organisation of inward investment promotion in Scotland

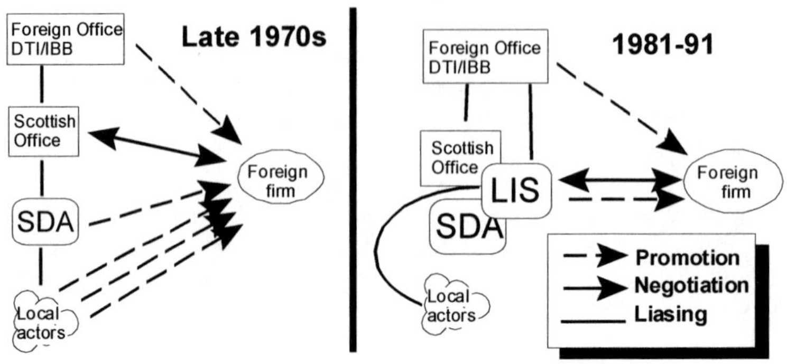

Source: Committee on Scottish Affairs 1980 vol. 2 pp. 1-66, IDS 1987a pp. 59f, Welsh Affairs Committee 1989 pp. 160-73.

From the perspective of foreign firms the post-1981 set-up would certainly seem to be much simplified, as illustrated by Figure 10.1, because coordination of Scottish efforts had been internalised within LIS. The work of the new joint entity was in the beginning overseen by the Secretary of State for Scotland together with senior executives from the two bodies, and in the late 1980s this evolved into a separate Steering Committee headed by the minister for industry at the Scottish Office,[58] at which point LIS also began to publish its own annual reports as a supplement to the FDI-related section of the main SDA report.[59] The first two directors of LIS both had a civil service background,[60] but this changed after LIS-head Iain Robertson had become Agency Chief Executive in 1987, because the new director of the inward investment body, professor Neil Hood of Strathclyde University, was an academic specialising in the study of inward investment and hence could be seen as firmly rooted in activities typical of the Agency rather than a department of central government.[61] LIS was physically based at the premises

58 Committee of Public Accounts 1981 p. 24; SDA 1983 pp. 76f, 1988 pp. 76f.

59 Locate in Scotland 1988ff.

60 *Glasgow Herald* 27.4.81, Iain Robertson 11.7.90.

61 Neil Hood had of course proposed a blue-print for organisational realignment to the Committee on Scottish Affairs strongly resembling the solution eventually adopted,

of the SDA in Glasgow, and its project-oriented style of operation was far removed from the bureaucratic rigour of mainstream government departments, but despite the high-profile nature of its activities and a growing number of overseas offices and representations, the size of the inward investment team remained fairly small throughout the 1980s.[62] In terms of human resources the new organisation seems to have been able to combine the strengths of its constituent parts in an effective way.[63] Basing promotional activities on the extensive information-gathering capacities of the Agency with its sectoral divisions and offices abroad made it possible to target industries and firms with overseas growth potential much more precisely, but having these professional members of the 'inward investment service class' backed up by the gravitas and authority of government ministers made it possible to use different strategies of persuasion,[64] and thus by integrating grant-negotiating civil-service experience into the organisation the simplicity and efficiency associated with the Irish-style one-door approach had been transferred to a Scottish setting.

If LIS had been an unforeseen result of a political conflict into which an enterprising academic was able to feed suggestions for professionalisation, it nonetheless "turned out to be a very good solution"[65] in the sense that subnational actors came increasingly to appreciate the professional efficiency of the new body in highlighting also their particular localities,[66] and relations to other British organisations appears to have become reasonably cordial.[67] This consensus of contentment would seem to indicate that a win-win situation had been created for key actors: the SDA could pursue inward investment as part of its sectoral and economic development strategies, the Scottish Office maintained

---

cf. Welsh Affairs Committee 1989 p. 16, Neil Hood 22.6.90, and the discussion above.

62 Welsh Affairs Committee 1989 pp. 160-73, Neil Hood 22.6.90. The total number of staff had reached 58 in the late 1980s, including the seven offices overseas (National Audit Office 1989 p. 5, SDA 1980-91).

63 SDA evidence to the Welsh Affairs Committee (1989 pp. 160-73), cf. the programmatic Hood & Young 1981, 1982.

64 The annual reports of both the SDA and LIS systematically exploit plant openings as photo opportunities, both as a way of having incoming investors endorse Scotland as location by their example, and as a vehicle for raising the profile of Scottish Office ministers.

65 Iain Robertson 11.7.90.

66 In the first years of the LIS set-up references can still be found in the SDA Annual Reports to the need for improving coordination (SDA 1982 pp. 70f, 1984 p. 82), but in the late 1980s even the New Towns were very supportive of the division of Labour within Scotland (Welsh Affairs Committee 1989 pp. 174-92).

67 Cf. Section 10.3.

control of discretionary regional grant assistance and hence ensured that UK-level coordination practices were observed with regard to financial assistance, and useful political publicity could be obtained by ministers being associated with 'good news' such as large-scale job creation – as long as LIS could continue to produce the only good that counted, namely incoming investors.

Changing Aims and Methods

Like in other policy areas that did not involve the SDA investing in productive assets, the parliamentary act gave few clues as to how the organisation might go about its tasks, except for the possibility of authorising the Agency to handle discretionary regional grants which was never used in practice before being removed by the incoming Conservative government.[68] Also in this field of activity the SDA was in other words given a proactive role in the development of aims and methods, and having officially assumed responsibility for inward investment in late 1976, the Agency gradually moved towards prioritising international promotion in statements of corporate strategy. The first annual report simply noted that despite indigenous industry being its main focus, the importance of incoming firms in the regional economy meant that it would be "endeavouring to secure a continuing flow of investment from sources outwith Scotland",[69] and only in 1978 was "promotion of Scotland as an attractive and profitable country to locate new industrial investment" described as "an important function of the Agency", and it was announced that research would be instigated to allow the SDA to contact individual companies "with attractive development and investment proposals".[70] Apart from emulating successful Irish practices, additional impetus for this targeted and proactive approach seems to have come primarily from the sectoral work initiated within Edward Cunningham's Strategic Planning Directorate, and in particular from the first electronics study which involved detailed examination of dynamic segments and firms within the industry.[71] The Agency had in other words begun to leave behind the approach inherited from the SCDI which primarily had tried to exploit networks of Scottish expatriates in the US and appears to have emphasised – alongside haggis, bagpipes and a wee Scotch – more traditional industrial strengths in e.g.

68 1980 Industry Act Section 8 Subsection 2.

69 SDA 1976 par. 44.

70 SDA 1978a pp. 26f, cf. SDA 1978 pp. 43f.

71 E.g. Edward Cunningham 13.6.90, John Firn 15.6.90. The former had himself publically stressed the importance of inward investment before it became priority business for the SDA (1977 p. 10).

engineering.[72] But despite this new strategic focus the rationale for pursuing FDI still remained largely implicit, although, speaking at the 1977 annual conference of the Regional Studies Association, Director of Strategic Planning Edward Cunningham did argue that "the potential of this source of investment in terms of employment, management techniques and technology cannot be ignored",[73] thereby underlining that the attraction of FDI was not just about quantitative expansion of economic activity but part of a much broader modernisation remit.

The change of government in 1979 saw promotion of inward investment elevated to a core function of the SDA via the new investment guidelines,[74] and through to the early 1990s the annual reports of the Agency maintained this impression through pictures of government ministers making announcements, shaking hands, breaking ground, revealing plaques, so that this aspect of Agency activity effectively replaced industrial investment – reported in a similarly detailed fashion in the 1970s – as its most visible and tangible activity.[75] Given the collaboration within LIS, it is hardly surprising that no evidence has been found of disagreement between the SDA and its sponsor department with regard to the aims and methods of promotional activities: comments in statements of corporate strategy were brief and general,[76] and the 1986 government review dedicated little more than two pages to the function which did *not* construct a market-failure rationale for promoting FDI but simply noted the "pressure to match the scale and professionalism of inward investment promotion of other potential host countries" before cursorily dismissing even considering alternative institutional set-ups and concluding that "we are impressed by the approach and professionalism of LIS".[77]

From this consensual perspective, the rationale for promoting inward investment included employment, sectoral change, introduction of new technologies and management practices, and often also knock-on effects like subcontracting opportunities for indigenous firms and the development of e.g. Just-In-Time capabilities in order to meet the exacting standards of foreign multinationals.[78] But at particular points in time additional, and even more specific, objectives came to the fore: in the

72 Sir William Gray 19.10.90, Gavin McCrone 20.6.90, cf. Alf Young 5.6.90.

73 Cunningham 1977 p. 10.

74 SEPD 1980a.

75 See the inward investment sections of SDA 1980-91, cf. Locate in Scotland 1988-91.

76 SDA 1981 p. 60, SDA 1984a, SDA 1987a.

77 IDS 1987a pp. 59-61.

78 SDA evidence to Welsh Affairs Committee (1989 pp. 160-73), National Audit Office 1989, Neil Hood 22.6.90, cf. the inward investment sections of SDA 1980-91, LIS 1988-91.

early- and mid-1980s a recurring theme was completion of the 'silicon chain', i.e. making it possible to source everything needed for high-tech products within the region,[79] and especially from the late 1980s onwards increasing emphasis was placed on efforts to enhance the quality of investments by encouraging foreign firms to let their Scottish subsidiaries undertake higher-order functions such as R&D or marketing.[80] In order to achieve these objectives, it was essential to be able to identify growing firms in dynamic sectors which might consider investing in Scotland, and in terms of firm-level selectivity the strategy of LIS would seem to gradually evolve along several lines:[81] the initial sectoral focus on electronics was broadened to include other sectors with growth potential, more attention was gradually given to expansion of the facilities of foreign firms already located in Scotland by encouraging 'intrapreneurialism',[82] and the emphasis on the US as a source of investment was supplemented by promotional efforts aimed at firms in the Far East and continental Europe. The marketing strategy of LIS in other words clearly revolved around identifying what *specific* gains *individual* investors could expect from a Scottish venture, and being able to pursue this from early-stage enquiries to the negotiations finally bringing the project to the region. This tailorised form of promotion was, moreover, supplemented by more general efforts to build a general image of Scotland conducive to inward investment: foreign firms already operating were continuously used to underwrite the professionalism of LIS and the usefulness of the region as a location for productive activities,[83] in the early 1980s defensive attempts to dispel notions of the region as plagued by industrial strife were seen as important,[84] and the mid-1980s saw the rise of the more forward-looking *Silicon Glen* brand which projected an image of Scotland as a hot-bed for high-tech economic activities to the outside world and actors within the region.[85]

All in all the analysis would seem to suggest that the basic assumptions underlying the promotional strategies of the SDA remained stable

---

79 SDA 1985 p. 77, cf. Iain Robertson 11.7.90.

80 SDA evidence to Welsh Affairs Committee (1989 pp. 160-73), Neil Hood 22.6.90, Iain Robertson 11.7.90.

81 SDA 1982 p. 71, 1983 p. 77, 1984 pp. 81ff, 1985 p. 78, 1986 pp. 84ff, 1987 pp. 106f, 1988 pp. 74ff, 1989 pp. 47ff, 1990 pp. 20f, 1991 pp. 14ff; Locate in Scotland 1988-91, SDA evidence to Welsh Affairs Committee (1989 pp. 160-73), Iain Robertson 11.7.90, Neil Hood 22.6.90, Young 1984.

82 Hood & Taggart 1999.

83 See the inward investment and pictorial sections of SDA 1980-91, cf. Locate in Scotland 1988-91.

84 Committee on Scottish Affairs 1980 vol. 2 pp. 10, 59; SDA 1980 pp. 24f.

85 SDA 1983 pp. 76f, 1985 p. 7; IDS 1988b, John Firn 15.6.90, Douglas Adams 18.7.90, cf. Hargrave 1985, Dunford 1989, Young *et al.* 1988 p. 103.

from the late 1970s onwards – ensuring that foreign investors were aware of the specific attractions of Scotland compared to alternative investment locations – but that *within* this targeted and proactive approach to inward investment, concrete priorities in terms of sectors, firms, and countries of origin evolved in a gradual and uncontested manner.[86] In short, neither the turbulence surrounding the institutional set-up in the wake of the change of government in 1979 nor the ensuing close partnership in LIS with the Scottish Office lead to new strategic departures, but would instead seem to have reinforced existing thinking within the SDA. In order to be associated with the good news of incoming investment Conservative ministers backed a proactive and selective strategy for international promotion.

10.2. Implementation

The analysis of the implementation of the SDA's inward investment activities is undertaken in two steps: first the development of activities is pursued, and the nature of the policy instruments and modes of implementation examined in some detail, and then the outcomes are considered with regard to firm-level impact and institutional change.

Policy Instruments and Modes of Implementation

The SDA's strategy for international promotion has clearly been labour intensive – involving extensive research, individual approaches to potential investors, lengthy negotiations with firms considering to locate in Scotland – and thus expenditure figures should be able to give an indication of the development of the Agency's commitment of resources to this policy area.[87] Unfortunately, a consistent time series does not exist because until 1981/82 promotional expenditure was subsumed under a broader heading, but from the point where the joint body started operating the level of expenditure more than doubled in just five years and then stabilised in the late 1980s, as illustrated by Figure 10.2, mainly reflecting changing levels of activity at Glasgow headquarters.[88]

---

86 At the level of corporate strategy the analysis above does in other words lend support to the interpretation put forward by the two former LIS directors, Iain Robertson and Neil Hood, in their evidence to the Welsh Affairs Committee (1989 pp. 160-72).

87 Not much is known about the changing 'project load' of LIS, but according to internal LIS information the level of enquiries and visits to Scotland remained relatively stable from 1981 to 1986 (Dunford 1989 table 4.9).

88 Even during the inquiry of the Committee on Scottish Affairs in the early 1980s overseas promotional expenditure continued to increase: according to SDA evidence to the Committee on Scottish Affairs (vol. 2 p. 340) expenditure on international promotion in 1979/80 amounted to £m 0.863, i.e. in real terms 35% below the level recorded in 1981/82 (SDA 1982).

It is, however, also worth pointing out that in the 1980s the share of total SDA expenditure accounted for by inward investment promotion averaged only 5%,[89] and thus the political interest in this policy area did, much like in the case of industrial investment, not reflect the level of financial resources committed.

Figure 10.2 SDA expenditure on promotion

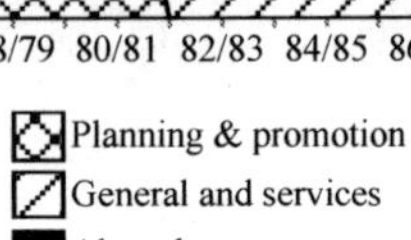

£m 1985/86.
Source: SDA 1977-91.

The nature of the policy instruments employed in the attraction of inward investment would seem to have varied between the different phases of the promotional process:[90] the *preparatory stage* depended primarily on the ability of the SDA to collect and process information by mobilising its organisational resources at home and abroad,[91] the *attractional stage* involved providing targeted firms with free information,[92] primarily tailored to the situation of the individual company but also in the form of general background information,[93] and the *negotiational stage* where the specific conditions associated with an inward

89 Calculated on the basis of SDA 1982-95.

90 For attempts to identify key stages of the promotional process, see e.g. Hood & Young 1981 pp. 40ff, Young *et al.* 1994b pp. 145ff, Brown & Raines 1999 pp. 32ff.

91 LIS evidence to the Welsh Affairs Committee (1989 pp. 160-73).

92 Promotional expenditure was overwhelmingly financed by public resources and receipts never exceeded 3% of expenditure (calculated on the basis of SDA 1982-91).

93 LIS evidence to the Welsh Affairs Committee (1989 pp. 160-73). The general promotional literature produced by the SDA with inward investors in mind was varied, ranging from specific information about labour costs and companies in particular sectors (e.g. 1981c, 1981d, 1986a, 1986b, 1987b, 1988b, 1988c, 1988d, 1989a, 1989c, 1990a, 1990b), to more general or image-oriented publications (e.g. 1978c, 1985b, 1986b, 1986c).

investment project would be clarified including discretionary regional grants from central government[94] and other SDA functions such as industrial property.[95] The policy resources central to the two first stages would appear to have been fairly constant from 1983/84 onwards in the sense that both expenditure and the organisational set-up were now stable, but the use of discretionary central government regional grants varied due to the influence of individual major projects and inevitably dwarfed the cost of the organisational and informational resources involved in promotion,[96] something which has undoubtedly helped to sustain the impression that inward investment was a prime strategic preoccupation of the Agency in the 1980s. While the policy instruments employed at each of these stages will of course have been important in their own right, it is generally claimed that the competitive position of LIS on the international market for footloose investment hinged on organisation as an underlying policy instrument in its own right,[97] because this enabled LIS to maintain its role as the 'company friend' throughout the two to four years it took for most projects to come to fruition, seeking to reduce some of the uncertainties involved in investing abroad.[98]

The modes of implementation through which these policy instruments have been put to use further illuminate the interaction between LIS and foreign investors. With regard to *project generation*, both the SDA and later LIS adopted a dual approach which primarily relied on proactive approaches to individual firms in targeted subsectors while at the same time being willing to react positively to out-of-the-blue enquiries from other firms considering Scotland as a possible investment location.[99] This occasionally opportunistic behaviour clearly affected the ability of the Agency to operate in a *selective* manner, because while proactivity could be targeted, reactive pursuance of additional projects

94 While the precise importance of such grants in the locational decision-making process surrounding FDI is disputed, they still figure prominently in accounts of negotiations with individual firms and thus must be seen as a policy instrument in its own right (see e.g. Benoit 1995, Brown & Raines 1999 pp. 15f, cf. Young *et al.* 1994b).

95 LIS evidence to the Welsh Affairs Committee (1989 pp. 160-73), Neil Hood 22.6.90.

96 In average SDA expenditure accounted for only 20% of total FDI-related expenditure in the period for which data is available (calculated on the basis of SDA 1982-87 and *Industry Act Annual Report* 1982-87).

97 LIS evidence to the Welsh Affairs Committee (1989 pp. 160-73), Neil Hood 22.6.90.

98 Neil Hood in evidence to the Committee of Public Accounts (1990 p. 6), cf. Phelps & Tewdwr-Jones 2001 p. 1256.

99 Gavin McCrone in evidence to the Committee on Scottish Affairs (1980 vol. 2 p. 7), Iain Robertson 11.7.90, Neil Hood 22.6.90.

meant that e.g. sectoral preferences could not be rigorously observed.[100] Moreover, the SDA gradually took an interest in foreign firms already operating in Scotland[101] – partly for opportunistic reasons as expansion gradually came to account for a larger share of new FDI within the region and indeed internationally[102] – but this new focus on existing subsidiaries was clearly additional to the traditional aim of attracting new green-field ventures, as suggested by the continued priority given to maintaining a physical presence abroad in the form of Agency offices. Within inward investment promotion, *project appraisal* was thoroughly discretionary:[103] the initial decision to pursue a particular project depended on an assessment of the 'seriousness' of the prospective investor, and in the negotional phase the willingness to contribute discretionary regional grant assistance and other forms of public support reflected the perceived value of the investment for the regional economy at large. Criteria like these were of course extremely flexible, but at least at the aggregate level quantitative indicators like employment and investment continued to prevail,[104] and while the SDA valued higher order functions like research and development, this did *not* imply that projects which only involved routine-type assembly functions were actively discouraged: upgrading the quality of foreign-owned plants within the region was seen as a long-term endeavour which involved aftercare through advisory services rather than stringent appraisal of first-round investment projects.[105] Also in other respects has project appraisal by LIS been based on the view that the customers decide: while the organisation became more relaxed about assisting prospective investors with contacts to relevant trade unions, union representation within foreign-owned plants was not actively encouraged,[106] and while sites in areas with high levels of unemployment would be suggested to investors, the locational preferences of the foreign investor would have to be respected.[107]

---

[100] An investigation by the National Audit Office noted that the share of jobs in non-targeted sectors had grown from around 25% in the early 1980s to around 50% in the late 1980s (calculated on the basis of National Audit Office 1989 p. 15).

[101] Neil Hood 22.6.90, Iain Robertson 11.7.90.

[102] Cf. e.g. Taylor 1986, Amin *et al.* 1994.

[103] Neil Hood 22.6.90, LIS evidence to the Welsh Affairs Committee (1989 pp. 160-73).

[104] National Audit Office 1989, cf. the 'key figures' in the annual reports of the SDA (1982-91).

[105] Neil Hood 22.6.90, Iain Robertson 11.7.90, LIS evidence to the Welsh Affairs Committee (1989 pp. 160-73).

[106] LIS evidence to the Welsh Affairs Committee (1989 pp. 160-73), Sir Robin Duthie 12.7.90, Iain Robertson 11.7.90, Neil Hood 22.6.90, Douglas Harrison 19.7.90, Helen Liddell 21.6.90, cf. STUC 1989 pp. 50f and Chapter 14).

[107] Especially the continued association of the Scottish New Towns with green-field high-tech ventures was often envied by actors in the old conurbations (see e.g. Ran-

In short, while the organisation of international promotion through LIS had evolved into a fairly sophisticated operation in terms of market intelligence and proactive targeting of individual firms, relatively general project appraisal criteria underlines that the options available to LIS were still limited by the basic fact that the organisation was trying to capture footloose projects. In order to influence the corporate investment strategies of multinational companies – which always had the option of placing their economic activities outwith Scotland – presenting itself as a flexible and accommodating location was deemed to be an essential prerequisite. In practice, therefore, these modes of implementation amounted to a two-stage strategy which first concentrated on attracting foreign capital and additional jobs to Scotland, and then, once a subsidiary had been established, attempted to attract higher-order functions and increase the integration of the operation through links with other firms in the region.

Inward Investment Promotion and Regional Economic Change

In the early 1980s the DTI-led Committee of Overseas Promotion established a common UK-wide standard for reporting the number of jobs and the amounts invested from abroad which applied to promoters of FDI both on the national and regional level, and to which LIS adhered.[108] Available figures suggest that the level of FDI activity in Scotland has gradually increased through the 1980s with regard to the number of investment projects, the number of jobs projected, and the amounts invested by the private sector,[109] with the total number of projected jobs recorded by LIS from 1981 to 1991 being nearly 84,000,[110] and thus the level of inward investment in Scotland continued to be significant throughout the 1980s. From a political perspective an aggregate figure like this is of course eminently useful, but when con-

---

dall 1985 pp. 262ff, Hargrave 1985 p. 92, Moore & Booth 1986b pp. 114f, Lever 1989, cf. evidence of the Scottish New Towns to Committee on Scottish Affairs (1980 vol. 2 pp. 213-44) and the Welsh Affairs Committee (1989 pp. 174-92).

108 Hill & Munday 1992 p. 538, Committee of Public Accounts 1990 p. 7. It has frequently been pointed out that a fundamental weakness of this approach is that the figures depend on individual projects being known to the reporting organisation: as applications for discretionary grant support would automatically bring a project to the attention of a government department, FDI in Assisted Areas was more likely to be registered than projects locating in e.g. the undesignated South East England, making inter-regional comparisons difficult and the aggregate national total unreliable (Hill & Munday 1992 p. 538, Stone & Peck 1996 p. 56). This bias was readily acknowledged by the DTI itself (e.g. IBB 1983a).

109 From the first to the last three-year period for which data is available the number of projects, jobs and private sector investment increased by 38, 44 and 97% respectively (calculated on the basis of SDA 1982-91).

110 Calculated on the basis of SDA 1982-91.

sidering the firm-level impact of public promotional activities it must still be treated with more than a little caution. *Firstly*, the reporting format included all foreign investments known to the various promotional bodies, irrespectively of whether they had been directly involved in them or not, undoubtedly in order to boost figures that could enhance credibility with foreign investors.[111] *Secondly*, the figures reported by LIS under both the original and the revised DTI rules were the number of jobs projected when the investment was being planned, but a detailed enquiry by the National Audit Office revealed that the employment eventually fell well short of projections, and hence in practice the Hoodian rule of thumb – in average around two-thirds of projected jobs are likely to materialise in practice – appears to have become generally accepted.[112] And *thirdly*, while the existence of a common UK reporting system plus the competitive nature of FDI promotion made the emergence of league-table style inter-regional comparisons possible,[113] establishing why Scotland appears to have been doing less well than e.g. Wales in terms of employment in foreign-owned firms in the 1980s would require in-depth examination on a historical case-by-case basis,[114] and as work along such lines is scarce, alternative lines of enquiry will have to be pursued. Fortunately, Scotland is one of the few regions within the UK where employment in foreign-owned manufacturing has been monitored relatively closely since the 1950s,[115] making it possible to track various forms of expansion and contraction in some detail and estimate the role of promotional bodies with regard to different types of inward investment projects.

Investment in foreign-owned firms may produce change in the regional economy along four different lines:[116] in the number of plants, in the ownership of existing facilities, in expansion of existing activities, or modernisation of existing operations. Despite the much-trumpeted Scottish one-door approach, the way in which these different forms of FDI were handled by public actors clearly differed: *new incoming* projects had in most cases LIS playing a major role in bringing the investment to Scotland,[117] *expansion* or *modernisation* of existing facili-

[111] See e.g. the IDS argument to the National Audit Office (1989 p. 7).

[112] Committee of Public Accounts 1990 p. iv, cf. the Scottish Office press release 7.4.89 (reprinted Committee of Public Accounts 1990 p. 2), and SDA 1989 p. 47.

[113] E.g. the rugby-style debate between Hill & Munday (1989) and Young (1989).

[114] See e.g. Hill & Munday 1992, Young *et al.* 1994a, 1994b; Benoit 1995, Stone & Peck 1996.

[115] On the Scottish database on foreign-owned firms, see Harris 1986 p. 26, Stone & Peck 1996 p. 68.

[116] Cf. Table 3.12.

[117] National Audit Office 1989 p. 7.

ties were handled by IDS with regard to discretionary regional grant assistance outwith the one-door approach[118] although such second-round projects may have been inspired by specific aftercare initiatives by the SDA, while *change of ownership* appears not to have been promoted proactively.[119] Given these organisational differences, illuminating the balance between various forms of FDI will therefore help illuminate the extent to which the SDA can have played an active part in facilitating inward investment. On the one hand the balance between opening of new plants and expansion of existing ones seem to have been skewed towards the former: in 1986-87 LIS-led projects account for 2/3 of projected jobs and more than 90% of planned foreign private investment.[120] On the other hand the overall balance between job-creating and 'jobless' FDI is also readily identifiable: one-third of the jobs 'claimed' by LIS over the period for which data is available would seem to be associated with modernisation of existing facilities,[121] although change of ownership did still account for a very significant of overall growth in foreign ownership in manufacturing in Scotland.[122]

All in all LIS would in other words clearly seem to have been involved in a significant part of inward investment into Scotland from the early 1980s onwards. While it is not known whether promotional activities have been decisive in making individual investors opt for Scottish locations, it has clearly been demonstrated that the joint body had been closely associated with the most conspicuous form of inward investment, the attraction of new green-field ventures.

Inward Investment and Institutional Change

The issues concerning the implications of inward investment promotion for the structure of the regional economy are neatly summed up in

118 National Audit Office 1989 p. 7.

119 Neil Hood 22.6.90. In individual cases the Agency may occasionally have had a role by stimulating international strategic alliances or using FDI as a form of new-model rescue.

120 Calculated on the basis of National Audit Office 1989 p. 7. The analysis of Stone & Peck of the development of the foreign-owned manufacturing sector in Scotland points in the same direction, with new openings accounting for more than twice as many additional jobs as expansion of existing plans in the period 1979-92 (calculated on the basis of Stone & Peck 1996 p. 60).

121 Calculated on the basis of SDA 1982-89. Until 1988/89 LIS published the same breakdown of its employment figures as the one used for discretionary grants by DTI/IDS, distinguishing between new jobs 'created' (new projects or expansions), and jobs 'safeguarded' through modernisation.

122 Stone & Peck (1996 p. 60) show that the increase in employment in foreign-owned manufacturing firms in Scotland for the 1979-92 period owed as much to acquisitions as to new incoming firms.

the work of three academics who also became involved in practical policy implementation.[123] In oft-quoted articles from the mid-1970s John Firn and Neil Hood highlighted the increasing degree of 'external control' within the Scottish economy and the associated 'branch factory syndrome' with foreign-owned units concentrating on routinised production,[124] and two decades later Stephen Young *et al.* contrasted this type of 'truncated' operation with the characteristics of, from a regional perspective, an ideal inward investment project in manufacturing, namely

> a single European facility with R&D, production and marketing responsibility for European or global markets, a large employer with a highly skilled, productive and high-wage work-force and high levels of input purchases locally to generate multiplier effects.[125]

In order to assess the possible institutional effects of Scottish promotional activities, two sets of issues will have to be explored:[126] on the one hand the 'static' effects concerning external control and sectoral change, and on the other hand 'dynamic' effects in the form of up-grading to higher-order functions within foreign-owned units and spill-overs to the rest of the regional economy through linkages and local sourcing.

The increasing importance of foreign-owned firms in the Scottish economy is well-documented: employment in foreign-owned manufacturing plants in Scotland rose steadily from 1950, reaching in absolute terms an all-time peak of nearly 117,000 persons in 1974 or 18% of the industrial workforce,[127] but during the economic crises of the following decade more than 30% of these jobs disappeared, and although the second half of the 1980s saw renewed expansion so that by 1992 more than 86,000 persons were employed in foreign-owned plants, indigenous activities had contracted even faster and external control in Scottish manufacturing now amounted to 24% in terms of employment.[128] High-tech activities such as electronics were around twice as common in FDI as in the Scottish economy at large,[129] and the increasing importance of the foreign-owned sector will therefore at the same time have pushed the structure of the regional economy in the direction of more 'modern'

---

123 While Firn and Hood both became SDA senior executives, Young provided extensive input to the work of both the Committee on Scottish Affairs in the early 1980s and the Welsh Affairs Committee in the late 1980s.

124 Firn 1975, Hood & Young 1976.

125 Young *et al.* 1994b p. 147. For a similar argument, see e.g. Amin *et al.* 1994 pp. 13ff.

126 The dichotomy is inspired by Young *et al.* 1994a pp. 658f.

127 Taylor 1986 pp. 20ff.

128 Taylor 1986 pp. 20ff, Stone & Peck 1996 pp. 59f.

129 Calculated on the basis of Taylor 1986 p. 27 and Scottish Office 1991 p. 49.

sectors, making computers one of the main exports by the early 1990s,[130] but one that depended on continued growth in imports of all manner of inputs from outwith the region.[131] The inflow of FDI in other words increased external control within the region while broadening and modernising its sectoral make-up, possibly creating a 'dual economy'[132] in the process where foreign plants were present in order to benefit from the resources available – e.g. labour and government grants – but had only limited spill-over effects on the indigenous sector of the economy.

The relationship between the foreign-owned firms and other economic actors within Scotland has often been studied with particular emphasis on the electronics industry, epitomising large-scale high-tech FDI projects, and although no coherent time series exists, parallel qualitative surveys have been undertaken at several points during the lifetime of the SDA. The available evidence would appear to suggest that foreign-owned electronics plants have become less rather than more closely integrated in the regional economy, with the share of local sourcing declining and that of overseas imports nearly doubling,[133] but there are, however, still some reasons for at least a degree of optimism: foreign-owned plants in other industries than electronics would often seem to have stronger linkages to the regional economy,[134] and as the likelihood of branch plants, even in electronics, having higher-order functions would seem to increase the longer the individual plant has operated in Scotland,[135] the decreasing propensity to source inputs locally may simply reflect the recent arrival of a large number of new firms in the second half of the 1980s.

The other feature traditionally used as a litmus test for the 'quality' of inward investment is R&D, and here the original 1970s survey by Hood & Young showed that while more than half of the US firms operating in Scotland undertook neither research nor development, a quarter of the sample actually did so at a fairly advanced level.[136] Later surveys seem to confirm this by indicating a preponderance of R&D which mainly adapts American technology to European markets, but still identified a significant minority of companies which undertook more basic forms of development activities.[137] While these results would

---

[130] Scottish Office 1991 pp. 24ff, cf. Brown *et al.* 1999 pp. 11f.

[131] Turok 1993.

[132] Cf. Young 1984 p. 102.

[133] Local sourcing decreased from 19 to 14% from 1980 to 1991, while overseas supplies increased from 32 to 55% (Firn & Roberts 1984 p. 306, Turok 1993 p. 406).

[134] Amin *et al.* 1994 pp. 42f, cf. Raines *et al.* 2001.

[135] Haug *et al.* 1983, cf. Young *et al.* 1994a, 1994b.

[136] Hood & Young 1976.

[137] Haug *et al.* 1983, Haug 1986, Young *et al.* 1988.

seem to suggest continuity rather than any clear-cut development trend, it is interesting to note that the share of highly qualified technical staff employed by electronics firms in Scotland nearly doubled from 1978 to 1989,[138] and although this changing occupational profile refers to both indigenous and externally-owned firms, this could warrant some cautious optimism with regard to the long-term development potential of the major FDI-dominated high-tech sector within the regional economy.

All in all the claim that Scotland occupied an 'intermediate' position with regard to the overall quality of inward investment – somewhere between the ideal-type high-quality projects and the assembly oriented branch plants – would seem to be warranted even in the electronics industry,[139] but as this position seems to have been established already in the mid-1970s, it is difficult to interpret this position as a consequence of the aftercare and supplier-development initiatives of the SDA in the 1980s, although this does of course not preclude that these policies have influenced individual firms. Establishing this would have required detailed firm-level enquiries, but even in the absence of this kind of research, it is still possible to draw conclusions about the political ramifications of the institutional changes brought about by the growing importance of inward investment within the regional economy.

As shown in Section 10.1, the reasons for promoting Scotland as a location for inward investment combined immediate goals like improveing the regional employment situation and more long-term endeavours such as modernising the economic structure and maximising the benefits from FDI through local linkages that would create multiplier effects and, hopefully, a permanent presence within the region. The political difficulty entailed in this dual strategy would seem to be that while in the 1980s the short-term quantitative results were clearly visible in the form of plant openings and job creation, the long-term impact of having FDI as a major element in regional development was still open to questioning. Although the modern nature of most foreign investment projects was obvious, the occasional, but very visible and well-publicised, closures and contractions could readily be construed as evidence of the fickleness of the commitment of multinational firms to Scotland, and the low-profile and gradual nature of quality enhancement found it difficult to match these very tangible phenomena. In short, the different nature of the two types of promotional efforts would seem to point towards a situation where 'the job numbers game' would be able to continue to prevail – unless, of course, key actors launched concerted efforts to

---

138 A survey by Kevin Morgan quoted by Phelps 1992 p. 15.

139 Randall 1985, Hargrave 1985, Haug 1986, Young *et al.* 1988, Young 1989, Amin *et al.* 1994, and, more sceptical, Turok 1993.

stress the long-term strategic role of inward investment at the expense of more immediate and tangible benefits.

10.3. The British Context: Inward Investment and Regional Development

In order to understand the position of the SDA in the wider UK policy network – already alluded to in the discussion of the authority to undertake international promotion – this section provides an analysis of the role of central government, both as a promoter in its own right and as a regulator of others, and sub-national actors in other British regions.

Central Government and Inward Investment Promotion

The general attitude of British governments towards inward investment had long been permissive or liberal in the sense that relative few restrictions had been placed on foreign investors,[140] but the active involvement of central government in attraction of FDI had evolved only gradually. In the 1970s joining the European Community did, however, trigger enquiries from multinational firms which saw Britain as a gateway to the continent, and in 1977 the handling of these was formalised through the formation of a dedicated unit within DTI, the Invest in Britain Bureau (IBB).[141] After the change of government in 1979 these international activities acquired a higher public profile: IBB began to issue its own annual reports,[142] and in the late 1980s inward investment promotion was presented as an integrated part of the White Paper *DTI – The Department for Enterprise*.[143] The organisational set-up did, however, remain unchanged till the early 1990s: an overseas marketing team was based in the world-wide network of embassies and consular offices maintained by the Foreign Office, and a DTI-based 'home team' handled enquiries, organising visits to relevant locations and only handing over to subnational actors once a specific location had eventually been decided.[144] The level of resources committed to promotional activities through the IBB cannot be followed in great detail, but the available information suggests that the around forty staff in the London office

[140] Bailey *et al.* 1994, Brown & Raines 1999 pp. 18f.

[141] Bailey *et al.* 1994 p. 124, IBB evidence to the Committee on Scottish Affairs (1980 vol. 2 pp. 67f), cf. IBB 1983a. For a 1990s European overview with historical perspectives, see Raines & Brown (eds.) 1999, cf. also Halkier & Danson 1997, Halkier, Helinska-Hughes & Hughes 2003.

[142] IBB 1983-91.

[143] DTI 1988 p. 18. Inward investment and the IBB also became a standard item in the Industry Act reports from the early 1980s onwards.

[144] IBB 1983-91, cf. Halkier, Helinska-Hughes & Hughes 2003.

were supplemented by an increasing number of full-time marketeers overseas, with thirteen persons working from consular offices in the US alone in the late 1980s and a staffing profile that gradually came to incorporate more commercial experience both at home and abroad – and, of course, could use senior government figures to add political clout to activities if necessary.[145] The costs of maintaining the domestic part of the operation also grew through the 1980s[146] but still constituted less than 30% of LIS total expenditure,[147] and although the latter included also the overseas operation and cost that would have been borne by e.g. the DTI regional offices in England, it is still clear that the relationship between UK and Scottish promotional efforts was fairly equal in terms of organisational and financial resources.

The overarching strategic goal of the IBB was to maintain Britain's position as "the preferred location" for FDI,[148] and its publications enumerated the well-known list of perceived benefits from inward investment in terms of job, trade, technology and modern management practices.[149] While the Bureau consistently claimed to have no preferences with regard to the location of individual projects within the UK,[150] in practice a focus on green-field manufacturing projects will have favoured areas designated for regional policy grant support, although the increasing interest in attracting foreign investment in services will have brought metropolitan areas like London more to the fore in the early 1990s.[151] From an early point its overseas activities appear to have operated in a proactive fashion in the sense that promotional events were supplemented by taking contact to individual firms, but it is also generally agreed that little systematic targeting of e.g. specific sectors took place.[152] In relation to individual firms the main functions of the IBB would in other words appear to have been dual: to further general awareness of the attraction of the UK as an investment location, and to

[145] Committee on Scottish Affairs 1980 vol. 2. pp. 68, 374f; Welsh Affairs Committee 1989 p. xxxvii, Martin & Tyler 1992, cf. IBB 1983a, 1983-91. The lack of commercial experience among the IBB civil servants in the early 1980s was readily acknowledged during the Committee on Scottish Affairs enquiry (cf. Section 10.1), but a commercial director was not appointed until two years later (IBB 1983).

[146] Committee on Scottish Affairs 1980 vol. 2. pp. 68, Welsh Affairs Committee 1989 p. xxxvii.

[147] Calculated on the basis of Welsh Affairs Committee 1989 p. xxxvii and SDA 1988.

[148] Used in the opening of Minister of State Norman Lamont's introduction to the first IBB annual report (IBB 1983 p. 1), and reappearing throughout the 1980s.

[149] See e.g. IBB 1983 p. 1, 1986 p. 1, 1991 p. 1.

[150] IBB 1983 p. 14, variations over this theme reappeared till the early 1990s.

[151] IBB 1983-91. The focus on services became particularly pronounced in 1990/91.

[152] Welsh Affairs Committee 1989 pp. xxxiif, cf. Hood & Young 1981, Young *et al.* 1994b.

handle incoming projects until the investor had developed clear preferences with regard to a particular locality within Britain.

In addition to its own promotional efforts through the IBB, the DTI also played a coordinating role through its central role in the UK regional policy network. The Committee on Overseas Promotion, comprising central government departments and their associated regional bodies, constituted a system of mutual information about promotional activities and specific enquiries,[153] and this coordination was backed not only by the general authority of the DTI but also by its role as handler of the largest regional grants.[154] Regional actors were in other words free to compete with each other for overseas investment both in general terms and with regard to individual projects, but while the playing field was certainly not even – the asymmetrical delegation of spending authority from the DTI meant that some regions enjoyed much greater grant-giving autonomy than others[155] – central government coordination did circumscribe the ability of subnational actors to conduct inter-regional bidding wars, and the introduction of what Raines has dubbed 'bounded competition'[156] therefore meant that the ability to mobilise other types of resources became even more important in the quest for foreign investment.

After a discreet, gradual and nearly under-cover, start in the 1970s, the following decade saw central government publically embrace inward investment. At the same time Britain retained its position as the 'preferred location' for FDI in Europe – attracting more than one-third of all foreign investment in Europe in the period 1984-89[157] – and projects by foreign-owned firms made up a significant share of central government expenditure on discretionary grant aid. Under these circumstances attraction of inward investment could therefore easily be construed as the primary regional development strategy of the Conservative government: a strategy which had the triple advantage of addressing the question of inter-regional disparities in a way very visible to the public eye while at the same time helping to modernise the industrial structure, and in which the use of financial incentives was less unacceptable because it helped to underpin Britain's position *vis-à-vis* international competition.

---

153 Committee on Scottish Affairs 1980 vol. 2 p. 338.

154 Cf. Section 8.3.

155 Regions with a territorial central government department were allowed considerably greater freedom than the regional offices of the DTI in England, and in practice the largest discretionary regional grants appear to have been awarded in Scotland and Wales (Raines 1998 pp. 11ff).

156 Raines 1998 p. 18.

157 Calculated of the basis of UNCTAD data reproduced by Brown & Raines (1999 p. 9).

New Actors, New Promotion?

For most of the sub-national actors that became active in regional development in Britain from the 1970s onwards, inward investment was a peripheral activity, with the main exception being major regional-level development bodies in industrialised regions.

The RDAs operating in rural areas predominantly focused on support for indigenous firms, to the extent that documents and reviews of the English Development Commission do not highlight the importance of international promotion at all.[158] The HIDB only engaged in promotion occasionally and on a scale which to some extent appears to resemble relocation of micro-firms rather than attraction of branch plants,[159] a strategic priority that will undoubtedly also have been informed by the short-lived major projects that had been brought to the Highlands in the 1960s.[160] The position of local authorities was more uneven:[161] especially in the early 1980s inward investment was seen by radical councils as creating dependency on foreign multinationals and therefore not promoted, while other localities opened offices, stationed staff abroad, and sent out missions to highlight their attractions to foreign companies. Through the 1980s the attitude to international promotion among local authorities generally became more positive, but at the same time efforts to coordinate activities at the regional level also increased: the UK-level Committee on Overseas Promotion required its regional members to coordinate sub-regional activities, and in Scotland, unlike England and Wales, the statutory right of local authorities to undertake independent promotional activities overseas was removed in 1982.[162] Few local actors are in other words likely to have devoted large-scale resources to activities outside the regional-level coordination framework, and as the European programmes introduced in the late 1980s were overwhelmingly oriented towards general infrastructure development and hence

158 See e.g. Development Commission 1984, cf. Rogers 1999 Ch. 10-11.

159 Committee on Scottish Affairs 1980 vol. 2 pp. 89-117, IDS 1987b p. 72, Welsh Affairs Committee 1989 p. xxxvi.

160 The two most prominent examples were the Invergordon aluminium smelter and a pulp factory in Fort William, the legacies of which were, respectively, the only central-government Enterprise Zone in the Highlands (Keating & Boyle 1986 pp. 58ff) and a rare opportunity to investigate the impact of organic pollution in the marine environment under controlled circumstances (Grethe Fallesen, personal communication).

161 Dicken & Tickell 1992, cf. Amin & Pywell 1989, Brunskill 1992 p. 447. In Scotland the early 1990s survey by McQuaid (1992) found that promotion, including place marketing for a domestic or local audience, accounted for 6% of local authority expenditure on economic development.

162 Hogwood 1987 pp. 15ff, cf. Section 10.1.

only indirectly of relevance to foreign investors,[163] we must turn to other regional-level actors in industrialised regions in order to find organisations for whom international promotion was a major policy activity.

The English regions, unlike Scotland and Wales, had not been endowed with multi-function RDAs, but central government still supported their efforts to attract inward investment by sponsoring what became known as Regional Development Organisations (RDOs), also in order to bring about more coordination of local authority efforts.[164] Formally partnership bodies which included the regional DTI office, local government and the private sector, RDO activities were predominantly funded by central government grants and included the setting up of representations overseas and the handling of investment projects within their area. RDOs were primarily established in areas otherwise designated for regional policy purposes, with the level of DTI financial support varying between in average 5% and 12% of SDA expenditure on promotion.[165] Perhaps because of the uneven picture presented by the RDOs, a so-called 'English Unit' was established within the DTI in 1989/90 in order to improve their position *vis-à-vis* the non-English RDAs,[166] but even though the RDOs did represent a one-door approach from the perspective of prospective investors and, like the IBB, were capable of operating in a proactive manner, they too lacked the targeting capacity of the LIS based on close links with an RDA with extensive resources and a much wider development remit. While the English RDOs were clearly an attempt to address the promotional consequences of the asymmetrical regional development set-up within the UK, their narrow remit, limited resources, and close links with the lead department in the national policy network meant that their room for manoeuvre was heavily circumscribed.

From the perspective of LIS, the WDA was undoubtedly its most important regional-level competitor within Britain, but both the Welsh organisational set-up and the promotional strategies pursued differed on several accounts. In terms of organisation the WDA operated in close partnership with other actors from the very start: taking over the sponsorship of the Development Committee for Wales from the Welsh

---

163 Brown *et al.* 1998 pp. 32ff, cf. Section 7.4.

164 See Hogwood 1987, Dicken & Tickell 1992, cf. Welsh Affairs Committee 1989 p. xxxviii.

165 Calculated on the basis of Hogwood 1987 p. 24, Dicken & Tickell 1992 p. 101, SDA 1984-91. The Northern Development Company, covering the North East England, consistently received by far the highest level of financial support from central government.

166 IBB 1990, cf. Dicken & Tickell 1992 p. 100.

Office meant working in conjunction with local authorities,[167] and in 1983 the arm's-length body joined forces with its sponsor department under the opaque name WINvest (Wales Investment Location), belatedly renamed Welsh Development International in 1990.[168] WINvest operated as a one-stop sales force and handler of individual enquiries, but apparently had neither the same level of integration with its sponsor department as its Scottish counterpart, especially in terms of negotiating discretionary regional grants,[169] nor the same success in bringing on board the local authorities which in Wales appear to have zealously guarded their right to promote themselves internationally.[170] In terms of financial resources and physical presence overseas, the Welsh operation would seem to have been broadly on the same level as LIS: the number of localities in which WINvest was represented gradually rose in the 1980s,[171] and the level of WDA expenditure on promotion was in average around 2/3 of its Scottish counterpart.[172] While the relative commitment of financial resources to inward investment attraction was broadly the same in both organisations,[173] the WDA would, however, seem to have attached greater strategic importance to this policy area right from the very beginning, probably reflecting the historical weakness of the indigenous sector in Wales.[174] In 1977 its first statement of strategy declared that the organisation would take "a major role" in promotional efforts,[175] and throughout the 1980s inward investment continued to overshadow activities oriented towards indigenous firms, and thus it is hardly surprising that a major policy innovation of the late 1980s, the *Source Wales* programme, focused on creating links between incoming and local firms and thereby maximise the long-term benefits from FDI within the region.[176] Promotion appears to have been proactive in the sense that prospective investors were approached in order to convince them of the attractions of locating in Wales,[177] but this targeting was not

167 WDA 1977a, Committee on Welsh Affairs 1980 p. 74.

168 Welsh Office 1987 Ch. 10, cf. Collis & Noon 1994.

169 Welsh Affairs Committee 1989 p. xi, cf. Young 1989 p. 62.

170 Evidence to the Welsh Affairs Committee by the WDA and Secretary of State Peter Walker (1989 pp. 96, 322f).

171 Welsh Affairs Committee 1989 pp. XLf.

172 Calculated on the basis of SDA 1983-91 and WDA 1983-91.

173 WDA expenditure on promotion amounted to 6.4% of total expenditure in the period 1985-91, the figure for the SDA was 5.4% (calculated on the basis of WDA 1986-91 and SDA 1986-91).

174 Morgan 1994.

175 WDA 1977a p. 9.

176 Cf. the discussion in Section 9.3 and 11.3.

177 WDA 1981a, Welsh Affairs Committee 1989 pp. 72ff, cf. Rees & Morgan 1991.

undertaken on the basis of long-term sectoral priorities, a form of selectivity the WDA had explicitly rejected in the mid-1980s because it would be too resource demanding in terms of front-end research and less useful than in a "more developed economy" like the Scottish.[178] Instead the Welsh approach seems to have been consciously 'opportunistic' in order to maximise the inflow of FDI and attract the greatest possible number of jobs – something which was greatly aided by the fact that Wales was close to major markets in South East England and continental Europe[179] – and thus the profile of the projects attracted to Wales differed from those coming to Scotland, with more high-volume consumer-oriented producers heading for the Principality and hence the branch-plant syndrome potentially being more exacerbated.[180] While Wales was seen to compete successfully with Scotland in quantitative terms, attracting 'more than its share' of new jobs,[181] their quality would still appear to be typical of especially peripheral regions where projects tended to be "more truncated, less autonomous and less well integrated"[182] than the ideal manufacturing project to some extent pursued by all organisations in the business of capturing foot-loose investment projects.

10.4. International Promotion and Regional Development

Promotion of inward investment could seem like an obvious example of a policy area that conforms to the external-revolution paradigm. Major changes clearly occurred: the strategic priority of promotion increased far beyond what many of the founding fathers of the SDA would have imagined, civil servants of the sponsor department became closely involved in implementation on an ongoing basis, and Scottish Office ministers systematically used individual projects to become associated with tangible good news like job creation in modern industries. Moreover, these changes would seem to stem from actions by the incoming Conservative government: the 1980 guidelines changed strategic priorities, and George Younger established LIS as a joint IDS-SDA operation. In short, the development of the SDA's promotion of Scotland as a location for foreign investment had involved major externally induced changes.

Such a conclusion can, however, only be sustained by concentrating on a limited number of aspects and episodes while ignoring features that

178 Welsh Affairs Committee 1989 pp. 85f.
179 Cf. Hill & Munday 1992.
180 E.g. Young 1989, Morgan & Henderson 1997, Phelps *et al.* 2003.
181 Hill & Munday 1992.
182 Young *et al.* 1994b p. 148.

make for a much more complex interpretation. Describing the origins of the LIS set-up as external would seem to require a very narrow interpretation of political fatherhood: of course the joint operation had been established by a Conservative Secretary of State, but the general notion of a one-door approach with a separate overseas presence had been presented as an, albeit probably unobtainable ideal, by the SDA both before and during the parliamentary enquiry, while the specific blueprint had been put forward by an enterprising academic and publically supported by the Labour party. At the same time this chapter has also demonstrated the existence of important continuities in the development of the policy area. *Firstly*, attraction of inward investment had actually been prioritised by the SDA in the late 1970s, not only through general promotional activities but also specifically in relation to its electronics sectoral initiative, and despite its high public profile in the 1980s, it never became more than one among many strategic thrusts in the policy profile of the organisation. *Secondly*, the proactive and selective targeting strategy – trying to make foreign investors an offer tailored to their particular circumstances in order to minimise the risk of it being refused – later became the preferred *modus operandi* of LIS, although this did of course not rule out opportunistic behaviour in responding to sudden openings on the international market for footloose investment projects. And *thirdly*, the project-driven nature of the joint body clearly bears the mark of the Agency rather than traditional civil service procedures, and thus in organisational terms the practices of the arm's-length body would seem to have continued to dominate operations. From the perspective of the private firms targeted, in other words, continuity will basically seem to have prevailed from the late 1970s onwards, and while changes did occur in the 1980s – sectoral priorities widened and more attention was devoted to existing foreign-owned facilities – these would seem to have been limited, incremental, and driven from within LIS rather than having been brought to it from the outside.

The key to understanding the relation between these very different interpretations of the same policy area would seem to be the nature of the LIS settlement, rightly in focus from an external-revolution perspective. Put simply, the new set-up embodied a compromise in terms of territorial politics which allowed the governing Conservative party to claim that specific Scottish interests had been safeguarded within the constitutional framework of the UK, had installed the IDS as the guardian of financial propriety so that the discretionary grants could not be used in bidding wars against other British regions, and had allowed the SDA to get on with refining its promotional approach while government ministers received much-needed positive publicity by being associated with job creation in modern high-tech industries. This convergence of interests in combination with the continued success of LIS in bringing new projects to Scotland probably explains the quiet and uncontroversial persistence of the experimental joined-up body, but, ironically, the ambitious nature of the promotional approach – seeing foreign investments as part of a more comprehensive strategy for regional development – also proved to be a potential Achilles' heel, because it increased expectations with regard to the quality and embeddedness of FDI projects. Perceiving its position *vis-à-vis* foreign investors as weakened by the fact that they could always decide to locate elsewhere, the Agency decided to pursue these goals not through stringent appraisal of new incoming firms but instead by focusing on second-round investments, a strategy involving a longer time-horizon which included measures with a much lower public profile. In this way strategic ambition could potentially backfire and create cracks in the otherwise increasingly pervasive consensus about the usefulness of importing "change from abroad".[183]

Compared to other British actors which adopted inward investment attraction as a strategic priority, the analysis has identified the SDA as a leader rather than a follower. The WDA probably embraced inward investment as a priority activity more single-mindedly and at a slightly earlier point in time, but in terms of promotional strategies the Scottish approach appears to have been the most innovative one, with its focus on selective in-depth analyses of industries and firms as the basis for proactive contacts to foreign firms, and perhaps this was also part of the reason why the DTI came to accept the existence of a separate Scottish effort overseas within the UK coordinating procedures. When Secretary of State George Younger announced the new institutional set-up, he claimed that the new institutional set-up would allow Scotland to make a fuller contribution to the task of maximising the flow of inward investment into the UK by exploiting "the fund of interest in and affection for Scotland built up abroad by emigration and historical association".[184] This may just have been a skilful piece of discursive engineering designed to soothe unionist feelings by alluding to perceptions of the traditional strength of the Scottish diaspora, but the point that LIS might bring something additional to the UK rather than simply lure away projects from other British locations may still, on the analysis above, have been true. Only, the Scottish contribution would appear to have less to do with the sentimental pulling power of tartan and a wee scotch than with the ability to mobilise information through extensive front-end research and having the organisational platform from which a mutually profitable dialogue could be maintained with individual prospective investors for a sustained period of time.

---

183 The title of the inward investment section of the SDA's 1985 annual report, carrying a large picture of the Conservative Prime Minister and a bemused worker.

184 Committee on Scottish Affairs 1981 p. 3.

CHAPTER 11

Advisory Services

Advisory services were not part of the traditional armoury of regional policy, but with the increasing importance of regionally-based economic development initiatives, strategic use of informational resources became more common as a means to influence private economic actors.[1] The original act of parliament gave the SDA the power to "provide or assist in the provision of advisory or other services",[2] and given the generally weak performance of the Scottish economy,[3] informational deficits could be expected to exist in many guises: firms in traditional sectors could need to know more about new market opportunities, firms in modern sectors could have difficulties in keeping up with the pace of technological change, small and growing firms could find it challenging to cope with organisational change, and potential entrepreneurs could need reassurance when facing the uncertainties of the market. Many private actors may not realise that additional information could assist them in achieving their goals, and they will therefore be reluctant to invest money and management resources in new knowledge, and public policies addressing informational deficits amongst private economic actors are therefore based on the assumption of policy-makers 'knowing better', but as they, like the firms themselves, cannot have perfect information about everything, advisory services will have to focus on particular areas of expertise. All in all this suggests that the use of advisory services to promote regional development will potentially entail at least two types of political issues: how can public actors 'know better' than the private firms they advise and, indeed, private providers of business advice? And because the informational needs of private economic actors differ, particular types of advisory services will support some forms of economic activity at the expense of other forms, and hence they are, like sectoral initiatives, inherently selective.

In the case of the SDA these general issues did, however, exist in a specific historical context in which great symbolic importance was

[1] See Bellini 2002, cf. Halkier & Danson 1997 and Section 2.1.

[2] SDA Act 1975 Section 4.

[3] Cf. Chapter 4.

attached to the relationship between public and private actors, and the advent of the Conservative Thatcher government in 1979 could therefore be expected to affect Agency advisory services in potentially contradictory ways: the *use of advice* was likely to increase because it could be construed as a relatively inexpensive policy instruments that interfered 'less' with the market process than e.g. financial subsidies, an increasingly *selective* focus on small and new firms could be expected with the emphasis on entrepreneurialism in the 1980s, and *implementation methods* that involved private sector actors were likely to come to the fore in order to avoid crowding out of consultants and other private providers of information-based business services. While changes along these lines would seem to be in accordance with the external-revolution perspective on the development of the SDA, alternative explanations for such changes – if they can indeed be verified – would, however, also seem to be possible:

- an increased use of advisory services could reflect Agency experience that informational policy instruments are useful when addressing specific problems of individual firms because of their interactive and knowledge-intensive nature,
- a focus on small and new firms could have developed because management resources here could be expected to be limited, and/or because this was an area in which the Scottish economy was perceived to be lagging behind more successful British regions, and
- increased use of external providers of advice could be a way to gain access to additional expertise or economise with scarce internal resources.

In short, an internal-evolution perspective could also seem to be plausible, and so there are plenty of reasons for investigating the development of the Agency's advisory services.

The analysis will be based primarily on published materials, especially the annual reports of the SDA, complemented by interviews with senior executives responsible for the advisory services of the organisation and the existing, limited but useful, academic literature.[4] Also this chapter falls in three parts: first the design of advisory services is explored (Section 11.1), then the implementation of key programmes is examined in some detail (Section 11.2), and finally the broader British policy context is surveyed, both with regard to developments within regional policy and the adjoining area of industrial policy (Section 11.3). On the basis of this it should be possible to draw conclusions with regard to the relationship between continuity and change in the

---

4 Especially Kirwan 1981, Lochhead 1983, Lever & Moore (eds.) 1986.

SDA's advisory services and their relationship to other policy areas, Agency corporate strategy, and wider British trends.

11.1. Design

Origins and Sponsorship of Advisory Services

In the 1970s proposals for a lowland RDA, improving the provision of advice for Scottish firms was not in itself a major argument, and thus this type of activity was not vested with an extensive baggage of economic expectations and political symbolism. Notions about the importance of informational resources were, however, by no means absent: both the SNP, the Scottish Labour party, and, not least, the SCDI research institute and the very indigenously oriented West Central Scotland plan had included informational instruments in the policy portfolio of their proposals.[5] Moreover, advisory services did in fact already play a role in policies for regional development in rural areas because the Development Commission had long been engaged in providing advice to small firms, and when the HIDB was established in 1965 the Board took over the Commission's responsibilities in the Highlands & Islands. And in the adjoining field of industrial policy a growing concern for the problems of small firms had led the Department of Industry to set up a network of information offices which gradually expanded from the early 1970s onwards.[6] Situated at the intersection of regional and industrial policy and with a geographical remit which included extensive rural areas, it was hardly surprising that advisory services became part of the SDA's policy profile.

The 1975 parliamentary act empowered the Agency to employ information, either directly or through intermediaries, in order to further its purposes,[7] but the act did not contain detailed rules regulating the use of this policy instrument, and as the Scottish Office did not issue additional guidelines, the initiative in developing these services was left with the arm's-length body itself.[8] This did, however, not mean that advisory services were completely ignored by the political sponsors of the organisation: as an increasing number of public providers of advice came to target SMEs in the mid-1980s, the question of coordination was taken up by a joint IDS-SDA high-level industrial policy 'study group',[9] and the 1986 government review was very critical about several aspects of

5 Cf. Table 5.2.

6 Lochhead 1983 pp. 12f, Barberis & May 1993 pp. 197ff.

7 SDA Act 1975 Section 4.

8 IDS 1987a p. 130.

9 Iain Robertson 11.7.90.

the Agency's use of advisory services to promote regional development.[10] Furthermore, in the 1980s the SDA also came to administer UK-wide advisory schemes in Scotland on behalf of central government departments,[11] and thus despite the absence of formal guidelines the sponsor department must have been able to influence the Agency's advisory services through coordination and its general powers of oversight.

Organisational and Informational Resources

For the perspective of organisational resources, the SDA's advisory services came to comprise both inherited activities and others created from first principles.

The incorporation of SICRAS, the Scottish implementing body of the Development Commission, meant that the Agency was operational from the very start in terms of advising small firms in rural areas. SICRAS became the Small Business Division of the new RDA, initially retaining its headquarters in Edinburgh and gradually extending its geographical remit until in 1981 advisory services were available to small firms throughout lowland Scotland.[12] Later the delivery of services became increasingly decentral, including local front-line offices as part of the Agency's area initiatives and culminating in the transfer of SBD activities to the new regional SDA offices in 1988,[13] and thus in terms of organisational resources, the structures taken over from SICRAS were in other words subject to continuous change for more than a decade. From the late 1970s onwards the inherited activities were gradually supplemented by a host of new advisory services, often relatively specialised and derived from or operating in conjunction with other Agency activities: the SBD began advising small firms on subcontracting opportunities, the concept of strategic management services was originally developed to support the industrial investment function, and many of the sectoral initiatives employed informational policy instruments.[14] It is therefore hardly surprising that provision of advice had not been made the responsibility of one particular unit but was instead undertaken by many different parts of the organisation,[15] and the development of new

---

[10] Cf. the discussion of advisory service strategies below.

[11] Cf. Section 11.2 and Table 11.1.

[12] SDA 1976 par. 15f, 1977 pp. 19ff, 1978 p. 51, 1979 pp. 42f, 1981 pp. 40f, cf. Lochhead 1983 pp. 29ff.

[13] SDA 1988 pp. 11ff, Iain Robertson 11.7.90, cf. Chapter 12.

[14] SDA 1976 par. 38, 1977a p. 6, 1981 pp. 27f, 41; Gerry Murray 30.7.90, cf. Chapter 9 and Table 11.1.

[15] In addition to the division which had provision of advice as their main responsibility – e.g. Management Development, Advisory Services, and Technology Transfer –

advisory services would thus appear to reflect the SDA's generally project-driven approach.

The balance between inherited and new advisory services in terms of manpower is difficult to establish on the basis of the staffing figures published by the Agency, but the more standardised services targeting the sizeable population of small firms appear to have required the largest input in terms of organisational resources.[16] In the early 1980s the SDA repeatedly complained about demand outstripping its ability to service (especially small) firms,[17] and through the 1980s a division of labour gradually evolved which saw the Agency retain highly specialised services closely linked to other policy activities,[18] while very standardised services for new and very small firms were hived off to a network of Enterprise Trusts. Although the trusts were formally private-sector led 'civic' initiatives, they delivered many of their services co-sponsored by the Agency and also continued the practice of employing the services of semi-retired businessmen as a way to gain access to relevant experience and added entrepreneurial credibility.[19] The reasons for outsourcing this particular service may have been pragmatic circumvention of staffing limits and/or a wish to involve private sector representatives in economic development work in order to strengthen its standing in the business community (and please its political sponsors), but either way around it effectively meant that scarce Agency resources could be deployed in other areas of activity. This did, however, not keep the organisation from undertaking other routine-type services on behalf of central government, e.g. signposting firms to relevant providers of advice, presumably because such functions involved access to additional funding, could readily be handled through its organisational network, and also helped to bolster the status of the SDA as the first port of call for firms looking for business advice.

---

other divisions such as Engineering, Scottish Resource Industries, and Small Business used information as a central instrument in pursuance of their goals (cf. Tables 6.1 and 11.1).

16 The SBD employed around ninety persons in average while the more specialised services amounted to little more than twenty all in all (IDS 1987a pp. 53f, 72; cf. Gerry Murray 30.7.90).

17 SDA 1981 p. 59, 1983 pp. 49ff, 1984 p. 8.

18 In the field of export promotion the SDA operated partly by supporting the long-standing efforts of the SCDI (Lochhead 1983 p. 44).

19 SDA 1979 pp. 42f, 1982 p. 30, 1983 p. 74, 1984 p. 56, 1985 p. 50, 1986 p. 6, 1987 pp. 62ff, 1989 pp. 89ff, cf. Lochhead 1983 p. 107, Keating & Boyle 1986 Ch. 7, Harding 1990, Metcalf *et al.* 1990.

Changing Aims and Methods

Tracking the development of the strategic thinking behind the use of advice as a policy instrument is a challenge in the sense that for more than a decade very little was said explicitly about the rationale for adopting this approach, either by the SDA or its sponsor department, and thus until the mid-1980s the analysis will primarily rely on what can be inferred from the development of specific advisory services by the organisation.

From the very beginning the Agency provided advice to small business in rural areas through SBD, and a major survey was launched in order to identify what the targeted firms saw as the most important problems they were facing – general aspects of business management such as finance, marketing and sub-contracting – so that activities could be adjusted accordingly throughout Scotland.[20] This suggests that the assumption about regional problems underlying the early advisory services was that economic development was hampered by a lack of information within small firms, that the firms themselves were able to identify their informational weaknesses, and that the best way of addressing this was through provision of advice by the Agency. Although the increased emphasis of the Labour government on small business was acknowledged,[21] the SDA also had its own reasons for developing a small-business strategy: this segment was seen as more readily influenced,[22] it would help to integrate inherited structures and policies into the new RDA, and it would ensure a clear economic dimension to the urban area initiatives that the organisation had recently been put in charge of by the Scottish Office.[23]

The change of government in 1979 did not result in the Scottish Office making major statements about the future role of the SDA's advisory services,[24] and the arm's-length body responded simply by 'getting on with it' in terms of developing existing and new services in which information was a central element. Small business – occasionally described in rather poetic ways that stand out from the otherwise 'hard-nosed' business-like rhetoric employed by the Agency[25] – continued to be a priority in its own right with a more comprehensive range of ser-

---

20 The findings were published in SDA 1978b.

21 SDA 1978a p. 12.

22 SDA 1978a p. 16, cf. Section 7.1.

23 Cf. SDA 1978a p. 12.

24 SEPD 1980a, cf. Sir Robin Duthie 12.7.90.

25 The SBD section of the SDA annual reports for 1979 and 1980 were captioned "Sowing and Growing" (1979 p. 42), and "The Seed Bed" (1980 p. 22).

vices and a more proactive approach,[26] but new advisory services were also introduced, targeting specific groups of larger firms in the context of especially sectoral initiatives.[27] It is therefore not surprising that the mid-1980s corporate strategy portrayed advice as a function on par with finance and property,[28] and thus this period saw the continuation of policy-making on the basis of the same assumptions as the late 1970s, although the problem of inadequate information had been extended beyond small enterprises, and the notion that firms would know their needs was questioned by the increasingly proactive approach.

The 1986 government review did, however, take a strong interest in the SDA's advisory services, and although no consultancy study had been undertaken to evaluate the experience of the first ten years, the review group not only constructed a rationale for these activities based on the market failure approach favoured by the Conservative government but also made some rather pointed recommendations with regard to changes in Agency practices. According to the reviewers the general rationale for the use of advisory services in regional policy was the occurrence of market failures with regard to information: especially small firms are unwilling or unable to pay for the services of private consultants, firms may be unaware that additional informational resources could be beneficial for business, and specialised services may not be available from private consultants. Such market failures could be addressed either by direct public provision or by stimulating the use of private provision through financial subsidies, and in order to avoid perpetuation of the underlying market failure the reviewers recommended that the role of private providers should be strengthened, both directly through subsidies and indirectly by increasing the charges for services provided by the Agency in order to avoid e.g. small firms getting accustomed to a price of information close to zero.[29] The Agency criticised these views both during the review process and, indeed, in public afterwards,[30] arguing that the use of private providers had in fact started well before the review, that the services provided were qualitatively different from those offered by e.g. private consultants and hence in effect created new markets,[31] and that e.g. the BOP approach in engineering was able to "communicate information and good practice

---

26 E.g. SDA 1981 p. 40, 1982 p. 54, cf. 1986 p. 50.

27 Cf. Table 11.1.

28 SDA 1984a, cf. the discussion in Section 7.2.

29 IDS 1987a pp. 55ff.

30 SDA Board Papers 16.2.77, National Audit Office 1988 p. 7, Iain Robertson 11.7.90, Edward Cunningham 13.6.90 – Policy Unit Head Frank Kirwan (1.6.90) accepted some of the points.

31 National Audit Office 1988 p. 7, cf. Edward Cunningham 13.6.90.

without interfering in the competitive process".[32] Conversely, the conditions on which information transfer to private firms should take place, i.e. the question of subsidisation, tended to be ignored by the Agency, and while the administration of central government schemes supporting the use of private consultants was continued, the organisation continued to cast itself not only as the subject initiating policies but also as the main provider of services.[33]

All in all it can be concluded that the assumptions underlying the strategies of the SDA with respect to advisory services either remained stable – informational weaknesses on part of private firms hampers regional economic and can be addressed by public policy – or became more radical over the years by being extended from small firms to larger ones, and, indeed, by acknowledging the possibilities of firms not realising that their growth is hampered by informational deficiencies. Only fairly late did the application of the market-failure approach provide an elaborate rationale for the use of advice in regional policy, but like in other policy areas the specific recommendations of the review group concerned primarily implementation-related issues – subsidisation and the use of private providers of advice – and these were not unconditionally accepted by the SDA. The arm's-length body would in other words appear to have enjoyed a relatively high level of operational freedom, but in order to illuminate how this have influenced its interaction with private economic actors, attention will now turn to the way in which the Agency implemented its information-based strategies.

11.2. Implementation

The analysis of the implementation the SDA's advisory services proceeds in three steps. Firstly the individual initiatives are introduced and levels of activity analysed. Then a detailed cross-service discussion of policy instruments and modes of implementation is undertaken and finally firm-level impact and institutional change will be considered, bearing in mind that individual services may have evolved in different directions.

32 SDA 1988a p. 29.

33 SDA 1988 pp. 37-52, 1989 pp. 29-38, 1990 pp. 20-30, 1991 pp. 12-20.

Table 11.1. Overview of SDA involvement in advisory services

Information deficiencies	*Focus of initiatives*	*Development of services (year of commencement, sponsor)*
General management	Finance and organisation	Small, originally rural (1976, SDA). New gradually to Enterprise Trusts (1981-, SDA)
	New firms and entrepreneurship	Small & new gradually to Enterprise Trusts (1981-, SDA)
	Corporate strategy development	SDA invested firms (1978), later larger firms (1980) and sectoral focus (1986, SDA)
Marketing & promotion	Trade promotion	First small (1977), then broader and sectoral (1987, SDA)
	Quality assurance and branding	First small (1983), then general and sectoral (1988, SDA)
Business Intelligence	Subcontracting	First small (1976), then sectoral (1981) and general (1989, SDA)
	Market Intelligence	First small (1983), then broader and sectoral (1985, SDA)
Production & technology	Application of standard technology	Small (1978), later broader (1983) and partly sectoral (1990, SDA)
	Transfer of new technologies	General, partly sectoral (1982, SDA)
Access to advisory services	Advisory services brokerage	Small Firms Information Service (taken over 1981, initiated in 1970s by DTI)
	Grants for small firms	Better Business Services (1985, European Commission & Scottish Office)
	Grants for SMEs	Consultancy Initiative (1988, DTI)

Source: SDA 1976-91.

Advisory Activity and Financial Resources

Over its nearly fifteen years of existence the SDA continuously employed advice in order to promote regional development, and Table 11.1 provides an initial overview, distinguishing between services on the basis of the business activity in which an informational deficit is assumed. It is immediately obvious that throughout the advisory services sponsored by the Agency covered a wide range of issues relating to the successful running of private firms, from general management practices, via production and technology, to key aspects of marketing in terms of information about opportunities and ways of pursuing these. In addressing all four types of informational deficiency, both general and more specific forms of support have been employed, and with the exception of business intelligence which by its very nature concern particular potential markets, the more specialised services have tended to emerge later than the more general ones.

With regard to *management and strategy*, the rural small-firm services inherited from SICRAS were gradually extended to urban areas, and the emphasis on new firms and entrepreneurship in the early 1980s increased demand for advice to an extent that made 'off-loading' the smallest and newest firms to the growing network of Enterprise Trusts an attractive option, not just for political but also for organisational reasons.[34] The more specialised services aiming to strengthen the strategic capabilities of larger Scottish companies were originally aimed only at SDA invested firms, but coverage was gradually extended and focused sectorally so that by the early 1990s the BOP scheme had taken on a high-profile life of its own.[35] *Marketing and promotion* was also quickly taken on board with regard to small firms in the form of support for participation in trade fairs and exhibitions in the UK and abroad, and in the mid-1980s human resource development in the form of marketing training courses was introduced.[36] A specialised promotional device was the *quality assurance and branding* schemes, originally introduced as an SBD service to help small firms live up to increasingly exacting standards but later functioning mainly as part of sectoral initiatives which combined certification schemes with an attempt to brand Made-in-Scotland food or textiles as high-quality on the back of their 'pure' nature.[37] Provision of *business intelligence*, was a more difficult area to engage in because information had to be very specific in order to be useful for private economic actors, but in 1983 a market research unit was established within the SBD, and provision of advice on market opportunities also became an integrated part of most sectoral initiatives.[38] A similar pattern can be observed with regard to advice on subcontracting opportunities, introduced early by the SBD and quickly taken up in relation to sectoral initiatives, starting with electronics al-

---

34 SDA 1977 pp. 19ff, 1978 p. 40, 1979 pp. 22 & 42f, 1981 p. 40, 1985 p. 50; IDS 1987a p. 53.

35 SDA 1978 p. 8, 1980 p. 28, 1981 pp. 27f, 1982 p. 42, 1983 p. 47, 1984 pp. 6 & 44, 1985 p. 46, 1986 pp. 45f & 58, 1988 p. 80, 1989 p. 35, cf. Section 9.2.

36 SDA 1977 pp. 19ff, 1978 p. 41, 1979 pp. 22 & 42f, 1981 p. 40, 1982 p. 54, 1985 pp. 50f, 1987 p. 50, 1988 pp. 37-41 & 77, 1989 pp. 31f, 1990 p. 25.

37 SDA 1983 p. 54, 1988 p. 47, 1989 p. 35, 1990 p. 24, cf. Section 9.2.

38 SDA 1983 p. 52, 1985 p. 46, cf. Section 9.2.

ready in 1981.[39] The SDA also engaged in forms of advice exclusively oriented towards *production and technologies*, first in the guise of initiatives creating awareness of the opportunities involved in applying new technologies,[40] and later technology transfer, a much more complex and uncertain area, became a priority, aiming to help Scottish firms access new technologies through licensing agreements, joint ventures and commercialisation of research undertaken in public research institutions.[41]

In addition to its own initiatives, the Agency also acted as a handling agent for a limited number of programmes sponsored by central government, aiming to improve the *access to advisory services* of private firms. In 1981 the DTI-sponsored Glasgow office of the Small Firms Information Service was incorporated so that the SDA now provided a signposting service which allowed firms to identify relevant public providers of advice,[42] but from the mid-1980s the emphasis shifted towards encouraging small firms to use external advisors by subsidising the expenditure incurred, working in parallel with central government and partly on the basis of European funding.[43]

In a labour-intensive activity like provision of advice, expenditure is an obvious indicator of changing levels of activity, and as illustrated by Figure 11.1, both expenditure on direct provision of advisory services and on other forms of informational support, including support for Enterprise Trusts and expenditure on schemes administered for central government, increased in real terms, especially in the 1980s, and as the overall level of SDA expenditure in the 1980s was fairly stable,[44] the relative position of advisory services within the organisation's policy profile was strengthened.[45] All in all this clearly demonstrates that the new emphasis on support for the use of private providers has been additional to rather than been replacing advice delivered by the Agency's own staff.

---

39 SDA 1976 par. 38, 1978 p. 40, 1979 pp. 22 & 42, 1981 p. 41, 1982 p. 54, 1983 p. 54, 1984 p. 59, 1989 p. 31, 1991 p. 13.

40 SDA 1978 p. 40, 1979 pp. 49f, 1985 pp. 39f & 51f, 1987 p. 66, 1990 p. 25.

41 SDA 1982 p. 42, 1983 p. 47, 1984 p. 38, 1986 pp. 38ff, 1987 pp. 49f.

42 SDA 1981 pp. 40f, 1982 p. 50, Wren 1996a p. 168, cf. Section 11.3.

43 SDA 1985 p. 51, 1986 p. 49, 1988 p. 48, 1991 p. 13; IDS 1988, Wren 1996a pp. 169 & 227, cf. Section 11.3.

44 Cf. Figure 6.7.

45 While its average share of SDA expenditure was around 1% in the late 1970s, the equivalent figure was more than 15% in the early 1990s (calculated on the basis of SDA 1977-91). This is also confirmed by the rising the number of company visits to (small) firms which more than doubled from the late 1970s to the late 1980s (calculated on the basis of SDA 1977-88).

Figure 11.1 SDA expenditure on advice

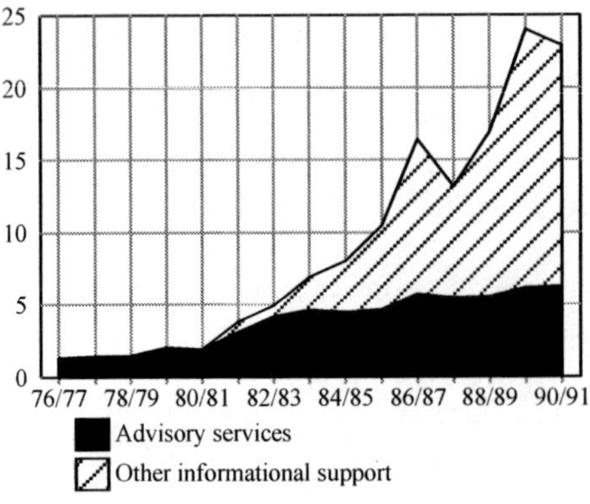

£m 1985/86.
Source: SDA 1977-91.

Policy Instruments and Modes of Implementation

It goes without saying that within advisory services information is the central resource which private economic actors are given access to, but still the conditions under which resources are made available must be considered, also because this may entail the transfer of resources other than information.

The advisory services operated by the Agency involved access to information on a variety of terms. At one extreme very standardised activities – e.g. the signposting service run on behalf of central government and some of the routine-like functions relating to small and new firms – could be said to provide information on a voluntary basis in the sense that it was up to the private actors to make the best of the input which came with 'no strings attached'. In contrast to this some of the more specialised services, e.g. the BOP management development schemes or the activities concerning technology transfer, appear to have offered access to information on a conditional basis, i.e. that firms accepted to commit considerable amounts of management time and to undertake certain tasks in order to be able to participate.[46] As illustrated by Table 11.1, the SDA's provision of advice came to include more and more specialised services over the years, and thus the relative importance of conditional access to information will have increased through the 1980s.

[46] Charles Fairley 26.7.90, Gerry Murray 30.7.90, Iain Shirlaw 1.6.90, Mike Sandys 31.5.90, SDA 1988a.

The Agency's advisory services were either free or heavily subsidised, at least from the early years to the mid-1980s,[47] and to the extent that similar services were available from private providers, this will have enhanced the attraction of Agency services because financial resources were used as an additional incentive to private firms. The 1986 government review singled this out as a major problem and recommended "a charging policy for directly-provided services designed to encourage private provision, and which takes account of the benefits to firms receiving the services",[48] and it could therefore be expected that the income generated through the fees charged would rise significantly from the late 1980s onwards. This did clearly not happen: while internally generated income reached its highest level after the review had been published, this expansion barely matched the extended scale of operations, whether measured in terms of the balance between income and expenditure or in relation to other activity indicators.[49] Despite criticism from its political sponsors, the advisory services of the SDA will have maintained their dual attraction to the very end, namely to provide useful information at very competitive prices.

The modes of implementation through which the Agency operated its advisory services entailed detailed rules about generation, selectivity and appraisal of projects, and also here differences within the broad range of information-based programmes are in evidence. With regard to *project generation*, the signposting service operated on behalf of central government from the beginning of the 1980s appears to have been reactive in that information was passed on to individual firms on their request. While all advisory services will to some extent have responded to enquiries, it is also generally agreed that those provided by the SDA were generally characterised by a relatively high, and indeed increasing, degree of proactivity. On the one hand the high-volume small-firm oriented and relatively standardised services provided by the SBD became increasingly proactive as the programme of annual visits to establish the needs of individual companies expanded.[50] And on the other hand the introduction in the 1980s of the more specialised services such as sector-specific BOPs and technology transfer also involved a

47 Page 1977 pp. 92f, Kirwan 1981 pp. 16f, Lochhead 1983 p. 106, IDS 1987a pp. 54f, cf. Charles Fairley 26.7.90, Gerry Murray 30.7.90, Edward Cunningham 13.6.90.

48 IDS 1987a p. 57.

49 At no point in the 1980s did internally generated income constitute more than 3.2% of expenditure under the advisory-service heading, and while the number of advisory visits more than doubled from the late 1970s to the early 1990s, the average level of internally generated income only increased 30% (calculated on the basis of SDA 1976-79 and 1988-91).

50 Lochhead 1983 p. 119, Jennifer Forbes 15.6.90.

more proactive approach, with groups of firms or individual companies being actively approached by the Agency.[51]

The advisory services of the SDA were inherently *selective* in the sense that particular forms of information had varying degrees of usefulness to individual firms.[52] *Size* of firm was the most common selectivity criteria, but although the grant schemes administered on behalf from central government from the mid-1980s onwards focused on SMEs, the Agency's own services did in fact move in the opposite direction: most forms of support started out by targeting small firms but subsequently came to include somewhat larger firms in the 1980s.[53] In terms of *sectors* some of the more specialised services were provided as integrated parts of SDA sectoral initiatives, while with regard to *area* selectivity, it was only the services oriented towards general management issues in small firms that until the late 1980s had a spatial dimension, originally because the inherited rural services were expanded to urban areas, and later as some areas were given additional attention through the direct organisational presence in particular localities of an area initiative office or an Enterprise Trust.

Finally, with regard to *appraisal* of individual projects, the advisory services implemented by the SDA fall in two categories. On the one hand the activities sponsored by central government would seem to operate on an automatic basis: signposting services were generally available 'on demand',[54] and the assessment of applications for grant support towards the use of private providers of advice appears to have been based on overt eligibility criteria such as firm size, what consultancy services could be supported etc.[55] On the other hand most of the Agency's own informational services would appear to involve a significant measure of discretion in the sense that the decision to proceed with advising a particular firm relied on judgements about its managerial qualities, through the discretion of 'frontline bureaucrats' or in the case of specialised services such as BOP through more formalised procedures.[56]

All in all it can be concluded that the trend with regard to modes of implementation was towards greater diversity, both in terms of genera-

---

51 Jennifer Forbes 15.6.90, Charles Fairley 26.7.90, Gerry Murray 30.7.90, Mike Sandys 31.5.90, Iain Shirlaw 1.6.90.

52 Cf. Table 11.1.

53 The SBD focused on firms employing less than 100 (Lochhead 1983, IDS 1987a p. 53).

54 Lochhead 1983 pp. 73f.

55 *Industry Act Annual Report* 1989 p. 61.

56 Charles Fairley 26.7.90, Gerry Murray 30.7.90.

tion, selectivity and appraisal of the suitability of individual firms as absorbers of new, economically useful knowledge – and towards patterns of interaction in which SDA priorities became increasingly important.

Advisory Services and Regional Economic Change

Assessing the impact of advisory services on individual firms is a demanding task because of the difficulties involved in tracking processes of knowledge development in individual firms,[57] but no systematic evaluation of Agency advisory services was ever published,[58] and thus it is only possible to address the question of firm-level impact in a rather general way, namely by focusing on the types of firms that were most likely to have benefited from the services provided.

As suggested by Table 11.1, the focus of the advisory services has exclusively been on promoting the expansion or modernisation of existing indigenous firms.[59] Although the importance of new firms and 'entrepreneurialism' became a conspicuous element of Agency parlance from the late 1970s onward and their public visibility increased through support for Enterprise Trusts in the 1980s, existing firms clearly remained a major target group, and another growth area became specialised forms of advice to existing firms about management practices, markets, and technologies. While the evidence available does not make it possible to ascertain the extent to which the firms advised actually did become more efficient, it has at least been possible to establish what firm-level impact the organisation did or did not try to achieve through its information-based activities.

Different views exist about the extent to which the SDA's advisory services brought about institutional change within the Scottish economy. On the one hand the 1986 government review alleged that subsidised provision of advice had been detrimental to private-sector providers of business services within the region, to the extent that even 'pump priming' of new services should be limited in order not to skew the future market for information.[60] On the other hand, the SDA and others have argued that there were indeed differences between what it provided itself and what was being – or could be – provided by private consultants:

---

[57] See e.g. Helmstädter *et al.* 2003.

[58] The 1986 review only refers to a 1984 survey which reported a positive impact of the Agency's advisory services, but no further details are given (IDS 1987a p. 56).

[59] This reflects the organisation of the book according to the 'bundling' of activities most commonly used by the SDA and its sponsor department, as promotion of inward investment were of course partly an advisory service aimed at economic actors outwith the region.

[60] IDS 1987a pp. 55ff.

- visibility through 'over-provision' is necessary because small-firm managers, a major part of the target group, are for many reasons – education, self-sufficiency, lack of time – less likely to make use of external sources of information,[61]
- it may be desirable to supply information that managers themselves have not recognised the need for, and this task of active market creation is less likely to be undertaken by private consultants primarily geared to exploiting existing opportunities,[62] and
- having a broad range of policy instruments enabled the SDA to employ its advisory services as part of an integrated approach which combined e.g. informational and financial resources in order to address the specific problems of particular firms.[63]

While the first argument is a general, and, indeed, contentious, one,[64] the second one hinges on the innovativeness and risk-willingness of private consultants within the regional economy, and the well-documented relatively small size of this sector in Scotland in the 1980s[65] could be cited in support of both interpretations: as evidence of the limited capacity of the private sector to take on the tasks performed by the Agency, or, conversely, as the result of 'crowding out' of private providers by public provision. Ultimately, the third point would appear to be the only one undisputed,[66] namely the institutional change brought about by the Agency's advisory services being able to integrate the use of informational and other policy instrument in support of private economic actors throughout lowland Scotland.

11.3. The British Context: Advisory Services and Regional Development

While it has been demonstrated that the SDA not only continued the advisory services inherited from its predecessor organisations but also, especially in the 1980s, developed new ones, it still remains to be seen whether these innovations were unique to the organisation or simply reflected general trends in the British regional policy network.

---

61 See e.g. Hull 1990, Amin & Thrift 1994, cf. Halkier & Damborg 2000.

62 Cf. Section 7.1.

63 A central assumption within the RDA approach to regional policy, cf. Section 3.4.

64 See e.g. Cooke & Morgan 1993, MacLeod 1997, Markusen 1999.

65 E.g. O'Farrell 1992. cf. Section 9.2, the SDA undertook a sectoral initiative in business services.

66 Lochhead 1983 p. 99, IDS 1987a p. 17, George Mathewson 9.7.90.

Central Government Regional and Industrial Policies

Reflecting the general strategic orientation of British regional policy towards addressing the urban consequences of industrial restructuring through financial subsidies for investment, provision of advice had traditionally only played a role as a means to support small firms in rural areas, and this situation continued for most of the period under review here. It was only in 1988 that expenditure for more 'intangible' purposes became eligible for central government regional aid: the Regional Enterprise Grant provided financial support for innovation and knowledge projects in very small firms,[67] and the Consultancy Initiative subsidised the use of external consultants by SMEs with a somewhat higher rate in the Assisted Areas than in the rest of Britain.[68] In practice the predominance of 'tangible' investment support did, however, continue through to the early 1990s,[69] and although the need to handle large numbers of grant applications had prompted the setting up of deconcentrated central government offices in the Assisted Areas, these offices never branched into advisory activities. Support for small firms became a priority in the adjoining field of industrial policy, especially from the late 1970s onwards,[70] and information-based measures were introduced, with a national network of Small Business Information Centres being gradually established from 1973 onwards. These centres provided basic advice and signposted small firms to other sources of information, and in 1987 the activities became part of the Small Firms Service before being off-loaded to the sub-regional Training and Enterprise Companies established in England and Wales in 1990.[71] In addition to this direct public provision, support was given to the setting up of Enterprise Trusts from the early 1980s, i.e. public-private partnerships which provided advice and other forms of support for new and very small firms.[72] Both these forms of informational support for small firms

---

67 DTI 1988 p. 30, cf. Wren 1996a pp. 186ff.

68 Wren 1996a pp. 187f.

69 Expenditure on the Regional Enterprise Grant accounted for only around 4% of central government regional subsidies in the period 1989-1991 (calculated on the basis of Yuill *et al.* 1989 & 1993, Wren 1996a p. 81), and the 'additional' Consultancy Initiative subsidy for firms located in Assisted Areas was of roughly the same size (calculated on the basis of *Industry Act Annual Report* 1991 p. 55 Tables 1 & 2).

70 With regard to small firms the Assisted Areas lagged behind relatively stronger regions like the South East of England, but national programmes of financial support introduced to strengthen small firms resulted in a geographically skewed take-up and hence in effect exacerbated existing inter-regional disparities (Mason & Harrison 1989, Barkham 1992, Mason 1992, Harrison & Mason 1993, Hart *et al.* 1993).

71 OECD 1977 pp. 141f, Mason 1987, Barberis & May 1993 pp. 199f, Smallbone *et al.* 1993.

72 *Industry Act Annual Reports* 1984-91, cf. Moore & Richardson 1988.

expanded in the 1980s, with preference given to direct provision through the Small Firms Service,[73] but compared to equivalent activities in Scotland, the level of expenditure would seem to be relatively low – in fact from 1983 onwards total SDA spending under its advisory services heading was 3% larger than central government expenditure on direct provision of advice to small firms in England[74] – and it therefore hardly surprising that evaluators found that private firms were rather unimpressed by the service provided through the DTI-sponsored network.[75]

All in all it can be concluded that although a move towards using informational instruments in central government already began in the early 1970s and gathered pace in the mid-1980s, supporting firms by improving their access to external sources of advice remained a marginal phenomenon within British regional and industrial policy.

New Actors, New Advisory Services?

Also actors situated above and below the national level to some extent engaged in advisory services in order to promote regional economic development, much in line with their counterparts elsewhere in Western Europe,[76] but again the British picture is a diverse one.

While the European Structural Funds supported information as a tool for regional development from the early 1980s by co-funding programmes such as Better Business Services in Scotland and equivalents elsewhere in Britain,[77] the sums involved in this were dwarfed by the amounts dedicated to e.g. infrastructure,[78] and this pattern of priorities recurred in the first post-1988 programming period.[79] Conversely, local authorities across Britain clearly saw advice as a useful policy instrument and dedicated significant shares of the economic development budget to informational support of especially new and firms, both directly by setting up their own advisory services and indirectly through sponsorship of e.g. Enterprise Trusts.[80]

---

73 In the late 1980s expenditure on the Small Firms Service was three times that on support for Enterprise Trusts (calculated on the basis of *Industry Act Annual Report* 1987-91).

74 Calculated on the basis of SDA 1983-91 and *Industry Act Annual Reports* 1983-91.

75 Smallbone *et al.* 1993.

76 See e.g. Bachtler 1993, Halkier & Danson 1997.

77 Lochhead 1983 p. 106, cf. Section 11.2.

78 Better Business Services constituted only 3% of ERDF expenditure in Scotland from 1985 to 1989 (calculated on the basis of *Industry Act Annual Reports* and *ERDF Annual Reports*).

79 See e.g. Hall Aitken Associates 1995 & 1996.

80 Lochhead 1983 pp. 68f, Todd 1985 Ch. 4, Keating & Boyle 1986 pp. 137ff, Moore 1988b, Totterdill 1989, Harding 1990, McQuaid 1992, Armstrong & Twomey 1993.

The rural RDAs provided advice along much the same lines as the SDA's Small Business Division, focusing on the needs of small firms and adopting an increasingly proactive approach over the years. Both the Development Commission and the HIDB provided much advice themselves, and in the Highlands central government schemes supporting the use of private consultants was administered by the RDA under arrangements similar to those which gave the SDA responsibility for the Scottish lowlands.[81] In contrast to this, the involvement of the industrial WDA in advisory services appears to have been rather more limited, despite having been vested with the same powers as the SDA and having inherited the Welsh arm of the Development Commission.[82] The WDA did extend the remit of its Small Business Unit to include urban areas,[83] but even at its peak in the late 1980s provision of advice still accounted for less than 5% of total expenditure,[84] most likely because advisory services for larger firms such as technology transfer and subcontracting were introduced at a much later stage than their Scottish equivalents,[85] and when even information-based services related to the organisation's main strategic focus, inward investment, was emerging rather late and partly prompted by the Welsh Office,[86] this could readily be interpreted as a long-standing WDA preference for policy instruments of a more tangible nature.[87]

Among the 'new' actors in regional policy it was in other words the minor ones, the two rural RDAs and the local authorities, that appeared to embrace the use of information as a means to influence the behaviour of private firms – while the 'big' players, the ERDF and the WDA, continued to give priority to other types of incentives.

11.4. Advisory Services for Regional Development

Advisory services was a policy area likely to have developed in accordance with the external-revolution interpretation. As a policy instrument, talking to managers is much less intrusive than owning shares in their firms, and as the discourse of the Thatcher government extolled the virtues of market competition, entrepreneurs setting up new firms were

---

81 IDS 1987b Ch. 7, 1988; Development Commission 1984, Tricker & Martin 1984, Rogers 1999 Ch. 10-12.

82 WDA Act 1975, WDA 1981a, Hogwood 1982 p. 53.

83 WDA 1981a, Welsh Office 1987 Ch. 8, cf. Howe 1996.

84 Calculated on the basis of WDA 1977-91.

85 Welsh Office 1987 Ch. 9, Welsh Affairs Committee 1989 pp. 85f, Advisory Council on Science and Technology 1990, cf. Morgan 1996, Morgan & Henderson 1997.

86 Welsh Office 1987 Ch. 9.

87 Cf. Chapters 10 and 12.

likely to be cast as heroes also in regional economic development, and it could therefore be expected that advisory services would expand after 1979 and become increasingly oriented towards small and new firms. The empirical analysis of the development of the advisory services of the SDA has, however, pointed in a different direction. Although some aspects could be interpreted as being in accordance with an external-revolution perspective, the overall impression is that this area of activity has combined constant evolution of inherited functions with creation of innovative programmes centred on making useful knowledge accessible to private economic actors.

In relation to the inherited responsibility for small firm development – which included a long-standing combination of access to finance and information – subsequent developments came to be an ongoing process of change: first the remit was extended from rural to urban firms, then the organisational presence across Scotland was gradually enhanced in order to be close to potential 'customers', before small business advice eventually was integrated into the new regional offices in the late 1980s – and throughout a range of information services was maintained which covered most aspects of business development, from general management issues via production technologies to marketing. The gradual transfer of responsibility for new and very small firms to the partnership-based Enterprise Trusts could be interpreted as an example of the ideological impact of the Conservative government, and indeed the timing – the process gathered pace in the early 1980s – would seem to point in this direction, but other possible rationales for this off-loading also exist: the limited capacity for in-house advice created by the staffing limits imposed on the SDA by the Treasury, the access to informational sources that would be credible in the eyes of small businessmen, the possibility of influencing the economic policies of co-sponsoring local authorities, and the political convenience of setting up separate organisations to deal with issues important to the party of government.

In addition to the inherited remit for provision of advice to small firms, the SDA did, however, also develop a series of innovative uses of information in regional development, often derived from other policy activities like industrial investments or sectoral initiatives. Moreover, these new advisory services were by and large the opposite of what one would expect a Thatcherite government to condone: they tended to focus on relatively large indigenous firms, predominantly but not exclusively within traditional sectors of the Scottish economy such as engineering, food and textiles, and some of them, notably the BOP management development programme, attempted to influence core patterns of behaviour within firms and hence may have had a much more lasting impact on private actors than e.g. a reactive grant scheme that does not ask questions about strategic priorities but merely subsidises existing management thinking.

Compared to other policy functions, the rationale for advisory services – apart from the underlying assumption of 'knowing better' – was only made explicit at a fairly late stage, namely when the 1986 government reviewers applied their general market-failure approach. Despite acknowledging the highly imperfect nature of markets for information and allowing for pump-priming policies that could alert private economic actors, service users as well as service producers, to the existence of new opportunities, the insistence of the reviewers on the need to avoid perpetuating market failures through provision of subsidised information did in fact become the only major controversy between the Agency and its political sponsors in this area of activity. This could of course be interpreted as an attempted external revolution, but if so, it appears to have been spectacularly unsuccessful. Although corporate rhetoric occasionally paid lip service to the need to stimulate private provision, nothing much happened with regard to increasing the charges paid by firms advised by SDA staff, and while this can hardly have gone unnoticed in the Scottish Office, perhaps the Agency's administration of central government schemes subsidising the use of private consultants may have been considered a sufficient degree of compliance, at least in political terms.

If the development of the advisory services of the SDA had not been driven by the discourse of their political sponsors, it can hardly be said to have replicated developments in other parts of the British regional policy network either. In short, compared to other public actors, the Agency's information-based activities were either a much more important part of its policy profile, and/or more innovative in terms of the services provided, e.g. by being a selective and proactive part of sectoral initiatives. All in all, despite a fairly low public profile, advisory services would seem to be a policy area in which the SDA did what RDAs were supposed to do, namely to develop policies targeting specific regional problems through measures that integrated a variety of policy instruments.

CHAPTER 12

Properties, Environments and Regional Development

Regional policy in Western Europe has not only been concerned with the 'software' of economic development – using access to financial or informational resources to promote particular localities – but has also employed 'hardware', i.e. the creation of attractive physical facilities which would allow private firms to organise production in a profitable and efficient manner. In Britain industrial estate companies started to provide subsidised factory space in the 1930s, attempts to clear away obstructive or unsightly legacies of early industrialisation acquired a regional dimension in the 1960s, and physical planning in the guise of e.g. new towns also became an integrated part of regional policy in many Assisted Areas.[1] As "some of the most potent symbols of and industrial age are to be found in its built environment",[2] substituting the 'dark satanic mills' and Victorian multi-storey factories in confined inner-city locations with facilities suitable for high-tech mass production could, however, also be interpreted as a symbolic repositioning of a locality in terms of temporal assumptions, leaving the past behind and demonstrating commitment to future-oriented industrial change. In other words, not despite but because of the visibility and endurance of much infrastructure, this form of public intervention could readily be seen as a tool of 're-imagineering' in order to change the internal and external perception of particular localities.[3]

Although the infrastructure-based functions of the SDA have not been surrounded by the same level of controversy as industrial investments or inward investment, their development has still been interpreted in rather different ways. From an external-revolution perspective on the development of the SDA it has been stressed that[4]

---

1 Cf. Section 4.3 and Chapter 5.

2 Pratt & Ball 1994 p. 1.

3 For a discussion of place promotion, see Therkelsen & Halkier 2004.

4 Barnekov *et al.* 1989 Ch. 7, Midwinter *et al.* 1991 pp. 193f, Danson *et al.* 1992.

- new guidelines for the industrial property function lead to a more commercial approach in the 1980s, culminating in the privatisation of a large part of Agency factories,
- the environmental activities became increasingly oriented towards projects with a potential for economic afteruse rather than as parks or other public amenities, and
- the spatially targeted 'area projects' moved away from having a social dimension in the late 1970s towards being industrial and economic in the 1980s.

As all of these developments could be construed as making Agency activities more 'market conform' and were either directly instigated by or agreeable to the Conservative government, this brings infrastructure activities in line with an external-revolution interpretation. In contrast to this, other authors have implied that many of the practices cited as inherently Thatcherite had in fact either existed, or at least been foreshadowed, well before the change of government in 1979, and hence the degree of change would seem to have been much more limited and not driven primarily by external political pressure.[5] In addition to the recurring issues concerning the relationship between public and private actors and between the SDA and its political sponsors, there are, however, other issues stemming from the characteristics of the policy instruments employed which are also likely to have influenced the development of Agency policies and how they were interpreted: being indivisible, immobile and long-lasting[6] meant that physical infrastructure would be tied to a particular site and hence an obvious object of inter-locality strife, but at the same time the more indirect link between public expenditure and additional economic activity made others dispute the relevance of infrastructure and especially environmental improvement in the context of regional development.[7] Again the extent and timing of change is in other words disputed, and thus also in this policy area a detailed enquiry is called for.

The analysis of the SDA's infrastructural policies will be based on published materials, especially the annual reports of the Agency, complemented by interviews with senior executives within the organisation and parliamentary enquiries investigating aspects of especially property-based development initiatives. A central feature of this policy area is the diversity of measures involved, but fortunately handling this diversity is to some extent facilitated by the existence of a sizeable body of litera-

5 Moore & Booth 1986e, Wannop 1990, cf. Keating & Boyle 1986 Ch. 1, Keating 1988b Ch. 4, Boyle 1993.

6 See Forslund & Karlson 1991 pp. 28ff, Spence 1992.

7 E.g. Douglas Harrison 19.7.90, Craig Campbell 16.7.90.

ture produced by academics interested in planning and urban policy[8] that makes it possible to draw upon the findings of previous research more than has been the case in other areas of SDA activity. Again the text falls into three parts: firstly the design of infrastructure-based policies is explored (Section 12.1) and then their implementation is examined (Section 12.2), both distinguishing between functions inherited from predecessor organisations and new ones added after 1975. Finally the broader British policy context is surveyed, both in terms of central government policies and the activities of RDAs and other sub-national actors (Section 12.3), and on the basis of this it will be possible to draw conclusions with regard to the relationship between continuity and change in the SDA's use of infrastructural measures and its relationship to other policy areas and the British policy network at large.

12.1. Design

Inherited Infrastructural Activities

In the 1970s proposals for setting up a lowland RDA, the provision of economic infrastructure and improvement of the quality of the physical environment can be found in some, but by no means all, of the blueprints circulated: the tripartite SCDI had repeatedly emphasised the importance of industrial property and the West Central Scotland Plan had highlighted the importance of improving the urban environment, but these issues were dwarfed by the critical importance accorded to industrial investment in the proposals emanating from STUC and Labour.[9] Central government had, however, been active in both industrial property and derelict land clearance from a regional development perspective for decades,[10] and the inclusion of infrastructural activities in the remit of the new organisation was therefore hardly surprising, especially when it involved devolution of powers from the British to the Scottish level and facilitated an integrated approach to regional development.

The 1975 parliamentary act defined provision of sites and premises as functions to be used in pursuance of economic development, and it specified that improvement of the environment included "bringing derelict land into use or improving its appearance".[11] With regard to

8 Wannop 1984, 1990, 1995; Lever & Moore (eds.) 1986, Keating & Boyle 1986, Donnison & Middleton (eds.) 1987, Keating 1988b, Boyle 1993.

9 See Chapter 5, in particular Table 5.2.

10 Cf. Section 4.3.

11 SDA Act 1975 Section 2. The Agency could pursue both these functions throughout Scotland, but in practice they were seldom used in the area covered by the HIDB: the economic history of the Highlands had left little industrial dereliction, and the facto-

industrial property, the Agency inherited the portfolio of industrial estates that had hitherto been administered by SIEC, the Scottish Industrial Estates Corporation, on behalf of the Department of Industry in London, and hence this function was regulated by the Scottish Office as part of the British policy network.[12] Similar conditions prevailed with regard to clearing of derelict land where the Agency incorporated a small unit within the Scottish Office which had administered support for projects initiated by local government, again much in parallel with similar schemes operating in England and Wales,[13] while the more general powers of environmental improvement were subject to ministerial approval with regard to individual projects[14] and thus more of an 'internal' Scottish affair. In practice the sponsoring of environmental activities appears to have operated in a relatively informal way, with guidance being "a series of notes and letters" between the Scottish Office and the SDA, supplemented by a delegated spending limit of 1/4 million pounds on environmental improvement projects that remained unchanged from 1975 to 1991.[15] In contrast to this the Agency's property function was regulated through formal guidelines issued by the sponsor department[16] which specified the purpose of the activity, the setting of rents and other conditions for tenants, the rules governing acquisition and disposal of sites and factories, and the role of the Scottish Office in the annual planning cycle. Moreover, in 1980 new guidelines heralded the introduction of performance indicators in the guise of financial duties,[17] but despite repeated misgivings by the Committee of Public Account in the early 1980s, it was only in 1983 that a financial duty was eventually introduced for the SDA and its counterparts in England and Wales,[18] and the 1986 review of the Agency proposed to adjust the formula.[19] A very important change from the pre-1975 regime governing the use of industrial property for regional policy purposes in Scotland was the extent to which powers had been delegated from

ries inherited from SIEC were transferred to the HIDB (SDA 1977 p. 22, Lochhead 1983 p. 142).

12 SDA Act 1975 Sections 6 & 15, cf. SDA 1977 p. 21, Committee of Public Accounts 1983 p. v, IDS 1987a p. 128.

13 SDA Act 1975 Section 8, cf. Prestwich & Taylor 1990 214f.

14 SDA Act 1975 Section 7.

15 Ian Hart 18.10.91.

16 While the incoming Conservative government published the property guidelines (SEPD 1980b), the original ones are only hinted at in the annual reports of the Agency (e.g. 1977 pp. 20f).

17 SEPD 1980b p. 5.

18 Committee of Public Accounts 1983 pp. viif, 28.

19 IDS 1987a p. 43.

central government: while SIEC had essentially been an executive arm of the Department of Industry in London, the SDA could decide the location of factories and select tenants without having to enter into prolonged negotiations with their sponsors in the Scottish Office.

The increased capacity for decision-making with regard to industrial property was, however, also a major challenge in terms of organisational resources because the staff and structure inherited were geared to large-scale implementation – from day one the SDA had become "Scotland's biggest industrial landlord"[20] and inherited SIEC staff accounted for more than half of the Agency's employees until 1983[21] – rather than strategic planning, and the latter continued to have an 'SIEC-external' dimension well into the 1980s.[22] In contrast to this, the environmental functions were the responsibility of a small team – only one person constituted the inheritance in human-resource terms – which saw itself as a technical service unit for other SDA entities and, indeed, for external champions of land clearance projects in especially local government, and the abundance of derelicts sites in lowland Scotland in combination with extensive reliance on external project champions and private contractors to undertake the physical work meant that environmental projects could act as a 'budgetary sponge', soaking up unspent resources at relatively short notice.[23] All in all this positioned the land renewal function as highly specialised implementors rather than makers of innovative policies in their own right, and the absence of attempts to install additional dynamics in this part of the organisation suggests that this has been generally agreeable to the Agency.

Also in terms of aims and methods was environmental improvement in many ways less complicated than the property function, except, that is, with regard to the rationale of the function. While the 1975 Act clearly gave the Agency two separate objectives, economic development and environmental improvement,[24] both the West Central Scotland Plan, the blueprint in which environmental issues were most prominent, and the official discussion paper issued by the Scottish Office claimed that

---

20 A slogan used by the SDA until at least the early 1980s (Rich 1983 p. 274).

21 Calculated on the basis of SDA 1976-87 (no breakdown of staff available for later years), cf. George Mathewson 9.7.90, Ian Hart 18.10.91, Glasgow Herald 14.5.84.

22 During the first two years factory policy of the SDA was the responsibility of the Industry Directorate (cf. Table 6.1), in the early 1980s the new Chief Executive George Mathewson established a high-level property policy group which considered all projects proposed (IDS 1987a pp. 35f, Ian Hart 18.10.91), and it was publically stressed that property should be treated "as one component of enterprise development and not an end in itself" (SDA 1984 p. 46).

23 Ian Hart 18.10.91, IDS 1987a Ch. 9, cf. SDA 1986c.

24 SDA Act 1975 Section 1.

the two objectives were intimately linked because industrial dereliction was seen to damage the image of certain areas and undermine the confidence of residents, and hence environmental improvement would help to promote economic development by eliminating impediments to both inward and indigenous investment.[25] The way the derelict land clearance programme operated, through 100% grant support for projects brought forward by local authorities, indicates that the scheme was essentially about central government bearing the abnormally high costs of bringing former industrial sites back in use, whether as public amenities or as 'brownfield sites' for private economic activities, with the Agency as gatekeeper for locally-driven projects. In the late 1970s the strategic statements of the SDA seldom touched more than briefly on environmental activities, but when they did, it was stressed that the environmental powers would be used throughout Scotland despite the concentration of efforts in and around Glasgow and that priority would be given to sites with prospects of industrial afteruse.[26] The ambiguity concerning what would appear to be the dual rationale of environmental improvement, desirable in its own right and/or conducive to economic development, continued after the change of government in 1979: land renewal was officially extolled – like any other activity except industrial investment – as one of the SDA's "principal functions",[27] but this did not lead to the function being mentioned separately in any of the Agency's corporate strategies in the early and mid-1980s,[28] and while the importance of industrial afteruse was still being emphasised,[29] new post-clearance purposes were gradually added – most importantly housing in 1984[30] – and spatial targeting increased with around 60% of environmental efforts now to be concentrated in a limited number of prioritised localities.[31] The 1986 government review endorsed the general rationale of public intervention in this area "because private sector investors are unlikely to give as much weight to environmental benefits as do the community as a whole",[32] but at the same time it also insisted that more attention should be paid to the environmental remit as an end

25 West Central Scotland Plan Team 1974a p. 107, SEPD 1975, cf. Gavin McCrone 20.6.90.

26 SDA 1977 p. 27, 1978 p. 51, 1978a p. 26, 1979 pp. 48f; Ian Hart 18.10.91, cf. Section 12.2.

27 SEPD 1980a p. 1.

28 See SDA 1981 pp. 57-63, SDA 1984a.

29 SDA 1982 pp. 64ff, Alan Dale, 16.10.91.

30 SDA 1984 p. 52.

31 Ian Hart 18.10.91, Morison 1987 p. 174, Keating 1988b Ch. 7.

32 IDS 1987a p. 63.

in its own right,[33] and in the 1988-91 corporate strategy of the Agency this was translated into a prominent "urban renewal" priority,[34] supplemented by attempts to develop a systematic ranking of 'awfulness' so that project selection could take expected environmental impact into account rather than primarily the economic one.[35] All in all it is evident that while the methods employed to tackle environmental problems did not change much from the late 1970s to the early 1990s, the way this was legitimised in terms of expected benefits would seem to have varied: from the late 1970s onwards the emphasis on economic afteruse for a widening range of purposes dominated – first general image and manufacturing, then also housing and city centre ambience – and only in the late 1980s did environmental improvement aimed primarily at individual citizens via public amenities regain some prominence.

The aims and methods of the other inherited infrastructure function, industrial property, were not expanded on at great length by either the SDA or the Scottish Office in the late 1970s where the underlying regional development rationale for public provision seems to have been taken for granted, while the widely used marketing slogan "Scotland's biggest landlord" foregrounded public ownership as a key feature of the function.[36] In practice, however, two strategic dimensions were clearly visible in the late 1970s: emphasis shifted towards units more suitable for the needs of small and new firms, both in newly-built property and through the conversion of large units inherited from SIEC,[37] and the building of advance factories – industrial property constructed without a known tenant in mind – was deliberately used to demonstrate the commitment of the SDA to all parts of Scotland.[38] The new factory guidelines issued by the Conservative government in 1980 stressed that public provision should *not* take place "in those areas of Scotland where (the SDA) judges private sector activity is sufficient to meet the requirements of industry"[39] – encouraging private-sector involvement in the financing of new ventures and ownership of existing properties – and that the property function should operate on a commercial basis in terms of the rents charged, although the method through which this should be secured meant that rents would still not reflect the actual cost of the

---

33 The report of the review group discussed the possibility of creating a separate derelict land quango but came out in favour of retaining the function within the SDA (IDS 1987a pp. 65ff).

34 SDA 1987a, cf. Frank Kirwan 1.6.90.

35 SDA Board Papers March 1988, cf. IDS 1987a pp. 65ff.

36 SDA 1978 p. 25, Rich 1983 p. 277, James Williamson 14.6.90.

37 SDA 1976 par. 11, 1977 pp. 21f, 1978 pp. 24ff, 1979 pp. 39f.

38 SDA 1986 par. 11, cf. Alan Dale 16.10.91.

39 SEPD 1980b p. 1.

service provided.[40] These two lines of thinking merged and were elaborated on in the light of existing experience in statements of strategy from both the Agency and the Scottish Office in the early 1980s: the primary rationale for intervening was now seen as the insufficient private provision of industrial property which could be rented by small/new firms, unwilling to buy premises because of uncertain prospects of growth or unable to do so because of insufficient financial means, while the existence of the capacity to undertake customised building projects for e.g. inward investors was seen as a secondary objective, essentially as a way to increase the financial attraction of Scotland *vis-à-vis* other locations.[41] Like in environmental improvement having most of the activity in a limited number of priority areas became a goal in its own right, but despite unequivocal demands by the new Conservative government for more private sector involvement, it was only in 1985 after a lengthy internal strategic review had concluded that "direct provision [...] is viewed as only one (relatively low priority) option" that disposal of existing properties seemed to have become firmly entrenched as a corporate priority.[42] Unsurprisingly, the 1986 government review reiterated the argument about private-sector failure in the market for small industrial premises for rent and the need to encourage more private sector participation,[43] and the Agency presented this as an endorsement of its internal property strategy review which culminated in the early 1990s disposal of "mature and developed elements of its industrial property portfolio" as the completion of its transition from "asset manager" to "strategic property developer".[44] All in all the strategic development of the property function would seem to display an interesting combination of change and continuity: while the inherited role as manager of existing industrial estates was gradually downgraded in the 1980s – prompted by new Scottish Office guidelines and emerging after a prolonged internal process as official Agency policy just before the 1986 government review – the role as property developer was maintained and the main strategic change, more emphasis on property for small and new firms, had been firmly put in place well in advance of the change of government in 1979. On this reading the impact of the change of government in 1979 would seem to have concerned the question of

40 SEPD 1980b.

41 Committee of Public Accounts 1983 pp. 16ff, cf. SDA 1981 p. 60, Chris Aitken 18.10.91.

42 SDA 1985 p. 54, cf. STUC reports of a meeting with Chief Executive George Mathewson on the SDA's new property policy (1985 pp. 54f). Property disposal is not mentioned in the 1985-88 corporate strategy (SDA 1984a).

43 IDS 1987a pp. 38ff, cf. SDA 1987 p. 38.

44 SDA 1991 pp. 6f, cf. Alan Dale, 16.10.91, Chris Aitken 18.10.91.

permanent public ownership of economic infrastructure rather than the principle or, indeed, the capacity of the SDA to provide subsidised industrial property for particular types of private firms.

New Infrastructural Activities

In addition to developing the inherited functions, the SDA also developed new infrastructural activities, and this section outlines the aims and methods of four new approaches to regional development, beginning with high-tech property, then considering the high-profile area initiatives of the Agency, and finally turning to one-off projects and financial support for property-related developments in city centres.

From a property perspective a significant innovation was the move into specialised forms of *advance high-tech premises* such as science parks and incubator units. While the SDA and SIEC had always been willing to build so-called 'bespoke' factories for individual tenants to particular specifications, now the sustained attempt to encourage the development of more high-tech industry in Scotland was also supported by high-tech facilities built in advance of specific demand, located in the vicinity of universities and providing not only accommodation suitable for research and development purposes, but also ready access to advisory or financial services of relevance to academic would-be entrepreneurs.[45] The first plans for advance high-tech premises were announced in 1981, and the scale and geographical spread of the activity gradually increased through the 1980s, operating as a series of one-off projects in collaboration with universities and other research establishments and subject to the general supervision of the Scottish Office.

In terms of financial resources and public profile, the most significant infrastructural policy innovation was the so-called *area initiatives* which focused a range of different policy instruments – primarily property and environmental improvement but also advisory and other services – within a designated locality while at the same time coordinating Agency efforts with those of other public and private actors. The 1975 Act explicitly stated that "the Agency may exercise their functions in relation to Scotland or any part thereof",[46] and both the preceding Scottish Office discussion paper and the West Central Scotland Plan had specifically mentioned the severe unemployment and dereliction faced by parts of the Glasgow conurbation as something the proposed RDA would have to address in collaboration with other public actors.[47] No separate guidelines governed the implementation of individual area

[45] SDA 1981 p. 43, 1982 p. 34, 1983 p. 36, 1985 pp. 40f, 1988 pp. 79f, 1991 p. 26.

[46] SDA Act 1975 Section 2 Subsection 1.

[47] SEPD 1975 pp. 3f, West Central Scotland Plan Team 1974a.

initiatives, and it was only in the late 1980s that attempts were made to institute a systematic approach to evaluation of these activities.[48]

Table 12.1 Organisational features of SDA area initiatives

		Main implementor	
		Partnership	*SDA*
Initiator	*Scottish office*	GEAR Peripheral estates	Task forces
	SDA	Self-help initiatives City-centre coalitions	Integrated area projects

The organisation of area initiatives changed over the years, both with regard to relations with external partners and internally within the SDA,[49] and on the basis of the origins of the individual initiatives and the approach to implementation adopted, the various types of area initiatives can be divided into four groups, as illustrated by Table 12.1. The first area initiative, the Glasgow Eastern Area Renewal project (GEAR), was initiated by the Scottish Office in May 1976 shortly after Lewis Robertson had taken up his full-time position as the first chief executive of the new arm's-length body, and it was the only one which the Agency itself placed under the *comprehensive urban renewal* heading. Focusing on the run-down eastern parts of the city where environmental dereliction and unemployment had replaced traditional industries like textiles and heavy manufacturing as prominent local features, GEAR entailed the SDA entering into an open-ended cooperation with local authorities and other public bodies where each of the partners retained full responsibility for their statutory tasks, and the role of the Agency as the coordinator relied on its ability to persuade partners on the back of its considerable resources with regard to environmental renewal and property development.[50] Although GEAR was a one-off in terms of the level of resources involved,[51] a similar organisational form was adopted in the late 1980s for another spatially-targeted initiative, namely a series of *peripheral estates partnerships* announced by the

[48] IDS & SDA 1988. Some individual initiatives had been subject to extensive evaluation (see Donnison & Middleton (eds.) 1987 pp. xvf).

[49] For overviews, see Gulliver 1984, Keating & Boyle 1986 Ch. 5, Moore & Booth 1986e, Morison 1987 Ch. 6 & 7, Keating 1988b Ch. 4 & 7, McCrone 1991a.

[50] SDA 1980a.

[51] Cf. Section 12.2.

Scottish Office in a 1988 white paper on urban policy.[52] These partnerships targeted derelict low-amenity public housing estates suffering from high levels of unemployment, and again the SDA was to take the lead in bringing together other, predominantly public, actors under organisational forms that echoed GEAR, albeit on a more limited spatial scale and maintaining a much lower public profile.[53] *Task Force* was the name used for the second generation of area projects, and this approach was employed in two localities, Garnock Valley and Clydebank, which had both been hit by severe unemployment problems in the wake of closures of major local employers. Again the Scottish Office, first under Labour and then under the Conservatives, had initiated activities, and again commitment was initially open-ended in terms of time and resources – but unlike GEAR the SDA was now clearly the dominant partner, with other public actors playing a more marginal role.[54] *Integrated area projects* were initiated by the SDA in the early 1980s on the basis of preparatory studies, and this organisational framework involved contractual relations between the SDA as the coordinating body and other actors, not least local authorities, which clearly defined the resource commitments of each partner. In contrast to this the so-called *self-help projects*, also introduced in the early 1980s, were only initiated and prepared by the Agency but required extensive local participation from both public and private actors because, unlike the two previous categories, they were implemented not by a locally based SDA team but by enterprise trusts, something which gave private sector actors a greater role in implementation.[55] Along similar lines the SDA also promoted what could be termed *city centre coalitions* – *Glasgow Action* was a prime example – which involved the setting up of a steering committee consisting of local private and public notables, acting like a think-tank but relying on its constituent parts for any action that needed to be taken in order to improve e.g. the image of a particular locality.[56]

The various organisational forms adopted have generally been interpreted as reflecting the evolving positioning strategies of the Agency and its sponsor department. Using the SDA as the coordination vehicle in urban renewal, first in GEAR and then in the peripheral estates

---

52 IDS 1988c, cf. Boyle 1989,

53 SDA 1988 pp. 70ff, 1989 pp. 56ff, 1990 p. 37, 1991 pp. 27f; cf. Boyle 1989.

54 SDA 1979 pp. 54f, 1980 pp. 50f, 1981 pp. 48f.

55 SDA 1981 p. 62, 1982 pp. 55-63, 1983 pp. 55-67, 1984 pp. 62-73. The claim by Lever (1987 p. 255) that integrated area projects are "rescue operations after major industrial collapses" fails to take the Agency's distinction between this and the so-called task forces into account.

56 SDA 1985 p. 71, 1987 pp. 110f; Wannop & Leclerc 1987 p. 229, Keating 1988b Ch. 7, Boyle 1989, McCrone 1991a.

partnerships, has generally been seen as an attempt by the Scottish Office to limit the influence of local authorities, either for functional or political reasons.[57] Conversely, proactively initiating the time-limited and contract-based integrated area projects could be interpreted as a way for the Agency to defend itself against being used by the Scottish Office as a fire-brigade in the event of a firm closing down and producing high levels of unemployment in particular localities,[58] and the increasing involvement of private-sector actors in self-help initiatives through enterprise trusts had the dual advantage of economising with the SDA's limited human resources while at the same time being inherently popular with Conservative ministers. Within the Agency itself a dedicated directorate handled area projects from 1976 onwards, and originally staff had primarily been recruited from a recently abandoned new-town project, thereby bringing planning-oriented competences to the organisation that was otherwise dominated by industrial economists and engineers,[59] and thus a potential tension between spatial and Scottish-wide goals may have been inscribed in the internal structure of the organisation.

In terms of aims and methods the area initiatives revolved around two types of strategic assumptions. *Firstly*, the rationale for targeting policies at selected localities rather than Scotland as a whole not only changed but possibly also remained a matter of dispute between the Scottish Office and the SDA. While the former clearly targeted on the basis of perceived need – massive dereliction and high levels of unemployment – and invoked this as the basis for both GEAR and the two task-force initiatives,[60] the Agency insisted that its selection of areas for integrated projects and self-help initiatives was based on having identified a discrepancy between a current crisis and the underlying development potential.[61] According to the sponsor department the rationale for area projects was in other words largely social or political, taking visible action after the collapse of major industrial undertakings, while the arm's-length body presented area projects as targeting of 'underperforming' localities that currently did not make a full contribution to the growth of the Scottish economy, i.e. an economic rationale at the regional level. Although at the level of corporate rhetoric the Agency's perspective would ultimately appear to have been accepted in the 1986

57 Leclerc & Draffan 1984, Naylor 1984, Keating 1988b p. 97, Danson *et al.* 1988, Boyle 1989, Wannop 1990, Midwinter *et al.* 1991 pp. 63ff.

58 Midwinter *et al.* 1991 pp. 184ff, David Lyle 19.7.90, cf. SDA 1981 p. 62.

59 Leclerc & Draffan 1984, Keating 1988b Ch. 4, McCrone 1991a, Pickvance 1990.

60 SDA 1979 pp. 50f, *Glasgow Herald* 21.8.79, 30.8.79.

61 SDA 1981 p. 62, SDA 1984a.

government review,[62] the subsequent involvement of the organisation in urban policy initiatives would seem to suggest that the Scottish Office still saw social need as a legitimate reason for letting the SDA concentrate its efforts in particular parts of Scotland, even if this entailed a good deal of "moving economic activity around" within Scotland rather than a net gain for the region as a whole.[63] *Secondly*, the specific goals involved in individual area projects also evolved over time. GEAR did combine economic and social goals with its emphasis on land renewal, industrial property and improved housing,[64] but the contribution of the SDA itself was primarily economic in nature, and all the initiatives subsequently initiated or implemented by the SDA focused much more single-mindedly on economic development, originally with a predominantly industrial focus. Since the mid-1980s, however, also new post-industrial images of Scottish cities were also promoted through support for e.g. the local authority initiated "Glasgow's Miles Better" campaign,[65] and this was eventually complemented by the peripheral estates partnership which had a predominantly social remit.

From a broader strategic perspective the move towards a greater focus on city centres and image repositioning had, however, begun well before the mid-1980s, and in fact some important infrastructural anchors for subsequent re-imagineering efforts had been instigated even before the change of government in 1979. On the one hand a series of *major special projects* in and around Glasgow were undertaken on a one-off basis combining land renewal and/or property powers of the Agency on a very large scale: the quay-side Scottish Exhibition and Conference Centre was initiated in the late 1970s and undertaken in partnership with local authorities and private developers,[66] the centrally located site of the former St. Enoch Railway Station was cleared in preparation for a major retail development and multi-storey parking,[67] and preparations for the UK Garden Festival in 1988 involved clearing a derelict dock area and managing the event in partnership with other public actors while having private housing development as projected afteruse.[68] On the other hand a city-centre oriented scheme of 'concessionary finance' to catalyse property-based economic development, Leg-Up, was instituted in 1982, operating on a reactive and competitive basis under the aegis of a high-

---

62 IDS 1987a p. 86, cf. SDA Board Papers 16.12.87.

63 IDS 1987a pp. 87, cf. Frank Kirwan 1.6.90.

64 SDA 1980a.

65 Gulliver 1984, Moore & Booth 1986e, Midwinter *et al.* 1991 pp. 184-91, McCrone 1991a pp. 922ff.

66 SDA 1978a p. 27, 1979 p. 55.

67 SDA 1979 p. 55, Aitken & Sparks 1986.

68 SDA 1985 pp. 53ff, 1986 pp. 78ff, 1987 p. 84, 1988e, cf. Castledine & Swales 1988.

level steering committee with SDA and Scottish Office representatives. Achieving a high degree of private sector financing was clearly a key criteria for evaluating individual Leg-Up proposals, and in practice priority was given to service and retail developments which could serve as physical back-up of ongoing attempts to change the industrial image which clung to many urban areas in Scotland, not least, of course, Glasgow.[69] In many ways, Leg-Up could be seen as a parallel to the local-authority oriented derelict land clearance grant, aiming to provide financial incentives for private developers to improve the attractiveness of privately owned city centre properties, but operating much like the industrial investment function with the Agency expecting to benefit financially from successful projects.

Designing Infrastructural Policies

Having surveyed the resources and strategies involved in the infrastructural policies of the SDA, it is now possible to consider both the balance between continuity and change, and identify some initial pointers about the main drivers of policy development.

The first conclusion that can be drawn is simple but important, namely that this policy area is characterised by a high degree of diversity, ranging from the inherited programmes providing relatively standardised services to the much more varied initiatives developing after 1975 that comprised individual (high-tech and major special) projects as well as more programme-like ones like the area initiatives and the Leg-Up scheme, and having both private and public actors as targets. It is therefore hardly surprising that the balance between continuity and change has varied between the various functions. One of the inherited functions, environmental improvement, changed little with regard to the way in which it operated – financial support for local authority initiatives and in-house projects – and although the balance between its two statutory rationales tended to lean towards the economic, the social dimension never completely disappeared. In contrast to this the other inherited function, industrial property, saw two major changes between 1975 and 1991: virtually from the word 'go' the SDA reduced the size of advance factories significantly in order to cater for the needs of small and new firms in particular, and from the mid-1980s onwards the role of the Agency as a strategic property developer was emphasised at the expense of managing its portfolio of industrial states, eventually resulting in plans for selling existing factories to sitting tenants or private management companies. A similar picture can be found among the first

69 IDS 1987a pp. 35, 87; SDA 1987 pp. 88ff, Scottish Enterprise 1991, George Mathewson 9.7.90, Alan Dale 16.10.91, Sir Robin Duthie 12.7.90, cf. Zeiger 1985, Barnekov *et al.* 1989 Ch. 7, Keating & Boyle 1986.

wave of policy innovations introduced in the late 1970s: while neither the concept nor the partnership-based set-up surrounding the major special infrastructure projects changed from the Scottish Exhibition and Conference Centre to the adjacent Glasgow Garden Festival nearly a decade later, the strategic orientation of the area initiatives gradually changed from being predominantly oriented towards manufacturing towards targeting a broader spectrum of commercial activities.

All in all none of the individual functions does in other words display a clear-cut strategic dichotomy between a social-interventionist pre-Thatcher experience and a post-1979 era of commercialisation, but when the six functions are seen as a whole the importance of change does, however, increase. While the SDA generally appears to have attempted to concentrate its infrastructural activities in areas of relevance from an economic development perspective, thereby in effect backgrounding its more 'social' environmental renewal remit, a shift can be seen from the original concentration on facilities of direct relevance for individual manufacturing firms to a much wider focus which also encompassed university-related science parks, commercial services, and the use of city-centre improvement to boost the general image of particular localities. This shift of focus was clearly most visible in the late 1980s, but it is, however, paramount to stress that several of these elements – notably the focus on poor image as an impediment to economic development and the first two major special projects in Glasgow – had been part of Agency corporate thinking also before the change of government in 1979, and thus strategic change must at least have been partly internally driven. At the same time the major intervention of the Scottish Office, the new industrial property guidelines which called for a more commercial approach and a more limited role for the SDA in property management, were designed to publically signal a break with an interventionist past, and although the impact of this on the strategic thinking of the arm's-length body was only visible from the mid-1980s onwards, the centrality of industrial property as an inherited function associated with traditional forms of regional policy made the new guidelines a point of reference around which an external-revolution interpretation could be constructed.

12.2. Implementation

Inherited Infrastructural Activities

Capital investment was obviously a crucial element of the industrial property function, and expenditure figures should therefore provide important information about changing levels of activity. As indicated by Figure 12.1, in financial terms the heyday of the function was in the early 1980s where expenditure reached a level in real terms more than four times the average spent by SIEC in the first half of the 1970s, and then gradually declined to a level 'only' twice that of its predecessor in the late 1980s.[70] Provision of factory space did, however, remain one of the main areas of SDA activity, accounting for 41% of total gross expenditure, but its dominant position with in average more than half of annual outlays only lasted until 1982/83 and then fell to around 30% in the early 1990s.[71] Other indicators do, however, suggest that the function was changing rather than just being reduced: while the floor space under SDA management was steadily reduced from 1982/83 onwards,[72] the number of tenancies kept growing so that in 1987/88 nearly three times as many firms were housed by the Agency as in 1979/80.[73]

Figure 12.1 Gross expenditure on infrastructure

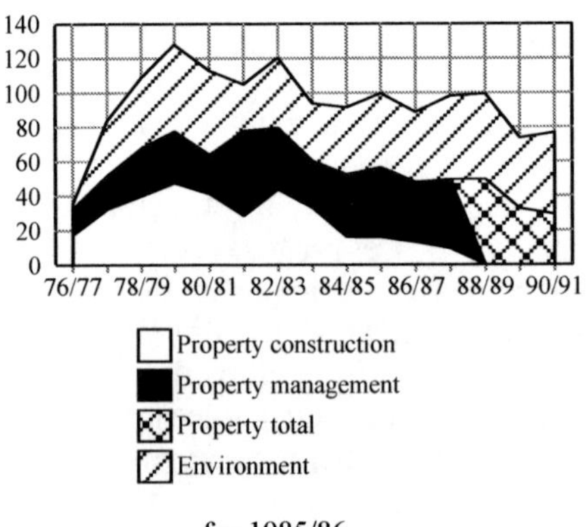

£m 1985/86.

Source: SDA 1977-91.

The impression of change is also sustained by a closer look at the policy instruments involved and the modes of implementation through which they were employed. In the 1980s both the Scottish Office and the SDA proclaimed that the arm's-length body should concentrate on being a strategic property developer rather than a manager of existing sites, but as illustrated by Figure 12.1 the opposite would seem to be

---

70 Calculated on the basis of SDA 1980-91, *Local Employment Act Annual Report* 1971-72, and *Industry Act Annual Report* 1972-75.

71 Calculated on the basis of SDA 1977-91.

72 By the end of the latest year for which data is available, 1987/88, floor space had been reduced by 37% compared to the peak in 1982/83 (calculated on the basis of SDA 1978-88).

73 Calculated on the basis of SDA 1980 and 1991.

happening, namely that the construction of new factories declined as a share of gross expenditure in the period for which data is available. At the same time it is, however, possible to detect a long-term trend towards a more proactive use by the Agency of its property building powers, with speculative building of advance units gradually increasing in importance and eventually accounting for around half the construction expenditure in the last five years for which expenditure breakdowns are available.[74] In terms of selectivity the most obvious trend was the drastic reduction in the size of the average factory unit owned by the SDA from around 4000 to less than 1000 square metres[75] and the preference given to localities covered by area initiatives.[76]

In terms of resource exchanges the 1980 property guidelines required that the property function should operate commercially in the sense that "rental should be not less than current market value",[77] but the available evidence suggests that moving in this direction was a rather slow process. Although total income from property rents etc. remained fairly stable for around a decade, as illustrated by Figure 12.2, this in effect constituted a gradual rent increase because the number of square metres owned was reduced from 1982/83 onwards,[78] but as many of the factories inherited from SIEC had long leases with low rents that could not be increased in the short term[79] and the method for determining a 'market-conform' valuation specified by the guidelines meant that rents were bound to be lower than the actual cost of provision,[80] in practice the notion of a 'commercial' approach implied 'limits to subsidies' rather than making a surplus on individual property investments.

The extent to which the SDA's industrial property function brought about institutional change in the Scottish economy is difficult to ascertain, especially in the absence of published in-depth evaluations of the function. It could be argued that making physical space for production available made sure that especially the development of small and new firms would not be hampered by lack of suitable premises, but the extent

---

74 Calculated on the basis of SDA 1977-91.

75 Calculated on the basis of SDA 1980-88. Although no data is available for the late 1970s, the unequivocal trend in the early 1980s clearly suggests that the size-selective policy of building and refurbishing primarily with small and new firms in mind have been adhered to right from the beginning.

76 SDA 1984 p. 49, IDS 1987a p. 87.

77 SEPD 1980b p. 5.

78 This was also reflected in the gradual upward movement of the financial duties for the property function throughout the 1980s (calculated on the basis of SDA 1984-91.

79 SDA 1982 p. 12, Alan Dale 16.10.91, cf. Glasgow Chamber of Commerce 1987 p. 7.

80 SEPD 1980b pp. 6f, Lochhead 1983 pp. 139f, Committee of Public Accounts 1982b p. 8.

to which its activities helped to rectify the failure of the property market that originally had prompted public intervention is less certain. On the one hand the Agency did dispose of a sizeable part of its property portfolio in the late 1980s and early 1990s, as illustrated by Figure 12.2: from the mid-1980s the level of disposals increased significantly – in real terms the average annual income from sale of industrial property increased more than four times from the late 1970s to the late 1980s[81] – before the final sale in the run-up to the Scottish Enterprise merger made figures go through the roof altogether, and thus by the early 1990s the role of the private sector in managing industrial property in Scotland would have increased significantly. On the other hand the fact that much of the Agency's activity was concentrated in a relatively limited number of priority areas made the possibility of sustaining expectations about cheap rents and hence maintained an impact on strategic segments of the market for industrial property.

Figure 12.2 Income from disposal of industrial property

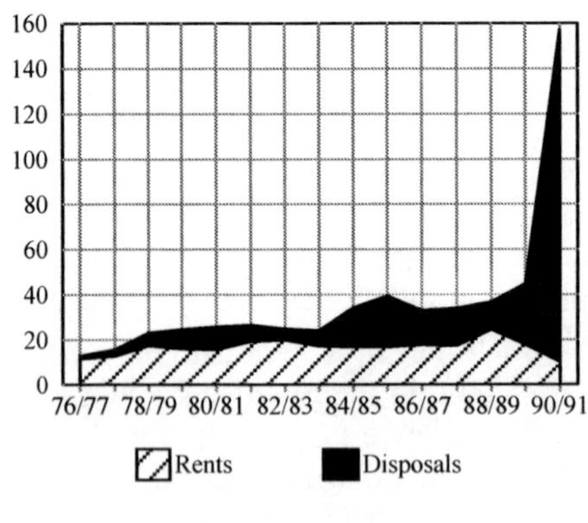

£m 1985/86.
Source: SDA 1977-91.

Also with regard to implementation the other inherited infrastructural function, environmental improvement, would seem to present a relatively uncomplicated picture. Although the number of projects varied,[82] the level of expenditure remained relatively stable from the late 1970s through to the early 1990s, and compared to the level of activity prior to the incorporation of the function in the SDA, this represented a very significant expansion because average annual expenditure was more

81 Calculated on the basis of SDA 1977-79, 1987-90.

82 The annual number of project varied from 70 to 448, averaging around 200 (calculated on the basis of SDA 1978-91).

than eight times that channelled through the Scottish Office in the first half of the 1970s.[83] Not just the volume of activity but also key aspects of implementation did, however, change after 1975 because although the generation of new projects remained predominantly reactive, some degree of proactivity was also introduced in relation to area initiatives, major one-off projects, or simply in order to soak up unspent Agency resources by functioning as a 'budgetary sponge'.[84] Moreover, the intended afteruse of the derelict land being treated changed away from being overwhelmingly public amenities towards including also projects with a future economic potential, originally as sites for industrial development but from the mid-1980s also for residential housing.[85] The outcomes produced by the function therefore fall in two categories: financial and technical support for a large number of environmental projects designed and physically delivered by other, primarily public, actors, and facilitation of other SDA projects in which environmental powers helped to deliver a cleaned-up building site and an enhanced ambience. In terms of institutional change this would seem to amount to a dual emphasis on creation of "customised space"[86] for development projects in localities dominated by brown-field sites left behind by earlier waves of economic activity on the one hand, and a more wide-ranging attempt to enhance the image of the region by removing dereliction and thereby increase confidence, both within the localities affected and among potential external investors. The difficulty, of course, is that while the former has a physical presence, the latter is a much less tangible form of change.

New Infrastructural Activities

In terms of expenditure the introduction of area initiatives, beginning with GEAR in 1976 and continuing in different organisational guises throughout the 1980s, was undoubtedly the most significant infrastructure-based policy innovation. Annual expenditure figures are, however, not stated under a separate heading in the accounts of the SDA, and therefore the analysis will have to rely on the financial summaries giving total figures for individual projects which appeared in the

83 Calculated on the basis of SDA 1977-91, *Local Employment Act Annual Report* 1971-72, and *Industry Act Annual Report* 1972-75.

84 Chris Aitken 18.10.91, cf. Lochhead 1983 pp. 147-53.

85 The 1986 government review produced expenditure breakdowns for the intended afteruse of land renewal expenditure relating to sites *not* owned by the SDA, showing that the relative importance of economic projects increased from 12% to 38% from 1981 to 1986 while recreation and amenities decreased from 71% to 48% in the same period (IDS 1987a p. 71).

86 Peck 1996.

Agency's annual reports until 1987.[87] As illustrated by Figure 12.3, all four types of area initiatives had clearly been partnerships in the sense that the financial contribution of the SDA was a minority one, ranging from 12-13% in comprehensive and integrated projects to 30% and 44% in self-help initiatives and task forces respectively, and with the exception of self-help projects and the final integrated one expenditure was predominantly infrastructural.[88] Clearly the area initiatives were an important part of the Agency's overall activity profile, but seen over the entire period 1976-87 for which figures are available they still fell spectacularly short of the 60% target the organisation had set itself because in practice the share of the targeted areas in total property and environmental expenditure constituted only 17% and 24% respectively.[89] The geographical targeting of activities ensured that the industrial property function operated primarily in a proactive manner through advance factories, and even the small business advisory services that otherwise tended to operate in a primarily reactive manner were used proactively with SDA staff visiting every firm in the targeted areas.[90] The geographical balance of the initiatives clearly showed that some priority was given to the Glasgow conurbation: five out of the nine initiatives for which expenditure data has been published were in the vicinity of the Clyde, and altogether they accounted for more than 70% of total SDA expenditure on this policy function.[91]

---

87 Financial information was only given for one self-help initiative, later ones appear to have been subsumed under general support for local Enterprise Trusts.

88 Calculated on the basis of SDA 1983 p. 68, 1987 p. 81. Overall 76% of total SDA expenditure on area initiatives was infrastructural.

89 Calculated on the basis of SDA 1977-87.

90 SDA 1980a pp. 12ff, Gulliver 1984.

91 Calculated on the basis of SDA 1983 p. 68, 1987 p. 81.

Figure 12.3 Breakdown of total expenditure by type of area initiative

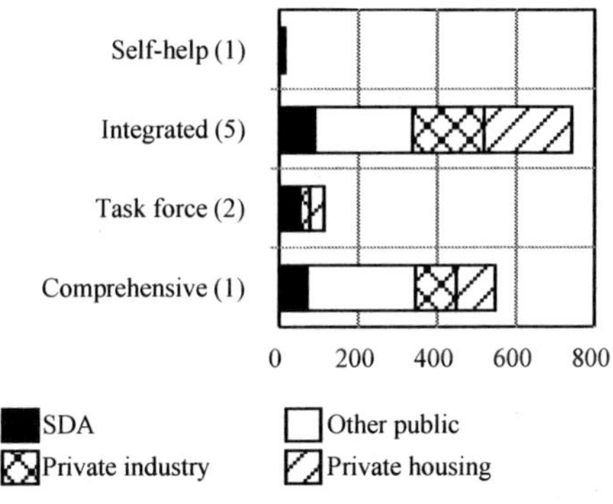

Current prices, £m.
Source: SDA 1983 p. 68, 1987 p. 81.

It is generally agreed – and indeed in line with what has been found in connection with infrastructural measures in other parts of Britain – that the main firm-level impact of individual area initiatives has been, in the words of Frank Kirwan, to "move activity about on the face of Scotland",[92] and as local government and other public bodies were subject to strict financial control from central government in the 1980s,[93] it seems reasonable to conclude that while prospects of access to Agency funding may have sustained levels of local authority expenditure in the designated localities,[94] this must have been "at the expense of other parts of their areas".[95] This does, however, not preclude that area initiatives can also have had more intangible effects such as improved implementation of SDA policies through closer and more proactive involvement of staff with particular localities or, indeed, may have achieved long-term image improvements both locally and with regard to Scotland. While the first argument was part of the reasoning behind the regionalisation of the Agency, the absence of systematic evaluations of the changing image of Scotland and its cities makes it difficult to draw firm conclusions with regard to the second point, although it is generally argued that especially the image of Glasgow was indeed transformed

92 Frank Kirwan 1.6.90, IDS 1987a p. 87, McArthur 1987, cf. Slowe 1981 p. 224, Pratt & Ball 1994.

93 See Midwinter *et al.* 1991 pp. 154ff.

94 Wannop 1984 p. 79, Wannop & Leclerc 1987, Wannop 1990, cf. Booth *et al.* 1982.

95 Keating & Boyle 1986 p. 100.

during the 1980s through a sustained combination of physical investment, major cultural events, and place marketing.[96]

Compared to the area initiatives, the other infrastructure-oriented policy innovations are fairly straight-forward with regard to their implementation characteristics. Both the advance high-tech premises and the major special projects were proactive SDA ventures, but while expenditure on science parks and similar facilities were subsumed under other headings in the accounts of the Agency, the cost of two of the major one-off initiatives are known – 14 million pound were spent on the Scottish Exhibition and Conference Centre[97] and more than 20 million on the Glasgow Garden Festival[98] – and as both these ventures involved sizeable contributions by other public and private partners, the SDA was in effect a major minority partner in both ventures.[99] Along similar lines Leg-Up could be seen as a way of handling a series of one-off projects involving partnership with external actors, and although the programme operated mainly in a reactive manner, it was also occasionally used proactively in connection with SDA-initiated projects such as the Scottish Exhibition and Conference Centre and the regeneration of Glasgow's Merchant City.[100] This function only accounted for only 5% of total infrastructural commitments by the Agency,[101] and the absence of controversy between the arm's-length body and its sponsor department[102] would seem to suggest that both its urban geographical selectivity and the crucial discretionary assessment criterion, sufficient private leverage, had generally been adhered to. Taken together this made Leg-Up a flexible instrument which could be used to support especially private investment in urban renewal, driven by commercial considerations and therefore in practice city-centre oriented, with the Agency sharing in financial risks and rewards and undertaking much of the 'due diligence' involved in assessing the viability of individual projects, and as such in effect a property-oriented complement to the industrial investment programme.

---

96 Gomez 1999, Belina & Helms 2003.

97 National Audit Office 1988 p. 9.

98 Calculated on the basis of SDA 1986-90.

99 National Audit Office 1988 pp. 9ff, cf. SDA 1983 p. 44; SDA 1988 p. 67, 1988e pp. 87ff.

100 Scottish Enterprise 1991 pp. 1, A1ff.

101 Calculated on the basis of SDA 1986-91. Expenditure figures for Leg-Up were not stated separately until 1985/86.

102 Zeiger 1985, IDS 1987a p. 87, National Audit Office 1988.

12.3. The British Context: Infrastructure and Regional Development

While it has been demonstrated that the SDA not only continued the infrastructural programmes inherited from its predecessor organisations but also became engaged in new ones, it remains to be seen whether these innovations were unique to the organisation or merely reflected broader British policy trends.

Industrial Property in British Regional Policy

Public provision of factory space for manufacturing had been part of British regional policy since the 1930s,[103] and this continued to be the case through to the early 1990, although a host of new actors arrived on the scene. On the one hand local government became increasingly involved in provision of premises for very small firms,[104] and on the other hand the organisational framework for activities sponsored by central government changed after the creation of the new RDAs in 1975. Hitherto industrial property had been managed by a small number of geographically-based industrial estates corporation which implemented policy-decisions taken in London by the department responsible for regional development, but from the mid-1970s decision-making capacity gradually moved away from central government: in Wales and Scotland the estate companies were merged into the new RDAs and hence became sponsored by territorial departments, and at the same time the policy-making powers of the various bodies gradually increased, although in England and Wales government departments did maintain a role in e.g. selection of tenants until the early 1980s.[105]

---

[103] Cf. Section 4.3.

[104] Fothergill *et al.* 1987 pp. 129ff, Lloyd & Rowan-Robinson 1988. In 1990/91 property accounted for only 8% of local government expenditure on economic development in Scotland (calculated on the basis of McQuaid 1992 p. 30).

[105] Committee on Welsh Affairs 1980 p. 82, King 1986 pp. 29ff, cf. Hogwood 1982.

Figure 12.4 Gross expenditure on industrial property

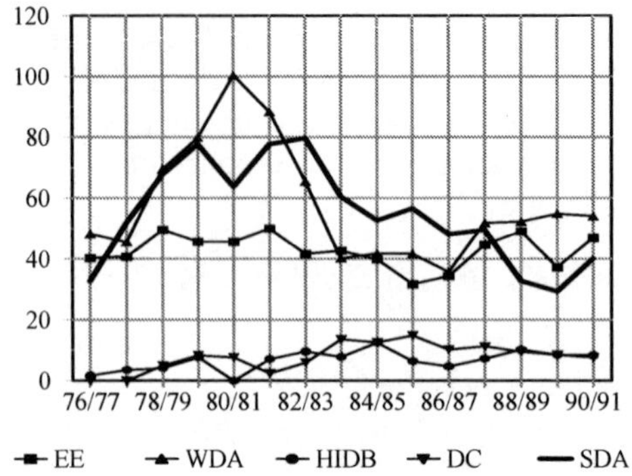

£m 1985/86.

Source: Calculated on the basis of SDA 1977-91, HIDB 1977-91, Development Commission 1977-91, Industry Act Annual Report 1977-91.

Within this revamped policy network broadly identical government guidelines governed financial aspects of public provision of industrial property – in particular rent levels and rent-free periods – in order to make sure that inter-regional competition for foot-loose investment projects was not skewed through excessive subsidies through low/no-rent agreements,[106] and the various bodies seem to have moved in the same directions in terms of their strategic priorities. Like its Scottish counterpart the WDA stressed the importance of provision of smaller units already in the late 1970s,[107] and English Estates took the same view at least from the early 1980s and also began to be involved in specialist property like science parks and incubators.[108] Moreover, also the modes of implementation were broadly similar in the sense that both reactive 'bespoke' building and proactive advance factories were given priority while the latter dominated in practice,[109] and spatial targeting of property activities also became a common phenomena: the WDA focused on localities hit by steel and coal closures,[110] while the Development Com-

106 Committee of Public Accounts 1983 pp. viif, Scottish Affairs Committee 1984 pp. 57ff, IDS 1987a Ch. 5, Welsh Office 1987 pp. 40ff.

107 Wilson Committee 1977f vol. 8 p. 17, Committee on Welsh Affairs 1980 pp. 72f. The Development Commission had of course been doing this in rural areas throughout (Chisholm 1984b, Rogers 1999 Ch. 3-4).

108 King 1986 pp. 31ff, Fothergill *et al.* 1987 pp. 120ff.

109 Fothergill *et al.* 1987 pp. 115ff, Welsh Office 1987 Fig. 6.1, IDS 1987b Ch. 5.

110 Committee on Welsh Affairs 1980 pp. xxvff, Rees 1997.

mission's general approach of giving priority to the rural localities with the greatest perceived need also applied to its property activities.[111]

The main difference between the various bodies sponsored by central government concern two related issues, namely the priority given to industrial property as part of the overall policy programme, and the way in which demands in the 1980s for a more 'commercial' approach was tackled. As illustrated by Figure 12.4, gross expenditure on property was generally higher in Wales and lowland Scotland than in the Assisted Areas in England catered for by English Estates, while, unsurprisingly, the smaller rural bodies lagged much further behind.[112] Moreover, in contrast to the constant level presented by English Estates, property expenditure by the two non-English industrial RDAs varied greatly, something which could suggest a greater importance of reactions to e.g. sudden plant closures in Scotland and Wales.[113] A rather different picture does, however, emerge when the relative importance of industrial property in the overall policy profile of the various bodies is considered: while the WDA comes across as being essentially a property-oriented development body and both the HIDB and DTI through English Estates have provision of premises as a stable, but secondary, activity – with around 66, 20 and 6% of expenditure devoted to property respectively[114] – in the 1980s the Development Commission upgraded while the SDA downgraded the function.[115] Moreover, all the arm's-length bodies engaged in public property provision were subject to similar pressures from their political sponsors for moving in direction of a more commercial style of operation in terms of the rents charged and, not least, reducing their role as estate managers through disposals,[116] but while their income-to-expenditure rate increased much in parallel in the first half of the 1980s, towards the early 1990s the SDA clearly stood out as the

---

[111] Chisholm 1984b p. 515, cf. Section 7.4.

[112] Over the entire period WDA and SDA property expenditure were respectively 55% and 28% higher than that of English Estates (calculated on the basis of SDA 1977-91, WDA 1977-91, *Industry Act Annual Report* 1977-91).

[113] The first peak in WDA property expenditure represents additional government funding in the wake of major steel closures (cf. Committee on Welsh Affairs 1980 pp. xxvff), and the highest figures for the SDA are recorded in the period in which area initiatives were most prominent.

[114] Calculated on the basis of WDA 1977-91, HIDB 1977-91, Industry Act Annual Report 1977-91.

[115] From the late 1970s to the early 1990s the property share of SDA expenditure fell from 57% to 26%, while its rural English counterpart expanded its role from 20% to 37% (calculated on the basis of the SDA 1977-91, Development Commission 1977-91).

[116] IDS 1987b Ch. 5, Welsh Office 1987 pp. 47ff, King 1986 p. 31.

organisation which was most successful in disposing of parts of its portfolio.[117]

Taken together this would seem to suggest that the use of public factory space for industrial development purposes was not simply governed by a uniform British approach but also reflected specific regional circumstances, especially concerning the priority accorded to property as a policy function.

Environmental Improvement in British Regional Policy

Improving the environment through the clearing of derelict land had been part of British regional policy from the 1960s,[118] but like in industrial property some aspects of the organisational set-up became more complex from the mid-1970s onwards. Again the setting up of the new RDAs in 1975 meant that central government was only directly responsible for clearing of derelict land in designated parts of England while in Wales and Scotland this responsibility had been transferred to the new arm's-length bodies.[119] Moreover, similar activities sponsored by the Department of the Environment as part of urban policy were gradually intensified, and thus in effect two parallel schemes were in operation in England while efforts in Wales and Scotland were unified by being channelled through the WDA and SDA. And finally both local and European actors became increasingly involved, especially in conjunction with one another as much ERDF funding in the 1980s was spent on projects which improved various aspects of the environment.[120]

Like in industrial property the purpose of activities seem to have been fairly similar across Britain, with the underlying rationale being social as well as economic, and the latter particularly prominent in the 1980s where corporate statements emphasised the importance of economic afteruse of land clearance. In practice, however, this priority was only pursued to a rather limited extent, and thus apart from the ability to 'customise space' for incoming investors, non-economic forms of afteruse seems to have dominated in England and Wales.[121] Throughout Britain land clearance schemes primarily operated in a reactive manner

---

[117] The average annual income as per cent of property expenditure in the period 1989-91 was 76% for the WDA, 88% for WDA and 219% for SDA (calculated on the basis of WDA 1977-91, *Industry Act Annual Report* 1977-91, SDA 1977-91).

[118] Cf. the discussion in Section 4.3.

[119] As dereliction was a problem mainly confined to industrial and urban areas, the non-industrial RDAs, environmental improvement were not central to either HIDB or the Development Commission (Gavin McCrone in evidence to the Scottish Affairs Committee 1984 pp. 62f).

[120] Goodstadt & Clement 1997, cf. Section 7.4.

[121] National Audit Office 1994 p. 3, cf. Prestwich & Taylor 1990 pp. 214ff.

– although also the WDA employed its environmental powers in support of major one-off projects like preparations for the Ebbw Vale garden festival in 1992[122] – and were subject to broadly similar government guidelines. Until the early 1980s local authorities in Scotland were, however, treated favourably compared to their counterparts south of the border, because in England the original policy of demanding co-funding from local authorities was retained while in Scotland all expenditure would be reimbursed,[123] and this, in conjunction with the rising prominence of area projects, may be part of the reason why the SDA was able to increase its environmental activities more than other organisations until the early 1980s: compared to the levels attained in the early 1970s, the activity increased in Scotland more than 800% while growth in Wales and England only averaged around 50%,[124] and although the difference is undoubtedly exaggerated by inclusion of urban funding in the Scottish figures that would have been channelled through the Department of the Environment in England, the difference would still seem to suggest that a considerable scope for regional variation existed with regard to the use of environmental powers for regional development purposes.

Urban Policy in Britain

The area initiatives of the SDA have often been interpreted as the Scottish version of urban policy,[125] and hence this policy area must be considered in order to complement the British context of the Agency's infrastructural activities.

From an organisational perspective urban policy in Britain changed significantly from the 1970s to the 1980s: originally the main emphasis had been on central government support for local authority projects, but the new Conservative government centralised efforts through the introduction of two complementary measures.[126] From 1980 onwards a number of Enterprise Zones were designated in (mainly) inner city areas where firms would be exempt from a number of taxes and levies and simplified physical planning procedures would operate. In 1981 this was complemented by the introduction of Urban Development Corporations (UDCs), quangos sponsored directly by the Department of the Environ-

122 National Audit Office 1994 p. 10.

123 Prestwich & Taylor 1990 pp. 214ff.

124 Calculated on the basis of *Local Employment Act Annual Report* 1970-72, *Industry Act Annual Report* 1973-91, WDA 1977-91, and SDA 1977-91.

125 IDS 1987a pp. 135f, cf. IDS 1988c.

126 For overviews of British urban policy, see Lever 1987, Turok 1987, Rhodes 1988 pp. 343-64, Barnekov *et al.* 1989, Lawless 1991, Healey 1991, Lawless & Haughton 1992, Atkinson & Moon 1994, O'Toole 1996.

ment which were given responsibility for urban renewal in a limited geographical area and extensive powers with regard to land-use, planning and economic development. Although resting on rather different assumptions – the market was being 'freed' in the Enterprise Zones while the UDCs involved extensive public intervention – both measures would seem to reflect a distrust in the ability of local government to handle large-scale urban problems, and in the early 1980s the new measures were subject to a good deal of controversy between the elected local councils and the unelected administrators who on behalf of central government were now in charge of implementing urban renewal measures locally. Moreover, while already the late 1970s had seen an increasing focus on economic rather than social issues in British urban policy, this hierarchy of priorities was intensified in the 1980s where direct involvement of the private sector through various partnership arrangements was seen as paramount, and individual measures revolved around property development,[127] image re-engineering and promotion of small businesses as "entrepreneurship in its highest form".[128] In many ways the macro-geography of the new urban policy measures is a particularly revealing feature: nearly thirty Enterprise Zones were established throughout Britain in the 1980s, including four in Scotland,[129] but the eleven UDCs were concentrated in England and Wales.[130] This would seem to suggest that while central government attempted to ensure some degree of inter-regional equality with regard to financially-oriented measures, the scope for variation was greater with regard to more complex activities which involved a number of different policy instruments.

12.4. Infrastructure for Regional Development

Having analysed the development of the infrastructural activities of the SDA and the wider British context in which they took place, it can be concluded that features which would seem to point towards an external-revolution interpretation are undeniably present. In the 1980s the industrial property increasingly focused on providing premises for small firms, including incubators for new high-tech firms, and ultimately the Agency's inherited role as a manager of industrial estates was eroded through large-scale disposal of properties in the early 1990s. Environmental improvement began to focus on improving the ambience of city

[127] The English counterpart of Leg-Up was administered by the Department of the Environment but differed by requiring local government participation in projects (Zeiger 1985, cf. Jones 1996).

[128] Atkinson & Moon 1994 p. 106.

[129] Wren 1996a pp. 134ff, Scottish Office 1991 p. 72.

[130] Atkinson & Moon 1994 p. 145.

centres rather than the traditional task of tackling industrial dereliction, and the spatial focusing of infrastructural activities through area initiatives moved from having comprehensive socio-economic goals, epitomised by GEAR launched in the late 1970s by the Labour government, to a much more single-mindedly economic approach which eventually saw the private sector take a leading role in urban regeneration. In short, a radically different approach that shifted focus away from manufacturing towards high-tech and service industries, introduced a commercial approach to the use of physical infrastructure, and relied increasingly on informational measures.

Such an external-revolution interpretation is, however, difficult to reconcile with other findings emanating from the empirical analysis. *Firstly*, it has been demonstrated that many of the features associated with the allegedly Thatcherite approach in the 1980s had in fact been present before the change of government in 1979: the emphasis on premises for small firms, the general focus on high-tech developments, the first two major special projects in Glasgow which foreshadowed the involvement in image-building and city centre development, the priority given to clearance of land with potential for economic afteruse, and the emphasis on economic development as the SDA's central contribution to area projects. In all of these cases the role of the new Conservative government must therefore have been, at best, to strengthen existing trends in Agency activities rather than having introduced them as completely new strategic departures. *Secondly*, other continuities would appear to be overlooked, especially when they occur in areas where change is evident at the level of corporate rhetoric: despite the official priority accorded to economic afteruse, most environmental renewal projects continued to produce public amenities rather than sites for commercial development, and although the size of the industrial property portfolio was reduced after the mid-1980s through gradually accelerating disposals, the capacity to construct new property for particular types of firms was retained. In short, the strategic core of the inherited infrastructural functions – to use public provision of industrial property and environmental improvement to influence the decisions of individual firms to locate or expand in Scotland – would appear to have been largely intact while at the same time enhancing the physical surroundings in practice remained an additional goal in its own right.

These continuities could of course have been overshadowed by genuinely new developments in the 1980s, but on closer inspection the significance of the latter would seem to have been rather limited, either because they remained marginal in relation to other infrastructural activities, or because they appear to have had little strategic impact. The Leg-Up scheme through which the Agency became a co-investor in urban property projects many of which were oriented towards enhancement of city centres belonged to the first category because of the relatively low levels of expenditure involved compared to the continued role of the organisation in industrial property and traditional forms of environmental renewal. Equally important, the increasing role of private sector actors in policy design and implementation would seem to belong to the second category: while city-centre partnerships like Glasgow Action did of course take on a life of their own, their dependence on public sector resources in effect made them vehicles of support for e.g. ongoing local government re-imagineering rather than the other way around, and the development potential of continuing to manage a broad portfolio industrial property would appear to be limited as long as the capacity to build new premises for selected types of firms remained intact.

All these 1980s innovations are of course in line with the general emphasis of the Conservative government on including non-industrial economic activities and private sector actors in bringing about market-oriented solutions to regional development problems, but it is also noticeable that the extent to which these priorities were translated into policy changes varied between the different parts of Britain. In many ways these changes appear to have been embraced more readily in Scotland than in Wales and, indeed, the English regions where organisational fragmentation and often tense relations with local government made it difficult to emulate the more integrated Scottish approach. And, most importantly, change was presented as part of a overarching attempt to reposition Scotland and its localities *vis-à-vis* the external world and as such may not have undermined, but perhaps even reinforced, the role of the public sector in regional economic development.

CHAPTER 13

Agency of Change? Reappraising SDA Policies

The previous chapters have analysed the development of the policies of the SDA on the basis of the conceptual framework developed in Chapter 3, and therefore it is now possible to address three of the central issues of this book, namely the balance between *continuity and change* in Agency policies, the relationship between *the SDA and its sponsor department* with regard to regional development activities, and the position of Agency policies *vis-à-vis* those of other actors in *the British regional policy network*. The corporate-level analysis in Chapters 6 and 7 had concluded that with regard to resources and strategies the development of the SDA would appear to have been characterised by gradual change, mainly driven by internal processes but at some junctions also affected by Scottish Office initiatives, but although sudden ruptures have clearly not been a recurring feature in the development of individual areas of Agency activity either, it is still necessary to keep the external revolution perspective in mind. Old inherited activities have evolved and new activities have been introduced, and in theory these changes could add up to such an extent that SDA policies could be considered an example of the 'gradual revolution' paradigm in the study of Thatcherism. For the purpose of this chapter, the policies of the SDA have been regrouped under eight headings, disaggregating the infrastructural functions which taken together account for the majority of Agency expenditure on the one hand, and highlighting the introduction of the two new financial instruments, development funding and Leg-Up, which share some key characteristics that differ from the headings under which they have previously been discussed. Table 13.1 provides a summary of the conclusions of the analyses of Agency policies, and two complementary features would seem to be in evidence.

Table 13.1 SDA Policies – Continuity and Change (part 1)

	SDA priority * *sponsor issues*	*Assumptions*	*Critical resources*	*Policy Instrument* * *implementation modes*	*Firm-level Impact* * *institutional change*
Industrial investment	Lower priority, stable activity * ownership, priority, viability	* regional competitiveness * SME equity gap * finance plus support, then initial scrutiny	Intelligence, external private network	Conditional finance * mainly reactive * discret. viability * mainly small, more high-tech	Indigenous * create regional venture capital network
Dev. funding & Leg-Up	Increasing from late 1970s * project scrutiny	* regional competitiveness * image, infrastructure * diverse city-oriented projects	Intelligence, external network	Conditional finance, organisation * pro-/reactive * discret. viability * city-centre focus	Modernise/new indigenous * improvement of regional position on image market
Sectoral strategies	High from late 1970s onwards * information procurement	* regional competitiveness * sectors/firms uncompetitive * tailorised diversification	External intelligence	Conditional/ voluntary information/ organisation * proactive * discretionary * high-tech and traditional	Indigenous, also incoming * create networks * modernise hierarchies * improve market positions
Advisory services	Gradually increasing * late 1980s: market failure perpetuation	* regional competitiveness * SME info gap * subsidised, tailorised diversification	Intelligence, in 1980s also private partners	Voluntary/ conditional information * re-/proactive * discretionary, automatic * small, sectors	Indigenous * increase access to market information, modernise hierarchies
Inward investment	Increasing from late 1970s * joint organisation since 1981	* regional competitiveness * sectoral structure, jobs * targeted one-door approach	Intelligence, Scottish network	Voluntary information, conditional finance * pro-/reactive * discretionary * high-tech, likely expanders	Incoming, mostly new * improve regional position on international FDI market

Table 13.1 SDA Policies – Continuity and Change (part 2)

	*SDA priority * sponsor issues*	*Assumptions*	*Critical resources*	*Policy Instrument * implementation modes*	*Firm-level Impact * institutional change*
Industrial property	High, less from mid-1980s * owner-ship, rent levels	* regional competitiveness * few small and incoming firms * subsidised provision	Organi-sation	Conditional organisation * pro-/reactive * discretionary * small, incom-ing	Indigenous, also incoming * improve regional position in market for customised space
Derelict land clearance	High, stable * late 1980s: balancing of goals	* physical attractiveness * image, urban dereliction * support/ organise renewal	Organi-sation	Conditional finance, organi-sation * re-/proactive * discretionary * space, afteruse	Diverse, diffuse * improvement of regional position on image market
Area initiatives	Decreasing from late 1980s * balancing of goals	* regio. comp., intra-regio. equality * underperforming localities * focus policies of partnership	Organi-sation	Conditional organisation * proactive * discretionary * space, small	Indigenous * spatial focus of infrastructu-ral policies

A recurring theme in the preceding chapters has been the existence of continuities which cut across the change of government in 1979, something which undermines radical versions of the external-revolution interpretation which present the return of the Conservatives to the Scottish Office as a watershed in the development of the SDA. In terms of objectives the vast majority of the 'original' policies revolved around improving regional competitiveness, either directly through private firms or indirectly via provision of infrastructure conducive to future economic development, and where non-economic – social and/or environmental – rationales were present, the balance between the two was shown to have been unresolved both in the 1970s and the 1980s. Moreover, also in the 1980s policy instruments remained predominantly conditional provision of resources so that firms had to comply with Agency criteria in order to gain access to finance, information or organisation, and in nearly all areas of activity a more or less pronounced element of proactivity continued to be present. Looking in more detail at individual policy areas, no clear-cut evidence of a more 'commercial' post-1979 approach to industrial investments has been found, the focus on small business – in advisory services, investment and industrial property – had been inherited and further developed in the late 1970s, the standard-setting first sectoral study in the high-tech field of electronics had been commissioned in April 1978, and the Agency intensified its promotional efforts by establishing its first overseas offices in early 1979 before the change of government. In short, if in the 1980s the arm's-length body focused on viable investments, built premises for small firms, promoted new high-tech ventures, and targeted multinational electronics firms as potential inward investors, then this will have been due to inertia rather than innovation.

It is, however, also important to stress that despite these important continuities, pervasive change has also characterised most areas of SDA policy from 1975 to 1991, and four types of innovations have been identified. *Firstly*, incremental application of existing concepts to adjoining areas, e.g. when one area or sectoral initiative leads to another, or when specialist advisory services develop on the basis of enquiries to generalist ones. *Secondly*, particular aspects of existing activities have been revamped, either as result of an internal learning-by-doing process such as the increasing emphasis on front-end scrutiny of industrial investment projects, or as a consequence of external demands for e.g. control with overseas promotional efforts or a reduced role for the Agency as a manager of industrial property. *Thirdly*, major new activities were introduced in addition to existing ones, again both through internal initiatives like the sector-based approach to industrial development, and through external ones such as the area-based approach and the Leg-Up scheme. And *finally*, while taken together these changes made the overall policy profile of the SDA more diverse – it is noticeable that in most areas of activity the same policy instrument was employed in multiple ways and hence several alternatives are given with regard to modes of implementation and firm-level impact – the analysis has also demonstrated an increasing trend towards integrated deployment of various functions as part of particular projects. Such an approach was of course inherited through the Small Business Division which had traditionally combined financial support and advice in rural areas, but from the late 1970s onwards this principle was extended through area and sectoral initiatives, major one-off projects, and via the joint LIS operation even came to include grant giving powers with regard to incoming firms that had originally been the preserve of the Scottish Office. In short, an increasingly diverse set of activities did not preclude an integrated approach to regional development from being employed consistently enough to become an unmistakable characteristic of the organisation.

At the level of individual policies changes have been found mostly to have originated within the SDA itself, especially with regard to incremental extension and/or development of existing services, while external impulses to change have been associated with a limited number of junctions in the development of Agency policies: on the one hand the introduction of new approaches to the ongoing regulation by the Scottish Office of the use of particular policy instruments, especially those involving direct transfer of financial and physical resources to individual firms, and on the other hand events further afield in the British policy network, like the introduction of Leg-Up in the wake of inner-city riots in England or the setting up of LIS after a parliamentary enquiry. A division of labour where most specific strategic initiatives were taken by the Agency itself and the sponsor department mainly attempted to influence the overall direction of policies is clearly in line with the rationale for setting up arm's-length RDAs, but as demonstrated in the preceding chapters, the strategic impact of high-profile external interventions in individual SDA activities have in fact been rather limited: the new industrial investment guidelines in 1980 largely reinforced existing efforts to invest in viable ventures by insisting on private co-funding, the setting up of LIS effectively made Scottish Office grant giving powers part of the Agency rather than the other way around, and – perhaps the most successful but also most belatedly implemented intervention by the sponsor department – the symbolic implications of the large-scale disposal of industrial property in the early 1990s would seem to be much larger than the strategic ones in view of the fact that the capacity of the SDA to construct new property was retained.

This clearly implies that also at the level of individual policies will a simplistic version of the external-revolution interpretation – Scottish Office interventions profoundly change strategic direction of arm's-length body – be impossible to sustain. At the same time it does, however, not preclude the possibility that the multitude of changes in individual policy areas plus a changing balance of activities could have added up to a substantial change of development strategy, and if this was found to have moved in directions central to Conservative ideological concerns or, potentially a less ambitious criteria, in parallel with policy developments in the rest of the British regional policy network, then it would be reasonable to speak of a 'gradualist external revolution'[1] which could have succeeded quietly because the Agency, the instigator of most policy changes, had tacitly accepted that new initiatives were more likely to be condoned if they appeared to be in tune with the government of the day.

---

1 A position close to the one adopted in the most recent texts by Danson (1999) and MacLeod (MacLeod & Jones 1999).

Table 13.2 Main dimensions of change in regional policy implementation 1975-91

Policy area	*SDA*	*English Regions*	*WDA*
Industrial investment	Priority stable Intensified appraisal Private partnership	Decreasing priority Discretionary appraisal gradually replace automatic More focus on incoming and small indigenous firms	Priority stable
Development funding & Leg-Up	Priority increasing City-centre focus	Priority increasing City-centre focus	Priority increasing City-centre focus
Sectoral strategies	Priority increasing Diversification	Absent (decreasing priority in industrial policy)	Late minor priority
Advisory services	Priority increasing Diversification More conditional More proactive More private partners	Limited expansion of standardised services	Limited expansion and diversification
Inward investment promotion	Priority increasing Integration of grants More proactive Sectorally selective	Priority increasing Limited integration More proactive	Priority increasing Integration of grants More proactive
Industrial property	Decreasing importance from mid-1980s Smaller units Rent increases and divestment in late 1980s	Less marginal priority Smaller units Rent increases in late 1980s	Variable high priority Smaller units Rent increases in late 1980s
Derelict land clearance	More proactive Integrated in major projects	Variable priority	Priority increasing in late 1980s
Area initiatives	Decreasing importance from mid-1980s More economic goals More private partners	Priority increasing from late 1970s (urban policy) Exclusion of public partners	Priority increasing from late 1970s (urban policy)

In order to illuminate the possible occurrence of such a 'gradualist external revolution', Table 13.2 provides a summary of policy changes identified in the previous chapters, and with regard to the SDA the overall picture emerging is one of two directions coexisting in the 1980s. On the one hand in most areas of activity – industrial investment, all information-based policies and environmental renewal – changes ran counter to what could be expected, because by becoming more proactive, more selective, more discretionary and more conditional, SDA policies increasingly exposed private firms to public priorities in the 1980s. On the other hand some changes did take place which do appear to be more agreeable from a Thatcherite perspective: the increasing emphasis on city-centre developments in area initiatives and the Leg-Up programme could be construed as being a shift away from the traditional focus in regional and urban policy on poor localities and manufacturing, and the growing involvement of private sector actors via industrial co-investments, advising Enterprise Trusts and privatisation of industrial estates could be interpreted as a dilution of public responsibility for regional development. Simply involving places and people cherished by the Conservative governments of the 1980s did, however, not in itself mean that policies necessarily changed in ways that eroded the Agency's capacity to purvey public priorities to actors in the Scottish economy. The emphasis on non-industrial activities was in line with the organisation's long-standing emphasis on sectoral modernisation and regional image improvement, and working with private partners to expand business support services or lower the threshold for venture capital projects changed the institutional environment faced by Scottish SMEs by drawing on private financial and informational resources – something which of course was likely to foreground private sector ways of doing business, much in line with the Agency's general commitment to improving the competitiveness of the region. Effectively this leaves us with the large-scale property disposal in the run-up to the Scottish Enterprise merger as the main potential standard bearer of Thatcherite policy change within the SDA, but, as argued in Chapter 12, because the capacity to produce new properties was retained this measure had rather limited strategic implications. In short, policy area by policy area changes strengthened rather than reduced the potential influence of public priorities from the late 1970s to the early 1990s, and in parallel with this the balance of the SDA's policy profile also shifted in the same direction. As illustrated by Figure 13.1, the policy areas expanding in the 1980s were those where public influence on the preferences of private economic actors was increasing: expenditure on industrial investment and information-based services not only grew nearly 160% in absolute terms from 1979 to 1991, but these areas also increased their share of total Agency spending from 14% to 37%.[2] All in all changes in SDA policies would seem to run counter to the liberal ideological

[2] Calculated on the basis of SDA 1980-91.

preferences of the governing Conservatives in the 1980s, and thus from the perspective of the policy activities of the SDA even a 'gradualist external revolution' interpretation has become difficult to sustain.

Figure 13.1 Distribution of gross SDA policy expenditure

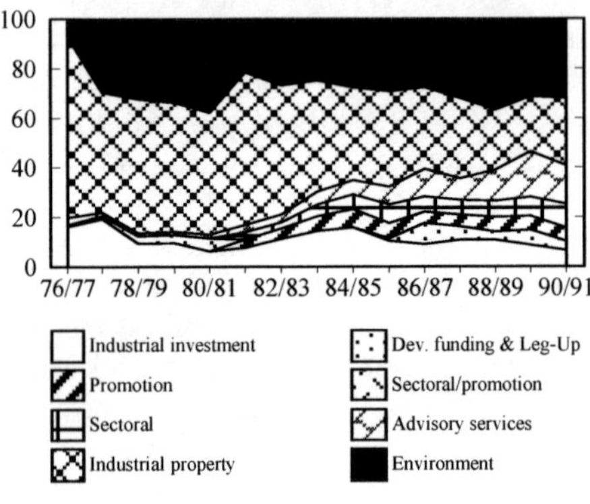

Per cent.
Source: SDA 1977-91.

This conclusion becomes even more remarkable when Scottish developments are seen in the context of the wider British regional policy network. Compared to its Welsh counterpart, the SDA comes across as having moved further away from its origins as a conglomerate of inherited policy function. The WDA did not develop new large-scale initiatives and diversify its policy profile to nearly the same extent, although some innovation would seem to be under way in the late 1980s, and this would seem to suggest that within the British regional policy network RDAs were allowed some degree of freedom to design strategies on the basis of regional preferences, much in line with the theoretical expectations about the role of this type of organisation in regional policy. England, however, offers an interesting contrast to the Scottish case: here central government played a dominant role through the DTI as the hub of an organisationally fragmented implementation structure in the Assisted Areas, the main financial policy instruments moved only gradually in the direction of the discretionary approach of the Agency, in advisory services and industrial property changes were limited, no counterpart of the sectoral initiatives existed at the regional level, and local government tended to be much less involved in, if not outright excluded from, the new-model urban partnerships established by central government. While the increasingly discretionary nature of the main financial policy instruments also enhanced the influence of public priorities in England, this aspect of the British-level approach was at the same time much more and much less radical than that of the SDA. More radical because discretion was vested with generalist civil servants and untempered by concertation with private sources of finance through a regional venture capital network, but at the same time also less radical because of the appraisal appears to have focussed on financial issues – minimising the public contribution to individual projects – rather than improving ventures by attaching e.g. organisational conditions to offers of financial support as practised by the Agency. On this reading it is certainly difficult to see the SDA as a follower of trends set by the DTI as the 'lead department' in the British regional policy network, but at the same time the long shift towards discretionary forms of central government support may also have functioned as a 'political cushion' for the Agency: the need to change the policies of the arm's-length body was perhaps less pressing when they already appeared to be close to the type of regional policy that government ministers themselves had declared they were aiming for.

We have, in other words, ended up with an interesting conundrum: in the 1980s a Conservative British government with self-proclaimed neoliberal leanings sponsored a Scottish arm's-length development body which not only managed to retain controversial policies but also gradually shifted its activities towards modes of implementation which implied more public influence in the regional economy. Understanding this paradox of the politics of regional policy in Scotland will, however, require that the focus of study is extended beyond the level of individual policies, and thus the following chapter considers the broader discursive environment in which the Agency and its sponsor department operated.

Part IV

The Politics of Scottish Regional Policy

CHAPTER 14

The Discursive Environment of the SDA

Focusing primarily on Agency activities and their relationship with the Scottish Office, the preceding chapters have established the balance between continuity and change with the SDA both on the corporate level and with regard to individual policy areas, and despite differences between policy areas and changes in the ways in which strategies were presented, the overall picture would clearly seem to be one that does not sit easily with the notion of an external-revolution perspective. Most of the policy areas in which the organisation became engaged have tended to evolve fairly gradually from the mid-1970s through to the early 1990s without the year 1979 shining through as a major watershed between an 'interventionist' Labour past and a 'market-oriented' Conservative future, and, indeed, without being subject to a series of major revamps that characterised central government regional grant schemes. This of course raises the next question, namely how this relatively high degree of continuity came about, and by combining the external-revolution and internal-evolution paradigms with the more general points made earlier about the possible impact of Thatcherism on public policy in Scotland,[1] some possible explanations could be:

- *Policy continuity despite political change:* a well-functioning insulation from political pressures through its position at arm's-length from a benevolently-minded sponsor department meant that party-political changes did not filter through to nearly the same extent as was the case in areas where policy was implemented directly by departments of central government (the RDA explanation, an internal-evolution perspective),
- *Policy change preceding government change:* the high-profile Conservative scepticism about the industrial investment function which dominated the political debate around the setting up of the organisation ensured that both the Agency itself and its political sponsors had embraced a commercial style of operation well before 1979, and thus the change of government eventually made little difference (the pre-

[1] See the discussion in Sections 2.2 and 4.1, and the contribution of Holliday (1992) in particular.

emptive adaption explanation, a variation on the external-revolution theme),

- *Political change in Scotland more limited than at the UK level:* the nature of the political pressures in Scotland differed substantially from those on the British level because of an extensive cross-party support for maintaining a high degree of preferential regionalism through proactive public sector involvement in economic development (the 'Scottish consensus' explanation), or
- *British policy changes preempts need for change:* the changing discourses and policy practices at the British level in the 1980s gradually brought mainstream central government policies closer to the selective approach the SDA had pursued since the late 1970s, and the need to bring the Agency 'into line' was therefore limited (the increasing policy-fit explanation).

Each of these explanations involve statements not just about policy practices but also about discourses on regional development, and hence in order to illuminate the possible sources of continuity and change in the development of the SDA, it will be necessary to consider the wider political environment in which the organisation operated. Both the arm's-length body and its sponsoring department will have been confronted by specific political demands and more general discourses concerning the nature of the regional problem and the relevance of the SDA in relation to these problems – entailing wider assumptions about the relationship between public and private actors or about the position of Scotland within the UK – and the eventual development of Agency policies will have reflected its ability to navigate this discursive terrain and make its policies appear attractive to its political sponsors without alienating other important actors in Scotland and beyond.

This chapter presents an analysis of the wider discursive environment of the SDA that will be primarily based on a combination of published and documentary sources, i.e. debates in the Westminster parliament and its committees, publications of key interest organisations and public debates as reported in the *Glasgow Herald* and business magazines supplemented to a limited extend by access to documentary sources. The text is organised according to two principles: first the Scottish and then the wider British political environments are explored, and in both cases the text is organised chronologically, beginning with the late 1970s under Labour, and then progressing through the Conservative 1980s before finally examining the debates on the proposal to merge the Agency into Scottish Enterprise.

14.1. Uncertain Commitments? The Scottish Environment 1975-79

Scottish issues played an unusually important part in British politics in the second half of the 1970s. The second general election in 1974 had resulted in a wafer-thin Labour majority in Westminster, and as the parliamentary dominance of the party north of the border was threatened by the rise of the SNP, the Labour government had plenty of party-political reasons for pursuing manifesto commitments such as special measures with regard to economic development – the SDA – and a constitutional adjustment of the union state which through devolution of power to an elected assembly would transfer decision-making authority in certain areas of public policy from London to Edinburgh.[2] Due to internal divisions in both the major parties – and especially within the governing Labour party – it proved difficult to find a formula that was acceptable both to Scottish aspirations and to the English majority in Westminster, and thus in a period of severe economic and fiscal crisis – unemployment nearly doubled from 1975 to 1979 and the IMF imposed strict conditions on British economic policy[3] – constitutional issues took up much parliamentary attention. In the short run this did, however, not result in changes in British governance, because the resulting compromise did not command sufficient electoral support in referenda held in Wales and Scotland, being resoundingly rejected in the former and failing to muster yes-votes from 40% of the Scottish electorate as stipulated by an anti-devolutionist amendment to the bill.[4] If Scottish issues were unusually prominent in British politics in the second half of the 1970s and Labour was clearly the dominant party north of the border, the position of the Secretary of State of State for Scotland, first Willy Ross and then from April 1976 Bruce Millan, was still a tenuous one, not just poised between territorial and functional policy networks but at the same time having to take an increasingly uncertain parliamentary situation into account as the government majority was gradually being undermined by by-election defeats.

The political environment into which the SDA was launched was in other words one that was likely to be dominated by two sets of issues: on the one hand the relationship between public and private actors in economic development which had been central in the industrial turmoils and strategic vagaries under the Conservative Heath government, and on

2 Cf. the discussion in Section 4.1.

3 Middlemas 1991 pp. 149-56.

4 On the 1970s devolution debate and its reverberations within the main political parties, see Keating & Bleiman 1979 Ch. 5, Kellas 1989 Ch. 8, Hutchison 1999 Ch. 4.

the other hand the possible reconfiguration of British territorial politics in order to accommodate pressures from the Scottish part of the union state. Still, it is, of course, hardly surprising that the public debate on the SDA in the late 1970s covered many of the themes which had been prominent when the new organisation was being established in 1974-75, but as implementation picked up speed and innovative functions introduced, new discursive dimensions were added because it was no longer only the intention of the Labour government and its arm's-length body that could be queried, but also the implications of concrete Agency activities.

The SDA and Territorial Politics

The level of financial resources allocated to the SDA functioned as a symbol of commitment to Scotland in general and Scottish economic development in particular, and as the policy instruments of the new organisation were a combination of inherited functions and innovative measures, both the overall level of funding and the additional amounts allocated to new functions like industrial investment and urban renewal acquired political salience.

The SNP repeatedly argued that the level of funding for the Agency was insufficient given the scale of the problem and the oil resources which could have been earmarked for this particular purpose,[5] and initially also the Conservatives, echoing their remarks during the Third Reading of the SDA bill, questioned the willingness of the Labour government to allocate adequate resources to the new organisation.[6] After Alick Buchanan-Smith had resigned from the Conservative front bench over the hardening of Margaret Thatcher's anti-devolutionary stand, the new Scottish Shadow Secretary Teddy Taylor shifted the focus from possible underfunding of the SDA towards probable overspending within the organisation,[7] but still the most substantial Conservative interventions came when in early 1979 the Labour government introduced a parliamentary bill which would significantly increase the spending limit of the Agency.[8] While Scottish Office minister Gregor MacKenzie argued that the activities of the new organisation were now beginning to gain pace and that raising the financial limit was "a firm demonstration of our confidence in the Agency's long-term future"[9] –

5 *Glasgow Herald* 7.2.76, Scottish Grand Committee 1976 col. 119, 1977 col 226; *HCPD* 22.6.78 vol. 952 cols. 737ff, SNP 1976a, 1976b, 1979a p. 5.

6 *HCPD* 21.10.75 vol. 898 cols. 401f, 5.2.76 vol. 904 cols. 1514f, cf. Conservative Central Office 1977 p. 539.

7 *Glasgow Herald* 5.2.77, 10.2.77, 11.2.77.

8 Cf. the detailed discussion in Section 6.3.

9 *HCPD* 14.3.79 vol. 964 col. 561.

but perhaps also a public indication that government would like to see a greater level of activity[10] – the Conservative George Younger proposed a much smaller increase of the financial limit in order to demonstrate support for the inherited activities and urban renewal.[11] During the Third Reading of the bill the Conservative amendment was eventually withdrawn after rather vague government assurances that the increase originally proposed would be used for a wide range of purposes and not just the investment function,[12] and thus ultimately cross-party agreement had been established with regard to the overall level of funding available for the Agency.

In terms of territorial politics the main issue in Scotland in the late 1970s was undoubtedly the successive proposals from the Labour government for some form of 'home rule' where Conservative opposition to devolution hardened after Thatcher became party leader.[13] The constitutional issue was now a major area of disagreement between the two union state parties, and from a party-political perspective this undoubtedly made it easier – or perhaps even important – for the Conservatives to maintain a critical-but-positive stance with regard to the SDA in order to signal a visible commitment to some form of preferential regionalism. As demonstrated above, after three years of practical experience with the SDA the leadership of the main political parties still agreed about the principle of having an arm's-length economic development agency in Scotland: although Labour and the SNP wanted to increase financial commitments significantly while the Conservatives derided Labour's great expectations for their "miracle machine"[14] and continued to be more than a little sceptical about the industrial investment function,[15] George Younger still stressed that – contrary to government innuendo to the opposite[16] – that his party had "absolutely no intention of abolishing the SDA".[17] While Conservative front bench spokesman Alex Fletcher repeatedly pondered the possibility of dividing

---

10 Wilson Committee 1977f vol. 6 p. 199, SDA 1978 p. 43, Robertson 1978 p. 31, Hogwood 1978, HCPD 14.3.79 vol. 964 cols. 558f, Lewis Robertson 20.7.90, Bruce Millan 23.7.90, Gavin McCrone 20.6.90.

11 Standing Committee B 1979 cols. 125ff.

12 *HCPD* 14.3.79 vol. 964 cols. 560ff. The comments of the English frontbench spokesman for the Conservatives, Norman Lamont, was rather more robust, dismissing the bill as "a rather grubby attempt to rustle up some votes from the Scottish nationalists" (*HCPD* 23.1.79 vol. 961 col. 261).

13 See Kellas 1989 Ch. 8.

14 *HCPD* 23.1.79 vol. 961 col. 246, Scottish Grand Committee 1978 col. 99, cf. Wood 1989 p. 124.

15 Cf. the discussion in the following section.

16 E.g. Scottish Grand Committee 1978 col. 57f, 107.

17 Standing Committee B 1979 col. 128.

the functions of the Agency between several separate bodies,[18] the platform on which the Tories fought the 1979 general election clearly announced their intention of maintaining the arm's-length body, albeit with a revised version of the industrial investment function,[19] and when the Conservative leader Margaret Thatcher campaigned in Scotland, she stuck rigorously to this line.[20] Interestingly, it was actually the Labour party that came closest to undermining the comparative advantage of Scotland when its 1979 manifesto promised the setting up of parallel RDAs in English regions experiencing similar economic problems,[21] thereby recognising the value of this new approach to regional policy but also responding to demands from Labour MPs from the north of England for a more evenhanded approach at the British level.[22]

Strategies and Policy Implementation

In the late 1970s the Labour Party in Scotland, in line with the Labour government generally, came to see SDA-style activities as the key to industrial renewal, in Scotland as well as on a wider UK level.[23] The persistently high levels of unemployment within the regional economy were seen as a symptom of the general decline of private industry in Britain which had weakened the indigenous sector, left failing private firms to be nationalised, and seen government rely primarily on incoming firms to create new employment and growth. In order to break out of this vicious circle, the Scottish party organisation argued, more attention should be given to the indigenous sector, in particular through public investments in private firms and provision of specialised advisory services, informed by strategic research into key sectors of the Scottish economy[24] – in short, a much broader view of the Agency than the original proposal which had revolved around the importance of public

18 Scottish Grand Committee 1978 col. 101, Standing Committee B 1979 col. 150, *HCPD* 23.1.79 vol. 961 cols. 246ff, 14.3.79 vol. 964 cols. 564ff.

19 Conservative Party 1979 p. 16, cf. Conservative Central Office 1978 p. 144, and the discussion below.

20 Thatcher 1999: Interview for the *Scottish Daily Express* 13.2.78, Speech to Glasgow Conservatives 19.2.79, Campaigning in West Aberdeenshire 26.4.79, Interview for *Glasgow Herald* 1.5.79.

21 Reproduced in Craig 1990 (p 288).

22 In 1976 a backbench bill which would have established RDAs in all the Assisted Areas in England (English Development Agencies Bill 1976) which was politely rejected by industry minister Gregor MacKenzie on the grounds that what was being proposed could already be achieved by combining NEB and Industry Act powers (*HCPD* 5.3.76 vol. 906 cols. 1801-11).

23 Cf. the discussion in Section 14.4.

24 Labour Party Scottish Council 1977, esp. pp. 4ff, 29ff.

ownership through a 'state holding company'.[25] Although the STUC continued to press for industrial investment to be given greater priority at the expense of environmental renewal,[26] the impression of overall appreciation of the efforts of the Labour government and the strategies of the SDA by organisations on the left is clearly in evidence. Nor did the Conservatives, on the other hand, produce an alternative blueprint for dealing with the problems of the Scottish economy: partly, of course, because their anti-decline strategy focused primarily on the need to bring about a new approach to macro-economic management,[27] but probably also because an organisation, the SDA, was at hand that could continue to function as a vehicle of preferential regionalism, provided that some adjustments were made to the way in which it operated.

The industrial investment function remained the most controversial activity of the arm's-length body, being the only policy area – apart from the inherited tensions among localities with regard to environmental projects and urban renewal[28] – that was subject to public debate on a regular basis in the late 1970s. Compared to the discursive patterns surrounding the setting up of the Agency in 1975, the relative importance of the issues had, however, changed, presumably as a result of the interpretations by various actors of the way in which the new function was being implemented. From the left came calls for the adoption of broader investment criteria which would allow the Agency both to invest in more risky ventures and to engage in rescues of firms experiencing financial difficulties,[29] although demands of the latter type may perhaps have been somewhat muted in order not to produce additional fuel for the persistent and vociferous criticism by the main opposition party.[30] The major issue raised by the STUC was, in other words, no longer the principle of extended state ownership but much more detailed operational aspects of the industrial investment function, and a rather similar change can be detected with regard to Conservative discourse on the SDA. During the public debates in 1975 the notion of the Agency as a vehicle of creeping back-door nationalisation had been prominent,[31] but Scottish business organisations took an increasingly pragmatic view

---

25 Cf. the discussion in Section 5.1.

26 STUC 1977 pp. 137, 141.

27 Cf. Section 2.2 and the discussion below.

28 Cf. Chapter 16.

29 STUC statement at meeting with Prime Minister Callaghan 2.9.77 (STUC 1978 pp. 83ff), STUC 1979 p. 417; cf. SDA 1977a, 1977b, 1977e, Sir William Gray 19.10.90, Lewis Robertson 20.7.90.

30 George Robertson 30.10.90, Helen Liddell 21.6.90, Sir Kenneth Alexander 25.7.90.

31 Cf. Section 5.2.

of the industrial investment function,[32] and in the late 1970s more emphasis was given to the themes that had been secondary in the original debate, namely the need to ensure that the Agency disposed of its equity holdings and operated in a commercial manner that did not result in unfair competition for existing firms.[33] Along these lines the 1978 conference of the Scottish Conservatives discussed a proposal for new investment guidelines which required the SDA to operate as a minority investor in conjunction with private sources of finance and demanded "that the SDA should 'sell out' within five years",[34] but these restrictive measures were later transformed into a more positive vision by Margaret Thatcher who – while applauding the "vigorous" regional policy of previous Conservative governments that had brought Ravenscraig and Linwood to Scotland[35] – envisaged a "revamped SDA providing support for cash-hungry infant firms",[36] and later added provision of short-term financial aid to firms with long-term prospects to the to-do list of the Agency.[37] This line of thinking was neatly summed up in the Conservative manifesto for the 1979 general election which stated that the SDA would be given

> new guidelines to ensure that aid and investment is channelled to assisting our overall aim of the creation and expansion of industries with a viable long-term future and with the potential for making profits and providing secure jobs. It must not be used to achieve any long-term increase in state control of industry or to support inefficient firms at the expense of their more efficient competitors.[38]

Statements like these clearly implied that the SDA and/or its political sponsors did not intend to invest along commercial lines in industries of the future, but with regard to the investment function Labour simply reiterated the need for proactive long-term investment by the Agency, also in situations involving considerable risk or exposing other firms to increased competition.[39] Although ministers had originally claimed that instituting a financial duty would ensure that the Agency would invest in

32 CBIS representative Douglas Hardie at the meeting of Scottish Economic Council 9.5.77 (according to SDA 1977e), cf. *Glasgow Herald* 4.10.76, Lewis Robertson 20.7.90, McCrone & Randall 1985 p. 236.

33 Scottish Grand Committee 1976 cols. 138f, 1977 cols. 197, 231; 1978 cols. 9, 54; *HCPD* 23.1.79 vol. 961 cols. 247ff, *Glasgow Herald* 20.4.78.

34 *Glasgow Herald* 25.4.78.

35 Cf. Chapter 5 and Section 14.4.

36 Thatcher 1999: Interview for the *Scottish Daily Express* 13.2.78.

37 Thatcher 1999: Speech to Glasgow Conservatives 19.2.79.

38 Conservative Party 1979 p. 16.

39 Scottish Grand Committee 1978 cols. 58f, cf. Millan 1981 p. 103.

viable projects,[40] the direct challenges to the investment guidelines from the Conservative frontbench were not responded to by government,[41] and hence the Conservative claim of non-commercial practices which could be linked to specific investment failures was being met by rather vague expressions of support for the principle of public proactivity. This could readily be construed as an acceptance of the underlying premise of the Tory position, namely that state ownership and financial support for industries of the past had in fact been central to the investment function under Labour, and thus, by implication, that the investment function would appear to have a future after a change of government only if changes were brought about in the ways in which the SDA operated.

Uncertain Commitments

Although the major parties continued to agree about a preferential regionalist approach to Scottish economic development, what this consensus about territorial politics should mean in practice had begun to look increasingly uncertain by the beginning of 1979. On the one hand the Labour government, presumably wanting high-profile success for the Agency to reflect positively on its political sponsors, seemed somewhat impatient with the speed at which the SDA moved, but at the same time the party also proclaimed its intention of reducing the advantage of Scotland by establishing parallel RDAs in the economically weak parts of England if it was still the governing party after the general election. The Conservative opposition, on the other hand, reiterated their intention to maintain the SDA as a major agent of regional change but also promised what was presented as a major revamp of the controversial investment function and repeatedly vented the possibility of splitting up the organisation along functional lines. Although the precise meaning of political demands for 'more proactivity' or 'more commercial investments' remained unclear, no matter which party would win the general election, it seemed likely to result in an urgent need for the SDA to adjust to a new environment in terms of the policies and politics of regional development.

14.2. Miraculous Reinvention? The Scottish Environment 1979-85

The general election in May 1979 was won by the Conservative party under the leadership of Margaret Thatcher. As the official parliamentary opposition in the late 1970s the Tories had increasingly sig-

---

40 Cf. the discussion in Section 5.2.

41 See especially Scottish Grand Committee 1978 and *HCPD* 23.1.79 vol. 961 cols. 247ff.

nalled general scepticism about state intervention and a preference for market-based solutions, but in 1979 the party conducted a fairly moderate campaign under the slogan "Labour isn't working" with a platform dominated by anti-inflation measures and the restoration of law and order after the industrial disputes of the 'winter of discontent'.[42] While the election did produce a comfortable Conservative majority in the Westminster parliament, the electoral trends within the union state did not move in parallel: in England the Tories increased their share of the vote from 38.9% to 47.2% while Labour's electoral support was reduced to 36.7%, but in Scotland the collapse of the SNP vote from 30.4% to 17.3% meant that while the Conservatives did increase their share of the vote from 24.7% to 31.4%, Labour grew nearly as much and maintained a clear lead with 41.6% of the votes cast.[43] In Scotland the electoral system translated these votes into an even more pronounced dominance of Labour with 44 MPs against 22 Conservatives,[44] and thus while in England – and therefore also at the UK level – 1979 was the year in which the Conservatives regained their dominant position with the voters, in Scotland it restored the electoral dominance of the Labour party. Like his predecessors, the new Conservative Secretary of State for Scotland, George Younger, found himself at the intersection between the Scottish territorial policy network and the functional British ones, but with the added complication that more than two thirds of the Scottish vote had been cast for other political parties.

Despite severe recession – unemployment figures grew steadily from 6.9% in 1979 to 14.2% in 1985[45] – the first half of the 1980s was a fairly stable period in Scottish politics, especially compared to the situation in England. In the early 1980s the opposition to the Thatcher government became increasingly divided when moderate Labour MPs created the centrist Social Democratic Party and allied themselves with the Liberal party. Yet, while the leftward drift of Labour, national unity during the war with Argentina, and the selling of public sector council houses to sitting tenants have often been credited with bolstering the electoral fortunes of the Conservatives,[46] at least two of these factors can hardly have made much of an impact in Scotland: relatively few council houses were sold north of the border, and the Scottish Labour party remained fairly moderate both at Westminster and in local government,[47] and these differences helped to produce a general election result in 1983

42 Cf. Section 2.2.

43 Brown *et al.* 1996 Table 7.1.

44 Kellas 1989 Table 15.

45 Balchin 1990 Table 2.7.

46 Cf. Section 2.2.

47 Hutchison 1999 pp. 147ff.

where, again, the difference between England and Scotland was conspicuous. At the British level the 1983 results reinstated the Conservative government with a much-enhanced majority on the basis of a decreasing share of the vote because opposition was divided, but in Scotland the rise of the SDP-Liberal Alliance was less pronounced and took place at the expense of all the other three parties, and thus Labour still accounted for more than half of the Scottish MPs.[48] At the Scottish Office this meant that the tenure of Secretary of State George Younger could continue, facing much the same parliamentary situation in Scotland but being able to govern on the basis of a greatly increased majority of English Tories at Westminster. Scotland was in other words no longer essential to the governing party in electoral terms, but would this affect the extent to which British political priorities were translated in policies by a government which had a long tradition for defending the union state?

The SDA and Territorial Politics: To Be or Not to Be?

From the perspective of the SDA, the first six months under the new government in 1979 proved to be anxious ones. Although the position of the Conservatives in opposition could be summarised as supportive of the inherited policy functions, critical of the investment powers and ambivalent about the merits of combining all these activities within one arm's-length body,[49] the general emphasis of the Thatcher government on fiscal prudence and its well-published scepticism about state intervention meant that question marks appeared to hover over the organisation, not least in the minds of its originators in the Labour party and, of course, within the organisation itself.[50] Shortly after the general election leading Conservative ministers, from the Prime Minister herself to Scottish Industry Minister Alex Fletcher, reiterated the pledge to retain the SDA with new investment guidelines,[51] but still the first major statement on regional policy by Sir Keith Joseph, the neoliberal intellectual that had become Secretary of State for Industry, would seem to be ambiguous with regard to the role of non-departmental bodies such as the RDAs. While it was explicitly pointed out that the "functions of the SDA [...] are unaffected by this announcement", at the same time it was also underlined that "we shall be examining the future role of these

48 Kellas 1989 Table 15.

49 Cf. the discussion in Section 14.1.

50 George Mathewson 9.7.90, Charles Fairley 26.7.90; the fate of the SDA in general and the investment function in particular was questioned by Labour only weeks after the election (*HCPD* 23.5.79 vol. 967 cols. 1042f).

51 Thatcher 1999: Speech to Scottish Conservative Conference 12.5.79; *HCPD* 23.5.79 vol. 967 cols. 1042f.

bodies in those parts of their present territories which [...] are to become non-assisted areas".[52] and thus not only the sponsoring Scottish Office but also the lead department in the UK regional policy network would apparently be involved in rethinking the role of the Agency. Already in the Autumn of 1979 it was clear that the new government found the organisation useful as a means of addressing localised employment crises in the wake of industrial closures,[53] but not before the publication of the long-foreshadowed revised investment guidelines in December was the detailed shape of the new-model SDA revealed: the organisation would be retained as an integrated multi-functional body, but its statutory purposes would be adjusted so that employment became less central, and the new guidelines made industrial investment an auxiliary function where private co-funding and early disposal of equity were required.[54]

One of the Agency's founding mothers, Helen Liddell, later described this outcome as "miraculous" in view of the general disposition of the Thatcher government *vis-à-vis* interventionist bodies,[55] and the actors directly involved and contemporary observers tend to point to three factors explaining the line taken by the Conservative government: external pressures through widespread support for the SDA in Scotland in general and within the consensual territorial policy network in particular,[56] the political usefulness of being seen to be doing something to address economic hardship within the region especially in times of recession,[57] and the pragmatic Conservatism of George Younger – generally described as "paternalist"[58], "patrician",[59] the "quintessential Tory aristocrat"[60] – who had become Secretary of State when the Thatcherite Shadow Scottish Secretary Teddy Taylor lost his parliamentary seat in the 1979 general election.[61] While Thatcher later praised the "robust patriotism" of Taylor and deplored the "standing up for Scot-

52 *HCPD* 17.7.79 vol. 970 cols. 483f.

53 *Glasgow Herald* 21.8.79, 26.10.79; STUC meeting with George Younger 14.11.79 (STUC 1980 p. 119).

54 Cf. Section 8.1 and the discussion below.

55 Helen Liddell 21.6.90. SDA Director Neil Hood used the term "amazing" (22.6.90).

56 Gavin McCrone 20.6.90, Sir William Gray 19.10.90, Sir Kenneth Alexander 25.7.90, George Robertson 30.10.90, Isobel Lindsay 6.6.90.

57 Edward Cunningham 13.6.90, Sir Robin Duthie 12.7.90, Charles Fairley 26.7.90, Harry Hood 25.7.90, Craig Campbell 16.7.90, Alf Young 5.6.90, cf. Booth & Pitt 1983.

58 Thatcher 1993 p. 620.

59 Midwinter *et al.* 1991 p. 25.

60 McCreadie 1982 p. 72.

61 Stevens 1990 p. 85, cf. Naughtie 1981, David Lyle 19.7.90.

land" practised by Younger,[62] the ability of the latter to engage even his more critical junior minister Alex Fletcher[63] in "mildly interventionist policies" at odds with the "punk monetarism" practised by government in the early 1980s[64] would, however, not seem to reflect a sudden change of balance within the Conservative party in Scotland caused by Taylor's defeat at the polls.[65] Under George Younger the new government did what it had officially proposed to do at least since the 1978 Scottish party conference and, indeed, specifically committed itself to doing in its 1979 election manifesto, namely to retain the SDA with revised investment guidelines, and thus the 'miracle interpretation' would either seem to pay too much attention to the oppositional rattling of sabres by Taylor and Fletcher, or, perhaps more likely given the timing of the research interviews, to assume the existence of a long-term Tory plan to get rid of a creature of 'Labour interventionism' culminating in the Scottish Enterprise proposal. As argued earlier, the Conservative party in Scotland had supported various forms of regional preferentialist measures in economic development for decades, and as most SDA functions were inherited and/or uncontroversial, the decision to retain the organisation as a vehicle for continued administration of long-standing policy programmes and flexible ad-hoc intervention in case of localised unemployment crises would after all seem to be rather unsurprising.

Having decided to retain the "reorganised"[66] or "restructured"[67] SDA, the Conservatives generally spoke of the organisation in positive terms in the first half of the 1980s, and the 1983 general election manifesto praised "the good work and experience of the SDA".[68] Also in this period adjustments of expenditure limits served as a symbol of government commitment to the Agency: the extended limit introduced by the Labour government just months before the 1979 election was removed in the beginning of 1980, only for a further extension of the ordinary limit to be introduced the next year,[69] accompanied by opposition la-

62 Thatcher 1995 p. 325, 1993 p. 619.

63 Lewis Robertson 20.7.90, Sir Robin Duthie 12.7.90, Edward Cunningham 13.6.90, David Lyle 19.7.90, George Robertson 30.10.90, cf. Naughtie 1985.

64 Aitken 1994. At the meeting between STUC and George Younger 17.9.79 the latter reportedly alluded to his position on the traditional Tory wing of the party when he insisted that "the Conservative government represented a wide spectrum of opinion" (STUC 1980 p. 111).

65 As suggested by Holliday 1992 p. 451.

66 Scottish Grand Committee 1981 cols. 142f.

67 Scottish grand committee 1984a cols. 132f, cf. *CBI News* 25.10.85 p. 23.

68 Conservative Party 1983 p. 20, cf. Conservative Central Office pp. 337f.

69 Cf. the discussion in Section 6.3.

ments about starvation of the arm's-length body[70] and government insistence that the significant increase in 1981 was an expression of "ministerial confidence in the Agency's future".[71] While Conservative backbench criticism became much more muted – largely confined to occasional interventions by a limited number of Thatcherites around Michael Forsyth[72] – the notion of government 'restructuring' was echoed in the repeated demands of the STUC for the Agency to be (at least) restored to its former position with regard to powers and funding in order to be able to "intervene more effectively in the process of re-investment and renewal of [...] industry",[73] and the SNP even demanded that the arm's-length body should be incorporated in a new Ministry of Industry and Development operating though Local Enterprise Boards throughout Scotland.[74] Apart from the occasional jibe about the Tories now embracing an organisation they used to oppose[75] – an allegation denied by both George Younger[76] and Margaret Thatcher[77] – the focus of Labour MPs was now primarily on the consequences of the macro-economic strategies of the Conservative government,[78] and thus despite disagreements about specific Agency activities, the basic question of the existence of the SDA as a vehicle of preferential regionalism effectively seemed to have become a non-issue in party-political terms,[79] if, indeed, it had ever seriously been one.

Strategies and Policy Implementation: New Industries for Old?

In terms of development strategies and policy implementation the early 1980s were, unlike the preceding period, *not* dominated by con-

---

70 E.g. Standing Committee E cols. 644ff, Scottish Grand Committee 1980 cols. 104ff.

71 Standing Committee A 1981 col. 38.

72 Scottish Grand Committee 1984a cols. 16, 28ff; cf. Adam Smith Institute 1983 p. 5.

73 Quoted from the 1983 Annual Congress resolution (reproduced STUC 1984 pp. 65f), cf. STUC 1980 pp. 105f, 1983 p. 67. The longstanding STUC interest in the SDA and its activities would, however, appear not to have been shared by Scottish trade unions in general, as a detailed study of the Agency (reported in STUC 1982 pp. 60-65) did only elicit very limited response when circulated to member unions in order to identify innovative policy proposals (STUC 1983 p. 84).

74 SNP 1983a pp. 6ff, cf. SNP 1982 p. 12, 1982a.

75 *HCPD* 19.11.81 vol. 13 cols. 437f.

76 Scottish Grand Committee 1981 cols. 141ff.

77 Thatcher 1999: Speech to Scottish Conservative Conference 13.5.83.

78 E.g. *HCPD* 29.11.83 vol. 49 cols. 783ff, 792ff, cf. the discussion below.

79 Sporadic proposals for parallel RDAs in English regions were vented by Labour backbenchers and ignored or dismissed by government (e.g. *HCPD* 17.7.79 vol. 970 cols. 1318f, 13.12.83 vol.50 col. 842, cf. Committee of Public Accounts 1981 pp. 7f), but parliamentary front-bench efforts in pursuance of manifesto commitments seemed rather half-hearted, cf. the discussion below.

flicts about the industrial investment function. Although the new investment guidelines were criticised for being too restrictive by both Labour[80] and, rather more insistently, the STUC,[81] business opinion seemed to accept that the function would henceforth primarily be brought into use in reactive rescue situations,[82] and the Conservatives seemed content with the new state of affairs which strengthened the involvement of both ministers and private sector actors in the industrial investment function and would preclude the SDA from extending public ownership and unfair competitive practices.[83] Instead of the retained but officially downgraded investment function, Labour primarily focused on two issues, namely what was seen as the complacency of the government *vis-à-vis* the state of the regional economy, and the way in which the Conservatives tried to pursue a development strategy based on attraction of inward investment.

Throughout the early 1980s the official parliamentary opposition concentrated on highlighting the growing problem of unemployment, constantly recounting unemployment figures and enumerating firms that had closed down or were about to do so, and, while welcoming new industries and inward investment, a central demand was more government attention to support for the retention and modernisation of existing firms in the regional economy.[84] A similar focus on the indigenous sector can be found in statements from the STUC[85] and the SCDI,[86] but, in sharp contrast to this, Secretary of State George Younger concentrated on presenting a picture of Scotland which focused on the growth of new enterprises and industries because "there is no future in clinging on to the old and out of date in Scotland".[87] In terms of policy this shift

80 E.g. Standing Committee E 1980 cols. 646ff, Standing Committee A col. 30, cf. George Robertson 30.10.90.

81 STUC 1981 p. 83, 1982 pp. 53, 62ff,; 1984 p. 29, 1985 pp. 198f, 1986a pp. 15f; *Scottish Trade Union Review* 18, 1983, pp. 10, 25.

82 Scottish Engineering Employers Association 1979 pp. 18f. A survey commissioned by the SDA showed that private businessmen tended to see the Agency primarily as a lender of last resort (Marplan 1983), reflecting both the prominence of the industrial investment function in public debates in general and the spin given to the new guidelines by the Conservative government.

83 Standing Committee E 1980 pp. 511ff, 674ff; Scottish Grand Committee 1985 cols. 161f. Scottish Thatcherites, including Michael Forsyth, never managed to get political mileage out of suggesting that perhaps the investment function was still not operating on a commercial basis in the mid-1980s (*Glasgow Herald* 4.7.79, *HCPD* 3.4.84 vol. 57 cols. 471f).

84 *HCPD* 19.11.81 vol. 13 cols. 431ff, 29.11.83 vol. 49 cols. 783ff; Scottish Grand Committee 1980 cols. 57ff, 1984 cols. 39ff, 1985 cols. 149ff.

85 STUC 1981 p. 83, 1982 pp. 53, 62ff; 1984 p. 29, 1985 pp. 198f, 1986a pp. 15f.

86 SCDI Executive Committee 6.6.83, 31.5.84.

87 *HCPD* 19.11.81 vol. 13 col. 447.

could best be facilitated through forward-looking attempts to attract high-tech inward investment and, in parallel with this, support for new small firms in localities hit by firm closures, and in 1985 Younger even claimed that the regional economy had been transformed to the extent that Scotland was now "widely recognised not as a museum of declining industry but as a new centre of high technology".[88] Also this message was reiterated on numerous occasions,[89] and towards the mid-1980s Labour's Scottish Shadow Secretary Donald Dewar claimed that the Conservative script was "pre-determined",[90] while his own, equally predictable but rather different, view of the Scottish economy was described by a Tory backbencher as evidence that Labour was "clinging to the corpse".[91] Both parties defined the nature of the regional problem in Scotland in terms of high levels of unemployment caused by the contraction of traditional industries, but the remedial policies favoured clearly differed. While Labour primarily concentrated on maintaining employment in existing firms – the effects of ill-judged macro-economic policies which translated into calls for rescues or investment in modernisation – the Conservatives concentrated on replacing lost jobs through support for new firms, incoming as well as indigenous. Whatever the long-term merits of both strategies in terms of economic development, these priorities associated Labour with a stream of bad news relating to declining industries of the past, while Tory ministers also had access to a stream of good news relating to growing industries of the future and even Margaret Thatcher began to praise 'Silicon Glen'.[92] Moreover, as the opposition was not too clear about the policy measures needed to tackle problems within the indigenous sector and government had well-known high-profile SDA remedies at their disposal in the form of FDI attraction and area initiatives in the wake of firm closures, the Conservatives seemed to have carved out a position for themselves where they could present themselves as practising regional preferentialists that were actively seeking to address longstanding problems within the regional economy.

Ironically, a key organisational vehicle for implementation of government strategy, Locate in Scotland, had only been created because Secretary of State George Younger decided to overrule the Conservative majority of the parliamentary committee investigating the attraction of

---

88 Scottish Grand Committee 1985 col. 166.

89 *HCPD* 29.11.83 vol. 49 cols. 772ff; Scottish grand committee 1984a cols. 47ff, 52f; 1985 cols. 156ff.

90 Scottish Grand Committee 1985 cols. 156ff.

91 *HCPD* 29.11.83 vol. 49 col. 798.

92 Thatcher 1999: Speech to the Glasgow Chamber of Commerce 28.1.83, speech to the Scottish Conservative Conference 13.5.83.

inward investment to Scotland.[93] It has been claimed that the attack on the overseas activities of the SDA was a backbench attempt to have a 'second go' at limiting the scope of Agency activities after the organisation had survived the change of government[94] – something Conservative committee member Ian Lang later seemed to suggest after having been promoted to Industry Minister at the Scottish Office[95] – but the discussions of the committee quickly came to focus on the organisational details of inward investment attraction, especially how to achieve efficient coordination between different actors within Scotland. Evidently both parties agreed that specific Scottish activities would be needed, and the issue that divided the committee was whether these efforts should operate through existing UK channels abroad or take the form of separate Scottish promotional efforts. The Conservative majority of the parliamentary committee had wanted to integrate Scottish activities into the overall UK efforts under the auspices of DTI-sponsored Invest in Britain Bureau in order to keep inter-regional competition under control, while Labour minority, along with large parts of the territorial policy network, had proposed an independent Scottish effort, including a permanent physical presence through a network of offices abroad. Ultimately the Secretary of State opted for a high-profile form of preferential regionalism in the shape of LIS and thereby maintained the position of the Scottish Office and, indeed, the party of government within the long-standing consensus about the need for special measures to address the economic problems of Scotland.

Predictable Reversions

In terms of political discourse the early 1980s saw the Conservative government reinterpret the meaning of the SDA. Any notion of state intervention or public enterprise disappeared, and instead the Agency was presented as an organisation that in the words of the newly elected Prime Minister Thatcher could help by "easing the transition from the industries and jobs of the past, to the industries and jobs of the future."[96] After inadvertently having helped government to acquire an organisational platform, LIS, through which its strategy of sectoral change could be pursued, the Labour opposition concentrated on highlighting what was presented as the economic failures brought about by government macro-economic policy, but the extent to which regional policy, including the SDA, would be part of the answer to the escalating crisis of

93 Cf. the discussion in Section 10.1.

94 E.g. Alf Young 5.6.90, cf. Section 10.1.

95 *HCPD* 29.10.87 vol. 121 col. 537.

96 Thatcher 1999: Speech to the Scottish Conservative Conference 12.5.79.

existing industrial firms remained unclear. In contrast to this the Conservative government had visible measures at their disposal – FDI attraction and post-closure area initiatives – which made it abundantly clear that the Tories remained within the long-standing preferential regionalist consensus. Strategic goals and policy priorities clearly had changed, but the underlying assumption about specific needs requiring special, additional, Scottish measures remained intact.

14.3. Due Credit? The Scottish Environment 1986-91

In many ways the late 1980s and early 1990s saw more turbulence in Scottish politics than the preceding period. The 1987 general election only slightly reduced the massive parliamentary majority of the Conservative government in the Westminster parliament, but again this result reflected diverging trends within the union state: in England the Tory share of the vote grew marginally while Labour recovered some of the terrain lost to the Liberal/SDP Alliance without getting more than 30% of the votes cast, while in Scotland the governing party recorded its lowest level of support since the Second World War while Labour nearly reached its pre-SNP level,[97] as illustrated by Figure 14.1. In terms of parliamentary strength the Conservative result in Scotland was, however, even worse, with only 10 MPs against Labour's 50, a 'Doomsday scenario'[98] which was not only seen as politically embarrassing but also greatly complicated the management of Scottish parliamentary business by government because of a shortage of suitable MPs to fill ministerial positions and serve on committees. Moreover, this problem had been exacerbated by the promotion of George Younger to Secretary of State for Defence in 1986, resulting in Malcolm Rifkind becoming Secretary of State for Scotland. While the industry portfolio at the Scottish Office remained with Ian Lang from 1986 until he eventually inherited the top Edinburgh job in 1990 after Rifkind had been made Transport Secretary in the first Major government, the prevalence of consensual Toryism seemed threatened by the inclusion of the Thatcherite Michael Forsyth in the Edinburgh ministerial team. The importance of these changes in government personnel was increased by the fact that after the 1987 general election many of the policy areas targeted for reform by the Conservatives were related to aspects of the welfare state that were subject to Scottish Office authority – and thus often fell within the Forsyth portfolio. A series of hard-fought controversies about e.g. education and local government (the 'poll tax') managed to further alienate not only popular opinion but also the 'elite'

97 Brown *et al.* 1996 Table 7.1.

98 An oft-used journalistic metaphor (Midwinter *et al.* 1991 pp. 95ff).

whose influence within the territorial policy network appeared to be threatened by reforms, making it possible to construe Thatcherism as an 'alien' (English) and 'extremist' attack on Scottish institutions and, ultimately, creating widespread support for some form of home rule that could function as a stronger bulwark than the Scottish Office against Westminster priorities.[99]

Figure 14.1 General election results in Scotland 1970-87

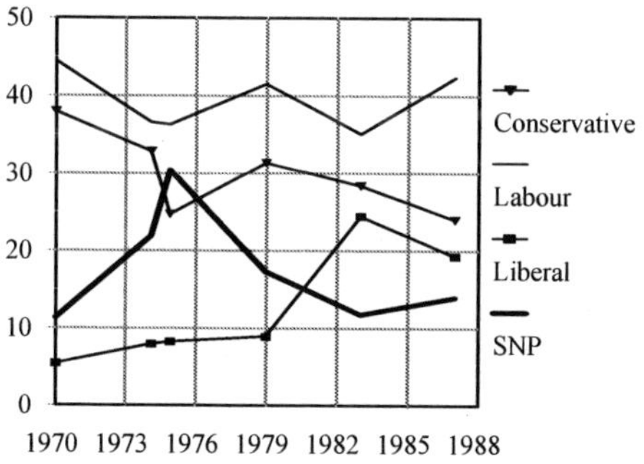

Per cent of votes.
Source: Brown *et al.* 1996 p. 146.

All in all the overall direction of politics and policy now seemed to be less certain than under George Younger in the first half of the 1980s: the pressure for reform from the UK level increased while at the same time the Conservative 'minority administration' at Scottish Office and its key initiatives proved to be not only unpopular but were also often actively resisted. Where would this leave the SDA which, at arm's-length from elected politicians, would appear to enjoy some degree of insulation from the political environment that had hitherto enabled the organisation to reposition itself when faced with new challenges?

Territorial Politics: The SDA as 'One of Us'

In 1986 and 1987 the SDA was subject to two parallel review processes: on the administrative level an in-depth evaluation of the rationale for and implementation of Agency policies was undertaken by the Treasury and the Scottish Office as part of the ongoing program of reviews of non-department public bodies,[100] and on the political level

---

[99] Midwinter *et al.* 1991 pp. 25f, McCrone 1991b, Holliday 1992, Keating 1996 pp. 211ff, Hutchison 1999 pp. 139-55.

[100] Cf. the discussion in Section 7.3.

parliament was due to consider extending the expenditure limit of the organisation which, for the first time ever, was coming close to being exhausted.[101] The administrative review concluded that although there was room for improved implementation procedures in a number of policy areas, overall the SDA "had adapted successfully to the change of approach to regional and industrial development under the present government".[102] Having passed the 'technical' test of demonstrating compatibility with the market-failure thinking in favour in the mid-1980s with the Conservatives, government proposed a substantial increase of the Agency's expenditure limit expected to fund activities through to 1992. Scottish Industry Minister Ian Lang motivated this with the "substantial and positive impact on the Scottish economy and environment" identified by the administrative review group and summarised the current role of the SDA as "working with the grain of our capitalist economy [...] [as] an engine of free enterprise, creative and catalytic".[103] While neither opposing the proposal or tabling amendments, Labour's Donald Dewar still characterised the bill as a "modest measure" and suggested that, although private co-funding was welcome, more public money would be needed if the SDA were to make a greater impact, especially in a situation where other forms of regional policy were being undermined by government policy.[104] The basic argument was the misfit between the needs of the regional economy, especially its indigenous sectors, and the resources being made available by government, and Labour's low-key opposition was easily surpassed by the SNP[105] and the STUC,[106] both of which combined demands for a substantial increase in funding for the SDA with proposals for strengthening economic development efforts in Scotland through the creation of new institutions or the strengthening of existing ones.

Apart from routinely demanding 'more of the same' with regard to funding, in terms of territorial politics the opposition concentrated their efforts in areas other than the SDA, namely campaigns against major plant closures and attempts to build a broad political coalition around renewed demands for Scottish devolution. The *ad hoc* anti-closure campaigns were clearly driven by trade unions, Labour and local authori-

---

101 Cf. Section 6.3.

102 IDS 1987a p. 11, cf. the discussion in Section 7.3.

103 *HCPD* 21.10.87 vol. 120 cols. 839ff. In his opening statement, minister Lang even took over two of the three 'Cs' from the early 1980s corporate lingo of the SDA (cf. Section 7.2).

104 *HCPD* 21.10.87 vol. 120 cols. 843-50.

105 *HCPD* 29.10.87 vol. 121 col. 486, SNP 1987a p. 11, 1988a.

106 STUC 1987a pp. 11-16, cf. 1987 p. 54.

ties,[107] while Conservatives were publically sceptical about the wisdom of preserving 'desert cathedrals' like the Ravenscraig/Gartcosh steel complex originally brought to Scotland through government inducement,[108] and Ian Lang taunted the opposition for instinctively thinking "that the only job worth having was bashing a bit of metal and developing industrial deafness".[109] Still, the governing party did, however, have good political reasons for being seen to make an attempt to retain what seemed to have become icons of Scottish industrialism, and in practice Scottish Office ministers and Tory MPs did – to some extent and with varying degrees of enthusiasm – give public support and sometimes managed to postpone redundancies.[110] In contrast to this, the Conservatives clearly made a point of opposing devolution and defending Scotland's existing position within the union state, repeatedly arguing that devolution was the 'slippery slope' towards full independence,[111] and thus the Tories continued to position themselves within the preferential regionalist tradition primarily through sponsorship of arm's-length development bodies: the party publically boasted about spending more on the SDA that the Labour government had done,[112] and despite English Tory misgivings about the Agency being "an engine of free enterprise liberally lubricated by public money",[113] the Conservative manifesto for the 1987 general election extolled SDA, LIS and HIDB as "our major instruments for economic regeneration and environmental improvement".[114]

Strategies and Policy Implementation: An Emerging Consensus?

Also in the late 1980s and early 1990s the state of the Scottish economy was interpreted very differently by the Conservative government and the increasingly vociferous opposition. Scottish Office ministers continued to present a picture dominated by positive messages about structural change and the growth of modern economic activities in high-tech industries and tradeable services,[115] while Labour and what Moore

[107] Dowle 1987, Smith & Burns 1988 pp. 263ff, Moore & Booth 1989 pp. 91ff.

[108] E.g. Scottish Grand Committee 1984a cols. 28ff, cf. Love & Stevens 1989.

[109] *HCPD* 25.1.88 vol. 126 col. 116.

[110] Lang 2002 p. 71, Macwhirter 1991, cf. Aitken 1997 p. 296.

[111] Brown *et al.* 1996 pp. 63f, Hutchison 1999 p. 141.

[112] Conservative Party 1987b p. 28.

[113] *HCPD* 29.10.87 vol. 121 col. 483.

[114] Conservative Party 1987a p. 33, cf. Conservative Central Office 1987 pp. 432f.

[115] See e.g. Scottish Grand Committee 1988 cols. 131ff. Scottish Industry Minister Ian Lang even started having produced a glossy publication, *Scotland – An Economic Profile* (Scottish Office 1988, 1991), in order to "dispel the myth" perpetuated by

& Booth have called the "defensive consensus"[116] focused on the continued threat of closure to many of the remaining large plants in traditional industries. The latter took the form of pressure on the Scottish Office to use political influence *vis-à-vis* e.g. the owners of Ravenscraig, the recently privatised British Steel, and in terms of discursive positioning this amounted to an attack of government for neglecting the plight of traditional industries. In the late 1980s a more thorough critique of the Conservative position did, however, begin to emerge on the basis of work undertaken for *The Standing Commission on the Scottish Economy*, an *ad hoc* body initiated in 1986 by the STUC and the Labour-controlled Strathclyde Regional Council, which had attracted much support in the territorial policy network, albeit less from business representatives and, of course, not the Scottish Office. Apart from highlighting the high levels of unemployment and the difficulties experienced by indigenous firms, this "proactive consensus"[117] also questioned the quality of the jobs and the long-term development prospects of the inward investment projects which were central to the modernisation strategy of the Conservative government.[118]

It is, however, interesting to note that while the analysis of the state of the Scottish economy of government and opposition contrasted sharply,[119] there now was a much higher degree of consensus about what policy instruments should be employed and which sectors of the economy targeted. On the one hand the Conservatives, safe in the knowledge that the SDA had "changed out of all recognition",[120] now explicitly endorsed that the Agency now also targeted indigenous sectors of the regional economy. In the words of Ian Lang

> the Agency recognises – and rightly – that Scottish industry, and not simply inward investors, is its client base. The Agency is now increasingly applying to the development of indigenous companies the targeted approach that has brought it success in the attraction of inward investment.[121]

On the other hand Labour's Donald Dewar not only commended the Agency's "honourable position […] and key place in deciding industrial

Labour-leaning media figures who "chose to dwell on the old heavy industries, whose decline had been under way for years and was now almost beyond redemption" (Lang 2002 p. 69).

116 Moore & Booth 1989 p. 91.

117 Moore & Booth 1989 p. 98.

118 Standing Commission 1988, 1989.

119 Compare e.g. Scottish Office 1988 with Standing Commission 1989 or STUC 1987a.

120 *HCPD* 29.10.87 vol. 121 col. 537.

121 *HCPD* 21.10.87 vol. 120 col. 841.

strategy"[122] but also stated that, contrary to what is suggested by Conservative "myth-making",

> I strongly support the efforts of the Agency to take areas in which it believes there can be genuine industrial growth in Scotland, to try to encourage excellence there and to attract firms that move in.[123]

This emerging consensus about a comprehensive overall development strategy did, however, not preclude different views about the way in which individual SDA activities should be conducted.

From a government perspective attraction of inward investment from abroad continued to be central to its attempt to promote sectoral change: the Scottish FDI record was routinely highlighted in Conservative publications,[124] and the abundance of photos of ministers opening new plants or announcing future investments in SDA and LIS publications[125] testifies to the systematic efforts of the Agency's political sponsors to become closely associated with the good news of job creation and economic modernisation. While in principle being more than a little sceptical about the impact of external ownership of Scottish industry,[126] STUC and the Standing Commission acknowledged the importance of FDI in the Scottish economy but questioned the quality of the projects being attracted to Scotland and emphasised the need to strengthen the links between incoming investment projects and local suppliers in order to make the new jobs more safe and create additional ones in existing firms.[127] The Conservatives, on the other hand, appeared to have taken to the new-model industrial investment function to the extent that rumours about an impending privatisation were dismissed out of hand along with opposition underestimates of the level of expenditure,[128] and the function was now described as "essentially an investment bank for prospectively or currently successful and expanding industry".[129] The Labour opposition accepted the usefulness of using public investment to 'pump prime' in order to attract additional private funding and mainly objected to the level of funding allocated to the function,[130] but the Standing Commission was highly sceptical about the commercial criteria according to which individual investment proposals were to be judged according to

122 *HCPD* 21.10.87 vol. 120 col. 844.
123 *HCPD* 21.10.87 vol. 120 col. 851.
124 E.g. Conservative Party 1987b, Conservative Central Office pp. 32f.
125 Cf. Section 10.1.
126 STUC 1987a p. 16, cf. STUC 1989.
127 STUC 1987 pp. 53f, Standing Commission 1989 pp. 46ff.
128 *HCPD* 25.11.87 vol. 123 cols. 320ff.
129 *HCPD* 29.10.87 vol. 121 col. 473.
130 *HCPD* 21.10.87 vol. 120 cols. 845ff.

the government's market-failure approach which created a "Catch 22" situation:

> Agency interventions must be increasingly commercial, but must simultaneously avoid areas where there is the prospect of the private sector filling gaps in market provision.[131]

Concurrently the STUC called for greater proactivity in industrial investment, returning to the early 1970s theme of equity investment as a means to exercise public influence on private economic actors and create accountability for firms in receipt of government financial support.[132] The main issue dividing government and opposition in this area would in other words seem to the question of the scale of activity desirable, where especially the Standing Commission sought to identify the elements of government thinking that would create 'paralysis through analysis' and limit the scope for Agency activity. All in all the late 1980s saw increasing agreement with regard to the strategic goals of regional policy in Scotland, but also a renewed prominence of debates about the ways in which policy instruments were being used, with government seemingly being less demanding *vis-à-vis* incoming firms than indigenous ones. As both of these aspects concerned the conditions under which support could be given to individual firms, the issue of the ideal relationship between public and private actors had in other words not disappeared from the discursive environment of the SDA.

Politics of the Scottish Enterprise Merger

In December 1988 the Conservative government published a white paper proposing the creation of a new arm's-length body, Scottish Enterprise, through a merger between the SDA and the Scottish part of the Training Agency, a public body responsible for training programmes aimed especially at the unemployed.[133] The document praised the achievements of the two organisations in the 1980s, including the "SDA's commercial and customer-oriented ethos" and its move towards becoming "a facilitator and enabler [...] rather than [...] a direct provider",[134] but increasing international competition and skills shortages caused by demographic decline made it paramount for government to coordinate and upgrade efforts in training and enterprise development by tailoring activities to the specific needs of different localities. As "employers need to be given a sense of ownership of the system of training and

131 Standing Commission 1988 p. 88.

132 STUC 1987a pp. 11ff, 1989 p. 47, 1990 p. 55.

133 IDS 1988a.

134 IDS 1988a p. 6.

enterprise creation",[135] the new organisation would be given a two-tier structure with a 'core' taking care of overall strategic issues and some functions best carried out at the Scottish level while most training and business development activities would be delivered by local agencies directed by boards with a majority of volunteers from the local business community.

The white paper contained a fairly low level of detail and thus came closer to a consultative document than a fully worked-up government proposal, and part of the explanation for this was undoubtedly that the idea had originally come to the Scottish Office from outside. Bill Hughes, chairman of the Scottish network for Enterprise Trusts providing advice to small and new firms,[136] had proposed a merger in the summer of 1988 and subsequently had it endorsed by Margaret Thatcher at the Scottish CBI conference in September as a "Scottish solution to respond to Scottish needs".[137] Having been launched by a prominent Thatcherite acting as private policy entrepreneur, the proposal gained momentum after having won the support of the Prime Minister,[138] but an air of inevitability did not prevent the white paper from receiving a rather critical response from political actors within the territorial policy network both with regard to some of its underlying assumptions and more practical aspects of the merger.[139]

Combining training and business development services within one and the same organisation divided respondents, with many local authorities, the STUC and some business organisations endorsing the principle.[140] Others, including the SCDI and Glasgow Chamber of Commerce,[141] argued that the nature of the services provided and the very different types of clients they targeted – primarily private firms on the business side and individual unemployed with regard to training – would make it difficult to achieve any real synergies, something which

[135] IDS 1988a p. 9.

[136] Cf. Section 11.1.

[137] Thatcher 1999: Speech to the Scottish CBI 8.9.88.

[138] For detailed accounts of the process which led government to adopt the merger proposal, see Danson *et al.* 1989b, Moore 1989, 1990.

[139] From the Scottish Office Library copies have been obtained of the responses from national interest organisations, local government, chambers of commerce and the SDA.

[140] Dumfries and Galloway Regional Council 1989, Fife Regional Council 1989, Tayside Regional Council 1989, Central Regional Council 1989, Edinburgh City District Council 1989, Junior Chambers Scotland 1989, Central Scotland Chamber of Commerce 1989, STUC quoted by Malcolm Rifkind (*HCPD* 9.1.90 vol. 164 cols. 839).

[141] SCDI 1989, Glasgow Chamber of Commerce 1989, Aberdeen Chamber of Commerce 1989, Glasgow Junior Chamber of Commerce 1989.

lead the tripartite SCDI to conclude that the rationale behind the merger was mainly "presentational", although the organisation of course welcomed the adding of training to the policy areas falling under the authority of the Scottish Office.[142] By far the most common target of criticism was the proposal to have two thirds of the board members of local agencies being from "the senior ranks of private-sector business",[143] with some respondents doubting that it would be possible to run major public policy programmes on the basis of "social philanthropy".[144] More importantly, not only local government but also business organisations argued that the omission of local government from the list of actors who would be expected to serve on the board of the local agencies would undermine effective policy coordination at the local level and/or the accountability of the new second-tier bodies.[145] While some criticised the 22 local agencies based on aggregated travel-to-work areas tentatively suggested in the white paper for being too small or conflicting with existing administrative borders,[146] especially business-oriented organisations but also the STUC stressed that having a sizeable strategic 'core' at the Scottish level was imperative,[147] and that the SDA name should be retained because it was a well-established brand, not least in relation to overseas investors.[148] In short, a diverse but largely critical message from across the territorial policy network that called for government to clarify the policy rationale for the organisational merger, argued for increased inclusiveness in the boards of the second-tier local agencies, and the retention of a sizeable strategic core.

---

[142] SCDI 1989, cf. Strathclyde Regional Council 1989.

[143] IDS 1988a p. 15.

[144] Edinburgh District Council 1989, cf. SCDI 1989, Borders Regional Council 1989.

[145] Borders Regional Council 1989, Fife Regional Council 1989, Strathclyde Regional Council 1989, Tayside Regional Council 1989, Lothian Regional Council 1989, Grampian Regional Council 1989, Edinburgh District Council 1989, Junior Chambers Scotland 1989, Central Scotland Chamber of Commerce 1989, Dundee & Tayside Chamber of Commerce and Industry 1989, Aberdeen Chamber of Commerce 1989, Edinburgh Chamber of Commerce 1989, Glasgow Junior Chamber of Commerce 1989.

[146] SCDI 1989, Grampian Regional Council 1989, Central Regional Council 1989, Edinburgh Chamber of Commerce 1989.

[147] SCDI 1989, STUC 1990 congress resolution (reproduced STUC 1991 p. 64), Scottish Engineering Employers' Association 1989a, Scottish Electronics Technology Group 1989, Junior Chamber Scotland 1989, Glasgow Junior Chamber of Commerce 1989, Fife Regional Council 1989.

[148] Strathclyde Regional Council 1989, Edinburgh City District Council 1989, Glasgow Chamber of Commerce 1989, Glasgow Junior Chamber of Commerce 1989, Junior Chambers Scotland 1989. Two business organisations were even sceptical about the term 'enterprise', noting that it was "rather overused" (Dundee & Tayside Chamber of Commerce and Industry 1989, cf. Glasgow Chamber of Commerce 1988 p. 3).

Probably unsurprisingly, the official response of the SDA largely moved in the opposite direction of this. The arm's-length body endorsed the principle of the merger adopted by its political sponsors but proposed "a more radical solution capable of immediate implementation": local Area Development Companies should be established immediately based on the existing regional structure of the Agency, each having "a strong, private sector-led board with a strategic remit for its area",[149] incorporating training functions on a basis similar to the one proposed for England by the Department of Employment,[150] and continuing the commercial SDA-style corporate culture with its emphasis on long-term partnership with private and public actors.[151] In short: take the government proposal, add some of the criticism of the territorial network and a considerable measure of self-perpetuation in the guise of avoiding interruption of important business and training services – and press fast forward.

At the parliamentary level, this pattern of response was presented by the Conservative government as endorsing the basic principles of the original proposal for a functional merger between training and business development activities under Scottish Office control with localised delivery through a Scottish-wide network, and although the principle of private sector dominance of the second-tier organisations was maintained, the willingness of local government to become involved with the local bodies was also welcomed.[152] The Labour front bench adopted a critical approach to the proposal, arguing that a merger would "undermine the central strengths of the SDA" and did not entail "a true partnership between the private sector, local authorities, trade unions and the wider community", with the alternative being closer collaboration between the SDA and a new Scottish Training Agency to be sponsored by the Scottish Office.[153] Although Scottish Industry Minister Ian Lang claimed that the opposition "dislike[s] the whole concept of enterprise" because "the independence of mind that enterprise engenders runs counter to Socialist philosophy", he saw "no major philosophical objection of principle to the proposal" but rather "a large number of Committee and niggling points"[154] and hence proceeded with the Bill that would mean that the SDA would cease to exist by the end of March 1991.

---

149 Cf. the discussion in Section 6.2.

150 Cf. the discussion below.

151 SDA 1989b.

152 Scottish Grand Committee 1989a cols. 1ff, *HCPD* 9.1.90 vol. 164 cols. 839ff.

153 Scottish Grand Committee 1989a cols. 11ff, *HCPD* 9.1.90 vol. 164 cols. 853ff.

154 *HCPD* 9.1.90 vol. 164 cols. 904-9.

To what extent did the emergence of and debate on the Scottish Enterprise proposal signify a change in the discursive environment of the SDA as the, until then, "major instrument for economic regeneration" in Scotland? In terms of timing it is of course tempting to think a crucial change must have taken place in the twelve months between the Third Reading of the SDA bill in late November 1987 which provided funding for five more years on the basis of a positive political evaluation, and the presentation of the Scottish Enterprise white paper in early December 1988 which heralded the end of the Agency. Moreover, the circumstances under which the Scottish Enterprise proposal became government policy – a Tory political entrepreneur shortcutting the territorial policy network by approaching the Prime Minister directly – would seem to suggest that the key change was imposition of the political authority of Margaret Thatcher from the UK level. While these circumstances are of course significant, it is, however, also important to place the proposal in the broader context of the development of the political discourse on regional development and policy in Scotland in order to avoid focusing exclusively on short-term developments at the expense of underlying continuities. While the endorsement by the Prime Minister was undoubtedly important in adding momentum to the merger proposal, other factors need to be considered in order to fully understand the enthusiasm with which Scottish Office ministers took Scottish Enterprise forward, including the extent to which the proposal seemed to herald a new approach to regional economic development in Scotland.

It is interesting to note that the Scottish Enterprise proposal focused primarily on the organisational aspects of the merger and only gave little attention to the policies that the new organisation would be pursuing: the division of labour between the two tiers of the organisation was largely left to be worked out later, the degree to which the planned local bodies would be able to influence training measures that were essentially British-wide schemes remained unclear, and, not least, the practical advantages for clients of having one provider of very different types of services seemingly having been given little thought. This certainly creates the impression of policy continuity for the main body of fairly standardised services hitherto provided by the merging organisations, with three important exceptions. *Firstly*, the main implication of the new two-tier structure was clearly an emphasis on tailoring business development and training services to the specific needs of individual localities. While this intention was radical in the context of training, dominated by standardised British programmes, it was hardly revolutionary in economic development where the reactive *ad hoc* area initiatives had already been succeeded by a regional implementation structure within the SDA, and the adding of a training element would seem to complement the long-standing priority given by the Conservative government to re-employment of human and other resources which had been made superfluous by the decline of traditional industries. *Secondly*, the future of Scottish-wide SDA activities like the sectoral programmes and specialised advisory services would seem to be uncertain because the size of the 'core' of the new organisation had not been resolved, something which could be construed as a change from the position adopted in 1987 where the Conservative government had attached great importance to sectoral programmes and commended their extension into traditional sectors of the Scottish economy. *Thirdly*, and also alluded to in the white paper[155] and underlined by the initial STUC enthusiasm for the merger, the merger could present an opportunity to revitalise training services by bringing a more traditional public body under the influence of the more action-oriented SDA, making the setting up of Scottish Enterprise seem more like a take-over than a merger between two equal parties. From a policy perspective the positive synergies of merging the two organisations were in other words more than a little unclear, and the overarching principle of localising service delivery would only be innovative in training but potentially a threat to some of the more specialised SDA activities. Indeed, the fact that the SDA proposal for a quick and comprehensive merger in accordance with the principles set out in the white paper was *not* adopted clearly suggests that non-functional considerations must have been more important than achieving a fast and smooth transition to an even more comprehensive form of regional policy in Scotland.

Turning now to changes in the wider policy environment of the SDA, the origins of the proposal, including the swift endorsement by the Prime Minister herself, could readily be construed as a 'Thatcherite conspiracy', in a sense the ultimate external-revolution interpretation: as the Scottish Office had repeatedly failed to bring the Agency's 'interventionist' activities under control, the entire organisation had to be terminated through the authority of the British level. Such an interpretation does, however, rest on the assumption that the Scottish Office merely transmitted the political preferences of London to Scotland, and although the importance of Scottish MPs to the Conservative majority at Westminster had become negligible after the 1987 general election, the tendency of Scottish Secretaries to 'stand up' to Prime Ministers for 'vital' Scottish interests continued – Margaret Thatcher clearly saw Malcolm Rifkind as belonging to this tradition[156] – and thus it is still necessary to account for the enthusiasm of Scottish Office ministers for the merger proposal. Four types of political considerations would seem to be relevant here, namely party-political positioning, territorial politics

155 IDS 1988a pp. 10, 12.

156 Thatcher 1993 p. 620.

vis-à-vis the UK, organisational politics relating to the sponsoring of arm's-length bodies, and the political usefulness of the SDA, each of which will be considered in the following. *Firstly*, from the perspective of party-political advantage, the proposal offered a rare chance to re-mould Scottish institutions in a high-profile way without being bogged down by drawn-out negotiations within the territorial policy network:[157] unlike most other welfare services both the SDA and the Training Agency were sponsored directly by central government, and thus the reforming credentials of Scottish Office ministers could be enhanced while the early backing of the STUC (attracted by prospects of improved training) and the interest of local authorities (wanting closer involvement in government-sponsored economic development activities) would make it more difficult for the main opposition parties to argue against basic principles of the proposal. *Secondly*, in terms of territorial politics the proposal brought an additional sizeable area of public policy, training, under the authority of the Scottish Office – something government and opposition agreed was a good thing – in a way that would preempt moves by the Department of Employment to extend its proposed English network of employer-led local training organisations north of the border by means of prime ministerial authority.[158] *Thirdly*, not only from an administrative but also from a political ministerial perspective the proposal would seem to strengthen the position of Scottish Office[159] because the core of the arm's-length became smaller and its relations with the second-tier local agencies more transparent because of their contractual nature. And *fourthly*, a specific assessment of the political usefulness of the SDA and its activities may also have made Scottish Office ministers inclined to accept the idea of the merger, and, interestingly, this line of argument can be found in two very different versions. The *'Agency-failure' view*, a version of the external revolution perspective, sees the idea of the merger as the result of the SDA as a legacy of 'Labour interventionism' being unable to make die-hard Thatcherites embrace the organisation, giving rise to a conspiracy to 'do something' about the SDA after the 1986 administrative review had disappointingly concluded that the arm's-length body operated in accordance with the market failure approach favoured by

---

[157] Cf. Holliday 1992.

[158] Although increased business influence on training policies had already been foreshadowed by the Conservative manifesto for the 1987 general election (Thomsen 1996 p. 217), timing would seem to suggest that the introduction of local networks may have been a Scottish invention, and thus the claim of the Scottish Conservatives that Scottish Enterprise was more than a "tartaned up" English proposal (Scottish Conservative Party 1989) would seem to be true.

[159] An interpretation that is difficult to substantiate but shared by especially SDA executives (George Mathewson 9.7.90, Alan Dale, 16.10.91, Macwhirter 1991.

government.[160] Several features of the proposal and the ensuing political debate clearly indicates that Scottish Enterprise was an attempt to shift the boundary between public and private actor roles in the governance of regional economic development in Scotland towards the latter. On the one hand the otherwise very publicity-oriented Conservative ministers insisted on including the 'e-word' in the name of the merged body despite the high international profile of the existing 'SDA brand'; something which could readily be interpreted as a discursive repositioning of two public bodies by employing a word that the Thatcher government had begun to use as a central metaphor to sum up the difference between themselves and their opponents,[161] and specifically used to headline the 1988 DTI white paper on regional policy.[162] On the other hand the importance attached to having a business majority on the boards of the local agencies, apart from being symbolic, also had a more material aspect in that it could be seen as a renewed attempt to limit the role of other public actors, local government, in economic development and training.[163] Both of these features would, in other words, seem to reflect either broader discursive developments at the British level or longstanding Scottish Office attempts to keep local economic policy under control, and thus their prominence do not necessarily reflect the belated ascendancy of a long-standing anti-SDA 'conspiracy'. Although a principled scepticism towards the Agency had existed as a discursive undercurrent since 1979 and the appointment of Michael Forsyth did strengthen the presence of Thatcherite thinking at the Scottish Office, this cannot explain the sudden change from the Agency-endorsing 1987 Bill to the Agency-termination 1988 white paper, and it is therefore better seen as an enabling factor, making it easier to adopt a proposal with other main attractions. A completely opposite interpretation of the political usefulness of the SDA, widespread among contemporary observers and actors involved[164] – and indeed implied by Scottish Industry Minister Ian Lang in his memoirs[165] – could be termed the *'Agency-success' view*, namely that Scottish Office ministers found that they as government sponsors were not given sufficient credit for the Agency's achievements in modernising the regional economy. Unlike the 'con-

---

160 George Mathewson 9.7.90, Gavin McCrone 20.6.90, Frank Kirwan 1.6.90, Neil Hood 22.6.90, Helen Liddell 21.6.90, Hood 1990 p. 65, cf. Danson *et al.* 1989b, MacLeod 1998a.

161 See e.g. Keat & Abercrombie (eds.) 1991 cf. the discussion below.

162 Cf. Sections 7.4 and the discussion below.

163 Moore 1990, Danson *et al.* 1992, cf. Chapter 12.

164 George Mathewson 9.7.90, Ewan Marwick 24.7.90, Danson *et al.* 1989a, Martin 1992b p. 146, MacLeod 1998a.

165 Lang 2002 pp. 78f.

spiratorial' Agency-failure theory, this avoids the problem of short-term timing because the Scottish Office and Ian Lang in particular had taken other measures to increase the political pay-off of inward investment promotion after the general election,[166] and the Scottish Industry Minister did indeed seem more adamant about the need for a symbolic change of name than his Secretary of State.[167] But as LIS would continue as a joint venture between the Scottish Office and the new arm's-length body and no additional concrete vehicles for ministerial photo-opportunities were envisaged, in practice the only way in which the position of elected politicians would be strengthened by the new set-up would have been indirectly through the disappearance of the strong 'SDA brand'.

All in all the discussion of the Scottish Enterprise proposal showed that in terms of territorial politics the underlying preferential regionalist consensus was still in place: both the Conservatives and Labour argued for special measures that they claimed would be to the benefit of Scotland. As demonstrated above the merger proposal lacked detail with regard to concrete policy developments: the synergies to be achieved were left in the penumbra, tailoring services to local business demand had already been attempted by the SDA and would therefore seem to pertain primarily to training, and thus the main probabilities were more adaptive training policies, possibly at the expense of fewer Scotland-wide strategic activities in low-profile and uncontroversial areas such as specialist advisory services and sectoral initiatives. This strongly suggests that the rationale behind the merger, despite government protestations to the contrary, was not primarily about upgrading policies. Instead it should be sought in other types of political considerations, primarily the party-political advantages to be accrued by Scottish ministers both within the Conservative party by demonstrating reforming zeal and *vis-à-vis* an opposition wrong-footed by the proposal's appeal to the interests of trade unions and local authorities. This was achieved through a symbolic organisational merger which strengthened the private-sector credentials of Scottish Office regional policy by replacing the old 'Agency brand' with a more 'enterprising' one while at the same time achieving an even broader appeal by situating itself in the longstanding tradition for preferential regionalism in Scottish politics by bringing training under the aegis of the Scottish Office. This was, however, not only as announced by Margaret Thatcher a "Scottish solution to respond to Scottish needs", but one that specifically increased the importance of private actors in Scottish regional policy, at least in terms of the discursive positioning of the new body. Like the SDA before it, Scottish Enterprise would therefore find itself navigating in a political environ-

[166] Cf. Section 10.1.

[167] See Scottish Grand Committee 1989a cols. 3 (Rifkind) and 39 (Lang).

ment where development and implementation of policies addressing perceived weaknesses of the regional economy was not the only criteria according to which the new organisation would be judged.

14.4. Demise or Reconfiguration? The British Environment 1975-91

Although certain themes tended to recur in Scottish discussions about the SDA, the multi-faceted activities of the Agency ensured that the organisation operated in a relatively varied discursive environment. In contrast the regional policies conducted by central government were of a much more unified nature, largely based on financial incentives in designated problem areas, and thus British-level debates tended to have a more monochrome nature.

From the 1960s and throughout the 1970s the rationale for the various regional policy measures gradually introduced by central government had been largely implicit, but the basic idea was clear enough. Assisted Areas were designated through a discretionary process in which the level of unemployment was an important parameter, and the remedial strategy involved a system of inducements designed to enhance the relative attractiveness of investing in the designated areas through financial support for key factors of production (machinery, buildings, wages) and restrictions on the expansion of industrial activities in wealthy regions through a system of planning controls. In terms of discourse this meant that two features of policy became central indicators of political commitment to problem regions, namely the geography of the Assisted Areas and the level of funding committed to regional policy purposes. The changes in the mid-1970s did, however, have limited party-political mileage, at least from a short-term perspective: the Labour government of Harold Wilson had been elected with a commitment to restoring the Regional Employment Premium, an automatic wage subsidy which the Conservative Heath government had scheduled for phasing out, only to end up terminating it in December 1976 as part of massive reductions of public expenditure in order to obtain IMF support for the crisis-ridden British economy,[168] and hence debates tended to focus on macro-economic policies rather than regional development strategies. The only major new Labour initiative with a regional dimension which was implemented was the setting up of the National Enterprise Board (NEB), but in the political debate the presence of a regional structure within the new organisation was completely overshadowed by discussions about the extent to which the NEB would become a vehicle for nationalisation of profitable firms or not. While

168 McCallum 1979 pp. 24-29.

Industry Secretary Tony Benn insisted on the need to "inject both the national interest and the regional interest, and the interests of working people, into the strategic decisions made by major industrial firms" when additional public funds were being made available for investment in order "to reverse the long decline in British manufacturing industry",[169] Michael Heseltine promised that the Conservatives would repeal what they saw as a left-wing attempt to create "the maximum powers for the Secretary of State" to give him "unfettered freedom to nationalise individual companies",[170] while Conservative backbenchers decried the NEB as "a means of turning this country into a Marxist Socialist state".[171] Perhaps the best indication of the lack of trust in the NEB's regional role was the 1976 attempt by Labour backbenchers to persuade government to establish RDAs in the English Assisted Areas like those having been established in Scotland and Wales, something which their own ministers declined, arguing that existing NEB and Industry Act powers were equivalent to those of the non-English arm's-length bodies.[172] The proposal for English RDAs was, however, later included in the Labour manifesto for the 1979 general election which also promised to continue "a strong policy of regional incentives",[173] thereby underlining the party's continued role as defender of the interests of regions with economic difficulties. In contrast to this, the commitment of the opposition to regional policy appeared to be at least uncertain in the late 1970s. On the one hand general scepticism about industrial subsidies had been voiced by market-oriented liberal intellectuals like Sir Keith Joseph, and significant reductions of regional policy expenditure by a new Conservative government had specifically been hinted at,[174] but on the other hand it was also pragmatically recognised that

> in the foreseeable future we cannot just cut off regional development grants, aids and policies. We shall maintain aid in certain highly deprived areas. We shall have a solid regional policy.[175]

Margaret Thatcher even spoke of having a "vigorous" regional policy,[176] something which of course did not rule out concentration of financial support in 'highly deprived areas' in order to achieve expenditure reductions.

---

[169] *HCPD* 17.2.75 vol. 886 col. 935.

[170] *HCPD* 17.2.75 vol. 886 col. 950.

[171] *HCPD* 18.2.75 vol. 886 col. 1208.

[172] English Development Agencies Bill 1976, *HCPD* 5.3.76 vol. 906 cols. 1801-11.

[173] Quoted from Craig 1990 p. 288.

[174] *HCPD* 12.12.78 vol. 960 cols. 535f, 564.

[175] *HCPD* 12.12.78 vol. 960 col. 543.

[176] Thatcher 1999: Interview with the Scottish Daily Express 13.2.78.

Having won the general election in May 1979, Prime Minister Thatcher declared that "government has a duty to mitigate the effects of industrial change" and stressed her party's long-standing support for regional aid,[177] and two months later the first statement on regional policy by Industry Secretary Sir Keith Joseph introduced a gradual programme of change which concentrated aid in fewer localities and resulted in useful reductions in public expenditure.[178] It was, however, also clear that the Conservatives now argued in ways that implied that regional policy primarily had a social rationale: jobs were transferred from one part of the country to another by "marginally improving the cash flow and profitability of companies that invest in Assisted Areas", including support for projects "that would have gone ahead in those areas anyway without the policy of regional assistance".[179] Moreover, the underlying importance of private economic actors and free markets was underlined:

> Regional differences will not be reduced simply by redistributing money from taxpayers: there needs also to be local enterprise and plenty of cooperation (between workforce and management) in making business competitive and profitable.[180]

The opposition condemned the cuts in regional aid as a "socially divisive policy" that undermined industrial investment and employment, and Labour promised to resist them "root and branch" while presenting the alternative as more resources, a shift away from automatic towards discretionary grants, strengthened location control and more funds for the NEB[181] – essentially a return to the policies pursued by the previous Labour government. While both the major parties were still committed to spatially discriminatory economic policies aimed at affecting the relative performance of particular localities, the differences between them was no longer only one of degree – downgrading or upgrading of localities in relation to the multi-tiered system of Assisted Areas – but also one of means – relying on financial incentives or location control and ownership – and underlying rationales. If there was an economic case for regional policy, then this government activity could be presented as a win-win situation where also the well-off regions gained through e.g. reduced congestion, but if the case was mainly a social one – taking work to the workers in order to reduce inter-regional migration

---

177 Thatcher 1999: Speech to the Scottish Conservative Party Conference 12.5.79.

178 *HCPD* 17.7.79 vol. 970 cols. 1302ff.

179 *HCPD* 24.7.79 vol. 971 cols. 364ff.

180 *HCPD* 17.7.79 vol. 970 col. 1307.

181 *HCPD* 24.7.79 vol. 971 cols. 378ff, 387.

– then its maintenance would require the well-off regions forgoing resources and hence a higher degree of solidarity at the British level.

Although the rhetoric of Conservative ministers often suggested a lack of commitment to this form of inter-regional redistribution of resources – Norman Tebbit famously advised the unemployed to get 'on yer bike'[182] and also the Prime Minister insisted that "there must be some mobility of labour"[183] – the second round of changes was argued along the same lines as the first one, albeit with a somewhat sharper edge. The Conservative government again declared that it was "firmly committed to an effective regional policy", but while the economic case was no longer clear-cut, there remained "a social case for regional industrial policy to reduce regional imbalances in employment opportunities". This lead to the introduction of a cost-per-job limit in automatic grants accompanied by an extension of the spatial coverage of the Assisted Areas and a reduction of the levels of support so that in financial terms the outcome was again projected to be a significant reduction of public expenditure.[184] During the early 1980s the Labour party had conducted an in-depth review of regional policy which took on board many of the findings of the growing body of research into two decades of British attempts to address the 'regional problem',[185] and the outcome of this was a scepticism about traditional central government programmes which concentrated on problem areas, ignored intra-regional inequalities and created dependency through encouragement of branch plants. According to Labour the alternative strategy was a regionally-based approach which involved RDAs under some form of public political scrutiny not just in traditional 'problem regions' but throughout Britain, a line that was used to challenge the Conservative government in 1983 and simply elicited the answer "no" from Industry Secretary Norman Tebbit when asked about the prospects of arm's-length development bodies in the English regions.[186] While the agreement about the need for some form of regional policy was maintained, the difference between the two main parties with regard to policy rationales and instruments had now widened, with the Conservatives seeing regional assistance as a central government social policy which shifted jobs to areas of high unemployment in the most economical way possible, and Labour arguing the case for their nation-wide RDA-based approach in

182 The original quote was rather more elegant, even by the forthright standards of Tebbit (cf. Parsons 1988 pp. 187f): "My father was unemployed in the 1930s and he did not riot. He got on his bike and looked for work".

183 Speech to the Welsh Conservatives 20.7.80 (quoted by Parsons 1988 p. 187).

184 *HCPD* 13.12.83 vol. 50 cols. 839ff, cf. DTI 1983a.

185 Compare Labour Party 1982 with e.g. Regional Studies Association 1983.

186 *HCPD* 13.12.83 vol. 50 col. 842.

terms of both economic and social need and insisting on tailoring measures to what was seen as the increasing diverse nature of development problems across the country.

The third round of changes in regional policy initiated by the Conservatives was part of a more comprehensive attempt to reshape the micro-economic policies of central government in the image of entrepreneurialism favoured by Margaret Thatcher and her Industry Secretary Lord Young. With regard to regional assistance this entailed termination of the automatic investment grants so that all grants became discretionary, focusing on support for inward investment, innovative indigenous firms and small firms in particular. In contrast to the early 1980s, the case for a "strong regional policy" was now argued in economic terms, as a means "to continue to improve the economy of the Assisted Areas" by changing

> the fundamental basis of the northern economy, so that alongside more competitive heavy manufacturing is the full range of service and manufacturing industries, as good and as strong as those in the south.[187]

Clearly the increased role for discretionary grants was potentially embarrassing for the Conservative government, but Industry Minister Kenneth Clarke insisted that applications would be "assessed objectively against published criteria to establish whether the project would go ahead without public money" and "it is the company that takes the initiative and the final decision whether to go ahead".[188] The response of Labour was a high-profile defence of automatic investment grants by front bench spokesman Tony Blair who deplored their scrapping as a departure from the longstanding cross-party commitment to regional policy as a means of reducing inter-regional differences in employment opportunities, insisted that automatic grants had more impact on company investment decisions and were cheaper for government to administer than discretionary ones, and claimed that the "concealed agenda" was "a massive reduction of resources to the hard-pressed regions".[189] This positive reevaluation of traditional forms of regional was retained in the Labour party's thorough policy review conducted after its defeat in the 1987 general election. Here the main opposition party favoured a dual approach to regional development: on the one hand it was stressed that "the instruments of regional policy must be revived to stimulate investment in the regions which are currently starved of capital", while

---

[187] *HCPD* 25.1.88 vol. 126 cols. 36f.

[188] *HCPD* 25.1.88 vol. 126 col. 42. The attempt to joke about civil servants having to 'pick winners' again as an attempt to alleviate criticism from Tory backbenchers and Labour jibes (cols. 42ff).

[189] *HCPD* 25.1.88 vol. 126 cols. 51-61.

on the other hand it would be important "to encourage each of our nations and regions to built the capacity to sustain their indigenous economies", a task in which the existing RDAs and a new network of "regional investment banks" would be crucial.[190] All in all the late 1980s comes across as an interesting period in British where, again, the main parties altered their position with regard to regional policy. The Conservative government, now embracing discretionary and relatively inexpensive regional programmes, provided for the first time in the 1980s an explicitly economic rational for its regional activities, while Labour reverted to a position close to the one the party had espoused in the manifesto for the 1979 general election, namely a combination of top-down and bottom-up measures and the use of a financial yardstick to compare the relative commitment of political parties to the goal of inter-regional equality. It was very much in this spirit that a senior Labour backbencher declared

> I am still puzzled why the Conservative government hates the regions. It is not as though they do not say the right things. They just seem to do things that undermine their protestations.[191]

Still, for the Labour party money seemed to be the currency that really counted in British territorial politics.

[190] Labour Party 1989 p. 15.

[191] *HCPD* 25.1.88 vol. 126 col. 107.

CHAPTER 15

The SDA and the Politics of Regional Policy

Investigating the development of the SDA in the context of British regional policy has turned out to be a task requiring extensive efforts with regard to both theoretical reflection and empirical analysis. In order to understand how and why organisations and policies have changed, an analytical framework had to be developed that was capable of accounting for the policy-making process in its entirety, from the minute details of interaction between public and private actors embodied in the ways policy instruments are employed, the strategies of different actors in pursuit of functional goals and organisational positioning, and the dominant assumptions about regions and regional development in the discursive terrain. This framework was then employed in the study of the origins of the proposal for an arm's-length development body for lowland Scotland, and the subsequent development of the Agency, focusing in particular on clarifying the nature and origins of continuity and change. The empirical analysis elucidated not only general features at the corporate level but also the characteristics of individual policy areas in the wide range of activities in which the organisation became involved, and throughout three aspects have been covered, namely statements of strategic intent, the interaction between the SDA and the actors targeted by policy measures, and the relationship between Agency activities and their political environment, from the sponsoring Scottish Office via the British regional policy network to prevailing ideological and territorial forms of discourse. A simple calculation shows that covering seven fields of enquiry (two corporate-level ones, five policy areas) with regard to three different aspects would result in 21 sub-studies, some larger than others but all of them needed to produce a comprehensive understanding of the development of the SDA in its specific British and wider European context. Fortunately, despite its multifaceted portfolio and the complex nature of its political environment, a relatively clear-cut pattern would seem to be emerging on the basis of the analyses in the preceding three parts of the book, and thus the task of this concluding chapter, therefore, is not so much to surprise the reader by last-minute revelations, but to draw together the arguments in order to, first and foremost, provide answers to the original research question, and then, finally and briefly, to suggest some more general points regarding the implications of the Scottish case for the study of the politics of regional policy in Britain and beyond.

15.1. Deconstructing Puzzles of the SDA

The first part of the empirical research question concerned the balance between continuity and change in the development of the SDA, and on the basis of the analyses it was concluded that the development of the strategies and policies of the SDA has been characterised by a combination of continuity and gradual change. Continuity in terms of the overall level of activity, with regard to the underlying assumptions about the nature of the regional problem – lack of competitive private firms causes unemployment – and concerning the policy instruments with which this problem should be addressed, namely conditional and discretionary use of financial, informational and organisational resources in order to increase the competitiveness of existing firms or bring about new incoming or indigenous enterprises with potential market clout. And gradual change in the sense that this approach was becoming diversified through the development of a range of tailorised initiatives selectively targeting specific forms of economic activity – particular sectors or types of firms – while at the same time policies became more proactive and began to involve ongoing network relations with especially private economic actors. In fact, the most obvious examples of radical change in the history of the SDA both concern the very existence of the organisation: suddenly in the mid-1970s it was decided to establish an arm's-length development body for lowland Scotland, and in the late 1980s, even more abruptly, it was decided to merge the organisation and form a new two-tier quango with an even broader remit covering training, economic development, and environmental improvement.

This combination of continuity and gradual change in terms of regional policy activities within an organisational framework subject to sudden reconfiguration constitutes a challenge for existing interpretations of the SDA. From an *external-revolution perspective* high degree of continuity and gradual change against the grain of liberal ideology constitute a conundrum, especially because the history of the Agency would seem to contrast sharply with developments of central government regional policy where objectives changed from inter-regional equality to regional competitiveness, the use of financial incentives shifted from automatic to discretionary ones, reducing both the extent of the areas designated for support and the level of expenditure in the process. Interpreted along these lines, the Scottish Enterprise proposal could be seen as the delayed revenge of Thatcherites frustrated by having been unable to control the policies of this free-roaming maverick left-over of 1970s interventionism. In contrast to this an *internal-evolution perspective* is of course easier to reconcile with the findings about the development of Agency activities, and it could be claimed that the combination of continuity and gradual change demonstrates that the arm's-length position of the organisation did provide shelter from a turbulent political environment. But at the same time the success of the RDA formula in Scotland makes it difficult to understand why it was not exported to England where regional policy continued to operate in a spatially selective and organisationally fragmented manner – and, of course, why a successful body like the SDA should be merged into oblivion through spatial fragmentation within a complex multi-tier structure with objectives and modes of operation so disparate that policy synergies were more than doubtful.

It could of course be tempting simply to combine the two perspectives, explaining the rise and fall of the Agency in purely party-political terms – an interventionist body was established to stave off the rising tide of Scottish nationalism and ultimately terminated by neoliberal Thatcherites exasperated by the inability of the Scottish Office to exert effective control – and letting internal learning-by-doing processes within the arm's-length body account for the continuity and gradual change in the Agency's strategies and policies in the intervening fifteen years. The empirical analysis above does, however, suggest four reasons why this composite interpretation should not be adopted. *Firstly*, seeing the creation of the SDA purely as an anti-SNP measure is a truncated interpretation in the sense that while nationalist success at the polls was undoubtedly an important factor behind the setting up of the Agency as a high-profile gesture of preferential regionalism, the concrete form that this initiative took cannot be separated from the intense debates taking place in the territorial policy network in Scotland which influenced the options available to the Labour government in its hour of political need. *Secondly*, the assumption that the Thatcher government always wanted to get rid of the SDA for ideological (liberal) reasons but only managed to succeed in doing so in the late 1980s through the Scottish Enterprise proposal ignores the obvious reasons why Conservatives, and of course especially those at the Scottish Office, would want to maintain the organisation: the Scottish Tories had long been part of the preferential regionalist consensus amongst the unionist parties north of the border, under the leadership of George Younger they were unequivocally placed among the moderates in the governing party, and most of the activities of the SDA were seen as useful and uncontroversial. *Thirdly*, interpreting the adoption of the Scottish Enterprise proposal purely in terms of ideological frustration omits other features that made Scottish Office ministers adopt the proposal, including the way in which specific policies operated and the increasingly marginalised position of the Conservative party in Scottish politics. And *finally*, the assumption that the arm's-length body was essentially left to 'get on with it' by its political sponsors underplays the occasional importance of political interventions and the ongoing sponsorship relation with the Scottish Office, and hence fails to appreciate the varying degrees to which particular policy areas have been subject of political and administrative scrutiny, and the role of Agency ingenuity in reinventing itself and its policies in ways agreeable to their political sponsors through institutional or discursive re-engineering.

In short, even what appeared to be the strong points of the two existing interpretations turned out to be, at best, partial contributions to the understanding of continuity and change in regional policy in Scotland, and thus it is necessary instead to develop an alternative rendering, capable of accounting for the interaction between politics and policies in the development of the SDA.

15.2. Agency Reconstructed: A History of the SDA in 1941 Words

The history of the SDA is best told as a story in four parts, a division that reflects the general approach of the Scottish Office to the existence of an arm's-length economic development body for lowland Scotland. As illustrated by Table 15.1, this creates a history in which support for the setting up of an arm's-length body dedicated to lowland Scotland first grew, then remained strong under both parties of government, and then, possibly, seemed to wane in the late 1980s after the launch of the Scottish Enterprise merger proposal.

The *genesis phase* in the early 1970s is a curious one in the sense that the parliamentary Labour Party that eventually legislated for the SDA to be established was actually the last major political grouping to embrace the notion of an arm's-length development for the Scottish lowlands. Many blueprints had been produced by various actors in the territorial policy network, with different profiles but falling broadly in two groups: on the one hand variations on the theme of an urban RDA modelled loosely on the Highlands & Islands Development Board, and on the other hand the left-wing proposal for a Scottish Enterprise Board which emphasised the importance of using public ownership as an instrument in regional policy. All such proposals had been dismissed or ignored by both the Conservative Heath Government and the official parliamentary opposition who preferred to maintain and expand traditional British-level regional grant schemes, and it was only after the nationalist electoral upsurge at the February 1974 general election – *and* the high-profile appearance of the West Central Scotland Plan and the Conservative proposal for a Scottish Development Fund – that the new Labour minority government announced its intention to set up a new RDA. The specific design of the new organisation involved amalgating existing, and at this point uncontroversial, activities with a new high-profile industrial investment function which, through the parallel proposal for a British NEB, became associated with extension of public ownership as a goal in its own right. The historical context and the policy profile proposed made it in other words possible to question the new organisation not just in terms of its possible contribution to the development of the Scottish economy, but also on the basis of its position in the wider discursive terrain with regard to the relationship between Scotland and the rest of Britain on the one hand, and relationship between public and private actors on the other.

Table 15.1 Phases of SDA development

		Support for lowland RDA	
		Uncertain	*Strong*
Party of government	*Labour*	Genesis (early 1970s)	Formation (1975-79)
	Conservative	Merger (1988-71)	Diversification (1979-87)

The first years of the SDA's existence could be seen as the *formative phase*[1] where the reshaping of inherited functions began and the development of new ones initiated. Both tasks constituted major challenges, as much of the organisational inheritance was characterised by limited strategic capabilities and the new functions were either untried in a regional context – industrial investment and sectoral initiatives – or potentially controversial within the region because they trespassed on areas of activity like urban renewal and inward investment promotion in which sub-regional actors had hitherto played a dominant role. The emphasis of the organisation itself, and perhaps even its Labour sponsors at the Scottish Office, was on promotion of economic activities that would become viable in their own right, and the emphasis on modernisation of firms and industries was only partly tempered by the introduction of area initiatives in local unemployment black spots, because while this instituted an additional intra-regional criteria of success in response

[1] An expression borrowed from Hood 1991a p. 7.

to a perceived political need to be seen to be doing something in the wake of widespread industrial decline, it still included additional economic rationales such as long-term improvement of the image of the region. Meanwhile, the period from 1975 to 1979 was a symbolic turning point in British regional policy because after more than a decade of expansion expenditure was being drastically reduced in response to severe economic crisis and budgetary restraints imposed by the IMF. Moreover, it was also a tumultuous time in political terms, with the new Conservative leadership under Margaret Thatcher increasingly questioning the traditional assumptions of the postwar consensus about the role of the public sector in the economy while at the same time mounting a vociferous defence of the constitutional status quo within the British union state through opposition to Labour proposals for devolution for Scotland and Wales. The SDA was inherently implicated in both these issues because of the way its origins and activities positioned it in the British discursive terrain: simply by existing the organisation was an example of preferential regionalism – giving Scotland something that English regions with similar problems did not get – and through parallels to the NEB the industrial investment function could readily be construed as being a vehicle of backdoor nationalisation. In practice, however, the SDA appears to have been somewhat sheltered by being seen as relevant mostly from the perspective of Scottish politics, so although the role of the investment function was a focus of attention for the opposition, this tended to be a more limited form of critique – patient public investment should not disadvantage competing private firms – that seemed to accept the potential usefulness of the function and did certainly not dispute the principle of giving Scotland preferential treatment compared to other parts of Britain.

After the change of government in 1979 the Conservatives did what they had promised during the election campaign, namely retain the SDA with revised guidelines for especially the industrial investment function. The new guidelines introducing demands akin to those in the 1972 Industry Act about commercial appraisal, disposal of equity holdings, and, most importantly, insisted on the importance of private sector co-investment, but in policy terms they merely strengthened existing trends within the Agency and thus their real significance was political, namely to serve as a symbolic marker of change having happened and thereby effectively reduced the role of the public-private dimension in future debates about the SDA. After the anxieties of the first months under the new government caused by rattling of sabres by radical elements in the governing party, the development of the arm's-length body over the next eight years is aptly summed up by the expression *diversification phase*:[2] the overall level of activity was maintained while existing policies were gradually developed and extended, new services introduced, and the Agency's proactive, selective and discretionary use of its policy instruments became even more pronounced. Moreover, the underlying assumptions of the organisation and its policies were explicitly endorsed by the Scottish Office after they had been reformulated in market-failure terms in connection with the 1986 government review, and thus in a period where British regional policy was seen as having primarily a social rationale in redistributing economic activity to areas of high unemployment and was gradually being reduced in terms of spatial coverage and expenditure, the competitiveness-oriented SDA maintained its levels of activity and was officially recognised as fully conforming with the thinking of the Conservative government on economic policy. Interestingly, the two major political debates about the arm's-length in this period both produced what appears to be agreement between the two major parties about the aims and means of regional policy. The controversy in the early 1980s about inward investment did not revolve around the fundamental question about whether specific Scottish efforts were called for but merely whether it was acceptable that these efforts were conducted outwith the common British setup or not, and the solution chosen by Secretary of State George Younger effective endorsed the view of the Labour minority of the Scottish Affairs Committee, thereby continuing in the tradition of open preferential regionalism but also running the risk of alienating more staunchly unionist Conservatives. On the other hand the early 1980s was characterised by the nature of the regional problem in Scotland becoming increasingly contested, with Labour focusing on the decline of traditional industries and the absence of preemptive government action, and government ministers extolling the successful modernisation of the regional economy through new incoming and indigenous firms. Towards the end of the period this 'tale of two Scotlands' would, however, seem to be replaced by cross-party agreement about the need to focus on both traditional and modern industries, on new firms as well as existing ones, and the symbolic apotheosis of this new near-consensual state of affairs was reached in the Autumn of 1987 where the spending limit of the Agency – previously alway a party-political battleground – was raised by parliament with both major parties praising its achievements.

Finally the *merger phase* was, unsurprisingly, characterised by continuity with regard to strategies and policies at the SDA while the organisation prepared for the impending organisational changes. As

2 Hood's expression "the flourishing period" (1991a) has too much of a "golden age" air to it.

argued in Chapter 14, the policy rationale for the merger proposal was given scant attention, as illustrated by the Agency's fast-track proposal being ignored by the Scottish Office, presumably because it did not contain the crucial political added-value for government ministers. The Thatcherite ideological connotations derived from the name and the personal intervention of the Prime Minister were of course important as an enabling factor, but the timing of the proposal – shortly after fulsome political endorsements of the SDA in 1987 that foreshadowed the enterprising 1988 white paper on British regional policy – would seem to suggest that interpreting the merger as the product of a longstanding ideological distrust is, at best, a rather incomplete view. Instead other political considerations seem to have been more important, including both short-term ones relating to ministerial self-advancement and wrong-footing the opposition, and more strategic ones such as being able to add a sizeable policy area like training to the responsibilities of the Scottish Office, a clear instance of preferential regionalism – putting a new and rather different perspective on Thatcher's words about "a Scottish solution to respond to Scottish needs". Moreover, this could be achieved in a manner that could be presented as increasing the accountability of public policy to the local business community while at the same time allegedly doing little damage to the existing policies of the SDA, and from the perspective of the politics of regional policy this last point is particularly interesting because it demonstrates that the information-based 'core services' like the sectoral initiatives were still seen as potentially dispensable and hence less important. These activities had been viewed with varying degrees of scepticism by the Scottish Office already in the 1986 review, unlike inward investment and industrial investment they were not mentioned specifically in the Scottish Enterprise proposal, and they were fairly marginal in the party-political debate, although some disquiet was expressed by the loss of strategic direction if 'the national core' of the new two-tier quango ended up being too small. After a decade of innovation and experimentation in new forms of bottom-up regional policy, policy instruments with tangible qualities still dominated the discursive terrain, but in retrospect this is hardly surprising. As demonstrated by the empirical analysis, throughout the period from the early 1970s into the 1990s policies that could be readily be construed as modifying *either* the boundary between public and private actors (industrial investment and property) *or* the relationship between Scotland and the rest of Britain (inward investment) had generated much more controversy than the new and more low-key information-based and network-oriented policy instruments. Despite wide-ranging changes in terms of the economic development initiatives being implemented, the politics of regional policy in Scotland would not seem to have change much by the early 1990s: preferential regionalism continued to be the shared starting point, and thus creating visible results, either through tangible policy measures or organisational change, remained crucial in the party-political contestations about regional policy in Scotland.

15.3. Scotland, Britain and Beyond

The reconstructed history of the SDA has more general implications along two lines which will be elaborated upon briefly in the following.

As a contribution to the growing literature on *Thatcherism*, the analysis started out from an 'implementation failure' perspective in the sense that scepticism about the tenability of the external-evolution perspective prompted closer scrutiny of the history of the Agency. It could be argued that the pervasive presence of continuities and gradual change running against the grain of liberal thinking is indeed evidence of a failure to implement Thatcherite thinking caused by e.g. the insulation provided by the arm's-length position of the organisation, but this presupposes that the Conservatives had actually wanted to change the Agency radically all along, something which is much less certain. In practice the Scottish Tories continued to adhere to significant measures of preferential regionalism also in the 1980s, culminating in the take-over of parts of Scottish operation of the British Training Agency from the Department of Employment in 1991, and although the Scottish Enterprise merger was couched in ideological discourse, the prominence of this form of territorial politics within the governing party would effectively seem to bring the argument of the book closer towards the 'ideational heritage' tradition which stresses the importance of ideational continuities within the Conservative party.

From a *regional studies* perspective the contribution of the book has been threefold. *Firstly*, developing and road-testing an analytical framework capable of spanning across different policy paradigms seems to be a timely contribution to the literature in a period characterised by multi-level governance and proliferation of policy initiatives within individual regions, especially because the framework recognises the possibility of policy implementation, corporate strategies and political discourse may develop unevenly and in different directions. *Secondly*, the detailed case-study of an arm's-length development body would seem to indicate that such arrangements may actually offer a considerable degree of operational freedom, if, and this is a big if, political sponsors, ministers and administrators alike, have decided to leave the development professionals to 'get on with it' and begin to focus on long-term economic change rather than short-term party-political advantage. And *thirdly*, the importance of accounting for the politics of regional policy in order to understand policy change or continuity has been underlined, including the likely centrality of particular types of general assumptions: about the position of regions *vis-à-vis* central government and the rest of the world on the one hand, and the relationship between public and private actors on the other. Even detailed political debates and administrative controversies about aims and methods in regional development entail positioning in relation to the wider discursive terrain with regard to issues which have a long-standing prominence in Europe and beyond because, after all, regional policy is still about making private economic actors adapt to public spatial priorities.

References

Bibliography

Abercrombie, Patrick & Robert H. Matthew (1949), *The Clyde Valley Regional Plan 1946*, Edinburgh, HMSO.

Aberdeen Chamber of Commerce (1989), *Scottish Enterprise*, Aberdeen, Aberdeen Chamber of Commerce.

Ackermann, Charbel & Walter Steinmann (1982), 'Privatized Policy Making: Administrative and Consociational Types of Implementation in Regional Economic Policy in Switzerland', *European Journal of Political Research* 10: 173-85.

Adam Smith Institute (1983), *Omega Report: Scottish Policy*, London, Adam Smith Institute.

Advisory Council on Science and Technology (1990), *The Enterprise Challenge: Overcoming Barriers to Growth in Small Firms*, London, HMSO.

Aitken, Ian (1994), 'Ian Lang's One-Nation Toryism', *New Statesman & Society* (310): 8.7.94.

Aitken, Keith (1997), *The Bairns o' Adam: the Story of the STUC*, Edinburgh, Polygon.

Aitken, P. & L. Sparks (1986), 'The Scottish Development Agency: A Case for Co-ordination', *Regional Studies* 20 (5): 476-80.

Albrechts, Louis & Eric Swyngedouw (1989), 'The Challenges for Regional Policy under a Flexible Regime of Accumulation', in Louis Albrechts *et al.* (eds.), *Regional Policy at the Crossroads – European Perspectives*, London, Jessica Kingsley, pp. 67-89.

Alden, Jeremy & Philip Boland (eds.) (1996), *Regional Development Strategies – A European Perspective*, London, Jessica Kingsley.

Alexander, Kenneth (1981), 'Developing the Highlands and Islands', in Christopher Blake &. Edgar Lythe (eds.), *A Maverick Institution*, London, Gee, pp. 72-85.

Alexander, Kenneth (1985), 'The Highlands and Islands Development Board', in Richard Saville (ed.), *The Economic Development of Modern Scotland 1950-1980*, Edinburgh, Donald, pp. 214-32.

Allen, Kevin (1979), 'Some National Approaches to Regional Inequality', *Studies in Public Policy* 41.

Allen, Kevin (1989), 'The Long-term Future of Regional Policy – A Nordic View', in NordREFO/OECD (ed.), *The Long-term Future of Regional Policy – A Nordic View*, København, NordREFO, pp. 11-39.

Allen, Kevin & Douglas Yuill (1990), 'Financial Incentives for the Promotion of Regional Development', in Hans-Jürgen Ewers & Jürgen Allesch (eds.), *Innovation and Regional Development*, Berlin, de Gruyter, pp. 169-86.

Allen, Kevin *et al.* (1986), *Regional Incentives and the Investment Decision of the Firm*, London, HMSO.

Allen, Kevin *et al.* (1988), 'Alternatives in Regional Incentive Policy Design', *Regional and Industrial Policy Research Series, EPRC, University of Strathclyde* 1.

Allen, Kevin *et al.* (1989), 'Requirements for an Effective Regional Policy', in Louis Albrechts *et al.* (eds.), *Regional Policy at the Crossroads – European Perspectives*, London, Jessica Kingsley, pp. 107-24.

Alonso, William (1989), 'Deindustrialization and Regional Policy', in Lloyd Rodwin & Hidehiko Sazanami (eds.), *Deindustrialization and Regional Economic Transformation*, Boston, Unwin Hyman, pp. 221-40.

Amin, Ash & C. Pywell (1989), 'Is Technology Policy Enough for Local Economic Revitalisation? The Case of Tyne and Wear in the North East of England', *Regional Studies* 23: 463-77.

Amin, Ash & Nigel Thrift (1994), 'Living in the Global', in Ash Amin & Nigel Thrift (eds.), *Globalization, Institutions, and Regional Development in Europe*, Oxford, Oxford UP, pp. 1-22.

Amin, Ash. *et al.* (1994), 'Regional Incentives and the Quality of Mobile Investment in the Less Favoured Regions of the EC', *Progress in Planning* 41 (1): 1-112.

Andersen, Niels Åkerstrøm (1994), 'Institutionel historie – En introduktion til diskurs- og institutionsanalyse', *COS-rapport* 10/94.

Andersen, Niels Åkerstrøm (1999), *Diskursive Analysestrategier*, København, Nyt fra Samfundsvidenskaberne.

Andersen, Niels Åkerstrøm & Peter Kjær (1996), 'Institutional Construction and Change: An Analytical Strategy of Institutional History', *COS-rapport* 1996/5.

Andersen, Ole Winckler & Jerome Davis (1986), 'Marked, kontrakt, hierarki: En introduktion til ny-institutionalismen', *Tidsskrift for Politisk Økonomi* 9 (3): 6-30.

Anderson, Jeffrey J. (1990a), 'Skeptical Reflections on a Europe of the Regions: Britain, Germany and the ERDF', *Journal of Public Policy* 10 (4): 417-47.

Anderson, Jeffrey J. (1990b), 'When Market and Territory Collide. Thatcherism and the Politics of Regional Decline', *Western European Politics* 13 (2): 234-57.

Anderson, Jeffrey J. (1992), *The Territorial Imperative – Pluralism, Corporatism and Economic Crisis*, Cambridge, Cambridge UP.

Anderson, Perry (1983), *In the Tracks of Historical Materialism*, London, Verso.

Anderson, Perry (1987), 'The Figures of Descent', *New Left Review* (151): 20-77.

Anderson, Perry (1994), 'The Invention of the Region 1945-90', *EUI Working Paper EUF* 94/2.

Armstrong, Harvey (1986), 'The Assignment of Regional Industrial Policy Powers', *Regional Studies* 20 (3): 258-61.

Armstrong, Harvey (1993), 'Subsidiarity and the Operation of EC Regional Policy in Britain', *Regional Studies* 27 (6): 575-83.

Armstrong, Harvey & Jim Taylor (1985), *Regional Economics and Policy*, Hemel Hempstead, Philip Allan.

Armstrong, H. W. & Jim Twomey (1993), 'Industrial Development Initiatives of District Councils in England and Wales', in Richard T. Harrison & Mark Hart (eds.), *Spatial Policy in a Divided Nation*, London, Jessica Kingsley, pp. 126-47.

Ashcroft, Brian (1978), 'The Evaluation of Regional Economic Policy – The Case of the UK', *Studies in Public Policy* 12.

Ashcroft, Brian (1982), 'Spatial Policy in Scotland', in Margeret Cuthbert (ed.), *Government Spending in Scotland – A Critical Appraisal*, Edinburgh, Harris, pp. 51-83.

Ashcroft, Brian (1983), 'The Scottish Regions and the Regions of Scotland', in Keith Ingham & James Love (eds.), *Understanding the Scottish Economy*, Oxford, Robertson, pp. 173-87.

Ashcroft, Brian & Jim H. Love (1988), Selective versus Automatic Regional Policy Assistance in Scotland – A Draft Interim Report, Paper, Fraser of Allander Institute, Strathclyde University.

Asheim, Bjørn (2001), 'Localised Learning, Innovation and Regional Clusters', in Åge Mariussen (ed.), *Cluster Policies – Cluster Development*, Stockholm, Nordregio, pp. 39-58.

Atkinson, Rob & Graham Moon (1994), *Urban Policy in Britain. The City, the State and the Market*, Houndmills, Macmillan.

Aughey, Arthur (1983), 'Mrs. Thatcher's Philosophy', *Parliamentary Affairs* 36: 389-98.

Bache, Ian *et al.* (1996), 'The European Union, Cohesion Policy, and Subnational Authorities in the UK', in Liesbet Hooghe (ed.), *Cohesion Policy and European Integration. Building Multi-level Governance*, Oxford, Oxford UP, pp. 294-320.

Bachtler, John (1993), 'Regional Policy in the 1990s – The European Perspective', in Richard T. Harrison & Mark Hart (eds.), *Spatial Policy in a Divided Nation*, London, Jessica Kingsley, pp. 254-69.

Bachtler, John (1997), 'New Dimensions of Regional Policy in Western Europe', in Michael Keating & John Loughlin (eds.), *The Political Economy of Regionalism*, London, Frank Cass, pp. 77-89.

Bachtler, John & Ivan Turok (eds.) (1997), *The Coherence of EU Policy. Contrasting Perspectives on the Structural Funds*, London, Jessica Kingsley Press.

Bacon, Robert & Walter Eltis (1976), *Britain's Economic Problem: Too Few Producers*, London, Macmillan.

Bailey, David *et al.* (1994), 'British Policy towards Inward Investment', *Journal of World Trade* 28 (2): 113-38.

Bain, Andrew D. & Richard G. Reid (1984), 'The Finance Sector', in Neil Hood & Stephen Young (eds.), *Industry, Policy and the Scottish Economy*, Edinburgh, Edinburgh UP, pp. 365-89.

Balchin, Paul N. (1990), *Regional Policy in Britain – The North-South Divide*, London, Chapman.

Balme, Richard & Laurence Bonnet (1994), 'From Regional to Sectoral Policies: The Contractual Relations Between the State and the Regions in France', *Regional Politics & Policy* 4 (3): 51-71.

Barberis, Peter & Timothy May (1993), *Government, Industry and Political Economy*, Buckingham, Open UP.

Barker, Anthony (ed.) (1982), *Quangos in Britain: Governments and the Networks of Public Policy-making*, London, Macmillan.

Barkham, Richard (1992), 'Regional Variations in Entrepreneurship: Some Evidence from the UK', *Entrepreneurship & Regional Development* 4 (3): 225-44.

Barnekov, Timothy *et al.* (1989), *Privatism and Urban Policy in Britain and the United States*, Oxford, Oxford UP.

Batt, Helge-Lothar (1996), 'Regionale und lokale Entwicklundsgesellschaften als Public-Private-Partnerships: Kooperative Regime subnationaler Politiksteuerung', in Udo Bullmann & Rolf G. Heinze (eds.), *Regionale Modernisierungspolitik. Nationale und internationale Perspektiven*, Opladen, Leske+Budrich, pp. 165-92.

Beckman, B. & A. Carling (1989), 'Förhandlingsekonomin i regionalpolitiken', *Arbetsmarknadsdepartementet Ds* 1989 (28).

Beesley, M. E. & G. M White (1973), 'The IRC: A Study in Choice of Public Sector Management', *Public Administration* 51 (1): 61-89.

Begg, Hugh M. (1983), 'Regional Policy. The Scottish Experience', *Occassional Paper, Department of Town and Regional Planning, University of Dundee* 83/13.

Begg, Hugh M. (1993), 'The Impact of Regional Policy Regrading in Scotland', in Richard T. Harrison & Mark Hart (eds.), *Spatial Policy in a Divided Nation*, London, Jessica Kingsley, pp. 85-107.

Begg, Hugh & Stuart McDowall (1986), 'Regional Industrial Policy', *The Scottish Government Yearbook* 1986: 211-24.

Begg, Hugh & Stuart McDowall (1987), 'The Effect of Regional Investment Incentives on Company Decisions', *Regional Studies* 21 (5): 459-70.

Belina, Bernd & Gesa Helms (2003), 'Zero Tolerance for the Industrial Past and Other Threats: Policing and Urban Entrepreneurialism in Britain and Germany', *Urban Studies* 40 (9): 1845-1867.

Bellini, Nicola (2002), *Business Support Services. Marketing and the Practice of Regional Innovation Policy*, Cork, Oaktree Press.

Bellini, Nicola & Francesca Pasquini (1998), 'Towards a Second Generation Regional Development Agency – The Case of ERVET in Emilia-Romagna', in Henrik Halkier, Mike Danson & Charlotte Damborg (eds.), *Regional Development Agencies in Europe*, London, Jessica Kingsley, pp. 253-70.

Benneworth, Paul (2001), *Regional Development Agencies – The Early Years*, RSA, Seaford.

Benoit, Sylvie (1995), 'Local Policies to Attract Mobile Investment: A Theoretical Survey with an Application to Two Sets of Local Organisations in France', in Paul Cheshire & Ian Gordon (eds.), *Territorial Competition in an Integrating Europe*, Aldershot, Avebury, pp. 222-43.

Benton, Ted & Ian Craib (2001), *Philosophy of Social Science. The Philosophical Foundations of Social Thought*, Palgrave, Houndmills.

Berrefjord, Ole *et al.* (1989), 'Forhandlingsøkonomi i Norden – En indledning', in Klaus Nielsen & Ove K. Pedersen (eds.), *Forhandlingsøkonomi i Norden*, København, DJØF, pp. 9-39.

Bevir, Mark & R. A. W. Rhodes (1998), 'Narratives of Thatcherism', in Hugh Berrington (ed.), *Britain in the Nineties – The Politics of Paradox*, London, Frank Cass, pp. 97-119.

Blackaby, Frank (ed.) (1979), *De-industrialization*, London, Heineman.

Bloch, Marc (1954), *The Historians Craft*, Manchester, Manchester UP.

Bogason, Peter (1985), 'Instrumental Choice and Effectiveness', *Arbejdspapir, Institut for Samfundsfag og Forvaltning, København Universitet* 1985 (5).

Bogason, Peter (1989), 'Nyinstitutionalisme set med politologiske briller', in Christian Knudsen (ed.), *Nyinstitutionalismen i samfundsvidenskaberne*, København, Samfundslitteratur, pp. 211-35.

Bogason, Peter (1992), *Forvaltning og stat*, Herning, Systime.

Bogason, Peter (1994), 'Nyinstitutionalisme, public choice og Bloomington-skolen', *GRUS* 44: 83-100.

Bogason, Peter (2000), *Public Policy And Local Governance. Institutions in Postmodern Society*, London, Edward Elgar.

Bölting, Horst M. (1976), 'Wirkungsanalyse der Instrumente der regionalen Wirtschaftsförderung', *Beiträge zu Siedlungs- und Wohnungswesen und zur Raumplanung, Universität Münster* 35.

Booth, Simon & Douglas Pitt (1983), 'A Paradox of Freedom. Intervention in the Scottish Economy', *Strathclyde Papers on Government and Politics* 23.

Booth, Simon *et al.* (1982), 'Organizational Redundancy? A Critical Appraisal of the GEAR Project', *Public Administration* 60: 56-72.

Borders Development (Scotland), Bill 1969, *HC Bills* 68.

Borders Regional Council (1989), *Scottish Enterprise*, Newton St Boswell, Borders Regional Council.

Bovaird, Tony *et al.* (1989), *An Evaluation of the Rural Development Programme Process*, London, Rural Development Commission.

Boyle, Robin (1989), 'Whose Responsibility? Analyzing Contemporary Urban Policy in Scotland, *The Scottish Government Yearbook* 1989: 126-38.

Boyle, Robin (1993), 'Changing Partners: The Experience of Urban Economic Policy in West Central Scotland 1980-90', *Urban Studies* 30 (2): 309-24.

Bredsdorff, Nils (2002), *Diskurs og konstruktion. En samfundsvidenskabelig kritik af diskursanalyser og socialkonstruktivismer*, København, Forlaget Sociologi.

Brown, Alice *et al.* (1996), *Politics and Society in Scotland*, Houndmills, Macmillan.

Brown, Ross & Philip Raines (1999), 'FDI Policy Approaches in Western Europe', in Philip Raines & Ross Brown (eds.), *Policy Competition and Foreign Direct Investment in Europe*, Ashgate, Aldershot, pp. 7-38.

Brown, Ross *et al.* (1999), 'Electronics Foreign Direct Investment in Scotland: Lessons for Nordrhein-Westfalen', *Regional and Industrial Policy Research Paper Series, EPRC, University of Strathclyde* 33.

Brunskill, Irene (1992), 'The Electronics Industry: Inward Investment versus Indigenous Development – The Policy Debate', *Environment & Planning C: Government and Policy* 10 (4): 439-50.

Budd, Leslie & Amer K. Hirmis (2004), 'Conceptual Framework for Regional Competitiveness', *Regional Studies 38* (9): 1015-28.

Bulpitt, Jim (1983), *Territory and Power in the UK*, Manchester, Manchester UP.

Burgess, Jacquelin A. (1982), 'Selling Places: Environmental Images for the Executive', *Regional Studies* 16 (1): 1-17.

Burton, Paul & Randall Smith (1996), 'The United Kingdom', in Hubert Heinelt & Randall Smith (eds.), *Policy Networks and European Structural Funds*, Aldershot, Avebury, pp. 74-119.

Buxton, Neil (1985), 'The Scottish Economy, 1945-79: Performance, Structure and Problems', in Richard Saville (ed.), *The Economic Development of Modern Scotland 1950-1980*, Edinburgh, Donald, pp. 47-78.

Cambridge Economic Policy Group (1980), 'Urban and Regional Policy with Provisional Regional Accounts, 1966-78', *Cambridge Economic Policy Review* 6 (2).

Cameron, Gordon (1979), 'The National Industrial Strategy and Regional Policy', in Duncan Maclennan & John B. Parr (eds.), *Regional Policy. Past Experiences and New Directions*, Robertson, Oxford, pp. 297-322.

Cameron, Gordon (1985), 'Regional Economic Planning – The End of the Line', *Planning Outlook* 2 (1): 8-12.

Cameron, Gordon & C. Mulvey (eds.) (1973), West Central Scotland – Appraisal of Economic Options I-III, Glasgow, Glasgow University.

Cameron, Greta & Mike Danson (2000), 'Partnership and Networking in Scotland', in Mike Danson, Henrik Halkier & Greta Cameron (eds.), *Governance, Institutional Change and Regional Development*, Aldershot, Ashgate, pp. 11-36.

Campbell, R. H. (1980), *The Rise and Fall of Scottish Industry, 1707-1939*, Edinburgh, John Donald.

Carr, Edward H. (1987), *What Is History? 2nd ed.*, Harmondsworth, Penguin.

Castledine, Pam & Kim Swales (1988), 'The Glasgow Garden Festival: A Wider Perspective, *Quarterly Economic Commentary* 14 (1): 58-64.

Central Regional Council (1989), *Scottish Enterprise*, Stirling, Central Regional Council.

Central Scotland Chamber of Commerce (1989), *Scottish Enterprise*, Polmont, Central Scotland Chamber of Commerce.

Chapman, Donald (1975), The Development Commission: Its New Role in the Rural Areas, Paper, RSA Conference "Planning for Areas of Population Decline", July 1975.

Charles, David R. & Jeremy Howells (1993), 'European Innovation and Regional Policies – Implication for the Periphery of the UK', in Richard T. Harrison & Mark Hart (eds.), *Spatial Policy in a Divided Nation*, London, Jessica Kingsley, pp. 232-53.

Chisholm, Michael (1984a), 'Regional Policy for the Late 20th Century', *Regional Studies* 18 (4): 348-52.

Chisholm, Michael (1984b), 'The Development Commission's Factory Programme', *Regional Studies* 18 (6): 514-17.

Chisholm, Michael (1987), Regional Development. The Reagan-Thatcher Legacy', *Environment & Planning C: Government and Policy* 5 (2): 197-218.

Christensen, Jørgen Grønnegård & Peter Munk Christiansen (1992), *Forvaltning og omgivelser*, Herning, Systime.

Christensen, Poul Rind (1990), 'Local Obstacles to Decentralization? – The Case of Industrial Policy in Denmark', *NordREFO* 1990 (6).

Christensen, Poul Rind *et al.* (1990), 'Firms in Network: Concepts, Spatial Impacts and Policy Implications', in Sven Illeris & Leif Jakobsen (eds.), *Networks and Regional Development*, København, Akademisk Forlag, pp. 11-58.

Christiansen, Peter Munk (1993), *Det frie market – den forhandlede økonomi*, København, DJØF.

Clarke, Peter & Clive Trebilcock (eds.) (1997), *Understanding Decline. Perceptions and Realities of British Economic Performance*, Cambridge, Cambridge UP.

Clarke, Allan & Allan Cochrane (1987), 'Investing in the Private Sector: The Enterprise Board Experience', in Allan Cochrane (ed.), *Developing Local Economic Strategies*, Milton Keynes, Open UP, pp. 4-22.

Clausen, Hans Peter (1963), *Hvad er historie*? København, Berlingske Forlag.

Cloke, Paul (ed.) (1992), *Policy and Change in Thatcher's Britain*, Oxford, Pergamon.

Coates, David (1994), *The Question of UK Decline: The Economy, State and Society*, Hemel Hempstead, Harvester Wheatsheaf.

Coates, David (1996), 'Introduction', in David Coates (ed.), *Industrial Policy in Britain*, Houndmills, Macmillan, pp. 3-29.

Coates, David & John Hillard (eds.) (1986), *The Economic Decline of Modern Britain. The Debate Between Left and Right*, Brighton, Wheatsheaf.

Cochrane, Allan (1990), 'Recent Developments in Local Authority Economic Policy', in Mike Campbell (ed.), *Local Economic Policy*, London, Cassell, pp. 156-73.

Cochrane, Allan & Allan Clarke (1990), 'Local Enterprise Boards: The Short History of a Radical Initiative', *Public Administration* 68 (3): 315-36.

Coffey, William J. & Mario Polese (1985), 'Local Development: Conceptual Bases and Policy Implications', *Regional Studies* 19 (2): 85-93.

Collin, Finn (2000), 'Kritisk realisme og socialkonstruktivisme: En kritisk af Roy Bhaskars videnskabsfilosofi', *GRUS* 60: 69-85.

Collin, Finn (2002), 'Kritik af diskursanalysens filosofiske forudsætninger', in Bøje Larsen & Kristine Munkgård Pedersen (eds.), *Diskursanalysen til debat. Kritiske perspektiver på en populær teoriretning*, København, Nyt fra Samfundsvidenskaberne, pp. 193-227.

Collis, Clive & David Noon (1994), 'Foreign Direct Investment in the UK Regions: Recent Trends and Policy Issues', *Regional Studies* 28 (8): 843-48.

Committee of Public Accounts (1980), 'SDA Accounts 1976-77, 1977-78 and 1978-79', *HC Papers* 736, 24.7.80.

Committee of Public Accounts (1981), 'Measuring the Effectiveness of Regional Industrial Policy', *HC Papers* 206, 30.4.81.

Committee of Public Accounts (1982a), 'SDA and WDA Accounts 1980-81', *HC Papers* 80, 26.4.82.

Committee of Public Accounts (1982b), 'HIDB Accounts 1980-81', *HC Papers* 301, 13.7.82.

Committee of Public Accounts (1983), 'Construction and Management of Factories by EIEC, SDA, WDA and HIDB', *HC Papers* 107, 7.11.83.

Committee of Public Accounts (1988), 'Regional and Selective Assistance', *HC Papers* 406, 11.7.88.

Committee of Public Accounts (1990), 'Locate In Scotland', *HC Papers* 217, 12.2.90.

Committee of Public Accounts (1995), 'The Welsh Development Agency', *HC Papers* 376, 21.6.95.

Committee on Scottish Affairs (1980), 'Inward Investment', *HC Papers* 769, 31.7.80.

Committee on Scottish Affairs (1981), 'Inward Investment: The Government's Reply to the Committee's 2nd Report of Session 1979-80', *HC Papers* 205, 5.3.81.

Committee on Welsh Affairs (1980), 'The Role of the Welsh Office and Associated Bodies in Developing Employment Opportunies in Wales', *HC Papers* 731, 30.7.80.

Conservative Party (1974), 'October 1974 Election Manifesto', in The Times: *Guide to the House of Commons October 1974*, London, Times Books, pp. 312-31.

Conservative Party (1979), *The Conservative Manifesto for Scotland 1979*, Edinburgh, Conservative Party.

Conservative Party (1983), *The Conservative Manifesto for Scotland 1983*, Edinburgh, Conservative Party.

Conservative Party (1987), *The Next Moves Forward – The Conservative Manifesto 1987*, London, Conservative Party.

Conservative Central Office (1950-89), *The Campaign Guide*, London, Conservative Party.

Cooke, Philip (1980), 'Discretionary Intervention and the WDA, *Area* 12 (4): 269-77.

Cooke, Philip (1990), 'Manufacturing Miracles: The Changing Nature of the Local Economy', in Mike Campbell (ed.), *Local Economic Policy*, London, Cassell, pp. 25-42.

Cooke, Philip & Kevin Morgan (1993), 'The Network Paradigm: New Departures in Corporate and Regional Development', *Environment & Planning D* 11: 543-64.

Cooke, Philip & Kevin Morgan (1998), *The Associational Economy*, Oxford, Oxford UP.

Cortell, Andrew P. (1997), 'From Intervention to Disengagement: Domestic Structure, the State, and the British Information Technology Industry 1979-90', *Polity* 30 (1): 107-31.

Craig, F. W. S. (1990), *British General Election Manifestos 1959-87*, Dartmouth, Aldershot.

Crichton, David (1984), 'The Textiles and Clothing Sectors', in Neil Hood & Stephen Young (eds.), *Industry, Policy and the Scottish Economy*, Edinburgh, Edinburgh UP, pp. 213-48.

Cromie, Stanley & Sue Birley (1994), 'Relationships Among Small Business Support Agencies', *Entrepreneurship & Regional Development* 6 (4): 301-14.

Cunningham, Edward (1977), *The Functional Role of the Scottish Development Agency*, RSA, London.

Cunningham, Edward (1978), *Regional Policy. The Role of a Development Agency*, Glasgow, SDA.

Curnow, R. & C. Saunders (eds.) (1983), *Quangos: The Australian Experience*, Alexandria, Hale & Iremonger.

Dahl, Ottar (1980), *Gruntrekk i historieforsknings metodelære*, Oslo, Universitetsforlaget.

Damborg, Charlotte & Henrik Halkier (1998), 'Regional Development Agencies in Denmark – Towards a Danish Approach to Bottom-up Regional Policy', in Henrik Halkier, Mike Danson & Charlotte Damborg (eds.), *Regional Development Agencies in Europe*, London, Jessica Kingsley, pp. 80-99.

Damesick, P. J. (1985), 'Recent Debates and Developments in British Regional Policy', *Planning Outlook* 28 (1): 3-7.

Damesick, P. J. (1987), 'The Evolution of Spatial Economic Policy', in P. J. Damesick & P. A. Woods (eds.), *Regional Problems, Problems Regions and Public Policy in the UK*, Oxford, Clarendon, pp. 42-63.

Damesick, P. J. & P. A. Woods (1987), 'Public Policy for Regional Development: Restoration or Reformation?', in P. J. Damesick & P. A. Woods (eds.), *Regional Problems, Problems Regions and Public Policy in the UK*, Oxford, Clarendon, pp. 260-66.

Danson, Mike (1980), 'The Scottish Development Agency', *Public Enterprise* (19): 12-14.

Danson, Mike (1997), 'Scotland and Wales in Europe', in Roderick Macdonald & Huw Thomas (eds.), *Nationality and Planning in Scotland and Wales*, Cardiff, University of Wales Press, pp. 14-31.

Danson, Mike & Greg Lloyd (1991), 'The Highlands and Islands Development Board: The Beginning or the End?', *Quarterly Economic Commentary* 16 (4): 76-79.

Danson, Mike, Henrik Halkier & Greta Cameron (eds.) (2000), *Governance, Institutional Change and Regional Development*, Aldershot, Ashgate.

Danson, Mike *et al.* (1988), The Changing Priorities of Development Agencies in Scotland, Paper presented to the Regional Studies Association Conference, Belfast 21-22.9.88.

Danson, Mike *et al.* (1989a), ''Scottish Enterprise'; Towards a Model Agency or a Flawed Initiative?', *Regional Studies* 23: 557-63.

Danson, Mike *et al.* (1989b), 'Scottish Enterprise: The Creation of a More Effective Development Agency or the Pursuit of Ideology?', *Quarterly Economic Commentary* 14 (3): 70-75.

Danson, Mike *et al.* (1989c), SDA, Economic Development and Technology Policy, Paper presented at UNECE Research Colloqium, Amsterdam 13-17 February 1989.

Danson, Mike *et al.* (1990a), 'Scottish Enterprise: An Evolving Approach to Integrating Economic Development in Scotland', *The Scottish Government Yearbook* 1990: 168-94.

Danson, Mike *et al.* (1990b), 'The Scottish Development Agency, Economic Development and Technology Policy', in Henk ter Heide (ed.), *Technological Change and Spatial Policy*, Amsterdam, Royal Netherlands Geographical Society, pp. 179-90.

Danson, Mike *et al.* (1992), 'Regional Development Agencies in the UK', in Peter Townroe & Ron Martin (eds.), *Regional Development in the 1990s – The British Isles in Transition*, London, Jessica Kingsley, pp. 297-303.

Danson, Mike *et al.* (1993), 'The Role of Development Agencies in Regional Economic Regeneration – A Scottish Case Study', in Richard T. Harrison & Mark Hart (eds.), *Spatial Policy in a Divided Nation*, London, Jessica Kingsley, pp. 162-75.

Danson, Mike *et al.* (1999), 'The European Structural Fund Partnerships in Scotland: New Forms of Governance for Regional Developments?', *Scottish Affairs* (27): 23-40.

Davenport, Michael (1983), 'Industrial Policy in the United Kingdom', in Gerard Adams & Lawrence R. Klein (eds.), *Industrial Policies for Growth and Competition*, Lexington, Lexington Books, pp. 331-57.

Davies, J. R. (1978), 'The Industrial Investment Policy of the SDA', *Quarterly Economic Commentary* 4 (2): 34-46.

Demko, George J. & Roland J. Fuchs (1984), 'A Comparison of Regional Development Policy Instruments and Measures in Eastern and Western Europe, in George Demko (ed.), *Regional Development Problems and Policies in Eastern and Western Europe*, London, Croom Helm, pp. 83-98.

Department of Energy (1974), 'United Kingdom Offshore Oil and Gas Policy', *Cmnd* 5696, 11.7.74.

Department of Industry (1976), 'Criteria for Assistance to Industry', *HC Papers* 596-II, 22.7.76.

Development Commission (1984), *The Next Ten Years*, Development Commission.

Development Commission (1958-91), 'Accounts', *HC Papers*.

Devine, Thomas M. (1996), 'Introduction', in Thomas M. Devine & Richard J. Finlay (eds.), *Scotland in the 20th Century*, Edinburgh, Edinburgh UP, pp. 1-12.

Diamond, D. R. & N. A. Spence (1983), *Regional Policy Evaluation: A Methodological Review and the Scottish Example*, London, Gower.

Dicken, Peter & Adam Tickell (1992), 'Competitors or Collaborators? The Structure of Inward Investment Promotion in Northern England', *Regional Studies* 26 (1): 99-106.

DiMaggio, Paul J. & Walter W. Powell (1991), 'Introduction', in Walter W. Powell & Paul J. DiMaggio (eds.), *The New Institutionalism in Organizational Analysis*, Chicago, Chicago UP, pp. 1-38.

Dintenfass, Michael (1992), *The Decline of Industrial Britain. 1870-1980*, London, Routledge.

Donnison, David & Alan Middleton (eds.) (1987), *Regenerating the Inner City: Glasgow's Experience*, London, RKP.

Dowle, Martin (1987), 'The Year at Westminster', *The Scottish Government Yearbook* 1987: 12-22.

Draper, Paul *et al.* (1988), *The Scottish Financial Sector*, Edinburgh, Edinburgh UP.

DTI (1983a), 'Regional Industrial Development', *Cmnd* 9111, Dec. 1983.

DTI (1983b), *Regional Industrial Policy: Some Economic Issues*, London, DTI.

DTI (1988), 'DTI – The Department for Enterprise', *Cm* 278, Jan. 1988.

Dumfries and Galloway Regional Council (1989), *Scottish Enterprise White Paper Response*, Newton Stewart, Dumfries and Galloway Regional Council.

Dundee & Tayside Chamber of Commerce and Industry (1989), *Comments on "Scottish Enterprise"*, Dundee, Dundee & Tayside Chamber of Commerce and Industry.

Dunford, Mick (1989), 'Technopoles, Politics and the Markets: The Development of Electronics in Grenoble and Silicon Glen', in Margaret Sharp & Peter Holmes (eds.), *Strategies for New Technology. Case Studies from Britain and France*, Hemel Hempstead, Philip Allan, pp. 80-118.

Dunford, Mick (1998), 'Regions and Economic Development', in Patrick le Galès & Christian Lequesne (eds.), *Regions in Europe*, London, Routledge, pp. 89-107.

Dunford, Mick & Diane Perrons (1986), 'The Restructuring of the Post-war British Space Economy', in Bob Rowthorn & Ron Martin (eds.), *The Geography of De-industrilisation*, Macmillan, Houndmills, pp. 53-106.

Dunford, Mick *et al.* (1981), 'Regional Policy and the Crisis in the UK – A Long-run Perspective', *International Journal of Urban and Regional Research* 5 (3): 377-410.

Dunleavy, Patrick (1989), 'The Architecture of the British Central State. Part I: Framework for Analysis', *Public Administration* 67: 249-75.

Dunleavy, Patrick (1993), 'The Political Parties', in Patrick Dunleavy *et al.* (eds.), *Developments in British Politics 4*, Houndmills, Macmillan, pp. 123-53.

Eaton, Fiona E. (1987), Examination and Analysis of the Role of SDA in Investment, Honours Dissertation, Department of Politics, Strathclyde University, Glasgow.

Edinburgh Chamber of Commerce (1989), *Scottish Enterprise – An Assesment of the White Paper Proposals*, Edinburgh, Edinburgh Chamber of Commerce.

Edinburgh City District Council (1989), *Scottish Enterprise White Paper*, Edinburgh, Edinburgh City District Council.

Eirug, Aled (1983), 'The Welsh Development Agency', *Geoforum* 14 (4): 375-88.

Eisinger, Peter (1988), *The Rise of the Entrepreneurial State. State and Local Economic Development Policy in the US*, Madison, Wisconsin UP.

Elbaum, Bernard & William Lazonick (eds.) (1986), *The Decline of the British Economy*, Oxford, Clarendon.

Elklit, Jørgen (1987), Noget om elite-interviews, samtidshistoriske interviews og vidnepsykologi, Department of Political Science, University of Aarhus.

Elmore, Richard F. (1987), 'Instruments and Strategy in Public Policy', *Policy Studies Review* 7 (1): 174-86.

English, Richard & Michael Kenny (eds.) (1999), *Rethinking British Decline*, Houndmills, Macmillan.

English Development Agencies Bill 1976, Bills 24.

ERDF (1975-88), Annual Report, Brussels, DG XVI.

Esmark, Anders (1998), 'Forandringen stiger. En kritik af forandringsbegrebet i ny-institutionalisme – og et systemteoretisk alternativ', *COS-rapport* 4/98.

EURADA (1995), *Reflections on the Creation, Development and Management of RDAs*, Brussels, EURADA.

Evans, Brendan (1999), *Thatcherism and British Politics, 1975-1999*, Stroud, Sutton.

Evans, Eric J. (1997), *Thatcher and Thatcherism*, London, Routledge.

Ewers, Hans-Jürgen (1990), 'Regional Economic Development and Innovation-Oriented Measures: Summary and Perspectives', in Hans-Jürgen Ewers & Jürgen Allesch (eds.), *Innovation and Regional Development*, Berlin, de Gruyter, pp. 337-44.

Ewers, Hans-Jürgen & R. Wettmann (1980), 'Innovation-oriented Regional Policy', *Regional Studies* 14: 161-79.

Expenditure Committee (1972), 'Public Money in the Private Sector, vol. 1: Report', *HC Papers* 347, 6.7.72.

Fairclough, Norman (1991), 'What Might We Mean by "Enterprise Discourse"?', in Russel Keat & Nicholas Abercrombie (eds.), *Enterprise Culture*, London, Routledge, pp. 38-58.

Fairclough, Norman (1992), *Discourse and Social Change*, Cambridge, Polity.

Fairley, John (1999), 'Economic Development: The Scottish Parliament and Local Government', in John McCarthy & David Newlands (eds.), *Governing Scotland: Problems and Prospects. The Economic Impact of the Scottish Parliament*, Aldershot, Ashgate, pp. 103-17.

Field, G. M. & P. V. Hills (1976), 'The Administration of Industrial Subsidies', in Alan Whiting (ed.), *The Economics of Industrial Subsidies*, London, HMSO, pp. 1-22.

Fife Regional Council (1989), *Scottish Enterprise White Paper Response*, Glenrothes, Fife Regional Council.

Fine, Ben (1989), 'Denationalisation', in Francis Green (ed.), *The Restructuring of the UK Economy*, Hemel Hempstead, Harvester, pp. 225-42.

Fine, Ben & Laurence Harris (1985), *The Peculiarities of the British Economy*, London, Lawrence & Wishart.

Finnegan, Marie (1998), 'Regional Development Agencies and Policy in Northern Ireland – Strategies and Implementation', in Henrik Halkier, Mike

Danson & Charlotte Damborg (eds.), *Regional Development Agencies in Europe*, London, Jessica Kingsley, pp. 306-23.

Finnie, Helen M. (1982), 'The SDA – Expanding Services for Expanding Industries', *Scottish Bankers Magazine* 1982 (Nov.), 85-88.

Firn, John R. (1975), 'External Control and Regional Development: The Case of Scotland', *Environment & Planning A* 7: 393-414.

Firn, John R. (1980), 'Economic Policies for the Conurbations', in G. C. Cameron (ed.), *The Future of the British Conurbation*, London, Longman, pp. 250-76.

Firn, John R. (1982), 'Industrial Regeneration and Regional Policy: The Scottish Perspective and Experience', in L. Collins (ed.), *Industrial Decline and Regeneration*, Edinburgh, Department of Geography and Centre of Canadian Studies, University of Edinburgh, pp. 5-19.

Firn, John R. (1985), 'Industry', in Roger Smith & Urlan Wannop (eds.), *Strategic Planning in Action. The Impact of the Clyde Valley Regional Plan 1946-1982*, Aldershot, Gower, pp. 100-138.

Firn, John R. & David Roberts (1984), 'High-technology Industries', in Neil Hood & Stephen Young (eds.), *Industry, Policy and the Scottish Economy*, Edinburgh, Edinburgh UP, pp. 288-327.

Firn, John R. & Kim Swales (1978), 'The Formation of New Manufacturing Establishments in Central Clydeside and the West Midlands Conurbation', 1963-72, *Regional Studies* 12: 199-213.

First Scottish Standing Committee (1975), 'Scottish Development Agency (No.2), Bill [Lords]', *Parliamentary Debates*, July 1985.

Folmer, Hendrik (1986), *Regional Economic Policy: Measurement of its Effect*, Dordrecht, Nijhoff.

Forslund, Ulla & Charlie Karlson (1991), *Infrastrukturens regionale effekter*, Stockholm, ERU.

Fothergill, Steve *et al.* (1987), *Property and Industrial Development*, London, Hutchinson.

Freeman, Chris (1979), 'Technical Innovation and British Trade Performance, in Frank Blackaby (ed.), *British Economic Policy 1960-74*, London, Cambridge UP, pp. 56-73.

Funck, Rolf H. (1990), 'Reflections on Regional Development Policy and the Concept of Region', in Manas Chatterji & Robert E. Kuenne (eds.), *New Frontiers in Regional Science*, London, Macmillan, pp. 173-83.

Gaardmand, Arne (1988), 'Jobbet til manden eller manden til jobbet? Om dansk regionalpolitik 1945-85', in NordREFO (ed.), *Om regionalpolitikken som politikområde*, Helsinki, NordREFO, pp. 67-108.

Galès, Patrick Le (1994), 'Regional Economic Policies: An Alternative to French Economic Dirigism?', *Regional Politics & Policy* 4 (3): 72-91.

Galès, Patrick Le (1998), 'Conclusion – Government and Governance of Regions. Structural Weaknesses and New Mobilisations', in Patrick Le Galès & Christian Lequesne (eds.), *Regions in Europe*, London, Routledge, pp. 239-67.

Gamble, Andrew (1983), 'Thatcherism and Conservative Politics', in Stuart Hall & Martin Jacques (eds.), *The Politics of Thatcherism*, London, Lawrence & Wishart, pp. 109-31.

Gamble, Andrew (1988), *The Free Economy and the Strong State: The Politics of Thatcherism*, Houndmills, Macmillan.

Gamble, Andrew (1990), *Britain in Decline. Economic Policy, Political Strategy and the British State, 3. ed.*, Houndmills, Macmillan.

Gee, J. M. Alec (1991), 'The Neoclassical School', in Douglas Mair & Anne G. Miller (eds.), *A Modern Guide to Economic Thought*, London, Elgar, pp. 71-108.

Gibbons, Peter D. (1989), *The Use of Advanced Manufacturing Technology by Scottish Manufacturing Firms and the Need for a More Considered Approach to Manufacturing Strategy*, Glasgow, Strathclyde University MBA Project Report.

Gibbs, David (1989), 'Government Policy and Industrial Change: An Overview', in David Gibbs (ed.), *Government Policy and Industrial Change*, London, Routledge, pp. 1-20.

Giddens, Anthony (1984), *The Constitution of Society*, Cambridge, Polity.

Glasgow Chamber of Commerce (1982-91), *Annual Report*, Glasgow, Glasgow Chamber of Commerce.

Glasgow Chamber of Commerce (1989a), *Scottish Enterprise*, Glasgow, Glasgow Chamber of Commerce.

Glasgow Junior Chamber of Commerce (1989), *Scottish Enterprise*, Glasgow, Glasgow Junior Chamber of Commerce.

Gomez, María V. (1999), 'Reflective Images: The Case of Urban Regeneration in Glasgow and Bilbao', *International Journal of Urban and Regional Research* 22 (1): 106-21.

Goodstadt, Vincent & Keith Clement (1997), 'Environmental Improvement within Regional Economic Development: Lessons from Clydeside', in John Bachtler & Ivan Turok (eds.), *The Coherence of EU Regional Policy. Contrasting Perspectives on the Structural Funds*, London, Jessica Kingsley, pp. 160-176.

Goodwin, Mark & Simon Duncan (1986), 'The Local State and Local Economic Policy: Political Mobilization or Economic Regeneration, *Capital & Class* (27): 14-36.

Gordon, I. R. (1990), 'Regional Policy and National Politics in Britain', *Environment & Planning C: Government and Policy* 8 (4): 427-38.

Grahm, Leif (1988), *Att välja regionalpolitik. Strukturförändring, kris och nyorientering i nordisk samhällsplanering*, København, NordREFO.

Grampian Regional Council (1989), *White Paper: Scottish Enterprise*, Aberdeen, Grampian Regional Council.

Grant, Wyn (1982), *The Political Economy of Industrial Policy*, London, Butterworths.

Grant, Wyn (1993), *The Politics of Economic Policy*, Hemel Hempstead, Harvester Wheatsheaf.

Grant, Wyn & Stephen Wilks (1983), 'British Industrial Policy: Structural Change, Political Inertia', *Journal of Public Policy* 3 (1): 13-28.

Grassie, James (1983), *Highland Experiment. The Story of the HIDB*, Aberdeen, Aberdeen UP.

Green, E. H. H. (2002), *Ideologies of Conservatism. Conservative Political Ideas in the 20th Century*, Oxford, Oxford UP.

Green, Francis (ed.) (1989), *The Restructuring of the UK Economy*, Hemel Hempstead, Harvester.

Greenwood, John & David Wilson (1989), *Public Administration in Britain Today*, London, Unwin Hyman.

Greve, Carsten *et al.* (1999), 'Quangos – What's in a Name? Defining Quangos from a Comparative Perspective', *Governance* 12 (2): 129-46.

Griffiths, Dylan (1996), *Thatcherism and Territorial Politics: A Welsh Case Study*, Aldershot, Avebury.

Gudgin, Graham (1995), 'Regional Problems and Policy in the UK', *Oxford Review of Economic Policy* 11 (2): 18-63.

Gulliver, Stuart (1984), 'The Area Projects of the Scottish Development Agency', *Town Planning Review* 53 (3): 322-334.

Gustaffson, Jeppe & Janne Seemann (1985), *Små institutioner i store systemer – Tilpasning og påvirkning*, Aalborg, Aalborg Universitetsforlag.

Hague, Douglas & Geoffrey Wilkinson (1983), *The IRC – An Experiment in Industrial Intervention*, London, George Allan & Unwin.

Hague, Rod *et al.* (1992), *Comparative Government and Politics – An Introduction, 3rd ed.*, Houndmills, Macmillan.

Halkier, Henrik (1986), 'Dunkle kapitler af en 'stor' fortælling: Franske borgere mellem feudalisme og kapitalisme', in Henrik Halkier *et al.* (eds.), *Mod en ny samfundshistorie?*, Aarhus, DJH, pp. 61-75.

Halkier, Henrik (1987), 'Stalin eller kaos – Evolution og historie i kritisk samfundsteori', *GRUS* 22/23: 88-113.

Halkier, Henrik (1990a), 'Samfundsforskning og tekst – Det skotske tilfælde som metodisk problem', *Hermes* (4): 101-24.

Halkier, Henrik (1990b), 'The Harder They Come – 1789 as a Challenge to Marxism', *Science & Society* 54 (3): 321-51.

Halkier, Henrik (1991a), 'Den 'engelske' syge – Konsensus og konflikt i Storbritannien 1951-79', *Den Jyske Historiker* (54-55): 199-231.

Halkier, Henrik (1991b), *Difficult to Cure. The Political Economy of Industrial Decline in Britain*, Aarhus, Ms.

Halkier, Henrik (1991c), 'The Political Economy of Britain's Decline – Problems and Perspectives', *Culture & History* (9/10): 105-36.

Halkier, Henrik (1992), 'Development Agencies and Regional Policy: The Case of the Scottish Development Agency', *Regional Politics & Policy* 2 (3): 1-26.

Halkier, Henrik (1994), 'The Origins of The Scottish Development Agency – A Study in Regional Policy and Politics', in Ernst-Ullrich Pinkert, E.-U. (ed.), *Universalisme og interkulturel kommunikation*, Aalborg, Sprog og Kulturmøde 8, Aalborg Universitet, pp. 197-208.

Halkier, Henrik (1996), 'Institutions, Power and Regional Policy', *European Studies Series of Occasional Papers, European Research Unit, Aalborg University* 16.

Halkier, Henrik (2001), 'Regional Policy in Transition – A Multi-level Governance Perspective on the Case of Denmark', *European Planning Studies* 9 (3): 323-38.

Halkier, Henrik (2003), 'Discourse, Institutions and Public Policy Change – Theory, Methods and a Scottish Case Study', *Spirit Discussion Papers* 23.

Halkier, Henrik & Charlotte Damborg (1997), 'Networks, Development Agencies and Intelligent Regions – Towards a Framework for Empirical Analysis', *European Studies Series of Occasional Papers, European Research Unit, Aalborg University* 22.

Halkier, Henrik & Charlotte Damborg (2000), 'Development Bodies, Networking and Business Promotion – The case of North Jutland, Denmark', in Mike Danson, Henrik Halkier & Greta Cameron (eds.), *Governance, Institutional Change and Regional Development*, Aldershot, Ashgate, pp. 92-114.

Halkier, Henrik & Mike Danson (1997), 'Regional Development Agencies in Europe: A Survey of Key Characteristics and Trends', *European Urban and Regional Studies* 4 (3): 243-56.

Halkier, Henrik, Mike Danson & Charlotte Damborg (eds.) (1998), *Regional Development Agencies in Europe*, London, Jessica Kingsley Publishers.

Halkier, Henrik, Ewa Helinska-Hughes & Michael Hughes (2003), 'Governing Inward Investment – Emerging National and Regional Patterns in West and East European Countries', *European Studies Series of Occasional Papers, European Research Unit, Aalborg University* 34.

Hall Aitken Associates (1995), *Evaluation of the Eastern Scotland Community Support Frameworks 1989-93 (Final Report)*, Glasgow, Hall Aitken Associates.

Hall Aitken Associates (1996), *Ex Post Evaluation of Objective 2 Regions 1989-93, North Sea Group (Summary Report)*, Glasgow, Hall Aitken Associates.

Hall, Graham & Pam Lewis (1988), 'Development Agencies and the Supply of Finance to Small Firms', *Applied Economics* 20 (12): 1675-87.

Hall, Peter A. (1986), *Governing the Economy: The Politics of State Intervention in Britain and France*, Cambridge, Polity.

Hall, Peter A. (1992), 'The Movement from Keynesianism to Monetarism: Institutional Analysis and British Economic Policy in the 1970s', in Sven Steinmo *et al.* (eds.), *Structuring Politics: Historical Institutionalism in Comparative Analysis*, Cambridge, Cambridge UP, pp. 90-113.

Hall, Peter A. (ed.) (1989), *The Political Power of Economic Ideas: Keynesianism across Nations*, Princeton, Princeton UP.

Hall, Stuart (1983), 'The Great Moving Right Show', in Stuart Hall & Martin Jacques (eds.), *The Politics of Thatcherism*, London, Lawrence & Wishart, pp. 19-39.

Hall, Stuart & Martin Jacques (1983), 'Introduction', in Stuart Hall & Martin Jacques (eds.), *The Politics of Thatcherism*, London, Lawrence & Wishart, pp. 9-18.

Hall, Stuart & Martin Jacques (eds.) (1983), *The Politics of Thatcherism*, London, Lawrence & Wishart.

Hallin, Göran & Anders Malmberg (1996), 'Attraction, Competition and Regional Development in Europe', *European Urban and Regional Studies* 3 (4): 323-337.

Ham, Christopher & Michael Hill (1984), *The Policy Process in the Modern Capitalist State*, Brighton, Wheatsheaf.

Hamilton, F. E. Ian (1987), 'Multinational Enterprises', in William F. Lever (ed.), *Industrial Change in the United Kingdom*, London, Longman, pp. 167-95.

Hanf, Kenneth & Laurence J. O'Toole, jr (1992), 'Revisiting Old Friends: Networks, Implementation Structures and the Management of Inter-organizational Relations', *European Journal of Political Research* 21 (1-2): 163-80.

Hansen, Niles *et al.* (1990), *Regional Policy in a Changing World*, New York, Plenum.

Harden, Ian (1988), 'United Kingdom', in Tore Modeen & Allan Rosas (eds.), *Indirect Public Administration in 14 Countries*, Åbo, Åbo Academy Press, pp. 300-332.

Harding, Alan (1990), 'Public-Private Partnerships in Urban Regeneration', in Mike Campbell (ed.), *Local Economic Policy*, London, Cassell, pp. 108-27.

Hargrave, Andrew (1985), *Silicon Glen. Reality or Illusion? A Global View of High Technology in Scotland*, Edinburgh, Mainstream.

Harris, Richard (1993), 'Retreat From Policy – The Rationale and Effectiveness of Automatic Capital Grants', in Richard T. Harrison & Mark Hart (eds.), *Spatial Policy in a Divided Nation*, London, Jessica Kingsley, pp. 64-84.

Harrison, Richard & Colin Mason (1992), 'The Roles of Investors in Entrepreneurial Companies: A Comparison of Private Investors and Venture Capitalists', *Venture Finance Research Project Working Paper, University of Southampton* 5.

Harrison, Richard T. & Colin M. Mason (1993), 'The Regional Impact of National Policy Initiatives – Small Firms Policy in the UK', in Richard T. Harrison & Mark Hart (eds.), *Spatial Policy in a Divided Nation*, London, Jessica Kingsley, pp. 192-215.

Hart, Mark & Richard T. Harrison (1990), 'Inward Investment and Economic Change: The Future Role of Regional Development Agencies', *Local Economy* 5 (3): 196-213.

Hart, Mark & Richard T. Harrison (1992), 'Northern Ireland', in Peter Townroe & Ron Martin (eds.), *Regional Development in the 1990s – The British Isles in Transition*, London, Jessica Kingsley, pp. 117-26.

Hart, Mark *et al.* (1993), 'Enterprise Creation, Job Generation and Regional Policy in the UK', in Richard T. Harrison & Mark Hart (eds.), *Spatial Policy in a Divided Nation*, London, Jessica Kingsley, pp. 176-91.

Harvie, Christopher (1994), *Scotland and Nationalism. Scottish Society and Politics 1707-1994, 2nd ed.*, Routledge, London.

Haug, Peter (1986), 'US High Tecnology Multinationals and Silicon Glen', *Regional Studies* 20 (2): 103-16.

Haug, Peter *et al.* (1983), 'R&D Intensity in the Affiliates of US Owned Electronics Companies Manufacturing in Scotland', *Regional Studies* 17 (6): 383-92.

Haughton, Graham & Paul Lawless (1992), *Policies for Potential: Recasting British Urban and Regional Policies*, London, RSA.

Hay, Colin (1995), 'Structure and Agency', in David Marsh & Gerry Stoker (eds.), *Theory and Method in Political Science*, Houndmills, Macmillan, pp. 189-206.

Healey, Patsy (1991), 'Urban Regeneration and the Development Industry', *Regional Studies* 25 (2): 97-110.

Heclo, Hugh (1986), 'Industrial Policy and the Executive Capacities of Government', in Claude E. Barfield & William A Schambra (eds.), *The Politics of Industrial Policy*, Washington, American Enterprise Institute, pp. 292-317.

Hedetoft, Ulf (1995), *Signs of Nations. Studies in the Political Semiotics of Self and Other in Contemporary European Nationalism*, Aldershot, Dartmouth.

Hedetoft, Ulf & Hanne Niss (1991), 'Taking Stock of Thatcherism', *Publications of the Department of Languages and Intercultural Studies, Aalborg University* 4.

Heinelt, Hubert & Randall Smith (1996), 'Introduction', in Hubert Heinelt & Randall Smith (eds.), *Policy Networks and European Structural Funds*, Aldershot, Avebury, pp. 1-8.

Heinelt, Hubert & Randall Smith (eds.) (1996), *Policy Networks and European Structural Funds*, Aldershot, Avebury.

Helmstädter, Ernst *et al.* (2003), Notes on Empirical Research Work under EURODITE, Paper presented to the EURODITE meeting in Brussels 20/21 October 2003.

Henderson, David M. (1984), 'The Natural Resource-based Sector', in Neil Hood & Stephen Young (eds.), *Industry, Policy and the Scottish Economy*, Edinburgh, Edinburgh UP, pp. 249-87.

Henderson, Dylan (1998), 'Building Interactive Learning Networks: Lessons from the Welsh Medical Technology Forum', *Regional Studies* 32 (8): 783-87.

Henning, Roger (1983), 'Regional Policy: Implementation through Bargaining', *Scandinavian Political Studies* 6 (3): 195-210.

HIDB (1966-91), *Annual Report*, Inverness, HIDB.

HIDB Act 1965, *Acts*, Ch. 46.

Hill, Stephen & Max Munday (1989), 'Scotland v Wales in the Inward Investment Game: Wales' Triple Crown?', *Quarterly Economic Commentary* 17 (4): 52-55.

Hill, Stephen & Max Munday (1992), 'The UK Regional Distribution of Foreign Direct Investment: Analysis and Determinants', *Regional Studies* 26 (6): 535-44.

Hindley, Brian (1983), 'What is the Case for State Investment Companies?', in Brian Hindley (ed.), *State Investment Companies in Western Europe – Picking Winners or Backing Losers*? London, Macmillan, pp. 1-24.

Hjern, Benny & David O. Porter (1993), 'Implementation Structures. A New Unit of Administrative Analysis', (orig. 1981), in Michael Hill (ed.), *The*

Policy Process – A Reader, Hemel Hempstead, Harvester Wheatsheaf, pp. 248-65.

Hobsbawm, E. J. (1990), *Nations and National since 1780 – Programme, Myth, Reality, 2nd ed.*, Cambridge: Cambridge UP.

Hodgson, Geoffrey M. (1986), 'The Underlying Economic Crisis', in David Coates & John Hillard (eds.), *The Economic Decline of Modern Britain*, Brighton, Wheatsheaf, pp. 320-33.

Hodgson, Geoffrey M. (1989), *Economics and Institutions*, Cambridge, Polity.

Hogwood, Brian W. (1977a), 'Intergovernmental Structures and Industrial Policy in the UK', *Studies in Public Policy* 2.

Hogwood, Brian W. (1977b), 'Models of Industrial Policy: The Implications for Devolution', *Studies in Public Policy* 5.

Hogwood, Brian W. (1978), 'The Primacy of Politics in the Economic Policy of Scottish Government', *Studies in Public Policy* 14.

Hogwood, Brian W. (1981), 'In Search of Accountability: The Territorial Dimension of Industrial Policy', *Studies in Public Policy* 82.

Hogwood, Brian W. (1982a), 'The Regional Dimension of Industrial Policy', in Peter Madgwick & Richard Rose (eds.), *The Territorial Dimension in UK Government*, Houndmills, Macmillan, pp. 34-66.

Hogwood, Brian W. (1983), 'The Instruments of Desire: How British Government Attempts to Regulate and Influence Industry', *Public Administration Bulletin* 42 (Aug.), 5-25.

Hogwood, Brian W. (1986), 'Analysing Industrial Policy: A Multi-perspective Approach', *Public Administration Bulletin* 29: 18-42.

Hogwood, Brian W. (1987), The Tangled Web – Networks and the Territorial Dimension of Industrial Policy, Political Studies Association Annual Conference, University of Aberdeen, 7-9 April 1987.

Hogwood, Brian W. (1992), *Trends in British Public Policy. Do Governments Make Any Difference*? Buckingham, Open UP.

Hogwood, Brian W. (1995), 'The 'Growth' of Quangos: Evidence and Explanations', in F. F. Ridley & David Wilson (eds.), *The Quango Debate*, Oxford, Oxford UP, pp. 29-47.

Hogwood, Brian W. (1996), *Mapping the Regions: Boundaries, Coordination and Government*, Bristol, Policy Press.

Hogwood, Brian W. & Lewis A. Gunn (1986), *Policy Analysis for the Real World*, Oxford, Oxford UP.

Holliday, Ian (1992), 'Scottish Limits to Thatcherism', *The Political Quarterly* 63 (4): 448-59.

Holliday, Ian (1993), 'Organised Interests After Thatcher', in Patrick Dunleavy *et al.* (eds.), *Developments in British Politics 4*, Houndmills, Macmillan, pp. 307-20.

Hood, Christopher (1978), 'Keeping the Centre Small: Explanations of Agency Type', *Political Studies* 26 (1): 30-46.

Hood, Christopher (1983), *The Tools of Government*, London, Macmillan.

Hood, Christopher (1986), 'The Hidden Public Sector: The 'Quangocratization' of the World?', in Franz-Xaver Kaufmann *et al.* (eds.), *Guidance, Control and, Evaluation in the Public Sector*, Berlin, de Gruyter, pp. 183-207.

Hood, Christopher (1988), 'PGOs in the United Kingdom', in Christopher Hood & Gunnar Folke Schuppert (eds.), *Delivering, Public Services in Western Europe*, London, Sage, pp. 75-93.

Hood, Christopher (1994), *Explaining Economic Policy Reversals*, Buckingham, Open UP.

Hood, Christopher & Gunnar Folke Schuppert (1988), 'The Study of Para-Government Organizations', in Christopher Hood & Gunnar Folke Schuppert (eds.), *Delivering, Public Services in Western Europe*, London, Sage, pp. 1-26.

Hood, Christopher & Gunnar Folke Schuppert (eds.) (1988), *Delivering Public Services in Western Europe. Sharing Western European Experience of Para-government Organization*, London, Sage.

Hood, Christopher *et al.* (1985), 'From Growth to Retrenchment? A Perspective on the Development of the Scottish Office to the 1980s', *The Scottish Government Yearbook* 1985: 53-76.

Hood, Neil (1984), 'The Small Firm Sector', in Neil Hood & Stephen Young (eds.), *Industry, Policy and the Scottish Economy*, Edinburgh, Edinburgh UP, pp. 57-92.

Hood, Neil (1990), 'Scottish Enterprise: The Basis of a Scottish Solution to Scottish Problems', *Quarterly Economic Commentary* 16 (2): 65-75.

Hood, Neil (1991a), 'The Scottish Development Agency in Retrospect', *The Royal Bank of Scotland Review* (171): 3-21.

Hood, Neil (1991b), 'The Development and Implementation of a Strategic Role within Scottish Enterprise', *Quarterly Economic Commentary* 16 (3): 70-80.

Hood, Neil & Stephen Young (1976), 'US Investment in Scotland – Aspects of the Branch Factory Syndrome', *Scottish Journal of Political Economy* 23 (3): 279-94.

Hood, Neil & Stephen Young (1981), 'Government Policy and Inward Investment Attraction in Scotland', *Quarterly Economic Commentary* 6 (4): 39-47.

Hood, Neil & Stephen Young (1982), 'The Attraction of Inward Investment. The Report of the Select Committeee and the Government's Reply', *The Scottish Government Yearbook* 1982: 258-64.

Hood, Neil & Stephen Young (eds.) (1984), *Industry, Policy and the Scottish Economy*, Edinburgh, Edinburgh UP.

Hood, Neil & James H. Taggart (1999), 'Subsidiary Development in German and Japanese Manufacturing Subsidiaries in the British Isles', *Regional Studies* 33 (6): 513-28.

Hooghe, Liesbet (1996), 'Building a Europe with the Regions: The Changing Role of the European Commission', in Liesbet Hooghe (ed.), *Cohesion Policy and European Integration. Building Multi-level Governance*, Oxford, Oxford UP, pp. 89-126.

Hooghe, Liesbet (ed.) (1996), *Cohesion Policy and European Integration. Building Multi-level Governance*, Oxford, Oxford UP.

Howe, Joe (1996), 'A Case of Inter-agency relations: Regional Development in Mid-Wales', *Planning Practice and Research* 11 (1): 61-72.

Hucke, Jochen (1983), 'Implementation von Finanzhilfeprogrammen', in Renate Mayntz (ed.), *Implementation politicher Programme II – Ansätze zur Theoriebildung*, Opladen, Westdeutscher Verlag, pp. 75-98.

Hudson, Ray (1989a), *Wrecking a Region: State Policies, Party Politics and Regional Change in North East England*, London, Pion.

Hudson, Ray (1989b), 'Rewriting History and Reshaping Geography: The Nationalised Industries and the Political Economy of Thatcherism', in John Mohan (ed.), *The Political Geography of Contemporary Britain*, Houndmills, Macmillan, pp. 113-29.

Hudson, Ray (1994), 'Institutional Change, Cultural Transformation, and Economic Regeneration: Myths and Realities from Europe's Old Industrial Areas', in Ash Amin & Nigel Thrift (eds.), *Globalization, Institutions, and Regional Development in Europe*, Oxford, Oxford UP, pp. 196-216.

Hughes, Jim (1982), 'Policy Analysis in the HIDB', *Journal of the Operational Research Society* 33: 1055-64.

Hughes, Jim (1998a), 'Regional Development Institutions – Rural Development in the UK', in Henrik Halkier, Mike Danson & Charlotte Damborg (eds.), *Regional Development Agencies in Europe*, London, Jessica Kingsley, pp. 183-98.

Hughes, Jim (1998b), 'The Role of Development Agencies in Regional Policy: An Academic and Practitioner Approach', *Urban Studies* 35 (4): 615-26.

Hull, Christopher John (1990), 'Information and Consulting as Instruments of Regional Development Policy: The Role of Public Agencies', in Hans-Jürgen Ewers & Jürgen Allesch (eds.), *Innovation and Regional Development*, Berlin, de Gruyter, pp. 199-206.

Humlum, Jøren (1990), Politologisk tekstanalyse, Aarhus, Department of Political Science.

Hutchison, I. G. C. (1999), *Scottish Politics in the Twentieth Century*, Houndmills, Macmillan.

IBB (1983-91), *Annual Report*, London, IBB.

IBB (1983a), *Inward Investment and the IBB 1977-82*, London, IBB.

IDS (1987a), *1986 Review of the SDA. Report of the Review Group to the Secretary of State for Scotland*, Edinburgh, HMSO.

IDS (1987b), *Review of the HIDB. Report to the Secretary of State for Scotland*, Edinburgh, IDS.

IDS (1988a), 'Scottish Enterprise. A New Approach to Training and Enterprise Creation', *Cm* 534, Dec. 1988.

IDS (1988b), *Regional Development: Encouraging Enterprise in Scotland*, London, HMSO.

IDS (1988c), *New Life for Urban Scotland*, Edinburgh, IDS.

IDS & SDA (1988), 'Area Initiatives Evaluation Handbook, *ESU Research Paper* 16.

Ifversen, Jan (2000a), 'Tekster er kilder og kilder er tekster – Kildekritik og historisk tekstanalyse', *Den Jyske Historiker* (88): 149-74.

Ifversen, Jan (2000b), 'Begreber, diskurser og tekster omkring civilisation', in Torben Bech Dyrberg *et al.* (eds.), *Diskursteorien på arbejde*, Roskilde, Roskilde Universitetsforlag, pp. 189-220.

Imberg, David & Jim Northcott (1981), *Industrial Policy and Investment Decisions*, London, PSI.

Industry Act 1975, *Acts*, 68.

Industry Act 1980, *Acts*, 33.

Ingham, Geoffrey (1984), *Capitalism Divided? The City and Industry in British Social Development*, Houndmills, Macmillan.

Ingham, Keith & James Love (eds.) (1983), *Understanding the Scottish Economy*, Oxford, Martin Robertson.

Izushi, Hiro (1999), 'Can a Development Agency Foster Co-operation Among Local Firms? The Case of the WDA's Supplier Association Programme', *Regional Studies* 33 (8): 739-50.

Jackson, Peter M. (1992), 'Economic Policy', in David Marsh & R. A. W. Rhodes (eds.), *Implementing Thatcherite Policies: Audit of an Era*, Buckingham, Open UP, pp. 11-31.

Jenkins, Keith (1991), *Rethinking History*, London, Routledge.

Jenkins, Bill (1993), 'Policy Analysis – Models and Approaches' (orig. 1978), in Michael Hill (ed.), *The Policy Process – A Reader*, Hemel Hempstead, Harvester Wheatsheaf, pp. 34-44.

Jensen, Bernard Eric (1991), 'Det engelske imperium 1688-1850', *Den Jyske Historiker*: 27-70.

Jepperson, Ronald L. (1991), 'Institutions, Institutional Effects, and Institutionalism', in Walter W. Powell & Paul J. DiMaggio (eds.), *The New Institutionalism in Organizational Analysis*, Chicago, Chicago UP, pp. 143-63.

Jessop, Bob (1980), 'The Transformation of the State in Post-war Britain', in Richard Scase (ed.), *The State in Western Europe*, London, Croom Helm, pp. 23-93.

Jessop, Bob (1989), 'Conservative Regimes and the Transition to Post-Fordism: The Cases of Great Britain and West Germany', in Mark Gottdiener (ed.), *Capitalist Development and Crisis Theory*, Houndmills, Macmillan, pp. 261-99.

Jessop, Bob (1990), *State Theory – Putting the Capitalist State in its Place*, Cambridge, Polity.

Jessop, Bob (1991), 'Foreword: On Articulate Articulation', in René Bugge Bertramsen *et al.*: *State, Economy & Society*, London, Unwin Hyman, pp. xiv-xxvii.

Jessop, Bob *et al.* (1988), *Thatcherism. A Tale of Two Nations*, Cambridge, Polity.

Johnes, Geraint (1987), 'Regional Policy and Industrial Strategy in the Welsh Economy', *Regional Studies* 21 (6): 555-64.

Johnson, Björn (1992), 'Institutional Learning', in Bengt-Åke Lundvall: *National Systems of Innovation. Towards a Theory of Innovation and Interactive Learning*, London, Pinter, pp. 23-44.

Johnson, Björn & Bengt-Åke Lundvall (1989), 'Limits of the Pure Market Economy', in J. Bohlin *et al.* (eds.), *Samhällvetenskap, Ekonomi, Historia*, Göteborg, Daidalos, pp. 85-106.

Johnson, Nevil (1982), 'Accountability, Control and Complexity: Moving Beyond Ministerial Responsibility', in Anthony Barker (ed.), *Quangos in*

Britain. Government and the Networks of Public Policy, London, Macmillan, pp. 206-18.

Jones, Colin (1996), 'Property-led Local Economic Development Policies: From Advance Factory to English Partnerships and Strategic Property Investment?', *Regional Studies* 30 (2): 200-206.

Jones, J. (1986), 'An Examination of the Thinking behind Government Regional Policy in the UK since 1945', *Regional Studies* 20 (3): 261-66.

Jordan, Grant & Klaus Schubert (1992), 'A Preliminary Ordering of Policy Network Labels', *European Journal of Political Research* 21 (1-2): 7-27.

Jørgensen, Marianne Winther & Louise Phillips (1999), *Diskursanalyse som teori og metode*, Frederiksberg, Samfundslitteratur.

Junior Chamber Scotland (1989), *'Scottish Enterprise – A Response*, Glasgow, Junior Chamber Scotland.

Kavanagh, Dennis (1987), *Thatcherism and British Politics: The End of Consensus*? Oxford, Oxford UP.

Kearney, Hugh (1989), *The British Isles – A History of Four Nations*, Cambridge, Cambridge UP.

Kearney, Hugh (1991), 'Nationbuilding – British Style', *Culture & History* (9/10): 43-54.

Keat, Russel & Nicholas Abercrombie (eds.) (1991), *Enterprise Culture*, London, Routledge.

Keating, Michael (1988a), *State and Regional Nationalism. Territorial Politics and the European State*, Hemel Hempstead, Harvester-Wheatsheaf.

Keating, Michael (1988b), *The City that Refused to Die. Glasgow: The Politics of Urban Regeneration*, Aberdeen, Aberdeen UP.

Keating, Michael (1991), *Comparative Urban Politics*, London, Elgar.

Keating, Michael (1992), 'Regional Autonomy in the Changing State Order. A Framework for Analysis', *Regional Politics & Policy* 2 (3): 45-61.

Keating, Michael (1993), 'The Politics of Economic Development. Political Change and Local Development Policies in the USA, Britain and France', *Urban Affairs Quarterly* 38 (3): 373-96.

Keating, Michael (1995), 'Local Economic Development: Policy or Politics?', in Norman Walzer (ed.), *Local Economic Development: Incentives and International Trends*, Boulder, Westview, pp. 13-30.

Keating, Michael (1996), *Nations Against the State. The New Politics of Nationalism in Quebec, Catalonia and Scotland*, London, Macmillan.

Keating, Michael (1997), 'The Political Economy of Regionalism', in Michael Keating & John Loughlin (eds.), *The Political Economy of Regionalism*, London, Frank Cass, pp. 17-40.

Keating, Michael (1998), 'Is There a Regional Level of Government in Europe?', in Patrick Le Galès & Christian Lequesne (eds.), *Regions in Europe*, London, Routledge, pp. 11-29.

Keating, Michael & David Bleiman (1979), *Labour and Scottish Nationalism*, London, Macmillan.

Keating, Michael & Robin Boyle (1986), *Re-making Urban Scotland. Strategies for Local Economic Development*, Edinburgh, Edinburgh UP.

Keating, Michael & Liesbet Hooghe (1994), *The Politics of EC Regional Policy*, Paper presented at the ECPR Joint Session, Madrid 1994.

Keating, Michael & Arthur Midwinter (1983), *The Government of Scotland*, Edinburgh, Mainstream.

Kellas, James G. (1989), *The Scottish Political System, 4th ed.*, Cambridge, Cambridge UP.

Kenworthy, Lane (1990), 'Are Industrial Policy and Corporatism Compatible?', *Journal of Public Policy* 10 (3): 233-65.

Kern, Manfred (1990), 'Financial Tranfers versus Real Transfers: Competing Strategies for Regional Economic Development', in Hans-Jürgen Ewers & Jürgen Allesch (eds.), *Innovation and Regional Development*, Berlin, de Gruyter, pp. 187-97.

King, Norman (1986), *English Estates 1936-86. Novel and Unorthodox. Government-sponsored Industrial Estates in England over 50 Years*, Team Valley, English Estates.

Kirwan, Frank (1981), 'The SDA: Structure and Functions', *Studies in Public Policy* 81.

Kjørup, Søren (2001), 'Den ubegrundede skepsis – En kritisk diskussion af socialkonstruktivismens filosofiske grundlag', *Sosiologi i dag* 31 (2): 5-22.

Kooistra, Daphne H. (1998), 'Entrepreneurs and Business People in Urban Growth Coalitions – Place Attachment and Active Participation in Urban Economic Development', in Henrik Halkier, Mike Danson & Charlotte Damborg (eds.), *Regional Development Agencies in Europe*, London, Jessica Kingsley, pp. 271-87.

Koselleck, Reinhart (1990), 'Sprogændring og begivenhedshistorie', *Den Jyske Historiker* (50): 121-35.

Kramer, Daniel C. (1989), *State Capital and Private Enterprise. The Case of the UK National Enterprise Board*, London, Routledge.

Kristiansen, Søren Brøndum (1987), *Industripolitik – med særligt henblik paa danske erfaringer*, København, Samfundslitteratur.

Labour Party (1958), *Let Scotland Prosper. Labour's Plans for Scotland's Progress*, London, Labour Party.

Labour Party (1962), *Signposts for Scotland*, London, Labour Party.

Labour Party (1963-79), *Report of the Annual Conference*, London, Labour.

Labour Party (1966), *Time for Decision. Manifesto of the Labour Party for the 1966 General Election*, London, Labour Party.

Labour Party (1982), *Alternative Regional Strategy – A Framework for Discussion*, London, Labour.

Labour Party (1989), *Meet the Challenge – Make the Change. A New Agenda for Britain. Final Report of Labour's Policy Review for the 1990s*, London, Labour.

Labour Party Scottish Council (1972), *Scottish Industry. An Outline of Industrial Policy*, Glasgow, Labour Party Scottish Council.

Labour Party Scottish Council (1973), *Scotland and the National Enterprises Board*, Glasgow, Labour Party Scottish Council.

Labour Party Scottish Council (1977), *An Industrial Strategy for Scotland*, Glasgow, Labour Party Scottish Council.

Laclau, Ernesto & Chantal Mouffe (1985), *Hegemony and Socialist Strategy*, London, Verso.

Lagendijk, Arnoud (2003), 'Towards Conceptual Quality in Regional Studies: The Need for Subtle Critique – A Response to Markusen', *Regional Studies* 37 (6/7): 719-28.

Lagendijk, Arnoud & James Cornford (2000), 'Regional Institutions and Knowledge – Tracking New Forms of Regional Development Policy', *Geoforum* 31 (2): 209-18.

Lambooy, Jan G. & Ron A. Boschma (2001), 'Evolutionary Economics and Regional Policy', *The Annals of Regional Science* 35 (1): 113-32.

Lane, Jan-Erik (1995), *The Public Sector. Concepts, Models and Approaches, 2nd ed.*, London, Sage.

Lane, Jan-Eric & Svante O. Ersson (1994), *Politics and Society in Western Europe, 3rd ed.*, London, Sage.

Lang, Ian (2002), *Blue Remembered Years – A Political Memoir*, London, Politico's.

Larsson, Mats (1988), 'The History of Regional Policy in Britain and Sweden. A Comprehensive Study of the Link between Economic and Political Decisions', *Working Paper from CERUM* 88/25.

Lauria, Mickey (ed.) (1997), *Reconstructing Urban Regime Theory*, Thousand Oaks, Sage.

Lawless, Paul (1988), 'Enterprise Boards: Evolution and Critique', *Planning Outlook* 31 (1): 13-18.

Lawless, Paul (1991), 'Urban Policy in the Thatcher Decade: English Inner-city Policy, 1979-90', *Environment & Planning C: Government and Policy* 9: 15-30.

Lawless, Paul & Graham Haughton (1992), 'Urban Policy Initiatives: Trends and Prospects', in Peter Townroe & Ron Martin (eds.), *Regional Development in the 1990s – The British Isles in Transition*, London, Jessica Kingsley, pp. 303-8.

Leclerc, Roger & Donald Draffan (1984), 'The GEAR Project', *Town Planning Review* 55 (3): 335-51.

Lee, Clive (1995), *Scotland and the United Kingdom. The Economy and the Union in the 20th Century*, Manchester, Manchester UP.

Lee, Simon (1996), 'Manufacturing', in David Coates (ed.), *Industrial Policy in Britain*, Houndmills, Macmillan, pp. 33-61.

Leruez, Jacques (1982), *L'Écosse, une nation sans État*, Lille, PU de Lille.

Letwin, Shirley Robin (1992), *The Anatomy of Thatcherism*, London, Fontana.

Levacic, Rosalind (1993), 'Markets as Coordinative Devices', in Richard Maidment & Grahame Thompson (eds.), *Managing the UK. An Introduction to its Political Economy and Public Policy*, London, Sage, pp. 30-50.

Lever, William F. (1987), 'Urban Policy', in William F. Lever (ed.), *Industrial Change in the United Kingdom*, London, Longman, pp. 240-57.

Lever, William F. (1989), 'International Comparative Aspects of Government Intervention in Local Economic Policy', in David Gibbs (ed.), *Government Policy and Industrial Change*, London, Routledge, pp. 209-31.

Lever, William & Chris Moore (eds.) (1986), *The City in Transition. Policies and Agencies for the Regeneration of Clydeside*, Oxford, Clarendon.

Leys, Colin (1989), *Politics in Britain. From Labourism to Thatcherism*, London, Verso.

Lindblom, Charles E. (1977), *Politics and Markets. The World's Political-Economic Systems*, New York, Basic.

Linder, Stephen H. & B. Guy Peters (1989), 'Instruments of Government: Perceptions and Contexts', *Journal of Public Policy* 9 (1): 35-58.

Lipsky, Michael (1980), *Street-level Bureaucracy*, New York, Russell Sage.

Lloyd, Greg (1997), 'Structure and Culture: Regional Planning and Institutional Innovation in Scotland', in Roderick Macdonald & Huw Thomas (eds.), *Nationality and Planning in Scotland and Wales*, Cardiff, University of Wales Press, pp. 113-32.

Lloyd, Greg & Rowan-Robinson, J. (1988), 'Local Authority Responses to Economic Uncertainty in Scotland', *The Scottish Government Yearbook* 1988: 282-300.

Locate in Scotland (1988-91), *Annual Report*, Glasgow, LIS.

Lochhead, Elspeth (1983), *Regional Development Agencies in the UK: Their Role in Small Firm Development*, Glasgow, Centre for the Study of Public Policy, University of Strathclyde.

Logan, John R. & Harvey L. Molotch (1987), *Urban Fortunes: The Political Economy of Place*, Berkeley, California UP.

Lothian Regional Council (1989), *Scottish Enterprise White Paper*, Edinburgh, Lothian Regional Council.

Loughlin, John & B. Guy Peters (1997), 'State Traditions, Administrative Reform and Regionalization', in Michael Keating & John Loughlin (eds.), *The Political Economy of Regionalism*, London, Frank Cass, pp. 41-32.

Love, James & James Stevens (1988), *The Scottish Development Agency and Regional Policy*, Working Paper, Department of Economics, University of Strathclyde, Glasgow.

Love, James & James Stevens (1989), 'The Scottish Steel Industry', *The Scottish Government Yearbook* 1989: 51-69.

Love, Jim (1984), 'SDA Annual Report 1984 – Review', *Quarterly Economic Commentary* 10 (1): 82-84.

Ludlam, Steve & Martin J. Smith (eds.) (1995), *Contemporary British Conservatism*, Houndmills, Macmillan.

Lythe, Charlotte & Madhavi Majmudar (1982), *The Renaissance of the Scottish Economy*? London, Allen & Unwin.

MacDonald, Mary & Adam Redpath (1980), 'The Scottish Office 1954-79', *The Scottish Government Yearbook* 1980: 101-34.

Macdonald, Roderick & Huw Thomas (1997), 'Nationality and Planning', in Roderick Macdonald & Huw Thomas (eds.), *Nationality and Planning in Scotland and Wales*, Cardiff, University of Wales Press, pp. 1-13.

Mackay, Tony (1984), 'The Oil and Oil-related Sector', in Neil Hood & Stephen Young (eds.), *Industry, Policy and the Scottish Economy*, Edinburgh, Edinburgh UP, pp. 326-64.

MacLennan, M. C. (1979), 'Regional Policy in a European Framework', in Duncan Maclennan & John B. Parr (eds.), *Regional Policy. Past Experiences and New Directions*, Robertson, Oxford, pp. 245-271.

Maclennan, Duncan & John B. Parr (1979), 'Postscript', in Duncan Maclennan & John B. Parr (eds.), *Regional Policy. Past Experiences and New Directions*, Robertson, Oxford, pp. 323-29.

Maclennan, Duncan & John B. Parr (eds.) (1979), *Regional Policy. Past Experiences and New Directions*, Oxford, Robertson.

MacLeod, Gordon (1996), 'The Cult of Enterprise in a Networked, Learning Region? Governing Business and Skills in Lowland Scotland', *Regional Studies* 30 (8): 749-56.

MacLeod, Gordon (1997), ''Institutional Thickness' and Industrial Governance in Low-land Scotland', *Area* 29 (4): 299-311.

MacLeod, Gordon (1998a), 'Entrepreneurial Spaces, Hegemony and State Strategy: The Political Shaping of Privatism in Lowland Scotland', *Environment & Planning A* 31 (2): 345-75.

MacLeod, Gordon (1998b), 'In What Sense a Region? Place Hybridity, Symbolic Shape, and Institutional Formation in (Post-)Modern Scotland', *Political Geography* 17 (7): 833-63.

MacLeod, Gordon (1999), 'Place, Politics and 'Scale Dependence'. Exploring the Structuration of Euro-Regionalism', *European Urban and Regional Studies* 6 (3): 231-53.

MacLeod, Gordon & Martin Jones (1999), 'Reregulating a Regional Rustbelt: Institutional Fixes, Entrepreneurial Discourse, and the "Politics of Representation"', *Environment & Planning D* 17 (5): 575-605.

Macwhirter, Iain (1991), 'The Year at Westminster', *The Scottish Government Yearbook* 1991: 12-28.

March, James G. & Johan P. Olsen (1984), 'The New Institutionalism: Organizational Factors in Political Life', *American Political Science Review* 78 (3): 734-49.

March, James G. & Johan P. Olsen (1989), *Rediscovering Institutions: The Organizational Base of Politics*, New York, Free Press.

Mariussen, Åge (2001), 'Introduction', in Åge Mariussen (ed.), *Cluster Policies – Cluster Development*, Stockholm, Nordregio, pp. 9-22.

Marks, Gary (1996), 'An Actor-Centred Approach to Multi-Level Governance', *Regional & Federal Studies* 6 (2): 20-38.

Marks, Gary *et al.* (1996), 'European Integration from the 1980s: State-Centric v. Multi-level Governance', *Journal of Common Market Studies* 34 (3): 341-78.

Markusen, Ann (1999), 'Fuzzy Concepts, Scanty Evidence, Policy Distance: The Case for Rigour and Policy Relevance in Critical Regional Studies', *Regional Studies* 33 (9): 869-84.

Marplan (1983), *Report on a Survey of Certain Scottish Businessmen*, Glasgow, Hall Advertising.

Marsh, David (1995), 'Explaining 'Thatcherite' Policies: Beyond Unidimensional Explanation', *Political Studies* 43 (4): 595-613.

Marsh, David & R. A. W. Rhodes (1992), 'The Implementation Gap: Explaining Policy Change and Continuity', in David Marsh & R. A. W. Rhodes (eds.), *Implementing Thatcherite Policies: Audit of an Era*, Buckingham, Open UP, pp. 170-87.

Marsh, David & R. A. W. Rhodes (eds.) (1992), *Implementing Thatcherite Policies: Audit of an Era*, Buckingham, Open UP.

Martin, Ron (1985), 'Monetarism Masquerading as Regional Policy? The Government's New System of Regional Aid', *Regional Studies* 19 (4): 379-88.

Martin, Ron (1986), 'Thatcherism and Britain's Industrial Landscape', in Bob Rowthorn & Ron Martin (eds.), *The Geography of De-industrilisation*, Macmillan, Houndmills, pp. 238-91.

Martin, Ron (1989a), 'The New Economics and Politics of Regional Restructuring: The British Experience', in Louis Albrechts *et al.* (eds.), *Regional Policy at the Crossroads – European Perspectives*, London, Jessica Kingsley, pp. 27-51.

Martin, Ron (1989b), 'Regional Imbalance as Consequence and Constraint in National Economic Revival', in Francis Green (ed.), *The Restructuring of the UK Economy*, Hemel Hempstead, Harvester, pp. 80-100.

Martin, Ron (1989c), 'Deindustrialisation and State Intervention: Keynesianism, Thatcherism and the Political Economy of Thatcherism', in John Mohan (ed.), *The Political Geography of Contemporary Britain*, Houndmills, Macmillan, pp. 87-112.

Martin, Ron (1992a), 'Financing Regional Enterprise: The Role of the Venture Capital Market', in Peter Townroe & Ron Martin (eds.), *Regional Development in the 1990s – The British Isles in Transition*, London, Jessica Kingsley, pp. 161-69.

Martin, Ron (1992b), 'The Economy', in Paul Cloke (ed.), *Policy and Change in Thatcher's Britain*, Oxford, Pergamon, pp. 123-56.

Martin, Ron (1993), 'Reviving the Economic Case for Regional Policy', in Richard T. Harrison & Mark Hart (eds.), *Spatial Policy in a Divided Nation*, London, Jessica Kingsley, pp. 270-90.

Martin, Ron & J. S. C. Hodge (1983a), 'The Reconstruction of British Regional Policy: 1. The Crisis of Conventional Practice', *Environment & Planning C: Government and Policy* 1: 133-52.

Martin, Ron & J. S. C. Hodge (1983b), 'The Reconstruction of British Regional Policy: 2. Towards a New Agenda', *Environment & Planning C: Government and Policy* 1 (3): 317-40.

Martin, Ron & Peter Townroe (1992), 'Changing Trends and Pressures in Regional Development', in Peter Townroe & Ron Martin (eds.), *Regional Development in the 1990s – The British Isles in Transition*, London, Jessica Kingsley, pp. 13-24.

Martin, Ron & Peter Tyler (1992), 'The Regional Legacy', in Jonathan Michie (ed.), *The Economic Legacy 1979-1992*, London, Academic Press, pp. 140-67.

Martin, Stephen (1990), 'City Grants, Urban Development Grants and Urban Regeneration Grants', in Mike Campbell (ed.), *Local Economic Policy*, London, Cassell, pp. 44-64.

Martin, Steve *et al.* (1990), 'Rural Development Programmes in Theory and Practice', *Regional Studies* 24 (3): 268-76.

Maskell, Peter & Anders Malmberg (1999), 'The Competitiveness of Firms and Regions. 'Ubiquitification' and the Importance of Localised Learning', *European Urban and Regional Studies* 6 (1): 9-25.

Mason, Colin (1987), 'The Small Firm Sector', in William F. Lever (ed.), *Industrial Change in the United Kingdom*, London, Longman, pp. 125-48.

Mason, Colin (1992), 'New Firm Formation and Growth', in Peter Townroe & Ron Martin (eds.), *Regional Development in the 1990s – The British Isles in Transition*, London, Jessica Kingsley, pp. 150-60.

Mason, Colin & Richard T. Harrison (1989), 'Small Firms Policy and the 'North-South' Divide in the UK: The Case of the Business Expansion Scheme', *Transactions, Institute of British Geographers, New Series* 14 (1): 37-58.

Massey, Doreen (1984), *Spatial Divisions of Labour. Social Structures and the Geography of Production*, Houndmills, Macmillan.

McAleavey, Paul (1993), 'The Politics of European Regional Developent Policy: Additionality in the Scottish Coalfields', *Regional Politics & Policy* 3 (2): 88-107.

McAleavey, Paul (1995), 'ERDF Expenditure in the UK: From Additionality to "Subtractionality"', *European Urban and Regional Studies* 2 (3): 249-53.

McArthur, Andrew A. (1987), 'Jobs and Incomes', in David Donnison & Alan Middleton (eds.), *Regenerating the Inner City – Glasgow's Experience*, London, RKP, pp. 72-93.

McAteer, Mark & Duncan Mitchell (1996), 'Peripheral Lobbying! The Territorial Dimension of Euro-lobbying by Scottish and Welsh Sub-Central Government', *Regional & Federal Studies* 6 (3): 1-27.

McCallum, J. D. (1979), 'The Development of British Regional Policy', in Duncan Maclennan & John B. Parr (eds.), *Regional Policy. Past Experiences and New Directions*, Robertson, Oxford, pp. 3-41.

McCreadie, Robert (1982), 'Scottish Tories at Mid-term: A Review of Scottish Office Ministers and Policies 1979-81', *The Scottish Government Yearbook* 1982: 69-88.

McCrone, David (1991), ''Excessive and Unreasonable': The Politics of the Poll Tax in Scotland', *International Journal of Urban and Regional Research* 15 (3): 443-52.

McCrone, Gavin (1969), *Regional Policy in Britain*, London, George Allen & Unwin.

McCrone, Gavin (1985), 'The Role of Government', in Richard Saville (ed.), *The Economic Development of Modern Scotland 1950-1980*, Edinburgh, Donald, pp. 195-213.

McCrone, Gavin (1991a), 'Urban Renewal: The Scottish Experience', *Urban Studies* 28 (6): 919-38.

McCrone, Gavin & John N. Randall (1985), 'The Scottish Development Agency', in Richard Saville (ed.), *The Economic Development of Modern Scotland 1950-1980*, Edinburgh, Donald, pp. 233-44.

McDermott, Philip (1979), 'Multinational Manufacturing Firms and Regional Development: External Control in the Scottish Electronics Industry', *Scottish Journal of Political Economy* 26 (3): 287-306.

McGreevy, T. E. & A. W. J. Thomson (1983), 'Regional Policy and Company Investment Behaviour', *Regional Studies* 17 (5): 347-58.

McKillop, Donal G. & Robert W. Hutchinson (1990), *Regional Financial Sectors in the British Isles*, Aldershot, Avebury.

McQuaid, Ron (1992), *Local Authorities and Economic Development in Scotland*, Edinburgh, COSLA.

Metcalf, Hilary *et al.* (1990), 'The Charitable Role of Companies in Job Creation', *Regional Studies* 24 (3): 261-68.

Middlemas, Keith (1989a), *Power, Competition and the State, vol. 1: Britain in Search of Balance, 1940-61*, Houndmill, Macmillan.

Middlemas, Keith (1989b), *Power, Competition and the State, vol. 2: Threats to the Postwar Settlement, 1961-74*, Houndmills, Macmillan.

Middlemas, Keith (1991), *Power, Competition and the State, vol. 3: The End of the Postwar Era. Britain since 1974*, Houndmills, Macmillan.

Midwinter, Arthur *et al.* (1991), *Politics and Public Policy in Scotland*, Houndmills, Macmillan.

Milgrom, Paul & John Roberts (1992), *Economics, Organization and Management*, Englewood Cliffs, Prentice Hall.

Millan, Bruce (1981), 'Scotland, the Scottish Office and the UK Economy', in Christopher Blake &. Edgar Lythe (eds.), *A Maverick Institution*, London, Gee, pp. 97-109.

Miller, David (1990), 'The Future of Local Economic Policy: A Public and Private Sector Function', in Mike Campbell (ed.), *Local Economic Policy*, London, Cassell, pp. 195-211.

Minns, Richard (1992), *The Value Added by Community Interventions to Enhance the Access of SMEs to Capital, in the Context of Regional Policy*, Greater London Enterprise.

Mitchell, Jeremy (1993), 'Coordination by Hierarchy', in Richard Maidment & Grahame Thompson (eds.), *Managing the UK. An Introduction to its Political Economy and Public Policy*, London, Sage, pp. 5-29.

Modeen, Tore & Allan Rosas (eds.) (1988), *Indirect Public Administration in 14 Countries*, Aabo, Aabo Academy Press.

Molotch, Harvey (1976), 'The City as a Growth Machine: Towards a Political Economy of Place', *American Journal of Sociology* 82 (2): 309-32.

Moore, Barry & John Rhodes (1974), 'Regional Policy and the Scottish Economy', *Scottish Journal of Political Economy* 21 (3): 215-35.

Moore, Barry *et al.* (1986), *The Effects of Government Regional Economic Policy*, London, DTI.

Moore, Chris (1988a), 'Modifying the Market: Regional Economic Development Strategies in Scotland and England', *Regional and Industrial Policy Research Paper Series, EPRC, University of Strathclyde* 2.

Moore, Chris (1988b), 'Enterprise Agencies: Privatisation or Partnership?', *Local Economy* 3 (1): 21-30.

Moore, Chris (1989), 'The Hughes Initiative: The Blueprint for Enterprise?', *Local Economy* 3 (4): 237-43.

Moore, Chris (1990), 'Scottish Enterprise: Evolution or Revolution of Enterprise Training', *Regional and Industrial Policy Research Paper Series, EPRC, University of Strathclyde* 6.

Moore, Chris (1994), 'Ny-institutionalisme og regional udvikling', *GRUS* 43: 46-53.

Moore, Chris (1995), 'Scotland and the SDA', in Martin Rhodes (ed.), *The Regions and the New Europe*, Manchester, Manchester UP, pp. 229-46.

Moore, Chris & Simon Booth (1984), 'Unlocking Enterprise: Policy Innovation in Scotland', *Studies in Public Policy* 137.

Moore, Chris & Simon Booth (1986a), 'The Post-Industrial Synthesis: Policies for Enterprise in Clydeside', in William Lever & Chris Moore (eds.), *The City in Transition. Policies and Agencies for the Regeneration of Clydeside*, Oxford, Clarendon, pp. 62-75.

Moore, Chris & Simon Booth (1986b), 'Unlocking Enterprise: the Search for Synergy', in William Lever & Chris Moore (eds.), *The City in Transition. Policies and Agencies for the Regeneration of Clydeside*, Oxford, Clarendon, pp. 92-106.

Moore, Chris & Simon Booth (1986c), 'The Pragmatic Approach: Local Political Models of Regeneration', in William Lever & Chris Moore (eds.), *The City in Transition. Policies and Agencies for the Regeneration of Clydeside*, Oxford, Clarendon, pp. 107-19.

Moore, Chris & Simon Booth (1986d), 'From Comprehensive Regeneration to Privatisation: The Search for Effective Area Strategies', in William Lever & Chris Moore (eds.), *The City in Transition. Policies and Agencies for the Regeneration of Clydeside*, Oxford, Clarendon, pp. 76-91.

Moore, Chris & Simon Booth (1986e), 'Urban Policy Contradictions: The Market Versus Redistributive Approaches', *Policy and Politics* 14 (3): 361-87.

Moore, Chris & Simon Booth (1989), *Managing Competition. Meso-Corporatism, Pluralism, and the Negotiated Order in Scotland*, Oxford, Clarendon.

Moore, Chris & Jon Pierre (1988), 'Partnership or Privatisation? The Political Economy of Local Economic Restructuring', *Policy and Politics* 16 (1): 169-78.

Moore, Chris & Jeremy J. Richardson (1988), 'The Politics and Practice of Corporate Responsibility in Great Britain', *Research in Corporate Social Performance and Policy* 10: 267-90.

Moran, Michael (1988), 'Industrial Relations', in Henry Drucker *et al.* (eds.), *Developments in British Politics 2, Revised Edition*, Houndmills, Macmillan, pp. 279-94.

Morgan, Bob (1996), 'An Endogenous Approach to Regional Economic Development: The Emergence of Wales', *European Planning Studies* 4 (6): 705-15.

Morgan, Kevin (1985), 'Regional Regeneration in Britain: The Territorial Imperative and the Conservative State', *Political Studies* 33: 560-77.

Morgan, Kevin (1994), 'The Fallible Servant – Making Sense of the WDA', *Papers in Planning Research, Department of City and Regional Planning, University of Cardiff* 151.

Morgan, Kevin (1997), 'The Learning Region: Institutions, Innovation and Regional Renewal', *Regional Studies* 31 (5): 491-504.

Morgan, Kevin (1998), 'Regional Renewal – The Development Agency as Animateur', in Henrik Halkier, Mike Danson & Charlotte Damborg (eds.), *Regional Development Agencies in Europe*, London, Jessica Kingsley, pp. 229-52.

Morgan, Kevin & Dylan Henderson (1997), 'The Fallible Servant: Evaluating the WDA', in Roderick Macdonald & Huw Thomas (eds.), *Nationality and Planning in Scotland and Wales*, Cardiff, University of Wales Press, pp. 77-97.

Morgan, Kevin & Andrew. Sayer (1983), 'Regional Inequality and the State in Britain', in James Anderson *et al.* (eds.), *Redundant Spaces in Cities and Regions? Studies in Industrial Decline and Social Change*, London, Academic Press, pp. 17-49.

Morison, Hugh (1987), *The Regeneration of Local Economies*, Oxford, Clarendon.

Morris, Paul (1991), 'Freeing the Spirit of Enterprise: The Genesis and Development of the Concept of Enterprise Culture', in Russel Keat & Nicholas Abercrombie (eds.), *Enterprise Culture*, London, Routledge, pp. 21-37.

Mottershead, P. (1978), 'Industrial Policy', in Frank Blackaby (ed.), *British Economic Policy 1960-74*, London, Cambridge UP, pp. 418-83.

Mullard, Maurice (1992), *Understanding Economic Policy*, London, Routledge.

Nairn, Tom (1977), 'The Twilight of the British State', *New Left Review* (101/102): 3-61.

Nairn, Tom (1981), 'The Crisis of the British State', *New Left Review* (130): 37-44.

National Audit Office (1985), 'Investment Activities of the SDA, WDA and HIDB', *HC Papers* 230, 20.2.85.

National Audit Office (1988), 'SDA: Involvement with the Private Sector', *HC Papers* 478, 5.5.88.

National Audit Office (1989), 'Locate in Scotland', *HC Papers* 300, 7.4.89.

National Audit Office (1994), 'Land Reclamation in Wales', *HC Papers* 461, 16.6.94.

Naughtie, James (1981), 'The Year at Westminster', *The Scottish Government Yearbook* 1981: 26-38.

Naughtie, James (1985), 'The Year at Westminster', *The Scottish Government Yearbook* 1985: 23-29.

Naylor, Derek (1984), 'Local Economic Initiatives Study: SDA', *University of Birmingham: Centre for urban and Regional Studies, Research Memorandum* 99.

Neumann, Iver B. (1999), 'Diskursanalyse av politikk: Forutsetninger og metodeproblemer', *Statsvetenskaplig Tidskrift* 102 (2): 163-81.

Neumann, Wolfgang & Henrik Uterwedde (1986), *Industriepolitik: Ein deutch-französischer Vergleich*, Opladen, Leske & Budrich.

Neuperts, Helmuth (1986), *Regionale Strukturpolitik als Aufgabe der Länder. Grundlagen, Verknüpfungen, Grenzen*, Baden-Baden, Nomos.

Newton, Scott & Dilwyn Porter (1988), *Modernization Frustrated: The Politics of Industrial Decline in Britain Since 1900*, London, Unwin Hyman.

Nicholls, David (1988), 'Fractions of Capital: The Aristocracy, the City and Industry in the Development of Modern British Capitalism', *Social History* 13 (1): 71-83.

Nicol, William & Douglas Yuill (1982), 'Regional Problems and Policy', in Andrea Boltho (ed.), *The European Economy. Growth and Crisis*, Oxford, Oxford UP, pp. 409-45.

Nielsen, Henrik Kaare (1994), 'Civilsamfund og demokratisering', *GRUS* 44: 40-54.

Nielsen, Klaus & Ove K. Pedersen (1988), 'The Negotiated Economy: Ideal and History', *Scandinavian Political Studies* 11 (2): 79-101.

Niss, Hanne (1991), 'Thatcherism – An Introduction', *Publications of the Department of Languages and Intercultural Studies, Aalborg University* (4): 9-19.

Nolan, Peter (1989), 'The Productivity Miracle?', in Francis Green (ed.), *The Restructuring of the UK Economy*, Hemel Hempstead, Harvester, pp. 101-21.

NordREFO (1978), 'Målkonflikter i regionalpolitikken', *NordREFO* 9 (4).

North, Douglas C. (1990a), *Institutions, Institutional Change and Economic Performance*, Cambridge, Cambridge UP.

North, Douglas C. (1990b), 'A Transaction Cost Theory of Politics', *Journal of Theoretical Politics* 2 (4): 355-67.

O'Farrell, P. N. *et al.* (1992), 'The Competitiveness of Business Service Firms: A Matched Comparison between Scotland and the South East of England', *Regional Studies* 26 (6): 519-33.

O'Toole, Mo (1996), *Regulation Theory and the British State. The Case of the Urban Development Corporation*, Aldershot, Avebury.

OECD (1977), *Selected Industrial Policy Instruments – Objectives and Scopes*, Paris, OECD.

OECD (1989), *Regional Policy Developments in OECD Countries*, Paris, OECD.

OECD (1994), *Regional Problems and Policies in the UK*, Paris, OECD.

Olsson, Jan (1995), 'Den lokala näringspolitikens politiske ekonomi. En jämföranda kommuntypstudie', *Örebro Studies* 12.

Oscarsson, Gösta (1988), 'Sammanfattande analys og förslag till fortsatt forskning', in NordREFO (ed.), *Om regionalpolitikken som politikområde*, Helsinki, NordREFO, pp. 11-66.

Ostrom, Elinor (1986), 'An Agenda for the Study of Institutions', *Public Choice* 48 (1): 3-25.

Ougaard, Morten (2000), 'Det kommer an på hvor man begynder. Rhoy Bhaskars kritiske realisme som erkendelsesteoretisk strategi', *GRUS* 60: 17-34.

Overbeek, Henk (1990), *Global Capitalism and National Decline: The Thatcher Decade in Perspective*, London, Unwin Hyman.

Page, Edward (1977), The Transformation of Decisions into Activities: The SDA as a Case Study, Glasgow, MSc Dissertation, Department of Politics, Strathclyde University.

Page, Edward C. & Michael J. Goldsmith (1987), 'Centre and Locality: Explaining Crossnational Variation', in Edward C. Page & Michael J. Goldsmith (eds.), *Central and Local Government Relations*, London, Sage, pp. 158-68.

Parsons, Wayne (1988), *The Political Economy of British Regional Policy*, Routledge, London.

Parsons, Wayne (1995), *Public Policy. An Introduction to the Theory and Practice of Policy Analysis*, Aldershot, Elgar.

Paterson, Lindsay (1994), *The Autonomy of Modern Scotland*, Edinburgh, Edinburgh UP.

Payne, Peter L. (1985), 'The Decline of the Scottish Heavy Industries, 1945-83', in Richard Saville (ed.), *The Economic Development of Modern Scotland 1950-1980*, Edinburgh, Donald, pp. 79-113.

Payne, Peter L. (1996), 'The Economy', in Thomas M. Devine & Richard J. Finlay (eds.), *Scotland in the 20th Century*, Edinburgh, Edinburgh UP, pp. 13-45.

Peck, Frank (1996), 'Regional Development and the Production of Space: The Role of Infrastructure in the Attraction of new Inward Investment', *Environment & Planning A* 28: 327-39.

Pedersen, Ove K. (1990), '..Og der var 10 og 11 og mange fler!', *GRUS* 30: 97-122.

Pedersen, Ove K. *et al.* (1992), 'Private Policies and the Autonomy of Enterprise: Danish Local and National Industrial Policy', *Journal of Economic Issues* 26 (4): 1117-44.

Petr, Jerry L. (1984), 'Fundamentals of an Institutionalist Perspective on Economic Policy', in Marc R. Tool (ed.), *An Institutionalist Guide to Economics and Public Policy*, Armonk, Sharpe, pp. 1-17.

Phelps, Nick (1992), 'From Local Economic Dependance to Local Economic Development?: The Case of the Scottish Electronics Industry', *Papers in Planning Research* 136.

Phelps, Nicholas A. & Mark Tewdwr-Jones (2001), 'Globalisation, Regions and the State: Exploring the Limitations of Economic Modernisation through Inward Investment', *Urban Studies* 38 (8): 1253-72.

Phelps, Nick *et al.* (2003), 'Embedding the Multinationals? Institutions and the Development of Overseas Manufacturing Affiliates in Wales and North East England', *Regional Studies* 37 (1): 27-40.

Phillips, Louise (1996), 'Rhetoric and the Spread of the Discourse of Thatcherism', *Discourse and Society* 7 (2): 209-41.

Pickvance, C. G. (1990), 'Introduction: The Institutional Context of Local Economic Development: Central Controls, Spatial Policies and Local Economic Policies', in Michael Harloe *et al.* (eds.), *Place, Policy and Politics: Do Localities Matter*? London, Unwin Hyman, pp. 1-41.

Pieda (1992), *Strathclyde Integrated Development Operation 1988-92 – Interim Evaluation*, Edinburgh, Pieda.

Pinder, John (1982), 'Causes and Kinds of Industrial Policy', in John Pinder (ed.), *National Industrial Strategies and the World Economy*, London, Allanheild/Osmun, pp. 41-52.

Pollard, Sidney (1983), *The Development of the British Economy 1914-1980, 3rd ed.*, London, Edward Arnold.

Pollard, Sidney (1984), *The Wasting of the British Economy. Economic Policy 1945 to the Present, 2nd ed.*, London, Croom Helm.

Pottinger, George (1979), *The Secretary of State for Scotland 1926-76. Fifty Years of the Scottish Office*, Edinburgh, Scottish Academic Press.

Pratt, Andy C. & Rick Ball (1994), 'Industrial Property, Policy and Economic Development: The Research Agenda', in Rick Ball & Andy C. Pratt (eds.), *Industrial Property: Policy and Economic Development*, London, Routledge, pp. 1-19.

Prestwich, Roger & Peter Taylor (1990), *Introduction to Regional and Urban Policy in the UK*, Harlow, Longman.

Pugh, Martin (1994), *State and Society. British Political and Social History 1870-1992*, London, Edward Arnold.

Radice, Hugo (1978), 'On the SDA and the Contradictions of State Entrepreneurship', *Stirling University Discussion Papers in Economics* 59.

Raines, Philip (1998), 'Regions in Competition: Inward Investment, Institutional Autonomy and Regional Variation in the Use of Financial Incentives', *Regional and Industrial Policy Research Paper Series, EPRC, University of Strathclyde* 29.

Raines, Philip (1999), 'Financial Incentives', in Philip Raines & Ross Brown (eds.), *Policy Competition and Foreign Direct Investment in Europe*, Ashgate, Aldershot, pp. 65-88.

Raines, Philip (2001), 'The Cluster Approach and the Dynamics of Regional Policy-making', *Regional and Industrial Policy Research Paper Series, EPRC, University of Strathclyde* 47.

Raines, Philip & Fiona Wishlade (1997), 'Cross-European Perspectives on the Use and Control of Financial Incentives in Foreign Inward Investment Promotion', in Mike Danson (ed.), *Regional Governance and Economic Development*, London, Pion, pp. 155-72.

Raines, Philip *et al.* (2001), 'Growing Global: Foreign Direct Investment and the Internationalisation of Local Suppliers in Scotland', *European Planning Studies* 9 (8): 965-78.

Randall, John N. (1980), 'Central Clydeside – A Case Study of One Conurbation', in G. C. Cameron (ed.), *The Future of the British Conurbation*, London, Longman, pp. 54-71.

Randall, John N. (1985), 'New Towns and New Industries', in Richard Saville (ed.), *The Economic Development of Modern Scotland 1950-1980*, Edinburgh, Donald, pp. 245-70.

Rees, Gareth (1997), 'The Politics of Regional Development Strategy: The Programme for the Valleys', in Roderick Macdonald & Huw Thomas (eds.), *Nationality and Planning in Scotland and Wales*, Cardiff, University of Wales Press, pp. 98-112.

Rees, Gareth & Kevin Morgan (1991), 'Industrial Restructuring, Innovation Systems and the Regional State: South Wales in the 1990s', in Graham Day & Gareth Rees (eds.), *Regions, Nations and European Integration: Remaking the Celtic Periphery*, Cardiff, University of Wales Press, pp. 155-76.

Regional Studies Association (1983), *Report of an Inquiry into Regional Problems in the UK 1983*, Norwich, Geo Books.

Rhodes, R. A. W. (1988), *Beyond Westminster and Whitehall – The Sub-central Governments of Britain*, London, Unwin Hyman.

Rhodes, R. A. W. (1995), 'The Institutional Approach', in David Marsh & Gerry Stoker (eds.), *Theory and Method in Political Science*, Houndmills, Macmillan, pp. 42-57.

Rhodes, R. A. W. (1997), *Understanding Governance*, Buckingham, Oxford UP.

Rhodes, R. A. W. & David Marsh (1992a), 'New Directions in the Study of Policy Networks', *European Journal of Political Research* 21 (1-2): 181-95.

Rhodes, R. A. W. & David Marsh (1992b), ''Thatcherism': An Implementation Perspective', in David Marsh & R. A. W. Rhodes (eds.), *Implementing Thatcherite Policies: Audit of an Era*, Buckingham, Open UP, pp. 1-10.

Rich, David C. (1983), 'The Scottish Development Agency and the Industrial Regeneration of Scotland', *The Geographical Review* 73: 271-86.

Richardson, Harry W. (1984), 'Regional Policy in a Slow-growth Economy', in George Demko (ed.), *Regional Development Problems and Policies in Eastern and Western Europe*, London, Croom Helm, pp. 258-81.

Riddell, Peter (1991), *The Thatcher Era. And its Legacy*, Oxford, Blackwell.

Ridley, F. F. & David Wilson (eds.) (1995), *The Quango Debate*, Oxford, Oxford UP.

Robertson, Lewis (1978), 'The Scottish Development Agency, *The Scottish Government Yearbook* 1978: 15-31.

Rogers, Alan (1999), *The Most Revolutionary Measure*, Salisbury, Rural Development Commission.

Rokkan, Stein & Derek W. Urwin (1983), *Economy, Territory, Identity: Politics of West European Peripheries*, London, Sage.

Rosas, Allan (1988), 'Indirect Public Administration: General Comments', in Tore Modeen & Allan Rosas (eds.), *Indirect Public Administration in 14 Countries*, Åbo, Åbo Academy Press, pp. 31-38.

Rose, Richard (1982), *Understanding the United Kingdom: The Territorial Dimension in Government*, London, Longman.

Rose, Richard & Phillip L. Davies (1994), *Inheritance in Public Policy. Change Without Choice in Britain*, New Haven, Yale UP.

Ross, J. M. (1981), 'The Secretary of State for Scotland and the Scottish Office', *Studies in Public Policy* 87.

Ross, William (1978), 'Approaching the Archangelic?', *The Scottish Government Yearbook* 1978: 1-20.

Rothwell, Roy & Walter Zegveld (1985), *Reindustrialisation and Tecnology*, London, Longman.

Rowthorn, Robert & J. R. Wells (1988), *Deindustrialization and Foreign Trade*, Cambridge, Cambridge UP.

Rubinstein, W. D. (1994), *Capitalism, Culture and Decline in Britain 1750-1990*, London, Routledge.

Sabatier, Paul A. (1993), 'Top-down and Bottom-up Approaches to Implementation Research' (orig. 1983), in Michael Hill (ed.), *The Policy Process – A Reader*, Hemel Hempstead, Harvester Wheatsheaf, pp. 266-93.

Sagan, Iwona & Henrik Halkier (eds.) (2005), *Regionalism Contested. Institutions, Society and Territorial Governance*, Aldershot, Ashgate.

Salamon, Lester M. & Michael S. Lund (1989), 'The Tools Approach: Basic Analytics', in Lester M. Salamon & Michael S. Lund (eds.), *Beyond Privatization: The Tools of Government Action*, Washington, Urban Institute Press, pp. 23-49.

Savage, Stephen P. & Lynton Robins (eds.) (1990), *Public Policy Under Thatcher*, Houndmills, Macmillan.

Saville, Richard (ed.) (1985), *The Economic Development of Modern Scotland, 1950-80*, Edinburgh, John Donald.

Sawyer, Malcolm C. (1992), 'Reflections on the Nature and Role of Industrial Policy', *Metroeconomica* 43 (1-2): 51-73.

Sayer, Andrew (1984), *Method in Social Science. A Realist Approach*, London, Hutchinson.

SCDI (1973), *A Future for Scotland. A Study of the Key Factors Associated With Economic Growth in Scotland, and Proposals Necessary to Achieve Success in the 1980s*, Edinburgh, SCDI.

SCDI (1989), *Scottish Enterprise – An Alternative View*, SCDI, Edinburgh.

SCDI Executive Committee Papers (1970ff), Edinburgh, SCDI.

Scharpf, Fritz W. (1983), 'Interessenlage der Adressaten und Spielräume der Implementation bei Anreizprogrammen', in Renate Mayntz (ed.), *Implementation politicher Programme II – Ansätze zur Theoriebildung*, Opladen, Westdeutscher Verlag, pp. 99-116.

Scott, John & Michael Hughes (1980), *The Anatomy of Scottish Capital. Scottish Companies and Scottish Capital, 1900-1979*, London/Montreal, Croom Helm/McGill-Queen's UP.

Scott, John *et al.* (1980), 'Patterns of Ownership in Top Scottish Companies', in Ron Parsler (ed.), *Capitalism, Class and Politics in Scotland*, Westmead, Gower, pp. 59-71.

Scott, Peter (1996), 'The Worst of Both Worlds: British Regional Policy, 1951-64', *Business History* 38 (4): 41-64.

Scottish Affairs Committee (1984), 'Highlands and Islands Development Board', *HC Papers* 418, 9.5.84.

Scottish Affairs Committee (1985), 'HIDB – Report and Proceedings', *HC Papers* 22, 13.3.85.

Scottish Conservative Party (1989), *Briefing Notes: Scottish Enterprise, 20.2.89*, Edinburgh, Scottish Conservative Party.

Scottish Development Agency Act 1975, *Acts*, 69.

Scottish Development Department (1963), 'Central Scotland. A Programme for Development and Growth', *Cmnd* 2188, Nov. 1963.

Scottish Electronics Technology Group (1989), *Scottish Enterprise*, Glasgow, SETG.

Scottish Engineering Employers Assosication (1970-91), *Annual Report*, Glasgow, Scottish Engineering Employers Assosication.

Scottish Engineering Employers Association (1989a), *Response to 'Scottish Enterprise' White Paper*, Scottish Engineering Employers, Glasgow.

Scottish Enterprise (1991), *LEGUP Handbook*, Glasgow, Scottish Enterprise.

Scottish Grand Committee (1976-91), 'Scottish Estimates', *Parliamentary Debates* 1976-91.

Scottish Grand Committee (1984a), 'Regional Policy as it Affects Scotland', *Parliamentary Debates* 1984-85: 4.12.84.

Scottish Grand Committee (1989a), 'Scottish Enterprise', *Parliamentary Debates* 1988-89: 20.3.1989.

Scottish Office (1966), 'The Scottish Economy 1965 to 1970. A Plan for Expansion', *Cmnd* 2864, Jan. 1966.

Scottish Office (1988), *Scotland – An Economic Profile*, Edinburgh, HMSO.

Scottish Office (1991), *Scotland – An Economic Profile, 2nd ed.*, Edinburgh, HMSO.

Scouller, John (1987), *Industry and Public Policy, 2nd ed.*, Glasgow, University of Strathclyde.

Scouller, John (1989), *Industry and Public Policy, Supplementary Unit 1: Industrial Policy: Rationale, Methodology, and Practice*, Glasgow, University of Strathclyde.

SCRI (1974), *Economic Development and Devolution*, Edinburgh, SCRI.

SDA (1976-91), *Annual Report*, Glasgow, SDA.

SDA (1977a), *The Agency's Industry Strategy*, Glasgow, SDA.

SDA (1977b), *The Agency's Industry Strategy: Meeting with SEPD, 21.3.77*, SDA internal note.

SDA (1977c), 'Accounts 1975-76', *HC Papers* 504, 13.7.77.

SDA (1977d), *Evidence to the Committee to Review the Functioning of Financial Institutions (The Wilson Committee)*, Glasgow, SDA.

SDA (1977e), *Scottish Economic Council, 6.5.77: Discussion of Agency's Industrial Strategy*, Glasgow, SDA internal note.

SDA (1978a), *The Agency's Industrial Strategy – Interim Paper for 1979-81*, Glasgow, SDA.

SDA (1978b), *Small Manufacturing Firms in Scotland – A Survey of their Problems and Needs*, Glasgow, SDA.

SDA (1978c), *The Scottish Economy. A Concise Analysis*, Glasgow, SDA.

SDA (1979a), *The Electronics Industry in Scotland: A Proposed Strategy*, Glasgow, SDA.

SDA (1980a), *GEAR: Strategy and Programme*, Glasgow, SDA.

SDA (1981a), *Development Strategy for the Health Care Industry in Scotland, 1-3*, Glasgow, SDA.

SDA (1981b), *Royal Bank Can Go It Alone*, SDA Press Release.

SDA (1981c), *Electronics in Scotland: The Leading Edge*, Glasgow, SDA.

SDA (1981d), *Health Care in Scotland: The Natural Choice*, Glasgow, SDA.

SDA (1982a), *Scotland's Electronic Strategy, I-II*, Glasgow, SDA.

SDA (1982b), *Biotechnology. A Review of the Potential*, Glasgow, SDA.

SDA (1984a), *Corporate Strategy 1985/86-1987/88 – Summary for the Executive 5.12.84*, Glasgow, SDA.

SDA (1985a), *Electronics, the State of the Art in Scotland*, Glasgow, SDA.

SDA (1985b), *Scottish First. Innovation and Achievement*, Glasgow, SDA.

SDA (1986a), *Health Care in Scotland. The Front Runner*, Glasgow, SDA.

SDA (1986b), *Scotland – An Open Country*, Glasgow, SDA.

SDA (1986c), *The Changing Face of Scotland*, Glasgow, SDA.

SDA (1987a), *Corporate plan 1988-91: Statement of Corporate Strategy*, Glasgow, SDA.

SDA (1987b), *Pay and Conditions in Major Scottish Manufacturing Companies*, Glasgow, SDA.

SDA (1988a), *The Engineering Industry in Scotland: Securing the Future*, Glasgow, SDA.

SDA (1988b), *Scottish Electronics Industry Database 1988*, Glasgow, SDA.

SDA (1988c), *Health Care and Biotechnology Companies in Scotland*, Glasgow, SDA.

SDA (1988d), *The SBHIA Directory of Biotechnology and Healthcare in Scotland*, Glasgow, SDA.

SDA (1989a), *Electronics Subcontract Assembly in Scotland*, Glasgow, SDA.

SDA (1989b), *Scottish Enterprise. The SDA Response to the White Paper*, Glasgow, SDA.

SDA (1989c), *Electronics and Support Companies in Scotland*, Glasgow, SDA.

SDA (1990a), *Pay and Conditions in Major Scottish Manufacturing Companies*, Glasgow, SDA.

SDA (1990b), *Overseas and European Companies Manufacturing in Scotland (Excluding North America)*, Glasgow, SDA.

SDA/DCC (1977-89), *Development Consultative Committee Papers*, Glasgow, SDA.

SDA Board Papers (1987f), Glasgow, SDA.

Segal, N. S. (1979), 'The Limits and Means of 'Self-Reliant' Regional Economic Growth', in Duncan Maclennan & John B. Parr (eds.), *Regional Policy. Past Experiences and New Directions*, Robertson, Oxford, pp. 211-24.

SEPD (1976a), *Scottish Development Agency. Draft Industrial Investment Guidelines*, Edinburgh, SEPD.

SEPD (1976b), *SDA – Industrial Investment Guidelines*, Edinburgh, SEPD.

SEPD (1980a), *SDA – Industrial Investment Guidelines*, Edinburgh, SEPD.

SEPD (1980b), *SDA – Site Development and Factory Building Guidelines*, SEPD, Edinburgh.

Sevaldsen, Jørgen (1988), 'Forfaldets fascination. The Decline of Britain-debatten', *Historisk Tidsskrift* 88: 282-312.

Sharp, Margaret & William Walker (1991), 'Thatcherism and Technical Advance: Reform without Progress? Part II: The Thatcher Legacy', *Political Quarterly* 62 (2): 318-37.

Sharpe, L. J. (1993), 'The European Meso: An Appraisal', in L. J. Sharpe (ed.), *The Rise of Meso Government in Europe*, London, Sage, pp. 1-39.

Sidenius, Niels Christian (1984), 'Dansk Industripolitik – Beslutningersformer, Interesser og Statens Relative Autonomi', in Gorm Rye Olsen *et al.* (eds.), *Stat, Statskundskab, Statsteori*, Århus, Politica, pp. 118-64.

Sidenius, Niels Christian (1989), *Dansk Industripolitik – Nye løsninger på gamle problemer*? Århus, Hovedland.

Sked, Alan (1987), *Britain's Decline. Problems and Perspectives*, Oxford, Blackwell.

Sleegers, Wilfred (1998), 'Regional Development Agencies in the Netherlands – Twenty Years of Shareholding', in Henrik Halkier, Mike Danson & Charlotte Damborg (eds.), *Regional Development Agencies in Europe*, London, Jessica Kingsley, pp. 66-79.

Slowe, Peter M. (1981), *The Advance Factory in Regional Development*, Aldershot, Gower.

Smallbone, David *et al.* (1993), 'The Use of External Assistance by Mature SMEs in the UK: Some Policy Implications', *Entrepreneurship & Regional Development* 5 (3): 279-95.

Smith, Roger (1985a), 'The Industrialization of the Clyde Valley', in Roger Smith & Urlan Wannop (eds.), *Strategic Planning in Action. The Impact of the Clyde Valley Regional Plan 1946-1982*, Aldershot, Gower, pp. 5-16.

Smith, Roger (1985b), 'The Setting Up of the Clyde Valley Regional Planning Team and Agencies for Implementation', in Roger Smith & Urlan Wannop (eds.), *Strategic Planning in Action. The Impact of the Clyde Valley Regional Plan 1946-1982*, Aldershot, Gower, pp. 17-40.

Smith, Peter & Malcolm Burns (1988), 'The Scottish Economy – Decline and Response', *The Scottish Government Yearbook* 1988: 259-81.

SNP (1959-88), *Annual Conference papers*, Edinburgh, SNP.

SNP (1965a), *SNP and You. Aims and Policy of the SNP, 5th ed.*, Edinburgh, SNP.

SNP (1965b), *Jobs Could Be Created Now – News Release 17.10.65*, Edinburgh, SNP.

SNP (1965c), *Industrial Development Corporation for Scotland – Press Statement 10.4.65*, Edinburgh, SNP.

SNP (1965d), *Development Districts – News Release 25.10.65*, Edinburgh, SNP.

SNP (1966a), *Putting Scotland First – The SNP Manifesto March 1966*, Edinburgh, SNP.

SNP (1966b), *UK Budget 1966 – Letter to the UK Treasury*, Edinburgh, SNP.

SNP (1966c), *Make Scotland's Money Count for You – SNP / Willy Wolfe election leaflet*, Edinburgh, SNP.

SNP (1968a), *Action NOW*, Glasgow, SNP.

SNP (1973a), *The Scotland We Seek*, Edinburgh, SNP.

SNP (1974a), *"It Is His Oil" Poster*, Edinburgh, SNP.

SNP (1976a), *Why Regional Policies Won't Work*, Edinburgh, SNP.

SNP (1976b), *Policy Document: Industry*, Edinburgh, SNP.

SNP (1979a), *Return to Nationhood – 1979 Election Manifesto*, Edinburgh, SNP.

SNP (1982a), *The Scottish Jobs Campaign*, Edinburgh, SNP.

SNP (1983a), *SNP Manifesto 1983: Choose Scotland – The Challenge of Independence*, Edinburgh, SNP.

SNP (1987a), *Play the Scottish Card – SNP General Election Manifesto 1987*, Edinburgh, SNP.

SNP (1988a), *SNP Calls for a Radical Overhaul of the SDA – News Release 18.10.88*, Edinburgh, SNP.

Spence, Nigel (1992), 'Impact of Infrastructure Investment Policy', in Peter Townroe & Ron Martin (eds.), *Regional Development in the 1990s – The British Isles in Transition*, London, Jessica Kingsley, pp. 229-36.

Staeck, Nicola (1996), 'The European Structural Funds – Their History and Impact', in Hubert Heinelt & Randall Smith (eds.), *Policy Networks and European Structural Funds*, Aldershot, Avebury, pp. 46-73.

Standing Committee A (1980), 'Industry Bill', *Parliamentary Debates* 1980-81: 11.12.80.

Standing Committee B (1979), 'Industry Bill', *Parliamentary Debates* 1978-79: 1.2.79.

Standing Committee E (1979), 'Industry Bill', *Parliamentary Debates* 1979-80: Dec. 1979.

Standing Commission on the Scottish Economy (1988), *Interim Report, February 1988*, Glasgow, Standing Commission on the Scottish Economy.

Standing Commission on the Scottish Economy (1989), *Final Report, November 1989*, Glasgow, Standing Commission on the Scottish Economy.

Steiner, Michael & Thomas Jud (1998), 'Regional Development Institutions in Austria – Trends in Organisation, Policies and Implementation', in Henrik Halkier, Mike Danson & Charlotte Damborg (eds.), *Regional Development Agencies in Europe*, London, Jessica Kingsley, pp. 48-65.

Steinmo, Sven *et al.* (eds.) (1992), *Structuring Politics. Historical Institutionalism in Comparative Analysis*, Cambridge, Cambridge UP.

Stephen, Frank (1975), 'The Scottish Development Agency', in Gordon Brown (ed.), *The Red Paper on Scotland*, Edinburgh, EUSPB, pp. 223-32.

Stevens, Chris (1990), 'Scottish Conservatism – A Faillure of Organisation?', *The Scottish Government Yearbook* 1990: 76-89.

Stöhr, Walter B. (1986), 'Changing External Conditions and a Paradigm Shift in Regional Development Strategies?', in Michel Bassand, M. *et al.* (eds.), *Self-reliant Development in Europe – Theory, Problems, Actions*, London, Gower, pp. 59-73.

Stöhr, Walter (1989), 'Regional Policy at the Crossroads: An Overview', in Louis Albrechts *et al.* (eds.), *Regional Policy at the Crossroads – European Perspectives*, London, Jessica Kingsley, pp. 191-97.

Stoker, Gerry (1990), 'Government Beyond Whitehall', in Patrick Dunleavy *et al.* (eds.), *Developments in British Politics 3*, Houndmills, Macmillan, pp. 126-49.

Stone, Ian & Frank Peck (1996), 'The Foreign-owned Manufacturing Sector in UK Peripheral Regions, 1978-93: Restructuring and Comparative Performance', *Regional Studies* 30 (1): 55-68.

Stout, D. K. (1979), 'De-industrialization and Industrial Policy', in Frank Blackaby (ed.), *British Economic Policy 1960-74*, London, Cambridge UP, pp. 171-96.

Strathclyde IDO (1994), *European Structural Funds in Strathclyde 1988-93 – Information Pack*, Glasgow, Strathclyde IDO.

Strathclyde Regional Council (1989), *Scottish Enterprise – Review of the Government's Proposal*, Glasgow, Strathclyde Regional Council.

STUC (1961-89), *Annual Report*, Glasgow, STUC.

STUC (1990-91), *Report of General Council to Annual Congress*, Glasgow, STUC.

STUC Economic Committee Papers (1967-79), Glasgow, STUC.

STUC (1986a), *Scotland: A Strategy for the Future*, Glasgow, STUC.

STUC (1987a), *Scotland – A Land Fit for People*, Glasgow, STUC.

STUC (1989a), *Scotland's Economy. Claiming the Future*, Glasgow, STUC.

Sturm, Roland (1991), *Die Industriepolitik der Bundesländer und die europäische Integration*, Baden-Baden, Nomos.

Swales, Kim (1983), 'Industrial Policy', in Keith Ingham & James Love (eds.), *Understanding the Scottish Economy*, Oxford, Robertson, pp. 188-200.

Swales, Kim (1989), 'Are Discretionary Regional Subsidies Cost-effective?', *Regional Studies* 23 (4): 361-68.

Swales, Kim (1993), 'A Game Theoretic Approach to Subsidizing Employment', *Regional Studies* 27 (2): 109-20.

Taylor, A. C. C. (1986), 'Overseas Ownership in Scottish Manufacturing Industries 1950-1985', *Scottish Economic Bulletin* (33): 20-28.

Taylor, Jim (1992), 'Regional Problems and Policies – An Overview', in Peter Townroe & Ron Martin (eds.), *Regional Development in the 1990s – The British Isles in Transition*, London, Jessica Kingsley, pp. 288-96.

Taylor, Jim & Colin Wren (1997), 'UK Regional Policy. An Evaluation', *Regional Studies* 31 (9): 835-848.

Tayside Regional Council (1989), *Scottish Enterprise White Paper Response*, Dundee, Tayside Regional Council.

Temple, Marion (1994), *Regional Economics*, Houndmills, Macmillan.

Thatcher, Margaret (1993), *The Downing Street Years*, London, HarperCollins.

Thatcher, Margaret (1995), *The Path to Power*, London, HarperCollins.

Thatcher, Margaret (1999), *Complete Public Statements, 1945-90*, Oxford, Oxford UP.

Theakston, Kevin (1996), 'Whitehall, Westminster and Industrial Policy', in David Coates (ed.), *Industrial Policy in Britain*, Houndmills, Macmillan, pp. 159-81.

Thelen, Kathleen & Sven Steinmo (1992), 'Historical Institutionalism in Comparative Politics', in Sven Steinmo *et al.* (eds.), *Structuring Politics: Historical Institutionalism in Comparative Analysis*, Cambridge, Cambridge UP, pp. 1-32.

Therkelsen, Anette & Henrik Halkier (2004), 'Umbrella Place Branding. A Study of Friendly Exoticism and Exotic Friendliness in Coordinated National Tourism and Business Promotion', *SPIRIT Discussion Papers* 26.

Thiel, Sandra van (2001), *Quangos: Trends, Causes and Consequences*, Aldershot, Ashgate.

Thomas, Ian C. & P. J. Drudy (1993), 'Advance Factories in Peripheral Areas – The Case of Mid-Wales', in Richard T. Harrison & Mark Hart (eds.), *Spatial Policy in a Divided Nation*, London, Jessica Kingsley, pp. 108-25.

Thompson, Edward P. (1978), 'The Peculiarities of the English' (orig. 1965), in Edward P. Thompson (ed.), *The Poverty of Theory and Other Essays*, London, Merlin, pp. 35-91.

Thompson, Grahame (1990), *The Political Economy of the New Right*, London, Pinter.

Thomsen, Jens Peter Frølund (1991a), 'Realisme, diskursanalyse og magtperspektiver', *GRUS* (33): 57-77.

Thomsen, Jens Peter Frølund (1991b), 'A Strategic-Relational Account of State Interventions', in R. B. Bertramsen *et al.*: *State, Economy & Society*, London, Unwin Hyman, pp. 146-95.

Thomsen, Jens Peter Frølund (1994), 'Ny-institutionalismen', *GRUS* 43: 5-22.

Thomsen, Jens Peter Frølund (1996), *Governing Against Pressure. State British Politics and Trade Unions in the 1980s: Governing Against Pressure*, Aldershot, Dartmouth.

Thomsen, Robert C. (2001), *Selves and Others of Political Nationalism in Stateless Nations: National Identity-building Processes in the Modern Histories of Scotland and Newfoundland*, PhD Thesis, Department of English, University of Aarhus.

Thorelli, Hans B. (1986), 'Networks: Between Markets and Hierarchies', *Strategic Management Journal* 7 (1): 37-51.

Tickell, Adam *et al.* (1995), 'The Fragmented Region: Business, the State and Economic Development in North West England', in Martin Rhodes (ed.), *The Regions and the New Europe*, Manchester, Manchester UP, pp. 247-72.

Todd, Graham (1985), *Investing in Scotland: Public Policy and Private Enterprise*, London, Economist Intelligence Unit.

Toothill, J. N. (1961), *Inquiry into the Scottish Economy, 1960-61*, Edinburgh, SCDI.

Totterdill, Peter (1989), 'Local Economic Strategies as Industrial Policy: A Critical Review of British Developments in the 1980s', *Economy & Society* 18 (4): 478-526.

Townroe, Peter (1986), 'Regional Economic Development Policy in a Mixed Economy and a Welfare State', *Journal of Regional Policy* 6: 355-372.

Townroe, Peter & Ron Martin (eds.) (1992), *Regional Development in the 1990s – The British Isles in Transition*, London, Jessica Kingsley.

Townsend, Alan R. (1987), 'Regional Policy', in William F. Lever (ed.), *Industrial Change in the United Kingdom*, London, Longman, pp. 223-39.

Tricker, Mike & Steve Martin (1984), 'The Developing Role of the Commission', *Regional Studies* 18 (6): 507-14.

Turok, Ivan (1987), 'Lessons for Local Economic Policy', in David Donnison & Alan Middleton: *Regenerating the Inner City – Glasgow's Experience*, London, RKP, pp. 235-47.

Turok, Ivan (1993), 'Inward Investment and Local Linkages: How Deeply Imbedded is "Silicon Glen"?', *Regional Studies* 27 (5): 401-17.

UNCTAD (1990ff), *World Investment Report, UNCTAD*, Geneva.

UNIDO (1997), *Regional Industrial Development Agencies. Types, Tasks and UNIDO Assistance*, Vienna, UNIDO.

Urry, John (1990), 'Conclusion: Places and Policies', in Michael Harloe *et al.* (eds.), *Place, Policy and Politics: Do Localities Matter*? London, Unwin Hyman, pp. 187-204.

Vanhove, Norbert & Leo H. Klaassen (1987), *Regional Policy – A European Approach*, Aldershot, Gower.

Veggeland, Noralv (1983), 'Noen teoretiske rammer for en handlingsorientert regionalpolitikk', *NordREFO* 14 (2-3): 63-78.

Veggeland, Noralv (1992), 'The Circular Motion of Planning: Top-down and Bottom-up Problems Once Again', in Markku Tykkyläinen (ed.), *Development Issues and Strategies in the New Europe*, Aldershot, Avebury, pp. 221-25.

Velasco, Roberto (1991), *The Role of Development Agencies in European Regional Policy – A Report for the DG XVI of the European Commission*, Brussels, European Commission.

Veljanovski, Cento (1990), 'The Political Economy of Regulation', in Patrick Dunleavy *et al.* (eds.), *Developments in British Politics 3*, Houndmills, Macmillan, pp. 291-304.

Villumsen, Gert (1994), 'Douglass North – Endnu en økonomisk fagimperialist', *GRUS* 43: 36-45.

Waarden, Frans van (1992), 'Dimensions and Types of Policy Networks', *European Journal of Political Research* 21 (1-2): 29-52.

Wad, Peter (2000), 'Kritisk realisme', *GRUS* 60: 53-68.

Walker, Gesa & Herbert Krist (1980), 'Regional Incentives and the Investment Decision of the Firm: A Comparative Study of Britain and Germany', *Studies in Public Policy* 57.

Walker, Jim (1987), 'The Scottish Electronics Industry', *The Scottish Government Yearbook* 1987: 57-80.

Walker, Peter (1991), *Staying Power – An Autobiography*, London, Bloomsbury.

Walsh, James I. (2000), 'When Do Ideas Matter? Explaining the Successes and Failures of Thatcherite Ideas', *Comparative Political Studies* 33 (4): 483-516.

Wannop, Urlan (1984), 'Strategic Planning and the Area Projects of the Scottish Development Agency', *Regional Studies* 18 (1): 77-81.

Wannop, Urlan (1990), 'The GEAR Project. A Perspective on the Management of Urban Regeneration', *Town Planning Review* 61 (4): 455-74.

Wannop, Urlan (1995), *The Regional Imperative: Regional Planning and Governance in Britain, Europe and the US*, London, Jessica Kingsley.

Wannop, Urlan & Roger Leclerc (1987), 'The Management of GEAR', in David Donnison & Alan Middleton: *Regenerating the Inner City – Glasgow's Experience*, London, RKP, pp. 218-231.

Wannop, Urlan & Roger Smith (1985), 'Robustness in Regional Planning: An Evaluation of the Clyde Valley Regional Plan', in Roger Smith & Urlan Wannop (eds.), *Strategic Planning in Action. The Impact of the Clyde Valley Regional Plan 1946-1982*, Aldershot, Gower, pp. 210-240.

Warner, Gerald (1988), *The Scottish Tory Party. A History*, London, Weidenfeld & Nicholson.

Warwick, P. (1985), 'Did Britain Change? An Inquiry Into the Causes of National Decline', *Journal of Contemporary History* 20: 99-133.

WDA (1977-91), 'Accounts', *HC Papers*.

WDA (1977a), *A Statement of Policies and Programmes*, Pontypridd, WDA.

WDA (1981a), *The First Five Years*, Cardiff, Wales.

WDA Act 1975, *Acts* 70.

Webb, Keith & Eric Hall (1978), 'Explanations of the Rise of Political Nationalism in Scotland', *Studies in Public Policy* 15.

Welsh Affairs Committee (1989), 'Inward Investment into Wales and its Interaction with Regional and EEC Policies', *HC Papers* 86-I, 8.2.89.

Welsh Office (1987), *1986 Review of the Welsh Development Agency. Report of the Review Group to the Secretary of State for Wales*, Cardiff, Welsh Office.

Welsh Office & WDA (1995), *Financial Management and Policy Review of the WDA – Response*, Cardiff, Welsh Office & WDA.

Wenneberg, Søren Barlebo (2000), *Socialkonstruktivisme – Positioner, problemer og perspektiver*, København, Samfundslitteratur.

West Central Scotland Plan Team (1974a), *West Central Scotland – A Programme of Action, Consultative Draft Report*, Glasgow, WCSP Team.

West Central Scotland Plan Team (1974b), *West Central Scotland Plan. Supplementary Reports 1-5*, Glasgow, WCSP Team.

Wiehler, Frank & Thomas Stumm (1995), 'The Powers of Regional and Local Authorities and their Role in the EU', *European Planning Studies* 3 (2): 227-50.

Wigley, John & Carol Lipman (1992), *The Enterprise Economy*, Houndmills, Macmillan.

Wiener, Martin J. (1985), *English Culture and the Decline of the Industrial Spirit, 1950-1980*, Harmondsworth, Penguin.

Wilding, Richard (1982), 'A Triangular Affair: Quangos, Ministers, and MPs', in Anthony Barker (ed.), *Quangos in Britain. Government and the Networks of Public Policy*, London, Macmillan, pp. 34-43.

Wilks, Stephen (1984), *Industrial Policy and the Motor Industry*, Manchester, Manchester UP.

Wilks, Stephen (1985), 'Conservative Industrial Policy 1979-83', in Peter Jackson (ed.), *Implementing Government Policy Initiatives: the Thatcher Administration 1979-83*, London, RIPA, pp. 123-43.

Wilks, Stephen (1987), 'From Industrial Policy to Enterprise Policy in Britain', *Journal of General Management* 12 (4): 5-20.

Williams, Karel *et al.* (eds.) (1983), *Why Are the British Bad at Manufacturing?* London, RKP.

Wilson Committee (1977f), *Evidence on the Financing of Industry and Trade, 1-8*, London, HMSO.

Wilson, Edgar (1992), *A Very British Miracle – The Failure of Thatcherism*, London, Pluto.

Windhoff-Héritier, Adrienne (1987), *Policy-Analyse – Eine Einführung*, Frankfurt, Campus.

Winter, Søren (1994), *Implementering og effektivitet*, Herning, Systime.

Wishlade, Fiona (1996), 'EU Cohesion Policy: Facts, Figures and Issues', in Liesbet Hooghe (ed.), *Cohesion Policy and European Integration. Building Multi-level Governance*, Oxford, Oxford UP, pp. 27-58.

Wishlade, Fiona (1998), 'EC Competition Policy: The Poor Relation of EC Regional Policy?', *European Planning Studies* 6 (5): 573-97.

Wishlade, Fiona (1999), 'Incentives Regulation', in Philip Raines & Ross Brown (eds.), *Policy Competition and Foreign Direct Investment in Europe*, Ashgate, Aldershot, pp. 89-114.

Wishlade, Fiona *et al.* (1997), *Economic and Social Cohesion in the European Union: The Reginal Distribution of Member States' Own Policies*, Paper, EURRN Conference Frankfurt/Oder 20-23 September 1997.

Wolfe, Billy (1973), *Scotland Lives – The Quest for Independence*, Edinburgh, Reprograph.

Wong, Cecilia (2000), 'Local Development Networks in England', in Mike Danson, Henrik Halkier & Greta Cameron (eds.), *Governance, Institutional Change and Regional Development*, Aldershot, Ashgate, pp. 57-88.

Wood, Frances (1989), 'Scottish Labour in Government and Opposition, 1964-79', in Ian Donnachie *et al.* (eds.), *Forward! Labour Politics in Scotland 1888-1988*, Edinburgh, Polygon, pp. 99-129.

Wren, Colin (1996a), *Industrial Subsidies. The UK experience*, Basingstoke, Macmillan.

Wren, Colin (1996b), 'Grant Equivalent Expenditure on Industrial Subsidies in the Post-War UK', *Oxford Bulletin of Economics and Statistics* 58 (2): 317-53.

Wright, Mike *et al.* (1994), 'MBOs in the Regions: A Tale of Two Recessions', *Regional Studies* 28 (3): 319-26.

Wright, Vincent (1998), 'Intergovernmental Relations and Regional Government in Europe – A Sceptical View', in Patrick Le Galès & Christian Lequesne (eds.), *Regions in Europe*, London, Routledge, pp. 39-49.

Young, Stephen (1984), 'The Foreign-Owned Manufacturing Sector', in Neil Hood & Stephen Young (eds.), *Industry, Policy and the Scottish Economy*, Edinburgh, Edinburgh UP, pp. 93-127.

Young, Stephen (1989), 'Scotland v Wales in the Inward Investment Game', *Quarterly Economic Commentary* 14 (3): 59-63.

Young, Stephen & Neil Hood (1984), 'Industrial Policy and the Scottish Economy', in Neil Hood & Stephen Young (eds.), *Industry, Policy and the Scottish Economy*, Edinburgh, Edinburgh UP, pp. 28-56.

Young, Stephen & A. V. Lowe (1974), *Intervention in the Mixed Economy: The Evolution of British Industrial Policy 1964-72*, London, Croom Helm.

Young, Stephen & Alan Reeves (1984), 'The Engineering and Metals Sector', in Neil Hood & Stephen Young (eds.), *Industry, Policy and the Scottish Economy*, Edinburgh, Edinburgh UP, pp. 128-73.

Young, Stephen *et al.* (1988), 'Global Strategies, Multinational Subsidiary Roles and Economic Impact in Scotland', *Regional Studies* 22 (6): 487-97.

Young, Stephen *et al.* (1994a), 'Multinational Enterprises and Regional Economic Development', *Regional Studies* 28 (7): 657-78.

Young, Stephen *et al.* (1994b), 'Targeting Policy as a Competitive Strategy in European Inward Investors Agencies', *European Urban and Regional Studies* 1 (2): 143-60.

Yuill, Douglas & Kevin Allen (1981), 'Regional Development Agencies in Europe: An Overview', *ESU Research Paper* 4.

Yuill, Douglas (ed.) (1982), *Regional Development Agencies in Europe*, Aldershot, Gower.

Yuill, Douglas & Kevin Allen (eds.) (1982), *European Regional Incentives 1982*, Glasgow, EPRC.

Yuill, Douglas *et al.* (1981-99), *European Regional Incentives, 1st-18th ed.*, London, Bowker-Saur.

Zukin, Sharon (1985), 'Industrial Policy as Post-Keynesian Politics: Basic Assumptions in the US and France', in Sharon Zukin (ed.), *Industrial Policy – Business and Politics in the US and France*, New York, Praeger, pp. 3-47.

Østergård, Uffe (1986), 'Marianne – Fra Minerva til Brigitte Bardot', *Den Jyske Historiker* 38-39: 11-34.

Östhol, Anders, Bo Svensson & Henrik Halkier (2002), 'Analytical Framework', in Anders Östhol & Bo Svensson (eds.), *Partnership Responses – Regional Governance in the Nordic States*, Stockholm, Nordregio, pp. 23-39

Periodicals

CBI News

Department of Employment Gazette

Economic Trends (CSO, Regional Accounts)

ERDF Annual Report

Glasgow Herald

Glasgow Chamber of Commerce Journal

House of Commons Parliamentary Debates (HCPD)

Industry Act Annual Report

Local Employment Act Annual Report

Scottish Abstracts of Statistics

Scottish Business Insider

Scottish Trade Union Review

Interviews

Interviews with the persons listed below were conducted in Scotland during the summer of 1990 and the autumn of 1991. The interviews were open-ended and no uniform questionnaire was used; each of them covered topics of which the interviewee had particular knowledge as well as more general questions. Most of the sessions took place on location, i.e. in the office of the interviewee, all of them were tape-recorded, and the final product was more than 800 pages of transcript.

Douglas Adams, SDA executive, head of Economic Services 1982-84, head of Industrial Programmes Development 1984-86. With Templetons Unit Trust when interviewed.

Chris Aitken, property executive 1966-88, first with Scottish Industrial Estates Corporation, then with the SDA's Property Directorate. Regional Director Strathclyde West and South 1988-91. Director of Environment, Scottish Enterprise when interviewed.

Sir Kenneth Alexander, non-executive member of SDA board 1975-85. Economic advisor to the West Central Scotland Plan, chairman of HIDB 1976-80. Professor of economics at Aberdeen University when interviewed.

Craig Campbell, economist with the SCDI since 1970, Chief Economist when interviewed.

Edward Cunningham, SDA director 1977-90. First Director of Strategic Planning, then Director Planning & Projects Division, finally Director Industry and Enterprise Development. Chairman of Business Options Ltd, a corporate development consultancy, when interviewed.

Alan Dale, SDA executive 1976-90. Factory Building Division Head 1976-83, Director Property Development and Environment 1983-87, Group Director Regions 1988-90. Independent property consultant when interviewed.

Sir Robin Duthie, SDA chairman 1979-88; industrialist.

Charles Fairley, former head of SDA's Advanced Engineering Division. Independent consultant when interviewed.

John Firn, former deputy head of SDA's Projects & Planning Division. Independent consultant when interviewed.

Jennifer Forbes, SDA executive 1985-91. Advertising Manager, since 1990 Head of Marketing.

Sir William Gray, SDA chairman 1975-79. Former Labour politician, Lord Provost of Glasgow 1972-75; lawyer.

Douglas Harrison, STUC research assistant.

Ian Hart, SDA land renewal executive 1978-91. From 1989 head of Environmental Development. Head of Land Engineering, Scottish Enterprise when interviewed.

Harry Hood, SDA executive 1975-86, Head of Public Relations. Formerly editor of the Glasgow *Evening Times*.

Neil Hood, SDA advisor and director 1976-91. Non-executive director of Scottish Development Finance from 1984, Director LIS 1987-88. Responsible for implementing the Scottish Enterprise merger as Director Enterprise and Special Initiatives 1989-91. Economic advisor, SEPD 1979. Professor of Business Policy at Strathclyde University's International Business Unit.

Frank Kirwan, SDA executive 1984-87. An economist with Strathclyde University before joining the Agency, first in the Industrial Programme Development Division, later Economic Services Division, and as head of the Policy Unit 1986-87. With the Royal Bank of Scotland when interviewed.

Helen Liddell, former Head of STUC's economic department (1971-76), and Scottish Secretary of the Labour Party (1977-88). Executive director with the *Daily Record* when interviewed.

Isobel Lindsay, on SNP's National Executive 1970-89. Vice-chairman for publicity and policy in several periods, now Convener of the Campaign for a Scottish Assembly. Lecturer in Politics at Strathclyde University.

David Lyle, SDA Secretary 1979-91.

Ewan Marvick, with the Glasgow Chamber of Commerce since 1980; Secretary and Chief Executive since 1982.

George Mathewson, SDA Chief Executive 1981-87. Before joining the director with venture capitalists ICFC 1972-81. Director with The Royal Bank of Scotland when interviewed.

Gavin McCrone, civil servant at the Scottish Office since 1970. Chief Economic Advisor at the Scottish Office 1970, Head of SEPD/IDS 1980-87. Formerly economist at Oxford University. Head of Scottish Development Department when interviewed.

Gregor Mackenzie, Labour minister at the Department of Industry 1974-76, minister at the Scottish Office 1976-79.

Bruce Millan, Labour minister at the Scottish Office 1966-70, 1974-79. Minister of State 1974-76, Secretary of State 1976-79. EC commissioner for regional affairs when interviewed.

Gerry Murray, SDA executive 1979-91. Head of Industry Services Division.

Donald Patience, SDA executive 1982-91. Director of Scottish Development Finance, SDA's Investment Division. Formerly with venture capitalists ICFC and 3i.

George Robertson, SDA board member 1975-78. Scottish Organiser of the General & Municipal Workers Union 1969-78, Labour MP since 1978.

Iain Robertson, SDA executive 1983-90. Chief Executive 1987-90, prior to that Director, Locate in Scotland 1983-87. Formerly civil servant at IDS. With County NatWest when interviewed.

Lewis Robertson, SDA Chief Executive 1976-81. Industrialist, appointed to the Steering Committee formed to set up SDA in 1975. Specialist rescue Chairman for private companies when interviewed.

Mike Sandys, SDA executive. Head, Advanced Engineering Division 1987-91, formerly with the Electronics Division.

Iain Shirlaw, SDA executive 1984-91. Joined Health Care and Biotechnology Division; since 1985 Head, Scottish Resource Industry Division.

James Williamson, SDA executive 1980-81, 1987-90. On secondment from Civil Service 1980-81 as Personal Assistant to SDA Chief Executives. Head of Financial Services 1987, Head of Policy Unit 1988-91.

William Wolfe, SNP Chairman 1969-79, President 1980-82.

Alf Young, research officer with the Labour Party in Scotland 1976-79. Economics editor at *The Glasgow Herald* when interviewed.

Regionalism & Federalism

The contemporary nation-state is undergoing a series of transformations which question its traditional role as a container of social, political and economic systems. New spaces are emerging with the rise of regional production systems, movements for territorial autonomy and the rediscovery of old and the invention of new identities. States have responded by restructuring their systems of territorial government, often setting up an intermediate or regional level. There is no single model, but a range, from administrative deconcentration to federalization. Some states have regionalized in a uniform manner, while others have adopted asymmetrical solutions. In many cases, regions have gone beyond the nation-state, seeking to become actors in broader continental and transnational systems.

The series covers the gamut of issues involved in this territorial restructuring, including the rise of regional production systems, political regionalism, questions of identity, and constitutional change. It will include the emergence of new systems of territorial regulation and collective action within civil society as well as the state. There is no a priori definition of what constitutes a region, since these span a range of spatial scales, from metropolitan regions to large federated states, and from administrative units to cultural regions and stateless nations. Disciplines covered include history, sociology, social and political geography, political science and law. Interdisciplinary approaches are particularly welcome. In addition to empirical and comparative studies, books focus on the theory of regionalism and federalism, including normative questions about democracy and accountability in complex systems of government.

Series Titles

No.8 – Henrik HALKIER, *Institutions, Discourse and Regional Development. The Scottish Development Agency and the Politics of Regional Policy*, 2006, 598 p., ISBN 90-5201-275-X

No.7 – Jörg MATHIAS, *Federalism in Practice. Regional Economic Development and Welfare Reform in the Berlin Republic*, forthcoming, ISBN 90-5201-254-7

No.6 – Daniele CARAMANI & Yves MÉNY (eds.), *Challenges to Consensual Politics. Democracy, Identity, and Populist Protest in the Alpine Region*, 2005, 257 p., ISBN 90-5201-250-4

No.5 – Nicola MCEWEN, *Nationalism and the State. Welfare and Identity in Scotland and Quebec*, 2006, 212 p., ISBN 90-5201-240-7

No.4 – Carolyn M. DUDEK, *EU Accession and Spanish Regional Development. Winners and Losers*, 2005, 202 p., ISBN 90-5201-237-7

No.3 – Stéphane PAQUIN, *Paradiplomatie et relations internationales. Théorie des stratégies internationales des régions face à la mondialisation*, 2004, 189 p., ISBN 90-5201-225-3

No.2 – Wilfried SWENDEN, *Federalism and Second Chambers. Regional Representation in Parliamentary Federations: The Australian Senate and German Bundesrat Compared*, 2004, 423 p., ISBN 90-5201-211-3

No.1 – Michael KEATING & James HUGHES (eds.), *The Regional Challenge in Central and Eastern Europe. Territorial Restructuring and European Integration*, 2003, 208 p., ISBN 90-5201-187-7